TOURS SYMPOSIUM ON NUCLEAR PHYSICS IV

Previous Proceedings in the Series of Tours Symposia on Nuclear Physics

Year	Volume	Publisher	ISBN
1997	III	AIP Conf. Proceedings vol. 425	1-56396-749-9
1994	II	World Scientific Publishing	981-02-2156-8
1991	I	World Scientific Publishing	981-02-0892-8

Related Titles from AIP Conference Proceedings

549 Intersections of Particle and Nuclear Physics
Edited by Zohreh Parsa and William J. Marciano, December 2000, 1-56396-978-5

529 Capture Gamma-Ray Spectroscopy and Related Topics: 10th International Symposium
Edited by Stephen Wender, July 2000, 1-56396-952-1

518 Proton Emitting Nuclei: PROCON'99—First International Symposium
Edited by Jon C. Batchelder, May 2000, 1-56396-937-8

508 Hadron Physics: Effective Theories of Low Energy QCD
Edited by A. H. Blin, B. Hiller, M. C. Ruivo, C. A. Sousa, and E. van Beveren, March 2000, 1-56396-927-0

495 Experimental Nuclear Physics in Europe: ENPE 99, Facing the Next Millennium
Edited by Berta Rubio, Manuel Lozano, and William Gelletly, November 1999, 1-56396-907-6

481 Nuclear Structure 98
Edited by C. Baktash, September 1999, 1-56396-858-4

459 Heavy Quarks at Fixed Target
Edited by Harry W. K. Cheung and Joel N. Butler, February 1999, 1-56396-864-9

455 ENAM 98: Exotic Nuclei and Atomic Masses
Edited by B. M. Sherrill, D. J. Morrissey, and C. N. Davids, December 1998, 1-56396-804-5

412 Intersections Between Particle and Nuclear Physics: 6th Conference
Edited by T. W. Donnelly, December 1997, 1-56396-712-X

To learn more about these titles, or the AIP Conference Proceedings Series, please visit the webpage **http://www.aip.org/catalog/aboutconf.html**

TOURS SYMPOSIUM ON NUCLEAR PHYSICS IV

Tours 2000

Tours, France   4–7 September 2000

EDITORS
M. Arnould
ULB, Belgium
M. Lewitowicz
GANIL, France
Yu.Ts. Oganessian
FLNR-JINR, Russia

H. Akimune
M. Ohta
H. Utsunomiya
T. Wada
T. Yamagata
Konan University, Japan

Melville, New York, 2001
AIP CONFERENCE PROCEEDINGS ■ VOLUME 561

Editors:

M. Arnould
Institut d'Astronomie et d'Astrophysique
Université Libre de Bruxelles Campus
Plaine CP 226
B-1050 Brussels
BELGIUM

E-mail: marnould@astro.ulb.ac.be

M. Lewitowicz
GANIL
BP 5027
F-14076 Caen Cedex
FRANCE

E-mail: lewitowicz@ganil.fr

Yu.Ts. Oganessian
Flerov Laboratory of Nuclear Reactions
Joint Institute for Nuclear Research
141980 Dubna, Moscow Region
RUSSIA

E-mail: oganessian@sungraph.jinr.ru

H. Akimune, M. Ohta, H. Utsunomiya,
T. Wada, T. Yamagata

Department of Physics
Konan University
8-9-1 Okamoto
Higashinada
Kobe 658-8501
JAPAN

E-mail: akimune@konan-u.ac.jp
masaota@konan-u.ac.jp
hiro@konan-u.ac.jp
wada@konan-u.ac.jp
yamagata@center.konan-u.ac.jp

Authorization to photocopy items for internal or personal use, beyond the free copying permitted under the 1978 U.S. Copyright Law (see statement below), is granted by the American Institute of Physics for users registered with the Copyright Clearance Center (CCC) Transactional Reporting Service, provided that the base fee of $18.00 per copy is paid directly to CCC, 222 Rosewood Drive, Danvers, MA 01923. For those organizations that have been granted a photocopy license by CCC, a separate system of payment has been arranged. The fee code for users of the Transactional Reporting Service is: 1-56396-996-3/01/$18.00.

© 2001 American Institute of Physics

Individual readers of this volume and nonprofit libraries, acting for them, are permitted to make fair use of the material in it, such as copying an article for use in teaching or research. Permission is granted to quote from this volume in scientific work with the customary acknowledgment of the source. To reprint a figure, table, or other excerpt requires the consent of one of the original authors and notification to AIP. Republication or systematic or multiple reproduction of any material in this volume is permitted only under license from AIP. Address inquiries to Office of Rights and Permissions, Suite 1NO1, 2 Huntington Quadrangle, Melville, N.Y. 11747-4502; phone: 516-576-2268; fax: 516-576-2450; e-mail: rights@aip.org.

L.C. Catalog Card No. 2001088177
ISBN 1-56396-996-3
ISSN 0094-243X
Printed in the United States of America

CONTENTS

Preface xi
Organizing Committee xiii
Letters xiv
Schedule xviii
Group Photo xix

NUCLEAR ASTROPHYSICS (NAP)

STELLAR STRUCTURE AND EVOLUTION—HANDLING NUCLEAR REACTION NETWORKS

Mixing and Thermonuclear Processes in Stars 3
    N. Langer, A. Heger, F. Herwig, and S. Wellstein
Important Current Issues in Supernova Theory 13
    A. Burrows
Exploding and Non-exploding Stars: Coupling Nuclear Reaction
Networks to Multidimensional Hydrodynamics 21
    K. Kifonidis, T. Plewa, and E. Müller
Phase Transitions in Dense Matter and Recent Topics of Neutron Stars 33
    M. Yasuhira
Nucleosynthesis in Massive Stars Using Extended Adaptive
Nuclear Reaction Networks 44
    A. Heger, T. Rauscher, R. D. Hoffman, and S. E. Woosley
Collapse-Driven Supernovae and the r-Process with New Data
of Unstable Nuclei 48
    K. Sumiyoshi

NUCLEAR PHYSICS AND COSMOCHRONOLOGY

Improved Predictions of Nuclear Data: A Continued Challenge
in Astrophysics 53
    S. Goriely
Cosmology, Cosmic Chemical Evolution, and the Age Problem 64
    G. J. Mathews
The Re/Os Clock: Open Questions Regarding the Neutron Cross Sections 72
    A. Mengoni and F. Käppeler
Theoretical Nuclear Astrophysics in Tours: A Summary 76
    M. Arnould and K. Takahashi

NUCLEAR REACTIONS IN ASTROPHYSICS

Advances in Cross Section Measurements at Low Energies 85
    M. Junker (for the LUNA-Collaboration)
Nuclear Astrophysics with a Recoil Mass Separator....................... 92
    F. Strieder
Radioactive Beams in Nuclear Astrophysics.............................. 102
    M. S. Smith
The Trojan Horse Method in Nuclear Astrophysics....................... 116
    M. Aliotta, M. Lattuada, M. G. Pellegriti, R. G. Pizzone,
    C. Spitaleri, D. Miljanic, C. Rolfs, S. Typel,
    and H. H. Wolter
First Result of the Cross Sectional Measurement of ^{3}He-^{3}He Solar Reaction in OCEAN... 127
    T. Itahashi, N. Kudomi, K. Kume, K. Takahisa, S. Yoshida,
    H. Ejiri, H. Toki, Y. Nagai, M. Komori, and H. Ohsumi
New Measurements of Astrophysical (γ,n) Reaction Rates Using a Quasi-thermal Photon Bath from Bremsstrahlung.................. 137
    K. Vogt, M. Babilon, W. Bayer, K. Denefleh, J. Enders,
    D. Galaviz, T. Hartmann, C. Hutter, P. Mohr, S. Volz,
    and A. Zilges
Nuclear Reactions in the Static Burning of Stars 145
    C. Rolfs
Nuclear Reactions in Hot and Explosive Burning: A Summary 159
    H. Utsunomiya, M. S. Smith, and T. Kajino

ATOMIC CLUSTER PHYSICS (ACP)

Atomic Clusters and Atomic Nuclei...................................... 167
    C. Guet
Shell Effects on Fission of Metal Clusters 174
    M. Nakamura
Evaporation of Finite Systems: The Case of Atomic Clusters.............. 184
    C. Bréchignac, Ph. Cahuzac, B. Concina, J. Leygnier,
    and B. Villard
Time-Dependent Mean-Field Method in Electronic Systems.................. 196
    K. Yabana
Formation of New Materials in Fullerenes by Using Nuclear Recoil 204
    T. Ohtsuki, K. Ohno, K. Shiga, Y. Kawazoe, Y. Maruyama,
    K. Shikano, and K. Masumoto
Electronic Properties of Nanotube Based Materials 214
    S. Saito

PHYSICS FOR NUCLEAR TRANSMUTATION (PNT)

Radiochemical Measurements of Nuclear Data for Transmutation of Minor Actinides 223
    N. Shinohara

Basic Nuclear Research Involvment in the French Research Programme on the Back-End of the Nuclear Fuel Cycle 232
    H. C. Flocard

PHYSICS WITH EXOTIC NUCLEI (PEN)

HALO NUCLEI AND NUCLEAR CLUSTERS

Nuclear Halo and Molecular States 247
    N. A. Orr

Measurements of Interaction Cross Sections for Nuclei Far from Stability at Relativistic Energies 259
    T. Suzuki (for the GSI-RIKEN-Kurchatov-Comenius Collaboration)

Shell Structure of Neutron-Rich Light Nuclei: New Vista 269
    F. Azaiez

NUCLEAR SHELLS VERY FAR FROM STABILITY

Proton Drip-Line Nuclei Studied at Intermediate Energies 280
    B. Blank

Nuclear Structure in the Vicinity of Shell Closures Far from Stability 287
    H. Grawe, M. Górska, J. Döring, C. Fahlander, M. Palacz,
    F. Nowacki, E. Caurier, J. M. Daugas, M. Lewitowicz,
    M. Sawicka, M. Pfützner, R. Grzywacz, K. Rykaczewski,
    O. Sorlin, S. Leenhardt, F. Azaiez, M. Rejmund, K. Hauschild,
    and J. Uusitalo

SUPERHEAVY ELEMENTS (SHE)

EXPERIMENTAL DEVELOPMENTS I

Heaviest Elements (Synthesis and Decay Properties) 303
    Y. T. Oganessian

Recent Developments in the Synthesis of Superheavy Elements 323
    D. Ackermann

EXPERIMENTAL DEVELOPMENTS II

Fusion of Deformed Nuclei in the Vicinity of the Coulomb Barrier 334
    H. Ikezoe, S. Mitsuoka, K. Nishio, K. Satou, and S. C. Jeong
Production of Superheavy Elements at GANIL 344
    C. Stodel, N. Alamanos, N. Amar, J. C. Angélique, R. Anne,
    G. Auger, J. M. Casandjian, R. Dayras, A. Drouart,
    J. M. Fontbonne, A. Gillibert, S. Grévy, D. Guerreau,
    F. Hanappe, R. Hue, A. S. Lalleman, N. Lecesne,
    T. Legou, M. Lewitowicz, R. Lichtenthäler, E. Liénard,
    L. Maunoury, W. Mittig, N. Orr, J. Péter, E. Plagnol,
    G. Politi, M. G. Saint-Laurent, J. C. Steckmeyer, J. Tillier,
    R. de Tourreil, A. C. C. Villari, J. P. Wieleczko, and A. Wieloch
**Search for a Z=118 Superheavy Nucleus in the Reaction of
Kr Beam with Pb Target at RIKEN** 354
    K. Morimoto, K. Morita, I. Tanihata, N. Iwasa, R. Kanungo,
    T. Kato, K. Katori, H. Kudo, T. Suda, I. Sugai, S. Takeuchi,
    F. Tokanai, K. Uchiyama, Y. Wakasaya, T. Yamaguchi,
    A. Yeremin, A. Yoneda, and A. Yoshida
    (RIKEN SuperHeavy Experimental Group)

NUCLEAR CHEMISTRY OF HEAVY ELEMENTS

Spectroscopic Studies of Mass-Separated Heavy Nuclei 358
    M. Asai, M. Sakama, K. Tsukada, S. Ichikawa, H. Haba,
    I. Nishinaka, Y. Nagame, S. Goto, Y. Kojima, Y. Oura,
    H. Nakahara, M. Shibata, and K. Kawade
Gas Chromatographic Studies of the Heaviest Elements 368
    A. Vahle

PROPERTIES OF SUPERHEAVY ELEMENTS

The Performance of Mean-Field Models for Superheavy Elements 377
    P.-G. Reinhard, M. Bender, T. Bürvenich, T. Cornelius,
    P. Fleischer, and J. A. Maruhn
**Decay Properties of Heavy and Superheavy Nuclei Predicted
by Nuclear Mass Formulas** 388
    H. Koura
Microscopic Description of Damped Collective Motion 399
    S. Yamaji, H. Hofmann, and F. A. Ivanyuk

REACTION THEORY FOR SUPERHEAVY ELEMENTS

Reaction Theory for the Synthesis of Superheavy Elements 411
    T. Wada, Y. Aritomo, T. Ichikawa, M. Ohta, and Y. Abe

Fusion and Quasifission in Collisions of Heavy Nuclei **421**
    G. G. Adamian, N. V. Antonenko, A. Diaz Torres, W. Scheid,
    and Y. M. Tchuvil'sky

Production of Superheavy Elements in Cold Fusion Reactions **433**
    V. Y. Denisov

FISSION MODES

**The Limit of Nuclear Deformation and Fission Properties
of Heavy and Superheavy Elements** **443**
    Y. L. Zhao, I. Nishinaka, Y. Nagame, K. Sueki,
    and H. Nakahara

**Five-Dimensional Potential-Energy Surfaces and Coexisting
Fission Modes in Heavy Nuclei.** **455**
    P. Möller, D. G. Madland, A. J. Sierk, and A. Iwamoto

List of Participants. **469**
Author Index. **475**

PREFACE

Tours Symposium on Nuclear Physics IV (September 4 – 7, 2000) was devoted to the following five topics in the broad field of nuclear physics: NAP (Nuclear Astrophysics), ACP (Atomic Cluster Physics), PNT (Physics for Nuclear Transmutation), PEN (Physics with Exotic Nuclei), and SHE (Super-Heavy Elements). We organized Tours2000 in the turning year to the 21th century and have reached another highest standard to carry it over to the next symposium. We are grateful to the members of the Organizing Committee for their every effort in organizing Tours2000. Needless to say, participants were the key to the success of Tours2000. We thank all the participants for their contributions. We remark that the NAP program had three round table sessions for the first time in this series of symposia. The discussions were lively and entertaining, thanks to the moderators and participants. We hope that the round table session will become a unique feature of our symposium in the future.

Tours2000 was strongly supported by the following organizations: Ambassade de France (Mr. Maurice GOURDAULT-MONTAGNE, Ambassadeur; and Prof. Henri ANGELINO, Censeiller pour la Science et Technologie), Conceil Général d'Indre-et-Loire (Mr. Jean DELANEAU, Président; Michel TROCHU, Vice-President Conseiller Municipal de Tours; and Ms. Brigitte ESPAZE, Chargée de Mission), GANIL (Prof. Daniel GUERREAU, Directeur), Univesité Françoir Rabelais, Tours (Prof. Jacques GAUTRON, Président), Lycée Collège Konan de Touraine (Mr. Shuzo FUJIWARA, Directeur; and Ms. Miki CREOLA), and Tokyo Marine Insurance Company.

M. Ohta
H. Utsunomiya
Konan University

ORGANIZING COMMITTEE

Y. Abe (YITP, Japan)
H. Akimune (Konan, Japan)
M. Arnould (ULB, Belgium)
D. Guerreau (GANIL, France)
C. Guet (Grenoble, France)
A. Iwamoto (JAERI, Japan)
T. Kajino (NAO, Japan)
S. Kubono (CNS, Japan)
S.M. Lee (Tsukuba, Japan)
M. Lewitowicz (GANIL, Japan)
G. Münzenberg (GSI, Germany)
Y. Nagai (RCNP, Japan)
T. Nomura (KEK, Japan)
Yu. Ts. Oganessian (FLNR-JINR, Russia)
M. Ohta (Konan, Japan)
C. Rolfs (Bochum, Germany)
H. Utsunomiya (Konan, Japan)
T. Wada (Konan, Japan)
T. Yamagata (Konan, Japan)
K. Yabana (Tsukuba, Japan)

HOST INSTITUTE
Konan University

SUPPORTED BY
Ambassade de France au Japon
Conceil Général d'Indre-et-Loire
GANIL
Univesité Françoir Rabelais, Tours
Lycée Collège Konan de Touraine
Tokyo Marine Insurance Company

**AMBASSADE DE FRANCE
AU JAPON**

L'Ambassadeur

Tokyo, le 25 juillet 2000

En septembre 2000 se tiendra le 4ème symposium de Tours de physique nucléaire.

A cette occasion je salue la coopération originale qui existe entre la ville de Tours et l'université Konan grâce à ce symposium sur les éléments chimiques super-lourds, les noyaux exotiques, et l'astrophysique nucléaire. Elle s'inscrit dans l'ensemble plus vaste des relations franco-japonaises dans le domaine de la physique, relations anciennes et actives. Et elle inclut de plus une composante éducative grâce à l'existence du Lycée-collège Konan de Touraine fondé en 1991.

Je souhaite donc cette année un plein succès au symposium de Tours, dans un domaine où excellent Français et Japonais. Je suis sûr que ce symposium contribuera à l'approfondissement des liens scientifiques entre nos deux pays.

Maurice GOURDAULT-MONTAGNE

 Relations Internationales

KONAN UNIVERSITY 4th SYMPOSIUM ON NUCLEAR PHYSICS

September 2000

KONAN University 4th symposium on nuclear physics was held in TOURS from the 4th to the 7th of September 2000 to the utmost satisfaction of the french partners who are delighted with the renewal of this event.

It enriches and strengthens the tights between KONAN and TOURS Universities in the frame of the Agreement signed in April 1999 between the two partners and more particularly in the field of scientific cooperation.

It will be a pleasure to receive such an event in TOURS again to the mutual benefit of KONAN and TOURS Universities.

Jacques GAUTRON
Président de l'Université
François-Rabelais de Tours

Tours, November 23rd, 2000

The 4th edition of the International Symposium on Nuclear Physics of Tours was an important event for us as it was held on the occasion of both the year 2000 and the 10th anniversary of the foundation of the Konan school in Touraine.

If the turn of the century is mainly symbolic, the development of our relationship with the Konan Institution proves to be very effective and rewarding. Ten years of mutual cooperation not only strengthened the ties between our two regions, but also encouraged outstanding initiatives such as the Symposium while giving birth to true and long-lasting friendships.

I take this opportunity to express my special thanks to Pr. Ohta and Pr. Utsunomiya who, once more, made this event successful.

At the dawn of the new millenium, we hope to reinforce our partnership with Konan, particularly in the fields of sciences where the Konan University scholars already play an important role on the international scene.

We would be proud to host the Symposium again in Touraine in 2003 as we are convinced the research work you are all leading in a spirit of cooperation and tolerance is another step to building a more advanced but also wiser humanity.

Jean DELANEAU
Senator
Le Président,
Conceil Général d'Indre-et-Loire

LA NOUVELLE RÉPUBLIQUE DU CENTRE-OUEST
LUNDI 4 SEPTEMBRE 2000 - INDRE-ET-LORE

Symposium de physique nucléaire

Du 4 au 7 septembre, le 4° Symposium international de physique nucléaire se tient à l'hôtel de l'Univers, à Tours. Cette manifestation, qui a lieu tous les trois ans, réunit près d'une centaine de chercheurs qui viennent présenter les dernières découvertes et les plus récents de leurs travaux. La présence du professeur Oganessian est attendue avec intérêt par les physiciens, car il parlera de son travail sur les noyaux super lourds, un des sujets les plus actuels dans le domaine de la recherche fondamentale.

ヌーヴェル・レピュブリック紙 記事訳
2000.9.4.付

核物理シンポジウム

9月4日から7日まで、トゥール市のホテル・ユヴェールで第4回原子核物理に関する国際シンポジウムが開催される。3年ごとに行われるこの会議には、最近の発見や研究成果を発表する100人近い研究者が集まる。基礎科学の最近の話題であるオガネシアン教授による超重元素に関する発表に期待が寄せられている。

LA NOUVELLE RÉPUBLIQUE DU CENTRE-OUEST
JEUDI 7 SEPTEMBRE 2000 - INDRE-ET-LORE

La physique nucléaire en congrès

Depuis le début de la semaine, une cinquantaine de spécialistes mondiaux en physique nucléaire sont réunis, jusqu'à aujourd'hui, à Tours. C'est la quatrième édition de ce symposium international de physique nucléaire organisé par des professeurs du Ganil, en France, et de l'université de Konan à Kobé au Japon.

« Depuis 1991, date de l'implantation du lycée Konan en Touraine, nous avons voulu renforcer l'esprit de collaboration qui existe entre la Touraine et Kobé », commmente le professeur Ohta, de l'université de Konan, initiateur du symposium qui a lieu tous les trois ans.

« C'est un bon forum, explique le professeur Arnould, professeur d'astronomie et d'astrophysique, directeur de l'Institut d'astrophysique de l'université libre de Bruxelles. On peut discuter entre participants d'un mariage heureux entre astrophysique et la physique nucléaire. On se rend compte que l'infiniment petit influence l'infiniment grand. Ce genre de colloque nous permet de comprendre que nous ne connaissons que très peu de choses. »

ヌーヴェル・レピュブリック紙　記事訳

2000.9.7.付

核物理シンポジウム

今週始めより、核物理の分野で世界的に活躍する物理学者約50人がトゥールに集まりシンポジウムを開いている。フランスのガニール研究所と神戸にある甲南大学が主催する国際核物理シンポジウムは今回で第4回目を迎える。

「1991年にトゥレーヌ甲南学園がこの地に開校されて以来、トゥレーヌ地方と神戸の協力関係を築いてきた。」と、主催者の甲南大学太田教授は語る。このシンポジウムは、3年毎にこの地で開催されている。

また、ブラッセルの大学で天文学と宇宙物理学教える、宇宙物理研究所のアルヌー所長は、「天文物理と核物理の研究について有意義な意見交換ができるこの様なフォーラムは、すばらしい。極小の研究は、極限大の理解を助ける。この種のシンポジウムで、我々は如何に多くの分野が未知であるかがわかる。」と話した。

SCHEDULE

	Sept. 4 (Mon)	Sept. 5 (Tue)		Sept. 6 (Wed)	Sept. 7 (Tur)
09:00 – 10:30	NAP I-1	NAP III-1	ACP I	PEN I	SHE IV
	coffee break	coffee break		coffee break	coffee break
11:00 – 12:30	NAP I-2	NAP III-2	ACP II	PEN II	SHE V
12:30 – 14:00	lunch	lunch		lunch	lunch
14:00 – 15:30	NAP II-1	NAP IV / NAP V	PNT	SHE I	SHE VI
	coffee break			coffee break	
16:00 – 17:30	NAP II-2	Excursion		SHE II	
	coffee break			coffee break	
18:00 – 19:30	NAP II-3	Banquet		SHE III	

NAP: Nuclear Astrophysics
PNT: Physics for Nuclear Transmutation
SHE: Superheavy Elements
ACP: Atomic Cluster Physics
PEN: Physics with Exotic Nuclei

Chairpersons

NAP I-1	G. Mathews (Notre Dame, USA)	PNT	M. Ohta (Konan, Japan)
NAP I-2	S. Goriely (ULB, Belgium)	PEN I	D. Guerreau (GANIL, France)
NAP I-3	M. Arnould (ULB, Belgium)	PEN II	M. Lewitowicz (GANIL, France)
NAP II-1	M. Smith (ORNL, USA)	SHE I	M. Ohta (Konan, Japan)
NAP II-2	K. Takahashi (Heidelberg Germany)	SHE II	G. Chubarian (Texas A&M, USA)
NAP III-1	T. Itahashi (RCNP Japan)	SHE III	K. Morita (RIKEN, Japan)
NAP III-1	M. Junker (Gran Sasso, Italy)	SHE IV	P. Möller (Los Alamos, USA)
NAP III-1	H. Utsunomiya (Konan, Japan)	SHE V	A. Iwamoto (JAERI, Japan)
ACP I	N. Takahashi (Osaka, Japan)	ShE VI	Yu. Ts. Oganessian (FLNR-JINR, Russia)
ACP II	C. Guet (CEA Grenoble, France)		

NUCLEAR ASTROPHYSICS (NAP)

Mixing and Thermonuclear Processes in Stars

Norbert Langer[1], Alexander Heger[2], Falk Herwig[3] and Stephan Wellstein[4]

[1] *Astronomical Institute, Utrecht University, The Netherlands*
[2] *Lick Observatory, UC Santa Cruz, U.S.A.*
[3] *Dept. of Physics and Astronomy, University of Victoria, Canada*
[4] *Potsdam University, Potsdam, Germany*

Abstract. According to standard wisdom, six different nuclear burning stages, from hydrogen to silicon burning, can occur during the hydrostatic evolution of stars. However, several instabilities can lead to the mixing of layers with different compositions and produce non-standard composition mixtures. Those — when at high enough temperatures — burn due to nuclear reactions which are unimportant otherwise. We provide various examples of such occurrences, for different ratios of burning over mixing time scale.

INTRODUCTION

The vast majority of heavy elements is produced in stars, either during their hydrostatic nuclear burning processes or in explosive burning events (e.g., Woosley & Weaver 1995). Due to the relatively long time scale of the hydrostatic burning stages, they do not involve nuclei far from the valley of stability in the nuclear chart. Thus, both, from the stellar physics and the nuclear physics point of view, hydrostatic nuclear burning is "easy" compared to explosive processing. However, as we shall see in the course of this article, there is one property of stars disturbing this simplicity: often, the chemical stratification in a star is *unstable*, and as a consequence layers with different composition are mixed, which may open up non-standard nuclear reaction channels.

To illustrate the term non-standard, we recall the "standard" onion skin model for stellar structure (cf. Fig. 1), where the ashes of the various nuclear burning stages are layered above each other, well separated by sharp boundaries which usually coincide with active nuclear shell sources. Indeed, active nuclear shell sources provide strong entropy barriers which do not allow any mixing process to penetrate them. However, in response to expansion and contraction of the various layers in the course of evolution, shell sources may become weak or even extinct (e.g., the

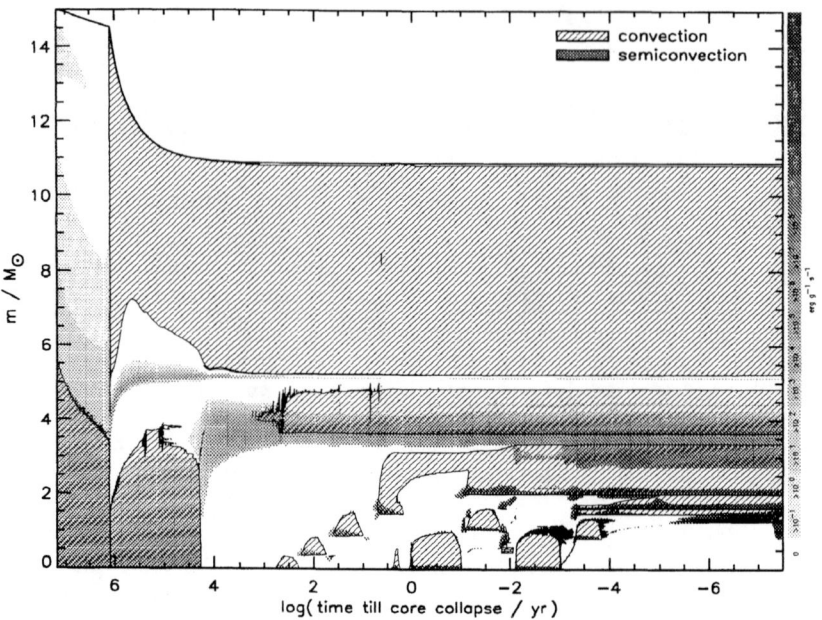

FIGURE 1. So called Kippenhahn-diagram, describing the evolution of the internal structure of a star with an initial mass of $15\,M_\odot$, from core hydrogen ignition until the onset of iron core collapse (cf. Heger et al. 2000). The X-axis gives the logarithm of the time (in years) with t=0 defined at the onset of iron core collapse. The Y-axis is a spatial coordinate; however, instead of a radius coordinate, the Lagrangian mass coordinate is used. Hatched areas are convectively unstable, and the convective cores of hydrogen, helium, carbon, neon, oxygen and silicon burning can be identified, successively. Gray shading denotes nuclear energy generation. After core burning, all burning stages develop shell sources, which give rise to the "onion skin model".

hydrogen burning shell in Fig. 1). In such a situation, the stage is set for mixing the ashes of neighbouring burning phases, which, if this mixture is exposed to high enough temperatures, gives rise to non-standard nuclear processes in stars.

MIXING PROCESSES

The list of instabilities which can lead to mixing processes in stars is very long (cf., Pinsonneault 1997); let us briefly consider the most important ones. Most important is certainly thermal convection. Fig. 1 shows that at any time a significant fraction of a massive star is convectively unstable; after the hydrogen burning stage even most of it. Convection is a dynamical instability, caused either by local nuclear energy production (note the nuclear energy production at the base of all convection zones in Fig.1, except for the convective envelope) or a low photon transmissivity.

While convection mixes on the local dynamical time scale, composition gradients in stars are typically such that light nuclei are lying on top of heavier ones. I.e., it costs energy to mix these layers in the gravitational potential of the star. Convection often is not able to mix in this situation; however, an instability confusingly named semiconvection may develop if the temperature stratification is superadiabatic. But, in contrast to convection, this instability acts only on the local thermal time scale, which is orders of magnitude larger than the local dynamical time scale.

Then, in particular as a result of mass transfer in binary systems, the situation may occur that heavy nuclei are placed above lighter ones in a star. This situation is unstable even if the temperature gradient is subadiabatic. So called thermohaline convection develops, which proceeds on the thermal time scale, like semiconvection.

Finally, all stars rotate, which can bring along a series of instabilities. Most important are the so called Eddington-Sweet circulations, slow currents which work towards thermal equilibrium in an ablated star with hotter poles and cooler equatorial regions, and the shear instability which may mix layers if they contain a sufficiently strong gradient in angular velocity (cf. Heger et al. 2000).

If any of these processes mixes elements into a nuclear burning zone, it is important to compare the time scale of this mixing process with the corresponding nuclear burning time scale. If the mixing time scale is short compared to the burning time scale — as it is, e.g., in a convective core — one may assume that the mixed region remains chemically homogeneous, and mixing and burning can be decoupled, as done in most stellar evolution models. If the mixing time is long compared to the burning time, obviously mixing and burning can be decoupled as well. However, if both time scales are comparable the situation is different, and a numerically costly coupled solution of the equations describing mixing and burning is necessary (Herwig et al. 1999). In all three cases, as well as when the burning starts only after the mixing is over, we may find non-standard nuclear fusion processes.

LONG OR SHORT MIXING TIME SCALES

Example 1: ^{13}C-Production in AGB stars

It is known since several decades that the main component of the s-process is produced in so called thermally pulsing AGB stars (cf., Gallino et al. 1998, and references therein), i.e. in red giants which consist of an electron-degenerate CO-core, an extended hydrogen-rich envelope, and two nuclear shell sources, the hydrogen and the helium burning shell, which operate very close together and alternately switch on and off. The s-process is supposed to occur in the helium burning shell using neutrons liberated by the ^{13}C(α, n)-reaction. The problem is: within standard assumptions, AGB stellar models do not have any significant amount of ^{13}C in the helium-rich layers (Gallino et al. 1998). Instead, one needs to invoke a non-standard mixing-process which transports protons into the helium burning region so that ^{13}C can be produced via proton capture on ^{12}C. I.e., not only the production

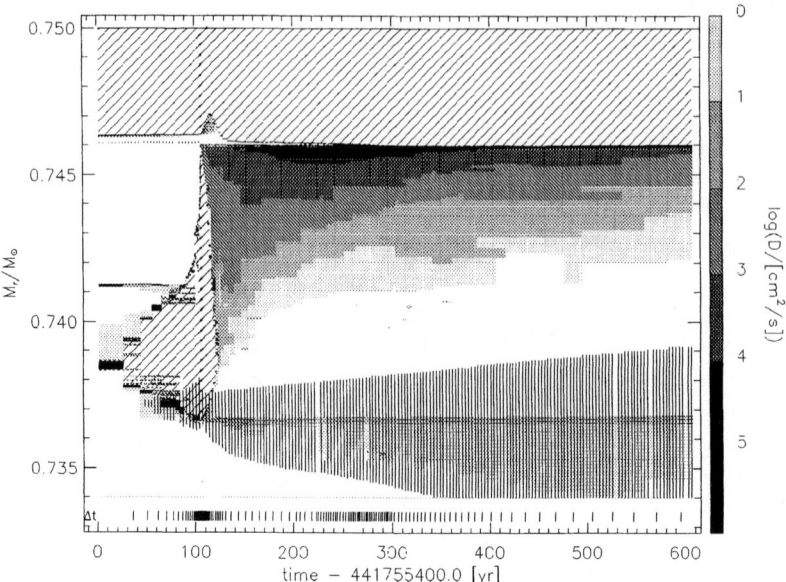

FIGURE 2. Section of the internal structure during and after the 25th thermal pulse of a rotating $3\,M_\odot$ sequence (Langer et al. 1999). Diagonal hatching denotes convection. The convective envelope extends down to $M_r \simeq 0.746\,M_\odot$. The pulse driven convective shell is located at $0.737\,M_\odot \lesssim M_r \lesssim 0.746\,M_\odot$ and $30\,\mathrm{yr} \lesssim t \lesssim 120\,\mathrm{yr}$. Vertical hatching denotes regions of significant nuclear energy generation, i.e., the hydrogen burning shell (at $M_r \simeq 0.746$ and $t \lesssim 100\,\mathrm{yr}$) and the helium burning shell ($0.734\,M_\odot \lesssim M_r \lesssim 0.739\,M_\odot$ and $t \gtrsim 40\,\mathrm{yr}$). Gray shading marks regions of significant rotationally induced mixing (see scale on the right side of the figure). Vertical marks at the bottom of the figure denote the time resolution of the calculation, where every fifth time step is indicated.

of ^{13}C, but the production of the whole main component of the s-process depends on non-standard mixing.

Very recently, two different approaches lead to significant ^{13}C production in AGB models: Herwig et al. (1997) found so called convective overshooting to be able to bring right amounts of protons into the helium-rich layers. Langer et al. (1999) have considered models of rotating AGB stars and found that shear mixing may also do the job (Figs. 2 and 3).

Mixing protons into helium burning layers may, however, involve further non-standard nuclear reactions, e.g. the production of fluorine or even of ^{15}N (cf. Jorissen & Arnould 1989). We note that this process may also occur in rotating massive stars with so far little explored consequences (Langer et al. 1997, Heger et al. 2000)

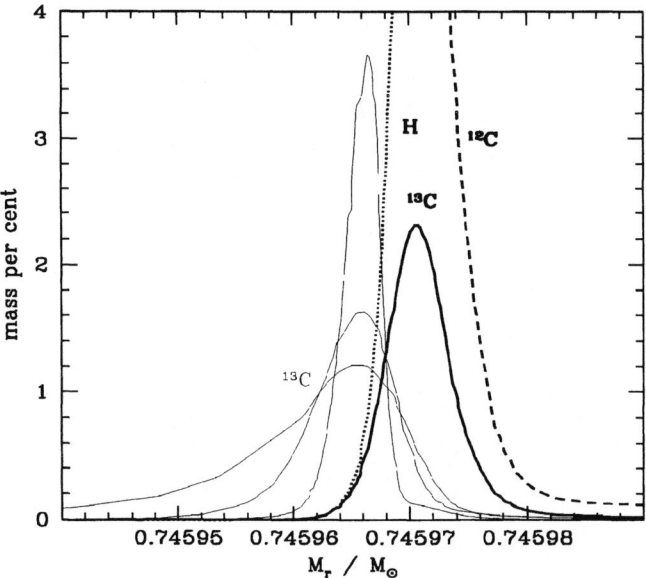

FIGURE 3. Chemical profiles at the location of the maximum depth of the convective envelope during the 25th thermal pulse (cf. Figure 2) of a rotating $3\,M_\odot$ AGB sequence (cf. Langer et al. 1999). The dotted and dashed lines mark the hydrogen and the ^{12}C mass fractions at $t = 1704\,\mathrm{yr}$, with $t = 0$ defined as in Figure 2. The fat solid line denotes the ^{13}C mass fraction at the same time. The three thin solid lines represent the ^{13}C mass fractions at $t = 2016\,\mathrm{yr}$, $t = 4155\,\mathrm{yr}$, and $t = 5139\,\mathrm{yr}$, with a later time corresponding to a smaller peak abundance. The maximum ^{13}C mass fractions of 3.6% occurs at $t = 2016\,\mathrm{yr}$. The ^{13}C peak moved inwards in the time interval from $t = 1704\,\mathrm{yr}$ to $t = 2016\,\mathrm{yr}$ due to continued proton captures on both, ^{12}C and ^{13}C.

Example 2: ^{26}Al-Production in Binaries

Most stars appear to be members of binary or multiple systems. The fraction of massive stars being members of *close* binaries — i.e. such in which mass overflow is expected to occur — is estimated to be of the order of 20...40% (cf., Podsiadlowski 1997, Mason et al. 1998). Langer et al. (1998) and Wellstein & Langer (1999) have studied the evolution and mixing processes in typical massive close binaries. Due to the transfer of most of the hydrogen-rich envelope of the primary to the secondary component, the primary becomes a helium star while the secondary evolves into a luminous blue supergiant. In both cases, the core masses evolve differently compared to single stars of the same initial mass (i.e., 20 or $18\,M_\odot$ in our example); the primaries' core masses are smaller, that of the secondaries larger.

The secondary star, which accretes the envelope of the primary during its core hydrogen burning evolution, is commonly assumed to evolve after accretion exactly like a single star of the corresponding new mass ($\sim 32\,M_\odot$ in our example).

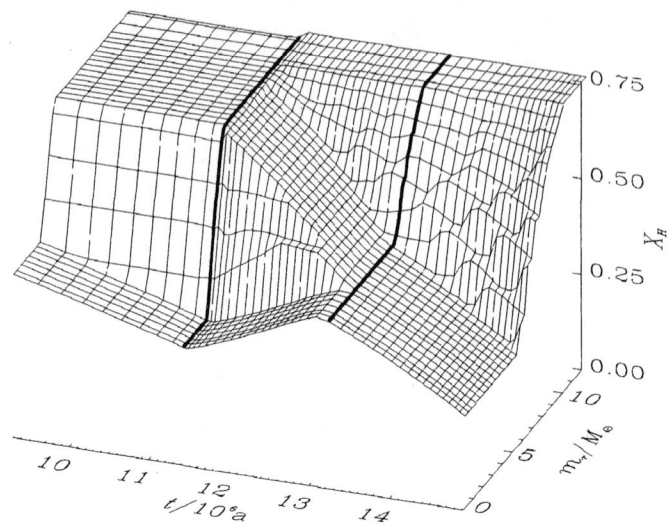

FIGURE 4. Hydrogen profile as function of time in a 12 M$_\odot$ star which accreted 8 M$_\odot$ at t$\simeq$ 11.4 10^6 a (left thick drawn line). At t$\simeq$ 13.1 10^6 a (right thick drawn line), the semiconvection controlled "rejuvenation process" is finished. The star has then adjusted its convective core size to its new mass (cf., Braun & Langer 1995). Note that were the accretion occurring later in the evolution of the 12 M$_\odot$ star, the time for rejuvenation might be too short.

However, Braun & Langer (1995) found that secondaries may retain significant structural differences compared to single stars (cf. Fig. 4), e.g. with the consequence that the supernova explosion occurs in the blue supergiant stage — as in the case of the progenitor of SN 1987A — rather than in the red supergiant stage.

Independent of this phenomenon, Langer et al. (1998) found that substantial differences in the synthesis of secondary CNO isotopes do occur in the secondary components of massive close binaries, due to the interplay between CNO burning and so called thermohaline mixing. This mixing process does occur in the whole hydrogen-rich envelope of the secondary from the surface down to the convective H-burning shell since the star accretes helium enriched matter from the primary component (i.e. matter with a higher mean molecular weight is lying above matter with lower mean molecular weight). As the time scale for thermohaline mixing — which is for the first time treated as a time dependent process in our models — and CNO processing at the bottom of the mixed zone are comparable, the production particularly of ^{13}C and ^{14}N is boosted in the whole H-rich envelope, increasing the yield of these isotopes by factors of the order of 2...3.

The synthesis of ^{26}Al in secondary components of close binaries was found to deserve special attention. The reason is that for ^{26}Al, as it is β-unstable with a mean life time of 1.03 10^6 yr ($\tau_{1/2}$ = 7.2 10^5 yr), not only the amount which is synthesised matters, but also the time of the synthesis if one wants to explain the observed γ-ray line emission from the decay of ^{26}Al in the Galaxy. We can only see the decay of ^{26}Al nuclei in the interstellar medium; the decay inside stars is unobservable. Therefore, the ^{26}Al which is observed should either be produced during supernova explosions or shortly before. From the spatial distribution of the γ-ray line emission (Prantzos & Diehl 1996) we know that it originates from massive stars. Supernovae are in fact the currently favoured production site, although the corresponding yields are very uncertain (Weaver & Woosley 1993, Woosley & Weaver 1995, Timmes & Woosley 1997).

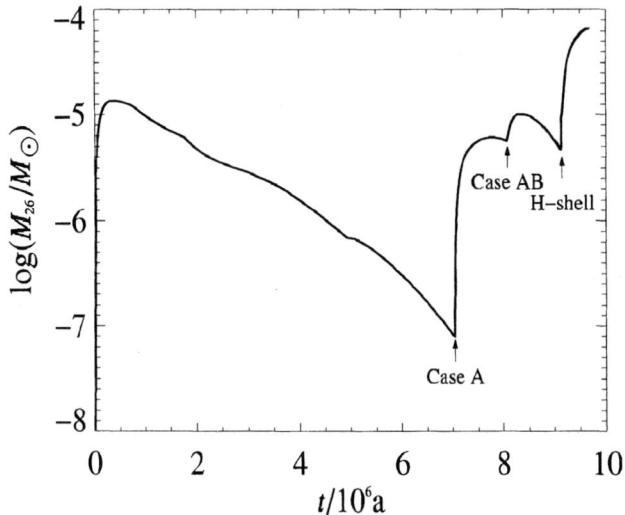

FIGURE 5. Evolution of the total mass of ^{26}Al inside the secondary component of a 20+18 M$_\odot$ close binary system as a function of time (in 10^6 yr). The beginning of the so called Case A and Case AB mass transfer phases is indicated, as well as the beginning of hydrogen shell burning which marks the end of the core helium burning evolutionary phase. When the star explodes as a supernova of Type SN 1987A ($t = 9.68\,10^6$ yr) it contains almost 10^{-4} M$_\odot$ of 26Al. Note that the initial metallicity of the stars is $Z_\odot/4$.

However, ^{26}Al is also produced during hydrostatic hydrogen burning, by proton capture on ^{25}Mg. Although very massive stars, through extremely strong stellar winds, can eject ^{26}Al generated during H-burning and contribute to the Galactic ^{26}Al (Langer et al. 1995, Meynet et al. 1997), the hydrogen burning contribution of less massive stars (say 10...30 M$_\odot$) is not considered as important since the major fraction of the ^{26}Al decays inside the star before it is released in the course of the supernova explosion.

Langer et al. (1998) found this to be different in massive close binary secondaries. Fig. 5 shows the time dependence of the amount of ^{26}Al generated by hydrogen burning in the interior of the secondary component of a 20+18 $M_\odot$ system. Obviously, the ^{26}Al mass, although 10^{-5} $M_\odot$ initially, would be of the order of 10^{-8} $M_\odot$ at the end of the evolution if no mass accretion would occur. However, since fresh fuel — i.e. also fresh ^{25}Mg — is mixed into the core due to the mass accretion process (cf. Fig. 4) the amount of ^{26}Al is increased by orders of magnitude at that time.

This would already be sufficient to produce as much as $\sim 10^{-6}$ $M_\odot$ ^{26}Al from this star of initially 18 $M_\odot$ — two orders of magnitude more than expected from single star calculations at $Z_\odot/4$ (cf. Langer et al. 1995). However, we find the H-burning shell source in the secondary component to be much more efficient than in corresponding single stars. This leads to the coupling of an extended convection zone to the hydrogen burning shell, and consequently to the enrichment of the whole convection zone with ^{26}Al. In the end, our secondary star contains almost 10^{-4} $M_\odot$ of ^{26}Al which will be liberated during the supernova explosion. As the chosen example appears to be a rather typical case, the Galactic ^{26}Al production may in fact be dominated by massive close binary systems.

COMPARABLE MIXING AND NUCLEAR TIME SCALES

Example 3: ^{7}Li-Production in Giants

As mentioned above, when mixing and nuclear time scale are comparable one needs to solve the equations describing both processes simultaneously in order to obtain the correct result. An example is the so called Cameron-Fowler mechanism for lithium production in red giants (Cameron & Fowler 1971). The mixing process involved is convection, which acts in the hydrogen-rich envelope of red giants. Under certain circumstances, the bottom of this convective envelope can be hot enough for thermonuclear processes to occur in it.

When the reaction ^{3}He$(\alpha,\gamma)^7$Be is enabled (with ^{3}He being produced by the pp-chain) we have a nuclear time scale involved which is of the same order as the convective mixing time scale (~ 1 yr), namely the β-decay of ^{7}Be to ^{7}Li ($\tau_\beta \simeq 50$ d). As the beryllium is transported out of the hot region into cooler layers by convection, the lithium which forms as its decay product may survive instead of being almost immediately destroyed.

The Cameron-Fowler mechanism may lead to considerable lithium production, and there is observational evidence for its operation in Li-rich red giants and AGB stars (cf. Charbonnel & Balachandran 2000).

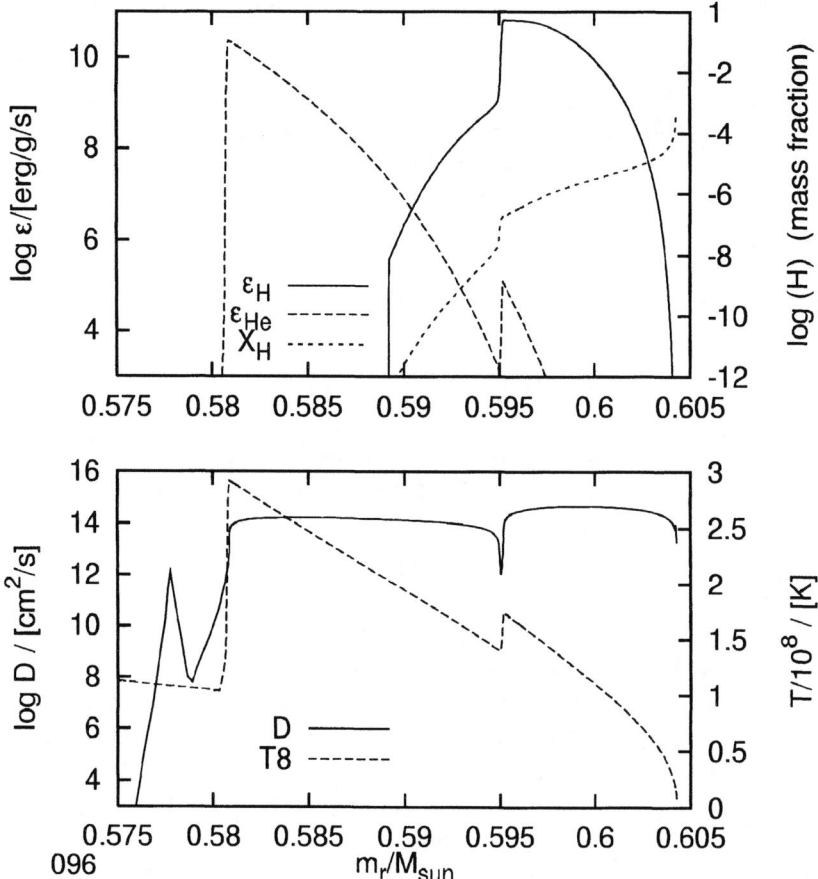

FIGURE 6. Profiles of various quantities in the nuclear burning region in post-AGB star during a late thermal pulse at the time of maximum energy generation (see text, and Herwig et al. 1999). The top panel shows nuclear energy generation rates due to proton and alpha captures, and the hydrogen mass fraction X_H. The bottom panel shows the diffusion coefficient D and the temperature in units of 10^8 K.

Example 4: Very Late Thermal Pulses

When AGB stars (see above) have lost all but a small fraction of their hydrogen-rich envelope due to their strong winds, they evolve within a short time scale into proto-white dwarfs. At this stage the hydrogen burning shell is practically extinct. However, there is evidence for some of these stars to undergo a so called "late thermal pulse" at this stage, i.e. the helium shell re-ignites violently. As the hydrogen burning shell is extinct, no entropy barrier prevents the convective mixing caused by this energetic event to penetrate into the remaining hydrogen-rich layers.

This is a most precarious situation, as convection brings the protons straight down into the helium burning layers. I.e., simultaneous hydrogen and helium burning occurs in the convection zone.

Fig. 6 describes this situation (cf. Herwig et al. 1999), which can not be modelled with standard stellar evolution codes. Protons survive up to temperatures of $\sim 2\,10^8$ K where they react with the products of helium burning. These, on the other hand, are mixed up into the cooler hydrogen-rich layers where they can react with protons. In the example shown in Fig. 6, all available protons are consumed, and the result is a hydrogen-free carbon-rich pre-white dwarf. We note that such objects are observed, as are stars which currently undergo a late thermal pulse.

Acknowledgements

This work has been supported by the NWO through grant No. 614.21.015, and by the Deutsche Forschungsgemeinschaft through grant No. La 587/15-2.

REFERENCES

1. Braun H and Langer N, 1995 A&A 297, 483
2. Cameron A G W, Fowler W A, 1971, ApJ 164, 111
3. Charbonnel C, Balachandran S C, 2000, A&A 359, 563
4. Gallino R, Arlandini C, Busso M, et al., 1998, ApJ 497, 388
5. Heger A, Langer N, Woosley S E, 2000, ApJ 528, 368
6. Herwig F, Blöcker T, Schönberner D, El Eid M F, 1998, A&A 324, L81
7. Herwig F, Blöcker T, Langer N, Driebe T, 1999, A&A 349, L5
8. Jorissen A., Arnould M, 1989, A&A 271, 161
9. Langer N, Braun H and Fliegner J, 1995, Astrophys. Space Sci. 224, 275
10. Langer N, Fliegner J, Heger A and Woosley S E, 1997 Nucl. Phys. A621, 457c
11. Langer N, Heger A, Braun H, 1998, in: *Atomic & Nuclear Astrophysics*, Proceedings of the 2nd Oak Ridge Symposium, T. Mezzacappa, ed., IOP Publishing, Bristol, p. 377
12. Langer N, Heger A, Wellstein S, Herwig F, 1999, A&A 346, L40
13. Mason B D, Gies D R, Hartkopf W I, Bagnuolo W G Jr, Brummelaar T T and McAlister H A 1998 ApJ 115, 821
14. Meynet G, Arnould M, Prantzos N and Paulus G 1997 A&A 320, 460
15. Pinsonneault M H, 1997, ARAA 35, 557
16. Podsiadlowski Ph 1997 in: *Evolutionary Processes in Binary Stars* eds. R A M J Wijers et al. NATO ASI Ser. C Vol. 477 (Dordrecht: Reidel) p 181
17. Prantzos N and Diehl R, 1996, Phys. Rep. 267, 1
18. Timmes F X and Woosley S E, 1997 ApJ 489, 160
19. Weaver T A and Woosley S E, 1993, Phys. Rep. 227, 65
20. Wellstein S, Langer N, 1999, A&A 350, 148
21. Woosley S E and Weaver T A. 1995 ApJS 101, 181

Important Current Issues in Supernova Theory

Adam Burrows*

*Department of Astronomy and Steward Observatory, The Univerity of Arizona, Tucson, AZ 85721 USA

Abstract.
With a kinetic energy of $\sim 10^{51}$ ergs, supernova explosions influence the birth of stars, are the source of the energetic cosmic-rays that irradiate us on Earth, and collectively, by their prodigious energy and momentum input during the birth of galaxies in the infant universe, may have helped shape the galaxies themselves. The neutrinos produced in great numbers during the supernova event are implicated as the source of their explosive power and their study has exercised supernova modelers for forty years, some of them productive. In this contribution, I summarize various new ideas and speculations concerning supernova physics and the role of neutrinos. In particular, I review possible systematics of nickel yields and explosion energies with progenitor mass, evidence for asymmetries in the explosion itself, evidence for kicks to neutron stars in their supernova cradle, and an example of an important (but hitherto neglected) neutrino production process. I also summarize the expected and dramatic signature in underground neutrino detectors of a galactic stellar collapse and explosion.

SETTING THE SUPERNOVA ENERGY SCALE

The basic, though provisional, facts of core–collapse supernova theory are that after the core of a massive star achieves the Chandrasekhar mass it becomes unstable to collapse, implodes to nuclear densities, bounces, and then drives a shock wave into the outer core. This shock wave stalls, sapped by prodigious neutrino losses and the dissociation of the nuclei through which it must pass. The delay before explosion, but after bounce, lasts perhaps 100 to 1000 milliseconds, during which time the region between the stalled accretion shock and the neutrinospheres is heated by the neutrinos issuing from the inner core. This heating not only supports the outer core from further collapse, but drives trans-sonic convection. This in itself makes the explosions, when they are finally launched, asymmetric and multi–dimensional. Rotation and asymmetries in the collapsing Chandrasekhar core might amplify this asphericity. The absorption in the outer part of this convective region of a fraction of the neutrinos emerging from the core eventually ignites and energizes the explosion and radiation pressure plays almost no role. The explosion is akin to a coronal ejection and is in fact followed by a classic wind from the protoneutron star. Hence, the pre–explosion structure is an accreting star bounded by an accretion shock, the post–explosion structure is a wind, and the supernova is the dynamical phase in between.

The accretion rate and its change and the binding energy that the explosion must overcome to eject the mantle vary with progenitor star. The density profiles have a steep drop off for the lower mass progenitors ($9 - 15 M_\odot$) and a shallow drop off for the higher mass progenitors ($\geq 20 M_\odot$). This translates into systematic trends in explosion characteristics such as nucleosynthesis, neutron star (or black hole) mass, explosion energy, blast asymmetries, ^{56}Ni yields, and kick velocity, as well as neutrino and gravitational wave signatures.

None of these systematics has been convincingly and quantitatively demonstrated, but there are intriguing hints. In particular, observations of SN1993J, SN1987A, SN1994W and SN1997D suggest that there is a systematic dependence of the ^{56}Ni yields and explosion energies with progenitor mass. The two recent Type IIp supernovae, SN1994W [1] and SN1997D [2], have very low ^{56}Ni yields ($\leq 0.0026 \, M_\odot$ and $\leq 0.002 \, M_\odot$, respectively) and long–duration plateaus. Hence, we can infer that SN1994W and SN1997D have large ejecta masses ($\geq 25 M_\odot$). The estimated explosion energy for SN1997D is a slight 0.4×10^{51} ergs. (SN1987A's explosion energy was $1.5 \pm 0.5 \times 10^{51}$ ergs, its ^{56}Ni yield was 0.07 $M_\odot$, and its mass has been estimated to have been $\sim 18 M_\odot$.)

The gravitational binding energy ($B.E.$) exterior to a given interior mass is an increasing function of progenitor mass, ranging at 1.5 $M_\odot$ interior mass from about 10^{50} ergs for a 10 $M_\odot$ progenitor to as much as 3×10^{51} ergs for a 40 $M_\odot$ progenitor [3,4]. This large range must affect the viability of explosion and its energy. It is not unreasonable to conclude, in a very crude way, that $B.E.$ sets the scale for the supernova explosion energy. In addition, a neutrino–driven explosion requires a neutrino–absorbing mass and there is more mass available in the denser core of a more massive progenitor. One might think that binding energy and absorbing mass partially compensate or that a more massive progenitor can just wait longer to explode, until its binding energy problems are buried in the protoneutron star and $\dot{M}$ has subsided. The net effect in both cases may be similar explosion energies for different progenitors, though the residue mass could be systematically higher for the more massive stars. However, if these effects do not compensate, the fact that binding energy and absorbing mass are increasing functions of progenitor mass hints that the supernova explosion energy may also be an increasing function of mass. Since $B.E.$ varies so much along the progenitor continuum, the range in the explosion energy may not be small. The amount of ^{56}Ni produced explosively also depends upon the mass between the residue and the radius at which the shock temperature goes below the explosive Si–burning temperature, a radius that depends upon explosion energy. Hence, the amount of ^{56}Ni produced may also increase with progenitor mass. Thermonuclear energy only partially compensates for the binding energy to be overcome, the former being about 10^{50} ergs for every 0.1 $M_\odot$ of ^{56}Ni produced.

Some of the ^{56}Ni produced may not be ejected. Fallback is possible and whether there is significant fallback must depend upon the binding energy profile. There is not much fallback for the lighter progenitors, perhaps for masses below 15 M$_\odot$, but there is significant fallback for the heaviest progenitors. The transition between the two classes may be abrupt. We base this conclusion on the miniscule binding energies and tenuous envelopes of the lightest massive stars and on the theoretical prejudice that the r–process, or some fraction of it, originates in the protoneutron star winds that follow the explosion for the lightest massive stars [5]. If there were significant fallback, these winds and their products would be smothered.

With significant fallback, the supernova may be in jeopardy and much of the ^{56}Ni produced will reimplode. There may be a narrow range of progenitor mass over which the supernova is still viable, while fallback is significant and both the mass of ^{56}Ni *ejected* and the supernova energy are decreasing. Above this mass range, a black hole may form. Hence, both low–mass and high–mass supernova progenitors may have low ^{56}Ni yields. These theoretical speculations have gained some support from the observations of SN1994W and SN1997D [6]. These two supernovae may reside in the fallback gap, above which might be found the black hole cut–off, perhaps near a progenitor ZAMS mass of 30–40 M$_\odot$. A major goal and theme of future supernova research will be the detailed exploration of these and/or other systematics using sophisticated numerical tools. Those tools must include complicated neutrino transport algorithms [7].

THE NEUTRINO SIGNATURE OF STELLAR COLLAPSE AND SUPERNOVA EXPLOSION

Neutrinos are the major signatures of the inner turmoil of the dense core of the massive star and they carry away the binding energy of the young neutron star, a full 10% of its mass–energy. The detection of collapse neutrinos, their "light curve" and spectra, will allow us to follow in real time the phenomena of stellar death and birth. The supernova, SN1987A, provided a glimpse of what might be possible, but it yielded only 19 events; we can expect the current generation of underground neutrino telescopes to collect thousands of events from a galactic supernova.

There is a broad consensus on the basic features of the neutrino light curve from a supernova [8], but it should be recalled that the luminosities and timescales for different massive star progenitors will be different. Generically, infall may last from 200 to 600 milliseconds during which time electron neutrinos will predominate. They will have roughly a capture spectrum that gradually hardens until shock breakout. The rise time of the associated luminosity depends upon the nuclear symmetry energy, but is approximately 5 milliseconds. The total energy radiated during this phase is roughly 10^{51} ergs. Bounce is almost immediately followed by the formation of the shock in the neutrino-opaque regions (at near 20 kilometers). The shock starts with a velocity near 50,000 km/s and so very quickly achieves the neutrinosphere (50-100 kilometers) and breaks out. Shock breakout is announced by a prodigious burst of electron neutrinos produced by electron capture on free protons newly liberated by shock dissociation of the infalling nuclei. The electron neutrino luminosity may achieve 10^{54} ergs s^{-1}. The characteristic time of the breakout burst is 3-10 milliseconds and the total energy radiated in electron-type neutrinos during breakout is $\sim 3 \times 10^{51}$ ergs. The magnitude of the latter will depend on the density structure of the collapsing Chandrasekhar core and will be higher for the more massive progenitors. During this phase, perhaps 10 events in both SuperK and in ICARUS can be expected from a collapse at 10 kpc.

During breakout, the matter is heated to such a degree that $\bar{\nu}_e$ neutrinos and ν_μ and ν_τ neutrinos and anti-neutrinos (hereafter "ν_μs") are thermally produced and radiated. The turn-on timescale of this component is less than one millisecond, but the initial luminosity of the $\bar{\nu}_e$s and the ν_μs depends upon the degree of degeneracy of the electrons near the neutrinospheres and the magnitude of the production sources, still poorly known. It is thought that the initial $\bar{\nu}_e$ neutrino luminosity is within about one order of magnitude of its peak value ($\sim$20-50 milliseconds after breakout). Even at such a level, at 10 kpc both SuperK and SNO will register 100's of $\bar{\nu}_e$ events per second, in SuperK perhaps a kilohertz. After the abrupt rise, the $\bar{\nu}_e$ neutrino luminosity rises further to approximately meet the falling ν_e luminosity. After 20-50 milliseconds, the two decay together as the light curve transitions to the longer-term protoneutron star cooling and neutronization phase. Similarly, the ν_μ neutrino luminosity per species achieves a value not more than 30% away from the electron neutrino luminosity.

The decay is gradual and there may be some quasi-periodic pulsation of the luminosities during this phase. However, the shock wave launched with such fanfare stalls into an accretion shock at 100-200 kilometers within 10-20 milliseconds of breakout. There is a delay to explosion, that may last between a hundred milliseconds and a second, during which time perhaps $\geq 10^{53}$ ergs of neutrinos may be radiated. When it comes, explosion should be accompanied by a decrease by about a factor of two over about 20 milliseconds in all the neutrino luminosities. This may be detectable. After explosion, the luminosities decay on timescales of seconds to a minute. Indeed, after as long as a minute, the event rate at 10 kpc in SuperK may still be as high as one per second. After breakout, the spectra of all the neutrino species first harden on timescales of hundreds of milliseconds, then soften, particularly after explosion, as the luminosity inexorably decays. The rise and fall timescales, as well as the explosion time, are not known theoretically with sufficient precision.

Hence, the important features detectors should key in on are: the infall rise, the breakout, the early $\bar{\nu}_e$ neutrino rise, the production of ν_μs, the signature of explosion, the rise and fall of the average neutrino energies, and the late-time persistence. In addition, if a black hole forms during the high-luminosity phase, the prediction is that the signal will stop within less than a millisecond [9]. Such a phenomenon will be detectable.

Given this generic neutrino light curve, can we use accurate timing of the features in the burst to triangulate on the supernova? This will depend upon the signal strength (and, hence, the distance). At the canonical distance of 10 kpc, with the forward-peaked SuperK ν_e neutrino events (~100-150), one should be able to achieve ~4-degree pointing without the aide of the network. If the initial $\bar{\nu}_e$ signal is indeed as abrupt as we believe and if it starts at a high luminosity, then initial count rates near one kilohertz in SuperK, SNO, and LVD/MACRO/ICARUS might enable the network to locate a supernova to within ~10 degrees at 10 kpc. The fact that the current detectors are all in the northern hemisphere is a problem, as is the possible fuzziness of the initial luminosity rise. Furthermore, there is general excitement that the network of neutrino telescopes now being established underground might indeed be able to announce, with whatever angular precision, the advent of a galactic supernova and allow the astronomical community the early warning it has never before enjoyed.

ASYMMETRIES OF SUPERNOVA EXPLOSIONS

There are many observational indications that supernova explosions are indeed aspherical. Fabry–Perot spectroscopy of the young supernova remnant Cas A, formed around 1680 A.D., reveals that its calcium, sulfur, and oxygen element distributions are clumped and have gross back–front asymmetries [10]. No simple shells are seen. Many supernova remnants, such as N132D, Cas A, E0102.2-7219, and SN0540-69.3, have systemic velocities relative to the local ISM of up to 900 km s^{-1} [11]. X–ray data taken by ROSAT of the Vela remnant reveal bits of shrapnel with bow shocks [12]. The supernova, SN1987A, is a case study in asphericity: 1) its X-ray, gamma–ray, and optical fluxes and light curves require that shards of the radioactive isotope ^{56}Ni were flung far from the core in which they were created, 2) the infrared line profiles of its oxygen, iron, cobalt, nickel, and hydrogen are ragged and show a pronounced red–blue asymmetry, 3) its light is polarized, and 4) recent Hubble Space Telescope pictures of its inner debris reveal large clumps and hint at a preferred direction [13]. Furthermore, radio pictures of the supernova SN1993J, which also has polarized optical spectral features, depict a broken shell. One of the most intriguing recent finds is the supernova SN1997X, which is a so-called Type Ic explosion. This supernova shows the greatest optical polarization of any to date (Lifan Wang, private communication). Type Ic supernovae are thought to be explosions of the bare carbon/oxygen cores of massive star progenitors stripped of their envelopes and some may be connected to a fraction of γ-ray bursts, for which jets have been inferred. As such, SN1997X's large polarization implies that the inner supernova cores, and, hence, the explosions themselves, are fundamentally asymmetrical. No doubt, instabilities in the outer envelopes of supernova progenitors clump and mix debris clouds and shatter spherical shells. The observation of hydrogen deep in SN1987A's ejecta [14] strongly suggests the work of such mantle instabilities. However, the data collectively, particularly for the heavier elements produced in the inner core, are pointing to asymmetries in the central engine of explosion itself.

NEUTRON STAR VELOCITIES IMPARTED AT BIRTH

Strong evidence that neutron stars experience a net kick at birth has been mounting for years. In 1993 [15,16], it was demonstrated that the pulsars are the fastest population in the galaxy ($<v> \sim 450$ km s^{-1}). Such speeds are far larger than can result generically from orbital motion due to birth in a binary (the "so-called" Blaauw effect). An extra "kick" is required, probably during the supernova explosion itself [17]. In the pulsar binaries, PSR J0045-7319 and PSR 1913+16, the spin axes and the orbital axes are misaligned, suggesting that the explosions that created the pulsars were not spherical [18,19]. In fact, for the former the orbital motion seems retrograde relative to the spin [20] and the explosion may have kicked the pulsar backwards. In addition, the orbital eccentricities of Be star/pulsar binaries are higher than one would expect from a spherical explosion, also implying an extra kick [21]. Furthermore, low–mass X–ray binaries (LMXB) are bound neutron star/low-mass star systems that would have been completely disrupted during the supernova explosion that left the neutron star, had that explosion been spherical [22]. In those few cases, a countervailing kick may have been required to keep the system bound. The kick had to act on a timescale shorter than the orbit period and the explosion orbit crossing time. Otherwise, the process would have been uselessly adiabatic. One is tempted to evoke as further proof the fact that pulsars seen around young (age $\leq 10^4$ years) supernova remnants are on average far from the remnant centers, but here ambiguities in the pulsar ages and distances and legitimate questions concerning the reality of many of the associations make this argument rather less convincing [23,24]. However, the ROSAT observations of the 3700 year–old supernova remnant Puppis A show an X–ray spot that has been interpreted as its neutron star [25]. This object has a large X–ray to optical flux ratio, but no pulsations are seen. If this interpretation is legitimate, then the inferred neutron star transverse speed is $\sim$1000 km s^{-1}. Interestingly, the spot is opposite to the position of the fast, oxygen–rich knots, as one might expect in some models of neutron star recoil during the supernova explosion. Whatever the correct interpretation of the Puppis A data, it is clear that many neutron stars are given a hefty extra kick at birth (though the distribution of these kicks is broad) and that it is reasonable to implicate asymmetries in the supernova explosion itself.

NUCLEON–NUCLEON BREMSSTRAHLUNG

A production process for neutrino/anti-neutrino pairs that has received but little attention to date in the supernova context is neutral-current nucleon–nucleon bremsstrahlung ($n_1 + n_2 \to n_3 + n_4 + \nu\bar{\nu}$). Its importance in the cooling of old neutron stars, for which the nucleons are quite degenerate, has been recognized for years [26], but only in the last few years has it been studied for its potential importance in the atmospheres of protoneutron stars and supernovae [27–29]. Neutron–neutron, proton–proton, and neutron–proton bremsstrahlung are all important, with the latter the most important for symmetric matter. As a source of ν_e and $\bar{\nu}_e$ neutrinos, nucleon–nucleon bremsstrahlung can not compete with the charged–current capture processes. However, for a range of temperatures and densities realized in supernova cores, it may compete with e^+e^- annihilation as a source for ν_μ, $\bar{\nu}_\mu$, ν_τ, and $\bar{\nu}_\tau$ neutrinos ("ν_μ"s). The major obstacles to obtaining accurate estimates of the emissivity of this process are an understandable reticence to include the full and proper nucleon–nucleon potentials, uncertainty concerning the degree of suitability of the Born Approximation, and ignorance concerning the true role of many–body effects [27,30,31].

CONCLUSIONS

Even after forty years of progress and development, we are far from a systematic and detailed understanding of the core–collapse supernova mechanism. To be sure, the subject has gotten much richer, the numerical tools have gotten much better, and many insights have been won. In addition, there are hints at connections between some supernovae and some gamma-ray bursts, providing yet another astrophysical context in which the neutrino and its interactions may be crucial. However, as we approach the new millenium the mind beggars at the number of basic questions with which we are still groping.

REFERENCES

1. Sollerman, J., et al., *Astrophys. J.* **493**, 933 (1998).
2. Turatto, M., et al., *Astrophys. J.* **498**, L129 (1998).
3. Burrows, A., Hayes, J., and Fryxell, B.A., *Astrophys. J.* **450**, 830 (1995).
4. Weaver, T.A. and Woosley, S.E., i*Astrophys. J. Suppl.* **101**, 181 (1995).
5. Mathews, G., Bazan, G., and Cowan, J., *Astrophys. J.* **391**, 719 (1992).
6. Nakamura, T., Umeda, H., Nomoto, K., Thielemann, F., and Burrows, A., *Astrophys. J.* **517**, 193 (1999).
7. Burrows, A., Young, T., Pinto, P., Eastman, R., and Thompson, T., *Astrophys. J.* **539**, 865 (2000).
8. Burrows, A., Klein, D., and Gandhi, R., *Phys. Rev.* D**45**, 3361 (1992).
9. Burrows, A., *Astrophys. J.* **300**, 488 (1986).
10. Lawrence, S.S. et al., *Astron. J.* **109**, 2635 (1995).
11. Kirshner, R.P., Morse, J.A., Winkler, P.F., and Blair, J.P., *Astrophys. J.* **342**, 260 (1989).
12. Strom, R., Johnston, H.M., Verbunti, F., and Aschenbach, B., *Nature* **373**, 590 (1994).
13. Pun, C.S.J., Kirshner, R.P., Garnavich, P.M., and Challis, P., *B.A.A.S.* **191**, 9901 (1998).
14. Wooden, D.H.et al., i*Astrophys. J. Suppl.* **88**, 477 (1993).

15. Harrison, P.A., Lyne, A.G., and Anderson, B., *Mon. Not. R. Astron. Soc.* **261**, 113 (1993).
16. Lyne, A. and Lorimer, D.R., *Nature* **369**, 127 (1994).
17. Fryer, C., Burrows, A., and Benz, W., *Astrophys. J.* **496**, 333 (1998).
18. Wasserman, I., Cordes, J., and Chernoff, D., (1998), in preparation.
19. Kaspi, V.M. *et al.*, *Nature* **381**, 584 (1996).
20. Lai, D., Bildsten, L., and Kaspi, V.M., *Astrophys. J.* **452**, 819 (1995).
21. van den Heuveli, E.P.J. and Rappaport, S., in *I.A.U. Colloquium 92*, eds. A. Slettebak and T.D. Snow (Cambridge Univ. Press), pp. 291–308 (1987).
22. Kalogera, V., *Pub. Astron. Soc. Pac.* **109**, 1394 (1997).
23. Caraveo, P., *Astrophys. J.* **415**, L111 (1993).
24. Frail, D.A., Goss, W.M., and Whiteoak, J.B.Z., *Astrophys. J.* **437**, 781 (1994).
25. Petre, R., Becker, C.M., and Winkler, P.F., *Astrophys. J.* **465**, L43 (1996).
26. Flowers, E., Sutherland, P., and Bond, J.R., *Phys. Rev.* **D12**, 316 (1975).
27. Hannestad, S. and Raffelt, G., *Astrophys. J.* **507**, 339 (1998).
28. Burrows, A., to be published in the proceedings of the 5'th CTIO/ESO/LCO Workshop *SN1987A: Ten Years Later*, eds. M.M. Phillips and N.B. Suntzeff, held in La Serena, Chile, February 24–28, 1997.
29. Suzuki, H., in *Frontiers of Neutrino Astrophysics*, ed. Y. Suzuki and K. Nakamura (Tokyo: Universal Academy Press, 1993), p. 219.
30. Raffelt, G. and Seckel, D., *Phys. Rev. Lett.* **69**, 2605 (1998).
31. Brinkmann, R. and Turner, M.S., *Phys. Rev.* **D38**, 2340 (1988).

Exploding and Non-exploding Stars: Coupling Nuclear Reaction Networks to Multidimensional Hydrodynamics

K. Kifonidis*, T. Plewa[†,*] and E. Müller*

Max-Planck-Institut für Astrophysik, Karl-Schwarzschild-Strasse 1, D-85741 Garching, Germany
[†]*Nicolaus Copernicus Astronomical Center, Bartycka 18, 00716 Warsaw, Poland*

Abstract. After decades of one-dimensional nucleosynthesis calculations, the growth of computational resources has meanwhile reached a level, which for the first time allows astrophysicists to consider performing routinely realistic multidimensional nucleosynthesis calculations in explosive and, to some extent, also in non-explosive environments. In the present contribution we attempt to give a short overview of the physical and numerical problems which are encountered in these simulations. In addition, we assess the accuracy that can be currently achieved in the computation of nucleosynthetic yields, using multidimensional simulations of core collapse supernovae as an example.

INTRODUCTION

Thermonuclear reactive flows are ubiquituous in astrophysics and occur in non-explosive environments as, e.g., in most (hydrostatic) stars as well as in explosive events, for which novae and supernovae are examples. Often they provide the energy which powers stellar outbreaks (as in the case of novae, X-ray flashes, and thermonuclear, i.e. Type Ia, supernovae) and even for stellar explosions where this is not the case (as e.g. in core collapse supernovae, which are driven by neutrino heating), the strong coupling of hydrodynamic advection and thermonuclear reactions is of utmost importance for the nucleosynthesis which accompanies these events. It is by a proper numerical modelling of this coupling through which a more detailed insight into the origin of the nuclear abundances in the solar system can be gained, which are themselves the result of a superposition of material which has been processed in explosive and non-explosive thermonuclear environments. By comparing the results of numerical models with the observed solar abundance pattern, on the other hand, one might also hope to learn more about the thermodynamic conditions in the otherwise unaccessible nucleosynthetic sites and events themselves.

The high precision with which nuclear abundances can be measured nowadays poses great demands on the accuracy of the numerical models, especially since it was convincingly demonstrated in recent years that due to the importance of hydrodynamic instabilities, rotation, and other effects, most of the nucleosynthetic sites do not possess spherical symmetry. Thus a reliable computation of the highly non-linear interaction of hydrodynamic advection and nuclear burning requires multi-dimensional numerical models. In the following sections we give a general overview of the methods which are currently employed for modelling thermonuclear flows and discuss some of the problems which are hereby encountered. Further reviews on reactive flow modelling can be found in [9], [10] and [11].

THE GOVERNING EQUATIONS

A rather wide range of astrophysical reactive flows, in which relativistic effects, viscosity and magnetic fields can be neglected, is described by the well-known (reactive) Euler equations. This system of non-linear partial differential equations which expresses the conservation of the total mass, momentum, total (i.e. kinetic + internal) energy and baryons of the fluid reads

$$\frac{\partial \rho}{\partial t} + \nabla \cdot (\rho \mathbf{v}) = 0 \tag{1}$$

$$\frac{\partial \rho \mathbf{v}}{\partial t} + \nabla \cdot (\rho \mathbf{v} \mathbf{v}) + \nabla P = \rho \mathbf{g} + \rho \mathbf{f}_{\text{add}} \tag{2}$$

$$\frac{\partial \rho E}{\partial t} + \nabla \cdot ([\rho E + P]\mathbf{v}) = \rho \mathbf{v} \cdot \mathbf{g} + \rho \dot{Q}_{\text{add}} + \rho \dot{Q}_{\text{nuc}} \tag{3}$$

$$\frac{\partial \rho X_i}{\partial t} + \nabla \cdot (\rho X_i \mathbf{v}) = \rho \dot{X}_i \tag{4}$$

$$\sum_i X_i = 1, \tag{5}$$

where ρ, $\mathbf{v}$, $E = v^2/2 + e$, and P have their usual meanings, X_i is the mass fraction of nucleus i, and $\rho \dot{X}_i$ as well as $\rho \dot{Q}_{\text{nuc}}$ are source terms due to nuclear transmutations. If self-gravity is important, the gravitational acceleration

$$\mathbf{g} = -\nabla \Phi \tag{6}$$

which appears in the source terms of Eqs. (2) and (3) and which depends on the gravitational potential, Φ, has to be obtained from a solution of the Poisson equation

$$\Delta \Phi = 4\pi G \rho. \tag{7}$$

In mathematical terms Eqs. (1–7) describe a mixed initial/boundary value problem due to the hyperbolic and elliptic nature of the Euler and Poisson equations, respectively. Given appropriate initial and boundary conditions, an equation of state relating ρ, P and e, and the additional source terms $\mathbf{f}_{\text{add}}$ and $\dot{Q}_{\text{add}}$, which in general will be problem-dependent, Eqs. (1–7) can be solved after an appropriate *flow representation* as well as a suitable *numerical scheme* have been adopted.

FLOW REPRESENTATIONS AND NUMERICAL SCHEMES

There are two primary approaches to solve the *homogeneous* part of the system of equations (1–5). In the Eulerian framework the system of conservation laws is solved on a grid which is *fixed in space* and the evolution of the flow is followed by advecting the fluid through the computational cells. The principal assets of this method are its straightforward extension from one to two or three spatial dimensions and the simplicity of its implementation on serial and parallel computer architectures. If an appropriate shock-capturing, finite-volume numerical scheme is used, it is equally straightforward to obtain strict numerical conservation of all physically conserved quantities and a sharp resolution of shocks. The major drawback is numerical diffusion. Consider the continuity equation (1) in its Eulerian form, which can be written as

$$\frac{\partial \rho}{\partial t} + \mathbf{v} \cdot \nabla \rho + \rho \nabla \cdot \mathbf{v} = 0, \qquad (8)$$

where the second term describes advection and the third term compression. Numerical diffusion is introduced into a numerical solution of this equation as a result of discretization errors of the $\mathbf{v} \cdot \nabla$ operator. There appears to be a simple remedy to this problem: using the comoving derivative $d/dt = \partial/\partial t + \mathbf{v} \cdot \nabla$ we can rewrite Eq. (8) in the frame comoving with the matter to obtain its *Lagrangian* form

$$\frac{d\rho}{dt} + \rho \nabla \cdot \mathbf{v} = 0. \qquad (9)$$

Note that in this frame the advection term $\mathbf{v} \cdot \nabla \rho$ has vanished. Therefore the Lagrangian approach is (in principle) not prone to numerical diffusion of mass (or composition). In Lagrangian methods each cell of the numerical grid represents a discretized fluid element which evolves subject to forces which are due to interactions with its neighbors and the time rate of change of the density of such a fluid element is solely determined by the compression (or expansion) that it experiences. Density interfaces (contact discontinuities) as well as composition discontinuities can be easily aligned with the boundaries of grid cells and do not have to be advected through the grid in the course of the calculation.

While this very desirable property of the Lagrangian approach has made it the method of choice for one-dimensional nucleosynthesis calculations, considerable difficulties are experienced when Lagrangian schemes are applied to multidimensional flows. Shear and vortices can severely distort a Lagrangian grid. The discrete approximation of differential operators over such a grid results in large errors in the

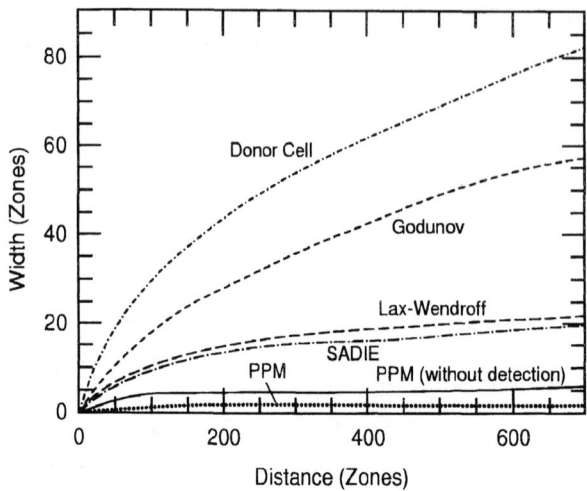

FIGURE 1. Comparison of the diffusivity of different advection schemes for the problem of the propagation of a contact discontinuity through an Eulerian grid. The curves give the width of the discontinuity (in grid zones) as a function of the number of zones it has propagated through the grid (adapted from [4] and [5]).

numerical derivatives, and in the extreme case that the grid lines cross (grid tangling) the calculations have to be stopped. Some remapping procedure to a new, more regular grid must then be applied which unavoidably introduces numerical diffusion to the solution. The distortion problem can be overcome if triangular instead of quadrilateral grids are used [11] or if (as in the Smoothed Particle Hydrodynamics, or SPH approach) no grid at all is adopted and instead the flow is sampled by a finite number of particles. In the former method considerable logic overhead is added in restructuring the deformed triangular grid, while in the latter case, due to the Monte-Carlo nature of the sampling, Poisson noise is introduced.

Due to the aforementioned drawbacks and due to significant progress in the development of accurate Eulerian schemes in the early 1980's, Lagrangian methods employing quadrilateral or triangular grids have not been used extensively in multidimensional calculations of astrophysical flows (see [18], [12] as well as [11] and the references therein for examples). On the other hand, the simplicity of SPH has made this method very popular for astrophysical (especially cosmological) simulations. Without attempting to escalate the very vigorous discussion, whether SPH or grid-based Eulerian schemes are to be prefered in astrophysical calculations (see e.g. [10]), we will argue below that, due to its Monte-Carlo nature, the SPH scheme appears to be rather unsuited for multidimensional nucleosynthesis calculations, especially in cases where hydrodynamic instabilities are known to be important.

Among Eulerian schemes, the so-called shock-capturing schemes have proven to be the most accurate ones for problems which involve discontinuities in the flow as shock waves (see [19] for details). The latter are very frequently encountered in explosive events, since in these cases the flows can attain supersonic speeds. Shock-capturing schemes derive their accuracy from a discretization of the hydrodynamic equations which closely mimics the physics of compressible flows by making use of the Riemann problem, i.e. the dissolution of an arbitrary flow discontinuity into a set of simple waves (shocks, contact discontinuities and rarefaction waves). Suitably constructed Riemann problems at the interfaces between adjacent computational cells are solved within each time step, from which the complete solution of the system of conservation laws is constructed. This allows one to avoid the use of large amounts of artificial viscosity in order to obtain a well-behaved numerical scheme in the vicinity of shocks. One of the most accurate shock-capturing schemes, which has been widely applied in astrophysics, is the (direct Eulerian) PPM scheme of [3], a second order extension of Godunov's original (and rather diffusive) first-order shock-capturing scheme [6]. In addition to its accurate treatment of shocks PPM includes a special detection and steepening algorithm to minimize numerical diffusion across contact discontinuities.

The superiority of shock-capturing schemes in computing compressible flows has been demonstrated e.g. in [19], and their performance for computing reactive astrophysical flows was studied in [4] and [13]. Fig. 1 shows a representative result from [4] in which PPM was compared to a number of older Eulerian schemes which were in wide-spread use until the mid 1980's. The figure shows the width of a contact discontinuity as a function of the number of zones that it has travelled across a numerical grid. Most Eulerian schemes tend to smear such fluid (and also composition) interfaces without limit, i.e. the width of the "discontinuity" tends to grow with time. Of all the schemes investigated, only PPM maintained a sharp resolution of the interface within two zones. Still however, numerical diffusion cannot be completely avoided in Eulerian calculations and its minimization necessitates an adequate spatial resolution in addition to an excellent advection scheme. This has led to the development of adaptive mesh refinement methods [1], which concentrate the computational effort in critical regions of the flow and thereby often allow for substantial savings in computer time.

ADDITIONAL PHYSICS

While the numerical problems encountered in solving the homogeneous part of the Euler equations are difficult to overcome, they represent only a part of the computational difficulties for a realistic simulation. The source terms, which are usually taken into account using the operator splitting technique [9], often require much more computer time than the solution of the hydrodynamic equations themselves. This holds, e.g. if large nuclear networks need to be evolved with the hydrodynamics or transport processes need to be taken into account (as e.g. neutrino transport in core collapse supernovae, see [15] and the references therein and A. Burrows, this volume). In some cases even phenomenological (sub-grid) models might have to be introduced. This is e.g. the case for turbulent combustion in thermonuclear supernovae, where a white dwarf is incinerated by a deflagration front whose propagation speed is impossible to compute in a direct simulation since this would require a resolution of the turbulent energy cascade down to the dissipation length scale [16]. Exacerbating the situation is the fact that stellar models, which serve as initial data for supernova simulations, might be affected by considerable uncertainties. In the absence of computational schemes and resources which allow for a consistent multidimensional treatment of stellar convection and rotation over stellar evolutionary time scales, one is forced to describe these phenomena by one-dimensional appoximations (see the contribution of N. Langer, this volume). Finally, uncertainties in nuclear reaction rates enter the calculations.

It is apparent that progress in only a *single* of the involved fields is *not* going to improve the accuracy of the desired nucleosynthetic yields considerably. In fact, a concerted effort in all areas appears to be required, since, as we will show in our example below, the different effects can conspire in falsifying the nucleosynthetic yields.

CORE COLLAPSE SUPERNOVAE: A CASE STUDY

Nucleosynthesis in core collapse supernovae is a good example for illustrating the aforementioned problems and we will start with a discussion of numerical diffusion using results of simple one-dimensional calculations. We subsequently address the complications introduced by convection in multidimensional calculations as well as by "additional physics", i.e. neutronization due to neutrino matter interactions. Finally we show how a multidimensional numerical failure, the so-called "odd-even-decoupling" phenomenon, an instability which appears to plague most shock-capturing schemes and whose effects have not yet been discussed extensively in the numerical astrophysics literature, can enhance neutronization by strengthening hydrodynamic convection and affect the nucleosythetic yields in multidimensional simulations.

Nucleosynthesis in a 15 $M_\odot$ star (1D)

In core collapse supernovae nucleosynthesis is triggered by a shock wave which forms after the collapse of the iron core of a massive star has proceeded to supranuclear densities. The shock, while initially powerful, stalls after a few milliseconds due to the energy losses from which it suffers while propagating through the outer iron core, but eventually ejects the outer stellar layers if heating by neutrinos from the collapsed core is able to overcompensate for the energy losses. In most nucleosynthesis calculations, however, these processes are not modelled in detail and instead a shock is initiated by simply depositing the typical observed supernova energy of $\sim 10^{51}$ erg near the center of a presupernova model.

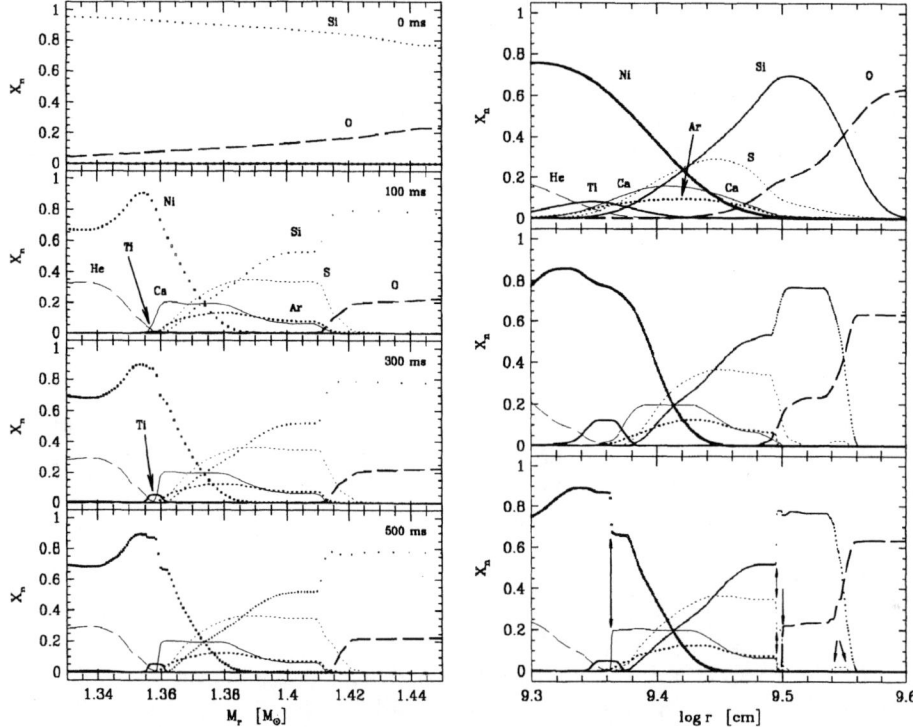

FIGURE 2. Left: Eulerian PPM calculation of explosive nucleosynthesis in the presupernova model of [20] using an α-nucleus network and the Consistent Multifluid Advection scheme (CMA) (from [13]). Right: Comparison of Eulerian PPM results using: 1st order advection for nuclear species (top), the FMA advection scheme for multifluid flows of [4] (middle), and the CMA scheme of [13] (bottom). Note the decreasing amount of diffusion and the sharp interfaces obtained with CMA (from [13]).

In the left panels of Fig. 2 we show snapshots of the mass fractions from the first 500 ms of such a calculation, focusing on the silicon-rich layers just outside the iron core in which explosive nucleosynthesis takes place. Only 100 ms after the start of the calculations explosive silicon burning has frozen out and has left behind a significant abundance of ^{56}Ni as well as ^{40}Ca, ^{36}Ar and ^{32}S. Particularly noteworthy for the following discussion is the nucleus ^{44}Ti. From one-dimensional Lagrangian nucleosynthesis calculations [17], [21] it is known that this isotope should be primarily synthesized in the innermost stellar layers which experience an α-rich freezeout. However, in the present Eulerian calculation a significant abundance of ^{44}Ti has formed in zones with mass coordinates around $1.36\,M_\odot$ (marked with arrows in the left panels of Fig. 2), i.e. at the interface of the regions enriched in ^{4}He and ^{40}Ca. This ^{44}Ti "bump" results from the reaction ^{40}Ca$(\alpha,\gamma)^{44}$Ti and the amount

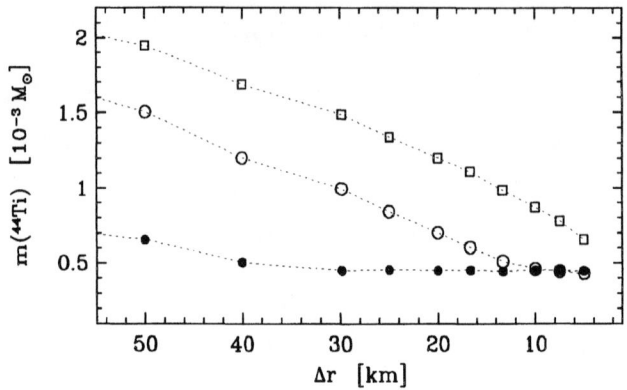

FIGURE 3. Final ^{44}Ti yield of the one-dimensional calculations shown in Fig. 2 as a function of radial resolution and different multifluid advection schemes. Top curve: 1st order species advection. Middle: FMA. Bottom: CMA (from [13]).

of ^{44}Ti thereby produced is very sensitive to numerical diffusion in this region. Consequently, the strength of the ^{44}Ti "bump" varies with the diffusivity of the numerical scheme which is used to advect the nuclear species. This is illustrated in the right panels of Fig. 2. The top right panel depicts results from a calculation with Godunov's first-order scheme. In this case, all mass fraction profiles are heavily smeared due to strong diffusion, as can be seen by a comparison to the middle right and bottom right panels which show results that were obtained with the FMA and CMA advection schemes of [4] and [13], respectively. Of all these schemes, CMA is the least diffusive since it is the only method which includes a detection and steepening algorithm for composition interfaces which was derived from PPM's original detection and steepening algorithm for contact discontinuities. Note the size of the ^{44}Ti bump for the three different runs. The more diffusive schemes produce much more ^{44}Ti. This is also illustrated in Fig. 3 which summarizes how the ^{44}Ti yield depends on the adopted advection scheme and the spatial resolution. While the CMA results (bottom curve) are already converged for a resolution of $\Delta r = 40$ km, FMA (middle curve) needs a resolution of about 10 km. The first-order scheme (top curve) would need much finer zoning than $\Delta r = 5$ km to yield results of comparable quality. Note also that, if a diffusive advection scheme and coarse resolution are used, the errors might be as large as a factor of four!

It should be pointed out, however, that ^{44}Ti is a somewhat extreme (though very important) example. The (relative) errors due to numerical diffusion are usually smaller for the more abundant nuclei. This is illustrated in Fig. 4 which shows the dependence of the yields of different α-nuclei on resolution in a 1D calculation from [8], in which no ad hoc energy deposition was adopted, but where the shock revival phase was followed in detail by including the effects of neutrino heating from a central light bulb neutrino source [7]. It can be seen from this figure that

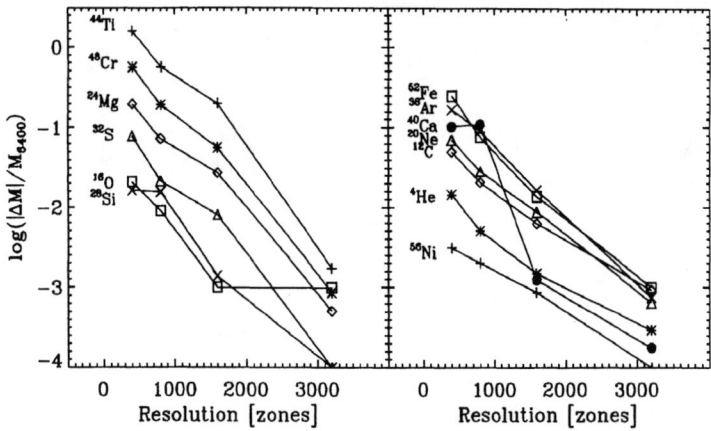

FIGURE 4. Accuracy of various elemental yields from one-dimensional shock-revival and explosive nucleosynthesis calculations in the post-bounce model of [2] (see also [20] for the presupernova model). The logarithm of the deviations of the elemental yields of various α-nuclei as compared to an essentially converged 6400 zone calculation ($\Delta r \leq 5\,\mathrm{km}$) is displayed as a function of the radial resolution (in grid zones)(from [8]).

individual elemental yields are more accurate than the ^{44}Ti yield by more than an order of magnitude. However, if one is aiming at a numerical accuracy of about 1% for all yields, about 3000 radial zones are required for this calculation. This amount of spatial resolution makes accurate multidimensional simulations very expensive.

Convection and ^{56}Ni synthesis in core collapse supernovae

Neutrino matter interactions play a crucial role in the explosion of core collapse supernovae. They heat the matter behind the stalling shock and thereby trigger the explosion. On the other hand they determine the electron fraction per baryon, Y_e, (or equivalently the ratio of protons to neutrons) in the ejecta and thus influence the nucleosynthetic yields. If the Y_e value of material that has been photodissociated by the shock is significantly reduced below 0.5 by neutrino/matter interactions, this matter will recombine mainly to neutron-rich nuclei after expanding and cooling. In that case nuclei with $Z = N$ like ^{56}Ni will not form in the ejecta. Fig. 5 which shows results of a two-dimensional simulation of shock revival illustrates this effect. In this calculation the luminosities of ν_e and $\bar{\nu}_e$ were such that $\bar{\nu}_e$ absorption on protons was favored against ν_e absorption on neutrons and thus the heated gas was neutronized [7]. The sharp division between this material which is visible in the bubbles behind the shock in Fig. 5 and the lepton-rich material farther out which has formed ^{56}Ni (the region enclosed by the white contour line in Fig. 5) is clearly visible. The negative entropy gradient that the heating has imprinted on the layers between the radius of maximum neutrino heating (close to the center) and the shock farther out, has also led to strong convective motions: bubbles of heated

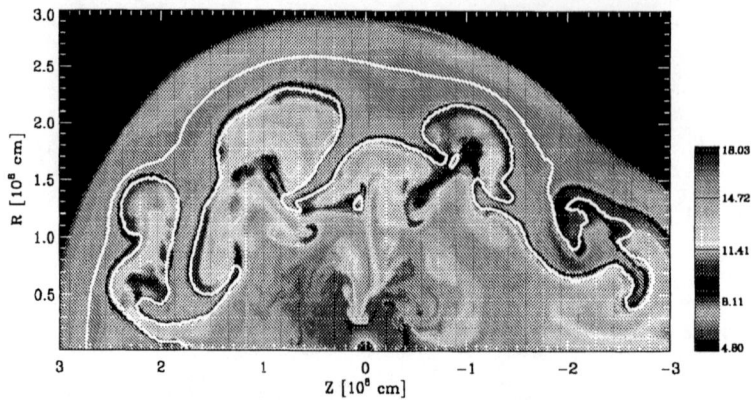

FIGURE 5. Distribution of the entropy (in units of k_B/nucl.) 320 ms after core bounce in a two-dimensional core collapse supernova simulation of a $15\,M_\odot$ star. The white contour line encloses the region in which the ^{56}Ni mass fraction exceeds 20% (from [8]).

deleptonized gas rise toward the shock while lower entropy flux tubes transport lepton-rich matter to deeper layers where it interacts with the neutrino fluxes much more efficiently. This interplay of convection and deleptonization is crucial for the ^{56}Ni yield. The shock is only able to heat a certain amount of lepton-rich material to temperatures in excess of the 5×10^9 K which are required for ^{56}Ni synthesis. If convection is strong enough to advect significant amounts of this matter close to the neutron star, where the gas will experience deleptonization, the ^{56}Ni yield will be lower than in a model with no or weak convection.

Multidimensional numerical failures: Odd-even decoupling

Quirk [14] has reported on a subtle flaw in a number of shock capturing schemes which becomes evident when calculating multidimensional flows with strong, grid-aligned shocks. He has dubbed this failure the "odd-even decoupling" phenomenon. The problem shows up only if a sufficiently strong shock is either fully or nearly aligned with one of the coordinate directions of the grid, and if, in addition, the flow is slightly perturbed. This can be due to either perturbations intentionally introduced in order to study *physical* instabilities, as it is done in all studies of convection in supernovae, or due to perturbations caused by other flow features. Many Riemann solvers show the tendency to allow these perturbations to grow without limit along the shock surface, thus triggering a strong rippling of the shock front as well as the post shock state. In supernova simulations these perturbations, whose amplitudes can exceed those of the seed perturbations by several orders of magnitude, enhance the growth of hydrodynamic instabilities. In case of neutrino driven convection they lead to large-scale overturn and angular wavelengths of convective bubbles which are significantly larger than in a "clean" calculation. This

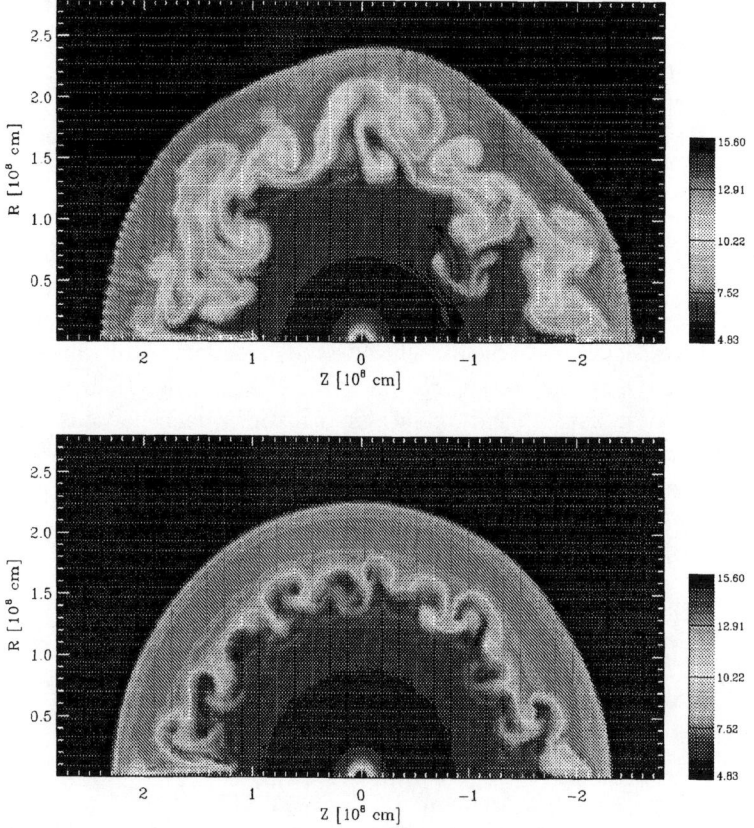

FIGURE 6. Top: Entropy distribution 208 ms after core bounce (in units of k_B/nucl.) in a two-dimensional supernova model showing odd-even decoupling. Bottom: Entropy distribution for an equivalent calculation in which odd-even decoupling has been suppressed (from [8]).

artificial enhancement of convection is demonstrated in Fig. 6 where the entropy distribution of a simulation exhibiting odd-even decoupling (top panel) is compared to one in which the numerical failure has been cured (bottom panel). A modification of PPM's original dissipation algorithms [3] was necessary for this purpose. Alternatively the hybrid Riemann solver method of [14] might be used. Note that the calculations have been carried out in spherical coordinates (r, θ) so that the (initially spherical) shock wave was fully aligned with the grid. However, cylindrical coordinates which have been used for plotting are indicated in the figures. The difference in the final ^{56}Ni yield for these two simulations was about 40%, the calculation not exhibiting odd-even decoupling showing the larger yield, as expected from our discussion in the previous section. This demonstrates that due to the strong coupling of neutrino physics and convection, the ^{56}Ni yield in multidimensional calculations is much more difficult to calculate correctly than in one spatial dimension. Therefore the error of at most a few percent which can be deduced for the latter case from Fig. 4 can be deceptive. The results shown in Fig. 6 also suggest that numerical noise, whatever its origin is, leads to a grossly overestimated efficiency of convection. Thus, schemes which are known to suffer from this problem (like SPH) do not appear to be suited for nucleosynthesis calculations in core collapse supernovae.

CONCLUSIONS

Realistic nucleosynthesis calculations in astrophysical contexts represent a challenge in many respects. The difficulties involve the numerical treatment of multidimensional hydrodynamic advection, complex physics in addition to hydrodynamics and burning, disparate length and time scales, realistic initial conditions and uncertainties in reaction rates. It is our conviction, that substantial efforts are required in *each* of these fields in order to obtain reliable yields in multidimensional nucleosynthesis calculations.

REFERENCES

1. Berger, M., and Colella, P., *Jour. of Comp. Phys.*, **82**, 64 (1989).
2. Bruenn S. W., *Nuclear Physics in the Universe*, Bristol: IOP, 1993, p.31
3. Colella, P., and Woodward, P. R., *Jour. of Comp. Phys.*, **54**, 174 (1984).
4. Fryxell B., Müller E., and Arnett W. D., *MPA Preprint*, **449**, 1 (1989).
5. Fryxell B., Müller E., and Arnett W. D., *ApJ*, **367**, 619 (1991).
6. Godunov S. K., *Mat. Sb.*, **47**, 271 (1959).
7. Janka H.-T., and Müller E., *A&A*, **306**, 167 (1996).
8. Kifonidis K., PhD thesis, Technische Universität München, (2000).
9. LeVeque R. J., *Computational Methods for Astrophysical Fluid Flow*, Berlin: Springer, 1998, pp. 84-101
10. Müller E., *Computational Methods for Astrophysical Fluid Flow*, Berlin: Springer, 1998, pp. 463-480
11. Oran, E. S., and Boris, J. P., *Numerical Simulation of Reactive Flow*, New York: Elsevier, 1987, ch. 10, pp. 358-394.
12. Pen, U.-L., *ApJS*, **115**, 19 (1998).
13. Plewa T., and Müller E., *A&A*, **342**, 179 (1999).
14. Quirk J. J., *Int. J. Num. Meth. Fluids*, **18**, 555 (1994).
15. Rampp M., and Janka H.-T., *ApJ*, **539**, L33 (2000).
16. Reinecke M., Hillebrandt W., and Niemeyer J. C., *A&A*, **347**, 739 (1999).
17. Thielemann F.-K., Nomoto K. I., and Hashimoto M., *ApJ*, **460**, 408 (1996).
18. Woodward, P. R., *ApJ*, **207**, 484 (1976).
19. Woodward, P. R., and Colella, P., *Jour. of Comp. Phys.*, **54**, 115 (1984).
20. Woosley S. E., Pinto P. A., and Ensman L., *ApJ*, **324**, 466 (1988).
21. Woosley S. E., and Weaver T. A., *ApJS*, **101**, 181 (1995).

Phase Transitions in Dense Matter and Recent Topics of Neutron Stars

Masatomi Yasuhira

*Department of Physics, Kyoto University,
Kitashirakawa-Oiwake-cho,
Kyoto 606-8502, Kyoto, JAPAN
e-mail:yasuhira@ruby.scphys.kyoto-u.ac.jp*

Abstract. Core of a neutron star consists of highly dense matter above normal nuclear density $\rho_0 \simeq 0.16 \text{fm}^{-3}$, where phase transitions are expected to take place. We review some phase transitions and recent topics of neutron stars.

INTRODUCTION

There exist many candidates of the phase transitions in highly dense nuclear matter: boson condensation (K^-,π), quark matter, hyperonic matter and so on [1]. They are often discussed with respect to neutron star (NS) physics, for example, maximum mass, cooling mechanism, glitch mechanism, delayed collapse, gamma ray burst and magnetar.

In this paper after the introduction why we believe the phase transitions in the core of a NS, We concentrate on three topics: the delayed collapse due to kaon condensation, the mixed-phase problem for first order phase transitions and strange stars as magnetar candidates.

Why do we believe the phase transitions?

The equation of state (EOS) of normal nuclear matter has been studied by many authors with various theoretical approaches. G-matrix and variational calculations (See ref. [1] and references therein.) are based on the microscopic theory, whose objective is to reproduce the properties of matter (saturation density, binding energy and nuclear incompressibility) based on the experimental data(2- or 3-body interaction). On the other hand, relativistic mean field theory is the effective theory and seems to be very useful method. The numerical table of EOS is submitted by Shen et al [2](also refer to talk by Prof. Sumiyoshi).

There have been suggested many phase transitions in a NS; e.g. nucleon or quark superfluidity, pion or kaon condensation, deconfinement transitions. In this paper we discuss the phase transitions beyond the normal matter by the following reasons: maximum mass and cooling mechanism.

Maximum Mass

Using the EOS for the normal matter, maximum mass of the NS is evaluated about $2.0 M_\odot$, however $M \simeq 1.35 \pm 0.04 M_\odot$ from the observation of NS from radio pulsars systems [3]. There is a large difference between observation and theoretical calculation. On the other hand, if there exists some phase transition, for example, kaon condensation, $M_{max} \simeq 1.4 - 1.6 M_\odot$ and it seems to be more plausible [4,5].

Cooling Mechanism

In the cooling scenario, the main contribution in the normal matter is the modified URCA reactions.

$$n + n \rightarrow n + p + e^- + \bar{\nu}_e,$$
$$n + p + e^- \rightarrow n + n + \nu_e.$$

This scenario based on the normal matter is called standard cooling scenario, and it has been suggested that we need extra cooling mechanisms to reproduce the observational data points consistently [6,7]. Each phase transition leads to the additional rapid cooling mechanisms. (The cooling curve including these mechanisms is called the non-standard scenario.) In the case of kaon condensation, K-induced URCA process exists,

$$n + \langle K^- \rangle \rightarrow n + e^- + \bar{\nu}_e,$$
$$n + e^- \rightarrow n + \langle K^- \rangle + \nu_e,$$

where $\langle K^- \rangle$ means the condensed kaon field [6]. The cooling curves for non-standard cooling scenario show the fast cooling, then with more additional process(heating or pairing), we can explain the observed surface temperature of NS.

We hereafter address three current issues about the phase transitions in neutron stars.

I DELAYED COLLAPSE

After supernovae explosions, protoneutron stars (PNS) are formed with hot, dense and neutrino-trapped matter. They usually evolve to cold ($T \simeq 0$) NS through two main eras: One is deleptonization era and the other the initial cooling era. In the deleptonization era, trapped neutrinos are released in about several seconds and PNS evolve to hot and neutrino-free NS. Then through the initial cooling era, they evolve to cold usual NS in a few tens of seconds.

However some of them may collapse to low-mass black holes during these eras by softening the EOS due to the occurrence of hadronic phase transitions [9]. This is called the *delayed collapse*. As a typical example, neutrinos from SN1987A were observed at Kaomiokande, but no pulsar yet, which suggests the possibility of the delayed collapse in SN1987A.

We consider here the possibility of the delayed collapse in the context of kaon condensation. Kaon condensation, one of the candidates of the hadronic phase transitions in high-density nuclear matter, is a kind of Bose-Einstein condensation. Fig.1 shows the mechanism of the occurrence of kaon condensation. As density increases, single particle energy for kaon ε_- decreases due to the attractive KN interaction in medium, while the electron chemical potential, which is equal to the kaon chemical potential in the beta equilibrium and neutrino-free matter, increases. When they become equal to each other, Bose-Einstein condensation of kaons occurs at a critical density ρ_c. Kaon condensation has been studied by many authors mainly at zero temperature since first suggested by Kaplan and Nelson [10]. We know that the kaon condensation gives rise to the large softening of EOS. Kaon condensation is the first order phase transition and thereby EOS includes thermodynamically unstable region. We applied the Maxwell construction to obtain the equilibrium curve for simplicity though, restrictly speaking, we need to take the Gibbs conditions(See Sect.II).

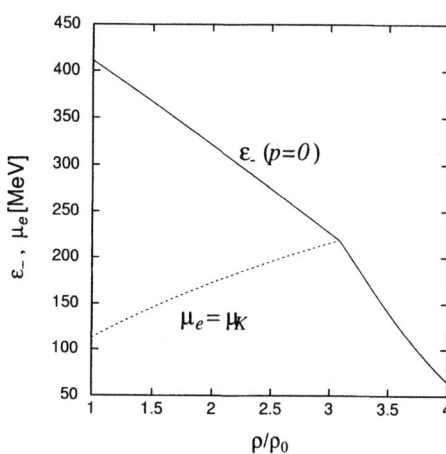

FIGURE 1. Mechanism of the occurrence of kaon condensation. Crossing point of chemical potential and single particle energy for K^-. represents a critical density.

Recently, to study the PNS, there appear a few works about kaon condensation at finite temperature [11,12] but there was no consistent theory based on chiral symmetry. Then we have presented a new formulation to treat fluctuations around the condensate based on chiral symmetry [13,14].

With thermodynamic potential in the reference [13,14], we can study the properties of kaon condensed state at finite temperature and then discuss some implications on the delayed collapse of PNS. We, hereafter, use the heavy-baryon limit for nucleons [13]. We show the phase diagram, EOS and then discuss the properties of PNS where thermal and neutrino-trapped effects are very important.

A Phase Diagram and EOS for kaon condensation

First we show the phase diagram in Fig.2. In the neutrino-trapped case we set $Y_{le} = Y_e + Y_{\nu_e} = 0.4$ where $Y_e(Y_{\nu_e})$ is the electron(electron-neutrino) number per baryon, while $Y_{\nu_e} = 0$ in the neutrino-free case. Both of the thermal and neutrino-

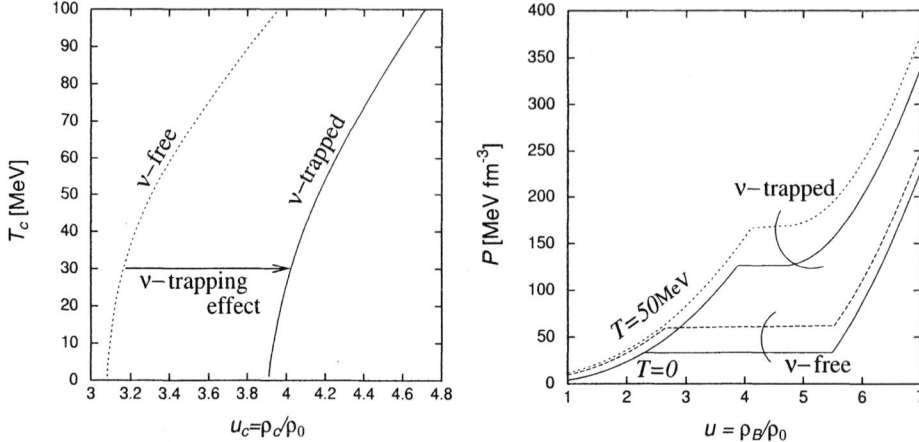

FIGURE 2. Phase diagram in density and temperature plane.

FIGURE 3. The EOS for kaon condensed matter.

trapped effects largely suppress the occurrence of kaon condensation. The reason for the latter case may be understood from the threshold condition, $\varepsilon_-(\rho_c) = \mu_K$. Chemical equilibrium holds the relation $\mu_K = \mu_e - \mu_{\nu_e}$: $\mu_{\nu_e} > 0$ in the neutrino-trapped case while $\mu_{\nu_e} = 0$ in the neutrino-free case, which means kaon chemical potential should be suppressed in the neutrino-trapped case. Then the occurrence of kaon condensation is suppressed through the suppression of number of kaons.

Next Fig.3 represents the EOS in density-pressure plane. Both of thermal and neutrino-trapping effects stiffen the EOS in the condensed phase. They seem to be more pronounced in the condensed state, especially around the critical density(see Fig.3), mainly through the rise of critical density.

In the realistic situation in PNS the isentropic condition is more relevant [11]. We reconstruct the isentropic EOS by evaluating the entropy as a function of temperature.

B Properties of PNS

Solving the TOV equation with the EOS, we can study the properties of PNS. In Fig.4 we show the gravitational mass versus central density for the neutrino-trapped and -free cases with entropy per baryon $S = 0, 1$ or 2. Both of the thermal and neutrino-trapping effects make the gravitational mass larger for the almost all of the central density. As the exception, at high central density in neutrino-free

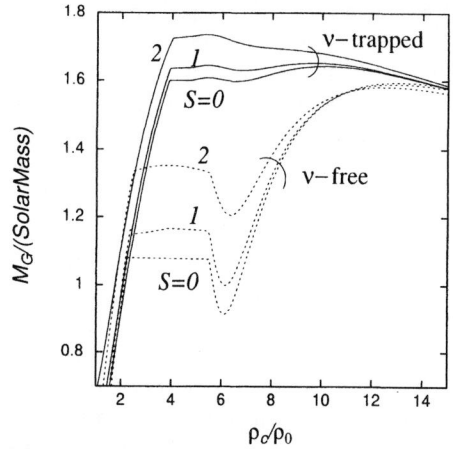
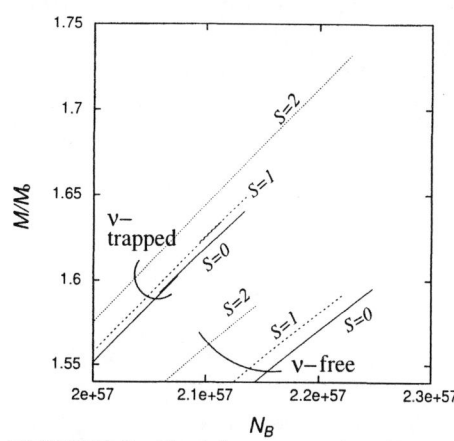

FIGURE 4. Central density versus gravitational mass.

FIGURE 5. Total baryon number N_B and gravitational mass for stable PNS.

case masses of hot NS are smaller than ones of cold NS. This is out of our intuition and the reason is as follows. Usually thermal effect enlarges the pressure and leads to the larger mass. At the same time, however, thermal effect enlarges the energy which contribute to gravitation. In the competition of the increase of pressure and gravitation, sometimes, mass of hot NS becomes smaller than one of cold NS.

In the neutrino-free case, once kaon condensation occurs in the core of a star, gravitational mass is little changed in the neighborhood of the equal pressure region in the isothermal EOS, then gravitationally unstable region (negative gradient part) appears. Therefore the neutron-star branch is separated into two stable branches: one is for stars with kaon condensate in their cores (right hand side from gravitationally unstable region) and the other consisting of only normal matter (left hand side from gravitationally unstable region). Thermal effect to the gravitational mass seems to be very large around the critical density because the EOS changes largely there. However the maximum mass, stars with which include kaon condensate in the core, is hardly changed and even decrease by the thermal effect as already discussed. As the temperature raises, gravitational mass grows largely in the normal branch but not in the kaon condensed branch.

In the neutrino-trapping case, thermal effect is large around the critical density and small for the heavy stars as well, but the situation is quantitatively different. For the $S = 0$ or 1 case we can see that the neutron-star branch is also separated by the gravitationally unstable region and the star with maximum mass exists in the kaon condensed branch. (Their central density $\rho_c \simeq 10\rho_0$) On the other hand, in the $S = 2$ case almost all of the stars with kaon condensate are gravitationally unstable, and the maximum-mass star whose central density $\rho_c = 5.4\rho_0$, still resides in the normal branch. For this reason the central density of the maximum-mass star is very different from those for $S = 0$ or 1.

To discuss the possibility of delayed collapse of PNS, the total baryon number N_B should be fixed as a conserved quantity during the evolution [15], under the assumption of no accretion. In Fig.5 we show the gravitational mass versus total baryon number for gravitationally stable PNS omitting unstable stars. Each terminal point represents maximum mass and maximum total baryon number. If an initial mass exceeds the terminal point in each configuration, the star should collapse into a black hole (not a delayed collapse but a usual formation of a black hole). We have shown the neutrino-trapped and -free cases; the former case might be relevant for the deleptonization era, while the latter for the initial cooling era. It is interesting to see the difference between the neutrino-free and -trapped cases: the curve is shortened as the entropy increases in the former case, while elongated in the latter case, where the remarkable increase in $S = 2$ and neutrino-trapped case results from that maximum mass exists in normal branch, different from other configurations with maximum mass in kaon condensed branch. These features are important for the following argument about the delayed collapse and maximum mass of the cold NS. The delayed collapse is possible if the initially stable star on a curve finds no corresponding point on other curves during the evolution through deleptonization or cooling with the baryon number fixed. Consider a typical evolution for example: A PNS is born as a neutrino-trapping and hot ($S = 2$) star after supernova explosion and evolves to neutrino-free and hot ($S = 2$) stage through deleptonization. Then through the cooling, the star evolves to be neutrino-free and cold ($S = 0$). We can clearly see the PNS with large enough mass can exist as a stable star at the beginning but cannot find any point on the neutrino-free and $S = 2$ curve. Therefore they must collapse to the low-mass black hole by deleptonization. It is to be noted that because the neutrino-trapped and $S = 2$ star never includes kaon condensate, its collapse is largely due to the appearance of kaon condensate in the core. Thus we may conclude that kaon condensation is very plausible to cause the delayed collapse in the deleptonization era.

On the other hand, in the initial cooling era after deleptonization delayed collapse does not take place because the NS on neutrino-free and $S = 2$ branch evolve to neutrino-free and cold branch and all of the stars seem to be able to find corresponding stable points in the each stage.

C Summary for delayed collapse

We have shown that the delayed collapse is possible in the deleptonization era due to the appearance of kaon condensation and the maximum mass of cold NS should be determined in neutrino-free and hot stage [16].

On the other hand, Pons et al. also studied kaon condensation in PNS matter [17] They concluded that the thermal effect is a key object to the delayed collapse and it is different from ours. We cannot give a clear reason for the discrepancy at present but it may originate from the difference of formulations: our discussion is based on the chiral Lagrangian and theirs, meson exchange model. Since the chiral model is based on the nonlinear Lagrangian the properties of the well developed kaon condensed phase become very different between both models.

In order to study the mechanism of delayed collapse and mass region which should collapse in more detail, we had better study the dynamical evolution beyond the static configurations. There neutrino opacity is important to determine the duration of the deleptonization [18], and in the kaon condensed phase neutrino opacities may become larger than in normal phase [19,20]. As another remaining issue, we will refine the EOS with the Gibbs conditions instead of the Maxwell construction.

II MIXED PHASE

For first order phase transition, for kaon condensation, quark matter, etc, there appears the thermodynamically unstable region in EOS. The Maxwell construction, which may be familiar for the liquid-gas phase transition for water, had been used as a standard method to get the proper EOS. Recently Glendenning pointed out that the Maxwell construction is not correct and we should impose the Gibbs conditions [21]. Here, we review some features due to the Maxwell construction and the results by Pons et al. [22].

A Maxwell Construction

The Maxwell construction imposes the conditions: $\mu_n^N = \mu_n^K$ and $P^N = P^K$, and can be written as equal-area rule,

$$\int_N^K V dP = 0. \qquad (1)$$

In Fig.6 we show the original EOS with thermodynamically unstable region and improved one by the Maxwell construction. As a result of the Maxwell construc-

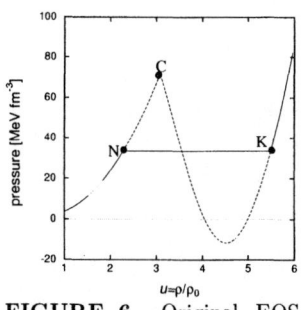

FIGURE 6. Original EOS and one in the Maxwell construction.

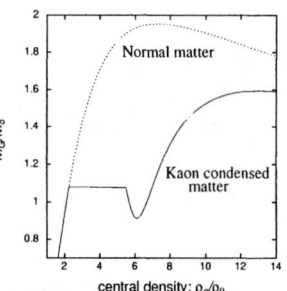

FIGURE 7. Mass-central density curve.

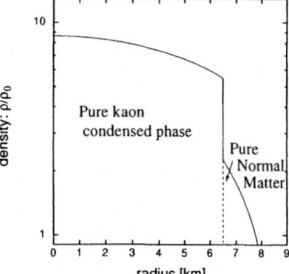

FIGURE 8. Interior structure of NS.

tion, there appears equal pressure region, which is called the mixed phase of kaon condensed matter and normal matter, only the volume ratio of two phases changes and properties of each phase (density, chemical potentials and so on.) never change. And the density gap leads to peculiar structure of NS. In Fig.7 mass-central density curve is shown. Once kaon condensation occurs at $\rho \simeq 2.3\rho_0$, density gap appears and gravitationally unstable region (region with negative slope) follows. Then picking up a NS, the core consists of two pure phases without the mixed phase (See Fig.8), which is known as a peculiar feature due to the Maxwell construction.

B Gibbs Conditions

In fact, however, for the first order phase transitions in nuclear matter, the Maxwell construction is not valid. Because there exist two chemical potentials, baryon and charge chemical potentials, we should use the Gibbs conditions [21].

The Maxwell construction is the same to the Gibbs conditions in case with only one chemical potential. On the other hand, there exist two chemical potentials in nuclear matter: baryon and charge chemical potentials which correspond to neutron and electron chemical potentials respectively. Then we need to use the Gibbs conditions,

$$\mu_n^N = \mu_n^K, \quad \mu_e^N = \mu_e^K, \quad P^N = P^K, \qquad (2)$$

between the phases of normal matter N and kaon condensed matter K.

Glendenning and Schaffner-Bielich discussed the EOS for kaon condensed matter imposing the Gibbs conditions [22] instead of using the Maxwell construction. Using the Gibbs conditions, compared to the case of the Maxwell construction, the EOS should be smoothed due to the appearance of mixed phase, the equal pressure region disappears and the mixed phase exist in the broad range of density. At the beginning of the mixed phase, the droplets consisting of kaon condensed matter appears in the normal matter. The densities for each phase can change at each point of the mixed phase. Charge neutrality is not achieved locally and the charge density is nonzero in each phase (Of course, globally achieved.); kaon condensed (normal) matter is negatively (positively) charged. This may be unfamiliar situation but simply understood as follows. In the mixed phase if there exists only baryon chemical potential, which is in the case of the Maxwell construction, baryon densities are not equal in two phases. On the other hand, in the case there exist two chemical potentials, as in the case of the Gibbs conditions, both of baryon and charge densities are not equal.

The properties of NS are also changed when we apply the Gibbs conditions. The gravitationally unstable region in the case of the Maxwell construction in Fig.7 disappears and the relation between central density and gravitational mass becomes to be smoothed. Here it is to be noted that the remarkable change appears for NS with comparably small mass (large radius) and behavior of heavy NS is little

	Sun	White Dwarf	NS	Magnetar
B[gauss]	10^3	10^8	10^{12}	10^{15}
R[cm]	10^{10}	10^8	10^6	10^6

TABLE 1. Magnetic fields and radii.

changed. Then our discussion about the delayed collapse in Sect.I may be hardly modified.

Interesting feature appears in the interior structure of NS in the case of the Gibbs conditions. The mixed phase may exist largely in the cores of NS. For a NS with $1.5M_\odot$, for example, a droplet phase exists largely and its radius is about 3[km] [22].

C Summary for mixed phase

In this section we have addressed the interesting issue about the mixed phase. The problem of the mixed phase seems to have two important aspects. One is the change of bulk properties of NS (mass, radius and so on.) through the modification of EOS and the other is of interior structure of NS due to the appearance of large extent of the mixed phase. The latter, especially, may give many implications to astrophysics. For example, Reddy et al. [19] discussed the coherent scattering of neutrinos in the mixed phase and they found that the mean free path of neutrinos becomes smaller in the droplet phase.

In the preceding discussion, we have ignored the finite volume effects: surface and Coulomb energy. They cannot be neglected to study the realistic EOS and they are expected to prevent the occurrence of the mixed phase [23]. Then the realistic EOS may exist between the EOS in the Gibbs conditions and the Maxwell construction.

III MAGNETARS AND STRANGE STARS

The magnetars are NS with strong magnetic fields ($B \sim 10^{14-15}$G derived from the P-$\dot{P}$ curve) and anomalous X-ray pulsars (AXP) and two pulsars in soft-gamma-ray repeaters (SGR) are known as the candidates [24]. The strong magnetic fields are out of the scaling law between radius and magnetic field $R^2 B \simeq const$, which can explain the magnetic fields of other stars(See Table 1). Then there is a possibility that they may originate from strong interaction. Recently there has been proposed on idea that the magnetars may be strange stars with complete spin alignment [25].

Once u,d,s-quark matter is formed, it may be more stable than normal nuclear matter at lower densities. Distinguished from the usual NS with core of quark matter, compact stars consisting mainly strange quark matter with or without thin crust are called strange stars.

As a very simple estimation, using non-interacting massless quarks in bag at $T=0$, energy per baryon of u,d-quark matter $\varepsilon_{u,d}$ and of u,d,s-quark matter $\varepsilon_{u,d,s}$ can be estimated.

$$\varepsilon_{u,d} \simeq 934\text{MeV}\frac{B^{1/4}}{145},$$
$$\varepsilon_{u,d,s} \simeq 829\text{MeV}\frac{B^{1/4}}{145},$$

with bag constant B. Then we can easily find that the u,d,s-quark matter is more stable than u,d-quark matter and nuclear matter. Of course we can use more complex model, with the one-gluon exchange and with heavy strange quark, and the result is similar [26]. A natural question may appear: if this is right, why nuclei consists of nucleon. But there is no contradiction because there is surface effect for light nuclei and no strange quark exists for heavy nuclei. A few strange quark in nuclear matter are not stable and it is almost impossible to produce enoughly many strange quarks at the same time through the higher-order weak interaction.

Based on the idea of strange stars, Tatsumi studied the possibility of magnetization of strange quark matter with relativistic one-gluon exchange interaction [25]. He concluded that magnetized strange quark matter may be stable in the low-density-region. At $\rho \sim 0.1\text{fm}^{-3}$ magnetization may occur in strange stars and can produce $B \sim 10^{15-17}$. He just suggested the possibility and realistic calculation(e.g. Hartree-Fock) is the future problem.

IV CONCLUDING REMARKS

We concentrated on 3 topics in this paper. We discussed the evolution of PNS from the view point of the nuclear theory (in the static limit) in Sect.I and found that the delayed collapse is possible in the deleptonization era due to the occurrence of kaon condensation. In Sect.II we reviewed the problem of the mixed phase, which may give important effects to NS physics: glitch, cooling scenario and so on. Then an interesting idea about the identification of magnetars as strange stars was introduced in Sect.III. Of course there exist many candidates of phase transitions and related astrophysical phenomena which we could not pick up here. And now we recommend a good and recent review by Heiselberg and Hjorth-Jensen [1].

At present it is not clear what kind of phase transitions really exist in NS. But the experimental data on the earth and the observation of signal from NS will limit the candidates and, sometime, determine what happens in NS, we hope.

ACKNOWLEDGEMENT

It is a great pleasure to thank the organizers of TOURS2000 for an enjoyable meeting. The author is grateful to T. Tatsumi, T. Muto and T. Takatsuka for the collaborations and helpful discussions.

REFERENCES

1. For a review article,
   H. Heiselberg and M. Hjorth-Jensen, Phys. Rep. **328** (2000) 237.
2. H. Shen, H. Toki, K. Oyamatsu and K. Sumiyoshi, Prog. Theor. Phys. **100** (1998) 1013, Nucl. Phys. **A637** (1998) 435.
3. S.E. Thorsett and D. Chakrabarty, *astro-ph/9803260*.
4. V. Thorsson, M. Prakash, J.M. Lattimer, Nucl. Phys. **A572** (1994) 693.
5. H. Fujii, T. Maruyama, T. Muto and T. Tatsumi, Nucl. Phys. **A597** (1996) 645.
6. S. Tsuruta, Phys. Rep. **292** (1998) 1.
7. D.N. Voskresensky, *astro-ph/0009093*.
8. D. Page, ApJ. **479** (1997) L43.
9. G. E. Brown and H. A. Bethe, Astrophys. J. **423** (1994) 659.
10. D.B. Kaplan and A.E. Nelson, Phys. Lett. **B175** (1986) 57; **B179** (1986) 409(E).
11. M. Prakash, I. Bombaci, M. Prakash, P. J. Ellis, J. M. Lattimer and R. Knorren, Phys. Rep. **280** (1997) 1.
12. V. Thorsson and P.J. Ellis, Phys. Rev. **D55** (1997) 5177.
13. T. Tatsumi and M. Yasuhira, Phys. Lett. **B441** (1998) 9.
14. T. Tatsumi and M. Yasuhira, Nucl. Phys. **A653** (1999) 133.
15. T. Takatsuka, Prog. Theor. Phys. **95** (1996) 901.
16. M. Yasuhira and T. Tatsumi, *nucl-th/0009090*.
17. J.A. Pons, S. Reddy, P.J. Ellis, M. Prakash and J.M. Lattimer, *nucl-th/0003008*.
18. J.A. Pons, S. Reddy, M. Prakash, J.M. Lattimer and J.A. Miralles, *astro-ph/9807040*.
19. S. Reddy, G. Bertsch and M. Prakash, Phys. Lett. **B475** (2000) 1.
20. T. Muto, T. Tatsumi, N. Iwamoto and M. Yasuhira, in preparation.
21. N.K. Glendenning, Phys. Rev. **D46** (1992) 1274.
22. N.K. Glendenning and J. Schaffner-Bielich, Phys. Rev. Lett. **81** (1998) 4564; Phys. Rev. **C60** (1999) 025803.
23. H. Heiselberg, C.J. Pethick and E.F. Staubo, Phys. Rev. Lett. **70** (1993) 1355.
24. C. Kouveliotou et al., *Nature* **393** (1998) 235, K. Hurley et al., Astrophys. J. **510** (1999) L111.
25. T. Tatsumi, *hep-ph/9910470*, Phys. Lett. in press.
26. For a review article, J. Madsen, *astro-ph/9809032*.

Nucleosynthesis in Massive Stars Using Extended Adaptive Nuclear Reaction Networks

A. Heger[1], T. Rauscher[1,2], R. D. Hoffman[3], and S. E. Woosley[1]

[1] *Astronomy Department,*
University of California, Santa Cruz, CA 95064, U.S.A.
[2] *Nuclear Theory and Modeling Group, L-414,*
Lawrence Livermore National Lab, Livermore, CA 94551-9900, U.S.A.
[3] *Departement für Physik und Astronomie,*
Universität Basel, CH-4056 Basel, Switzerland

Abstract. We present the first calculations to follow the evolution of *all* stable isotopes and their abundant radioactive progenitors in a finely zoned stellar model from the onset of central hydrogen burning through explosion as a Type II supernova. An extended adaptive nuclear reaction network is implemented that contains about 700 isotopes during hydrogen and helium burning and more than 2500 isotopes during the supernova explosion. The calculations were performed for 15, 20, and 25 $M_\odot$ Pop I stars using the most recently available set of experimental and theoretical nuclear data. We include revised opacity tables, take into account mass loss due to stellar winds, and implement revised weak interaction rates that significantly affect the properties of the presupernova core.

An s-process is present, which, along with the usual nucleosynthesis from advanced burning stages and the explosion, produces nearly solar abundances for most nuclei up to $A = 60$ in the 25 $M_\odot$ star. Between $A = 60$ and $A = 90$ we find that the s-process leads to an over-production of key nuclei by a factor $\sim 2-3$. Above $A = 90$ the s-process has makes little contribution, but we find the approximately solar production of many proton-rich isotopes above mass number $A = 120$ due to the γ-process.

INTRODUCTION

Massive stars with initial masses of more than $\sim 8\,M_\odot$ are the main source of oxygen and heavier elements. To understand the details of the nucleosynthesis in these objects numerical simulations using extended nucleosynthesis networks have to be performed, that follow the stellar evolution and the supernova explosion. Unlike previous studies who followed single zones with extended networks [1–3] or full stellar models with restricted networks (e.g., [4–6]), we are now able to follow the nucleosynthesis of all stable isotopes from hydrogen to bismuth and all their radioactive progenitors.

INPUT PHYSICS AND NUMERICAL METHOD

We employed a hydrodynamic stellar evolution code, KEPLER [7], to follow the stellar evolution from central hydrogen ignition to onset of core collapse. In the study presented here we used most recent input physics, including, e.g., OPAL opacity tables [8], mass loss, and the weak rate set of [9], which causes significant changes of the stellar structure at the late evolutionary phases [10,11]. The supernova explosion was simulated by a piston, as described in [4], that typically deposits $1.2 \cdot 10^{51}$ erg, or leads to the ejection of $\sim 0.1\,M_\odot$ of ^{56}Ni, both typical values for supernovae. The mass-cut between ejection and fall-back was then determined self-consistently from the dynamics of the explosion.

During the early stages of stellar evolution much smaller reaction networks ($\sim$ 500 isotopes) are necessary than at the late stages ($\sim$ 1500 isotopes), and, in particular, during the supernova explosion ($\sim$ 2500 isotopes). Some burning phases require more neutron-rich isotopes while during others proton-rich isotopes become important to properly follow the flow of the nuclear reactions. Also, the maximum reaction network needed may vary from star to star and therefore is difficult to determine before the calculation without including many unneeded isotopes. Due to these issues, and since huge nuclear reaction networks are computationally quite expensive, we implemented an adaptive nuclear reaction network algorithm that automatically adds and removes isotopes as needed. This cut the computational cost by a factor 3–10 and allowed for extended studies.

From neon to bismuth we used the strong nuclear reaction rates set of [12], below neon the rates of [13] were taken, and for α captures on self-conjugate nuclei we followed [14]. Experimental rates were used for neutron capture along the line of stability [15], the (α, γ) reaction on ^{70}Ge [16] and ^{144}SM [17], and for β^-, β^+, and α-decay [18], supplemented with theoretical β^- and β^+ rates [19]. For ^{180}Ta we implemented a temperature-dependent decay rate [20]. A more detailed description of the nuclear rate sets will be given in [21].

OVERVIEW OF RESULTS AND CONCLUSIONS

We calculated the evolution of 15, 20, and $25\,M_\odot$ stars of solar composition (Pop I stars) using the code described above. For the $25\,M_\odot$ star we also performed additional calculations using alternative reaction rate sets (NACRE [22] and different estimates for the ^{22}Ne$(\alpha,n)^{23}$Mg reaction [23,24]). Due to space restrictions in this conference proceedings we can only briefly present some results of the $25\,M_\odot$ star using the rate set of [12].

The production factor of all stellar ejecta, including wind and supernova explosion, with respect to solar abundances [25] is shown in Figure 1. Not included is any (r-process) contribution from the neutron star wind and the ν-process on nuclei

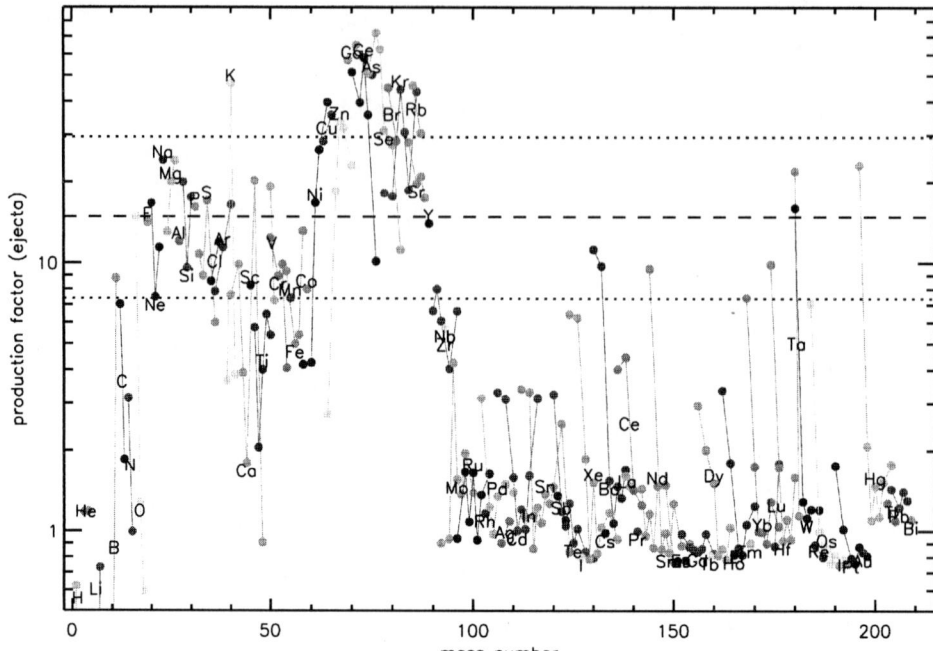

FIGURE 1. Decayed production factors as a function of mass number, A. The different elements are color coded and their isotopes (*dots*) are connected by *thin lines*. The *dashed line* indicates the production factor of ^{16}O, the "metal" with the biggest mass fraction, and *dotted lines* span a range of 0.3 dex around this value, showing the "range of good agreement" with solar abundance ratios.

with A or Z greater than 40. The range of acceptable agreement with solar abundance ratios, defined by a production factor similar to that of oxygen, is indicated by dotted lines.

Most isotopes from neon to nickel are produced at about solar production factors, while the iron group is slightly low. The trans-iron isotopes of $A = 60 - 90$ are overproduced by a factor 2 due to the s-process. This is similar to what has been found before in studies that used smaller reaction networks [4,13]. If assumed that less metal-rich stars have a correspondingly weaker s-process contribution, a discrepancy with solar abundance data is avoided. We also found weaker s-process in the lower-mass stars of our sample. Above $A = 90$ we only found a weak s-process that changes the isotope abundances by no more than a factor 2.

Above $A = 100$ the we found a significant production of proton-rich isotopes by the γ-process. In the mass ranges $A = 123 - 150$ and $A = 170 - 200$ the production factors were compatible with solar abundance patterns, but for $A = 100 - 123$ and $A = 150 - 170$ the production was low by a factor 2. Especially the rarest isotope, ^{180}Ta, however, was produced in good agreement with the sun. We note that in other models of our calculations a stronger γ-process was found. Its strength varied significantly with details of the stellar structure, in particular, the extent and radius of the neon-oxygen core. For more details we refer the reader to [21,26].

Acknowledgments. This research was supported, in part, by Prime Contract No. W-7405-ENG-48 between The Regents of the University of California and the United States Department of Energy, the National Science Foundation (AST 97-31569, INT-9726315), and the Alexander von Humboldt-Stiftung (FLF-1065004). T.R. acknowledges support by a PROFIL professorship from the Swiss National Science foundation (grant 2124-055832.98).

REFERENCES

1. Cowan, J.J., Cameron, A.G.W., and Truran, J.W., *ApJ* **294** 656, (1985).
2. Kratz, K.L., et al., *ApJ* **402**, 216 (1993).
3. Freiburghaus, C., et al., ApJ **516** (1999) 381.
4. Woosley S.E., and Weaver, T.A., *ApJS* **101** 181 (1995).
5. Thielemann, F.-K., Nomoto, K., and Hashimoto, M.-A., *ApJ* **460**, 408 (1996).
6. Chieffi, A., Limongi, M., and Stanerio, O., *Proc. 10th Workshop on "Nuclear Astrophysics"*, eds. Hillebrandt, W., and Müller, E., *MPA/P12*, p. 115, (2000).
7. Weaver, T.A., Zimmermann, G.B., and Woosley, S.E., *ApJ* **225**, 1021 (1978).
8. Iglesias, C.A., and Rogers, F.J., *ApJ* **464**, 943 (1996).
9. Langanke, K., and Martínez-Pinedo, G., *Nucl. Phys.* **A673**, 481, (2000).
10. Heger, A., Langanke, K., Martínez-Pinedo, G., and Woosley, S.E., *PRL*, (2000), submitted.
11. Heger, A., Woosley, S.E., Martínez-Pinedo, G., and Langanke, K., *ApJ*, (2000), in prep.
12. Rauscher T., and Thielemann, F.-K., *ADNDT* **75** (No. 1+2), 1. (2000).
13. Hoffman, R.D., Woolsey, S.E., and Weaver, T.A., *ApJ*, (2000), in press.
14. Rauscher T., Thielemann, F.-K., Görres, J., and Wiescher, M., *NP* **A675**, 695 (2000).
15. Bao, Z., et al., *ADNDT* **76**, 1 (2000).
16. Flüp, Zs., et al., *Z. Phys.* **355**, 203 (1996).
17. Somorjai, E., et al., *A&A* **333**, 1112 (1998).
18. Tuli, J.K., et al., *Nuclear Wallet Cards*, 5th ed., Brookhaven National Laboratory, USA, (1995); Kratz, K.L., priv. comm.; Thielemann, F.-K., priv. comm.
19. Möller, P., Nix, J.R., and Kratz, K.-L., *ADNDT* **66**, 131 (1997).
20. Belic, D., et al., *PRL* **83**, 5242 (1999).
21. Rauscher, T., Heger, A., Hoffman, R.D., and Woosley, S.E., *ApJ*, (2000), in prep.
22. Angulo, C., et al., *NP* **A656**, 3, (1999).
23. Costa, V., Rayet, M., Zappalà, R.A., and Arnould, M., *A&A* **358**, L67 (2000).
24. Käppeler, F., et al., *ApJ* **437**, 396 (1994); Wiescher, M., priv. comm.
25. Anders, E., and Grevesse, N., *Geochim. Cosmochim. Acta* **53**, 197 (1989).
26. Heger, A., Woosley, S.E., Hoffman, R.D., and Rauscher, T., *ApJ*, (2000), in prep.

Collapse-driven supernovae and the r-process with new data of unstable nuclei

K. Sumiyoshi

Numazu College of Technology (NCT)
Ooka 3600, Numazu, Shizuoka 410-8501, Japan [1]

Abstract. We have studied the ν-driven winds from the nascent proto-neutron star. We have found in the general relativistic hydrodynamical simulations using the relativistic EOS table that the short expansion time scale is realized in the cases of massive and compact proto-neutron stars. Having the trajectories of ejecta, we have performed r-process calculations to find that a successful r-process up to A $\sim$ 195 occurs in the ν-driven winds with a short expansion time scale.

I INTRODUCTION

Physics of supernovae and r-process nucleosynthesis are both fascinating problems where the interplay of nuclear physics and astrophysics takes the important keys. To attack these problems, one has to perform careful numerical simulations with the information of nuclear physics at extreme conditions, and then, one has to compare with observational signature of stellar evolution. Here I focus on our recent studies of the r-process in ν-driven winds in supernovae with two strong motivations from the recent progress in astronomy and nuclear physics.

The first motivation is what the recent observations of r-process elements in extremely metal-deficient stars [1,2] have exhibited. Metal-deficient stars, whose metal abundance is 10^{-3} or less relative to the solar system abundance, are thought to have formed as first or second generation stars in the early stage of the evolution of the Galaxy. The abundance patterns have been found to match very well with a scaled solar system r-process abundance pattern. These observations strengthen our interests to study the nucleosynthesis in ν-driven winds as a primary process. We would like to clarify whether the r-process occurs and how much amount of heavy elements are produced in ν-driven winds.

The second motivation is the recent advance in the physics of unstable nuclei which have been developed in radioactive nuclear beam facilities in the world [3,4].

---

[1] E-mail: sumi@la.numazu-ct.ac.jp

There are increasing experimental data of unstable nuclei in neutron-rich region, which can help to probe the neutron-rich matter in neutron stars and supernovae. Along this line of research, we have extensively studied the nuclear structure and nuclear matter in the relativistic many body framework [5–8]. Recently, we have completed the relativistic EOS table in order to apply to supernova simulations from these theoretical studies [9,10]. Reliability of the EOS table has been checked by comparing with the data of unstable nuclei, which ensures the validity of application to various simulations of supernova phenomena. In the light of this progress in physics of unstable nuclei, we perform the hydrodynamical simulations of ν-driven winds. I report in this article that a short expansion time scale is found in our hydrodynamical studies [11] to be a promising condition for successful r-process in ν-driven winds, as was previously proposed by Otsuki et al. in the semi-analytic studies of the winds [12].

II R-PROCESS IN ν-DRIVEN WINDS

A Relativistic EOS table

The relativistic EOS table, which has been completed recently by the present author and his collaborators [9,10], spans the wide range of density, composition and temperature relevant to supernova explosions. This EOS is based on the relativistic mean field theory, which is known to be a successful model for the studies of nuclear structure [5–7]. We have constructed the RMF theory based on the relativistic Brückner-Hartree-Fock theory [13]. The RMF theory has been carefully polished by incorporating the experimental data such as masses and radii [8,14] in order to set up the theoretical parameters. There have been accumulated successful applications to a variety of studies in nuclear physics and astrophysics. The relationship between EOS and neutron-skin thickness has been discussed [15]. The relativistic EOS table has been applied to numerical simulations of core collapse [16] and neutrino emission from proto-neutron stars [17]. The EOS table is available for open use upon request to the author.

B Condition of r-process in ν-driven winds

The ν-driven wind occurs in supernovae shortly after the core bounce. At this time, there is a hot neutron star containing neutrinos at the center. A part of neutrinos emitted as supernova neutrinos heats the surface material in the nascent neutron star. As a result, a small portion of material is ejected forming a ν-driven wind. During the expansion of ejected material, nucleosynthesis starts from neutrons and protons at high temperature above 1 MeV. At around 0.5 MeV, α-process goes on through three-body reactions to produce a small amount of seed elements. At around 0.2 MeV, r-process takes place if neutrons are still abundant enough.

Woosley et al. have indicated a profound implication that the r-process could occur if neutron-to-seed abundance ratio is high enough at the end of the α-process in the ν-driven winds with high entropy per baryon $\sim$400 k_B [18]. In their case the density is low, because $S \sim T^3/\rho$, so that the three-body reaction ($\alpha\alpha n \to {}^9\text{Be}$) creates not very large amount of seeds, leaving plenty of free neutrons. Consequently the neutron-to-seed abundance ratio becomes very high. However, in other studies, entropy has been found to be too low ($\sim$100 k_B) [19,20]. In addition, ν-process is found to reduce neutron fraction significantly to unfavor the r-process [21].

It has recently been proposed in the semi-analytic studies of the ν-driven winds [12] that a high neutron-to-seed abundance ratio is realized in the winds of short expansion time. If the expansion time scale is short enough, there is little time for the three-body reaction to create many seed elements, leading to high neutron-to-seed abundance ratio. Moreover, one can avoid harmful ν-process due to short time period [12,22]. Since these findings in the semi-analytic studies are quite favorable for the successful r-process, we tried to establish the validity by carrying out hydrodynamical simulations of the ν-driven winds [11].

C Numerical simulations

Numerical simulations have been carried out by solving general relativistic hydrodynamics [23] with neutrino heating and cooling being taken into consideration. The relativistic EOS table has been adopted for these simulations. We follow the hydrodynamics of surface layers of a proto-neutron star. Having trajectories of density and temperature of ejected material, we calculate nucleosynthesis by solving r-process reaction network [24]. We have studied the dependence on the mass

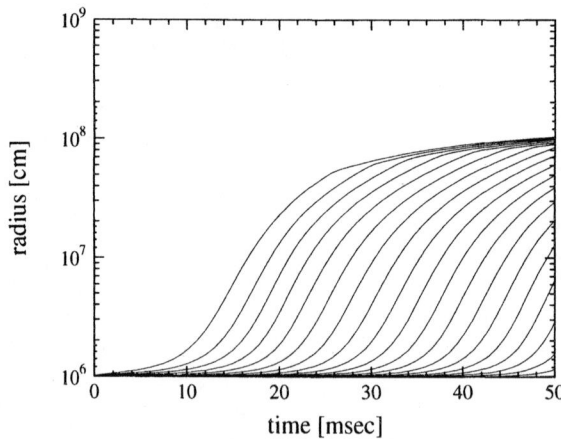

FIGURE 1. Calculated mass trajectories in neutrino-driven wind.

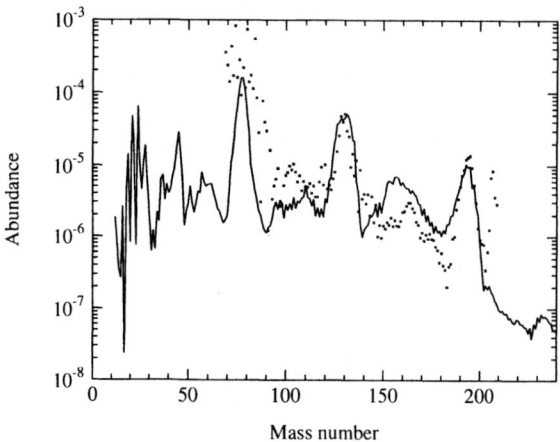

FIGURE 2. Calculated r-process abundance.

and radius of neutron star, and neutrino spectrum and luminosity as a first step. We are currently linking with supernova simulations.

Figure 1 demonstrates mass trajectories as a function of time in the case of the neutron star of $2M_\odot$ and 10 km with a total ν-luminosity of 6×10^{52} [erg/s]. The material is heated up by neutrinos, and the surface layers are ejected successively. In this particular case, the expansion time scale is 5 msec and the entropy per baryon is 160 k_B. Picking up a trajectory from numerical simulations, we have calculated the r-process nucleosynthesis with an extended nuclear reaction network including very light neutron-rich nuclei [24]. Figure 2 shows the final abundance yields of r-process calculation. We can observe that the heavy elements up to third peak around A $\sim$ 195 are successfully reproduced. Relative height of second peak is also nicely reproduced in the present work. In the current study, we have seen that massive and compact neutron stars are preferable. We have also found that a new nuclear-reaction flow runs through the light neutron-rich nuclei near the drip line and changes dramatically the final abundances. This new nuclear-reaction flow can change the final heavy element abundances by orders of magnitudes [24].

III SUMMARY

By performing hydrodynamical simulations with the relativistic EOS table, which has recently been completed, we have found that the r-process is possible in ν-driven winds with short expansion time scale. Further simulations of the ν-driven winds are now underway by taking care of the evolution of proto-neutron star and ν-spectrum in order to derive the total amount of r-process yields in supernovae from various progenitor masses.

ACKNOWLEDGMENTS

The author is grateful to M. Terasawa, K. Otsuki, H. Shen, H. Suzuki, S. Yamada, G. Mathews, H. Toki, and T. Kajino for continuous, fruitful collaborations. The numerical simulations have been performed on the supercomputers at RIKEN and KEK, Japan. The author is grateful for the travel support by Numazu College of Technology.

REFERENCES

1. C. Sneden, A. McWilliam, G. W. Preston, J. J. Cowan, D. L. Burris and B. J. Armosky, *Astrophys. J.* **467**, 819 (1996).
2. C. Sneden, J. J. Cowan, I. I. Ivans, G. M. Fuller, S. Burles, T. C. Beers and J. E. Lawler, submitted to *Astrophys. J.*, (2000); astro-ph/0003086.
3. I. Tanihata et al., *Phys. Rev. Lett.* **55**, 2676 (1985).
4. I. Tanihata et al., *Phys. Lett.* B **289**, 261 (1992).
5. K. Sumiyoshi, D. Hirata, H. Toki and H. Sagawa, *Nucl. Phys.* A **552**, 437 (1993).
6. Y. Sugahara and H. Toki, *Nucl. Phys.* A **579**, 557 (1994).
7. K. Sumiyoshi, H. Kuwabara and H. Toki, *Nucl. Phys.* A **581**, 725 (1995).
8. D. Hirata, K. Sumiyoshi, I. Tanihata, Y. Sugahara, T. Tachibana and H. Toki, *Nucl. Phys.* A **616**, 438c (1997).
9. H. Shen, H. Toki, K. Oyamatsu and K. Sumiyoshi, *Nucl. Phys.* A **637**, 435 (1998).
10. H. Shen, H. Toki, K. Oyamatsu and K. Sumiyoshi, *Prog. Theor. Phys.* **100**, 1013 (1998).
11. K. Sumiyoshi, H. Suzuki, K. Otsuki, M. Terasawa and S. Yamada, *Pub. Astron. Soc. Japan* **52**, 601 (2000).
12. K. Otsuki, H. Tagoshi, T. Kajino and S. Wanajo, *Astrophys. J.* **533**, 424 (2000).
13. R. Brockmann and R. Machleidt, *Phys. Rev.* C **42**, 1965 (1990).
14. T. Suzuki et al., *Nucl. Phys.* A **658**, 313 (1999).
15. K. Oyamatsu, I. Tanihata, Y. Sugahara, K. Sumiyoshi and H. Toki, *Nucl. Phys.* A **634**, 3 (1998).
16. K. Sumiyoshi, H. Suzuki, S. Yamada and H. Toki, in preparation for submission to *Nucl. Phys.*, (2000).
17. K. Sumiyoshi, H. Suzuki and H. Toki, *Astron. Astrophys.* **303**, 475 (1995).
18. S. E. Woosley, J. R. Wilson, G. J. Mathews, R. D. Hoffman and B. S. Meyer, *Astrophys. J.* **433**, 229 (1994).
19. J. Witti, H.-Th. Janka and K. Takahashi, *Astron. Astrophys.* **286**, 841 (1994).
20. Y.-Z. Qian and S. E. Woosley, *Astrophys. J.* **471**, 331 (1996).
21. B. S. Meyer, G. C. McLaughlin and G. M. Fuller, *Phys. Rev.* C **58**, 3696 (1998).
22. M. Terasawa, T. Kajino, S. Wanajo, G. J. Mathews and K.-H. Langanke, in preparation for submission to *Astrophys. J.*, (2000).
23. S. Yamada, *Astrophys. J.* **475**, 720 (1997).
24. M. Terasawa, K. Sumiyoshi, T. Kajino, I. Tanihata and G. J. Mathews, submitted to *Astrophys. J.*, (2000).

Improved predictions of nuclear data: a continued challenge in astrophysics

S. Goriely[1]

Institut d'Astronomie et d'Astrophysique
Université Libre de Bruxelles
Campus de la Plaine, CP 226
1050 Brussels – Belgium

Abstract. Although important effort has been devoted in the last decades to measure reaction cross sections and decay half-lives of interest in astrophysics, most of the nuclear astrophysics applications still require the use of theoretical predictions to estimate experimentally unknown rates. The nuclear ingredients to the reaction or weak interaction models should preferentially be estimated from microscopic or semi-microscopic global predictions based on sound and reliable nuclear models which, in turn, can compete with more phenomenological highly-parametrized models in the reproduction of experimental data. The latest developments made in deriving the nuclear inputs of relevance in astrophysics applications are reviewed. It mainly concerns nuclear structure properties (atomic masses, deformations, radii, etc...), nuclear level densities, nucleon and α-optical potentials, γ-ray and Gamow-Teller strength functions.

I INTRODUCTION

Stong, weak and electromagnetic interaction processes play an essential role in nuclear astrophysics (for a review, see [1]). The thermonuclear reactions of astrophysical interest mainly concern the capture of nucleons or α-particles at relatively low energies (far below 1 MeV for neutrons and the Coulomb barrier for charged particles)[2]. β-decay rates, as well as electron or positron captures are also crucial for our understanding of specific scenarios in stellar evolution (e.g presupernova and supernova models) and nucleosynthesis (e.g the r-process). Although important effort has been devoted in the last decades to measure decay half-lives and reaction cross sections, most of the relevant rates remain unknown and must be estimated theoretically. In addition, given astrophysical applications (e.g the r- or p-processes of nucleosynthesis) often involve a large number (thousands) of unstable nuclei, so that only *global* approaches, i.e a unique description in the whole (N, Z)-plane, are

---

[1] S.G. is FNRS senior research assistant
[2] Spallation reactions induced by the interaction of primary particles with relative energies in excess of some tens of MeV per nucleon are not considered here.

to be considered. For these reasons, when the nuclear ingredients to the reaction (e.g Hauser-Feshbach) or weak interaction (e.g the Quasi-Particle Random Phase Approximation or QRPA) models cannot be determined from experimental data, use is made preferentially of microscopic or semi-microscopic global predictions based on sound and reliable nuclear models which, in turn, can compete with more phenomenological highly-parametrized models in the reproduction of experimental data. The selection criterion of the adopted model is fundamental, since most of the nuclear ingredients in reactions rate calculations need to be extrapolated in an energy and mass domain out of reach of laboratory measurements, where parametrized systematics based on experimental data can fail drastically.

In most of the applications, thermonuclear reaction mechanisms are satisfactorily described by the statistical model of Hauser-Feshbach[3]. Regular updates as well as improvements are brought to the Hauser-Feshbach codes dedicated to astrophysical applications (e.g. the code MOST [2]). These concern mainly the ground-state description, the nucleon- and α-nucleus optical potentials, the nuclear level density (NLD) and the γ-ray strength function. The latest developments in these fields are discussed in Sect. II. Similarly, the description of weak interaction processes in extreme stellar conditions requires a continued effort in the develpment of global microscopic models, as discussed in Sect. III.

II NUCLEAR REACTION RATES

A Ground state properties

Among the ground state properties, the atomic mass $M(Z,A)$ is obviously the most fundamental quantity and enter all chapters of nuclear astrophysics. Their knowledge is indispensable to estimate the rate and energetics of any nuclear transformation. The calculation of the reaction and decay rates also requires the knowledge of other ground state properties, such as the deformation (β_2), density distribution (ρ_q), single-particle level scheme ($\epsilon_\nu, j_\nu, \pi_\nu$), pairing force and phase space (G_q, Λ), shell correction energies (δW_q), ..., as sketched in Fig. 1. The importance of estimating these properties reliably should not be underestimated. For example, the NLD of a deformed nucleus at low energies (typically at the neutron separation energy) is predicted to be significantly (about 30 to 50 times) larger than of a spherical one due principally to the rotational enhancement. An erroneous determination of the deformation can therefore lead to large errors in the estimate of radiative capture rates.

Until recently the atomic masses were calculated on the basis of one form or another of the liquid-drop model, the most sophisticated version of which is the "finite-range droplet model" (FRDM) [4]. Despite the great empirical success of

---

[3] As discussed in details in [3], the validity of the Hauser-Feshbach predictions has to be questioned for radiative neutron captures by exotic n-rich nuclei, and obviously for reactions involving light elements ($A \lesssim 20$).

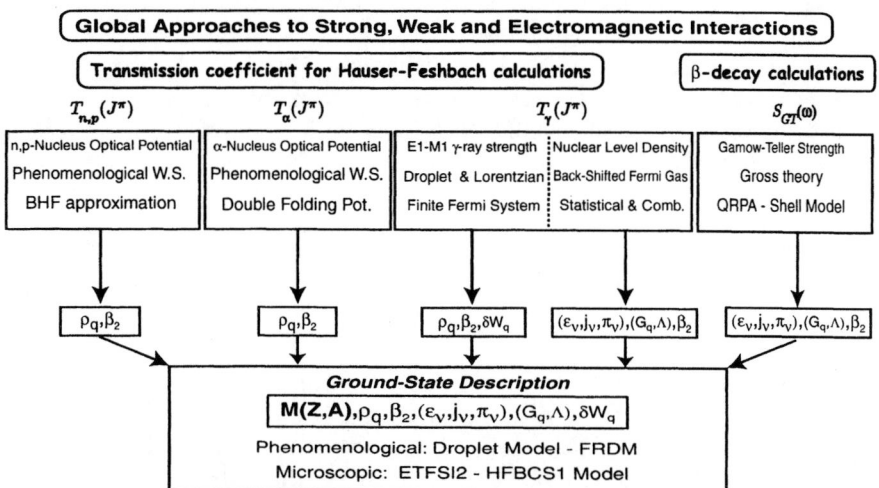

FIGURE 1. Ground-state quantities of relevance in the Hauser-Feshbach and β-decay rates calculations. Global phenomenological and microscopic theories used to estimate the different transmission coefficients T and Gamow-Teller strength function, as well as their dependence to ground state properties are listed. Details are given in the text.

this formula (it fits the 1888 $Z \geq 8$ masses with an rms error of 0.689 MeV), there is still an obvious need to develop a mass formula that is more closely connected to the basic nuclear interactions. Two such approaches can reasonably be contemplated at the present time, one being the non-relativistic Hartree-Fock (HF) method, and the other the relativistic Hartree method, also known as the relativistic mean-field (RMF) method. Progress in the HF and RMF mass models has been slow, presumably because of the computer-time limitations that arose in the past with deformed nuclei. Nuclear forces are traditionally determined by fitting to the masses (and some other properties) of less than ten or so nuclei. The resulting forces give rise to rms deviations from experimental masses well in excess of 2 MeV. This is far from reaching the level of precision found by droplet-like models (around 700 keV).

The result is that the most microscopically founded mass formulas of practical use were till recently those based on the so-called ETFSI (extended Thomas-Fermi plus Strutinsky integral) method [5]. The ETFSI method is nothing else than a high-speed macroscopic-microscopic approximation to the HF method based on Skyrme forces, with pairing correlations generated by a δ-function force that is treated in the usual BCS approach (with blocking). The macroscopic part consists of a purely semi-classical approximation to the HF method, the full fourth-order extended Thomas-Fermi method, while the microscopic part (based on what is called the Strutinsky-integral form of the Strutinsky theorem) constitutes an attempt to improve this approximation perturbatively, and in particular to restore the shell corrections that are missing from the ETF part. In the latest version of the ETFSI mass model (ETFSI2), eleven parameters are found to reproduce the 1719 experimental masses of the $A \geq 36$ nuclei (the $N = Z, Z \pm 1$ nuclei subject to Wigner-term anomalies are not included) with an rms deviation of 709keV [6]. The precision of ETFSI mass table is therefore comparable with the one obtained by the droplet-like formula. The ETFSI model remains an approach of the macroscopic-microscopic type, although it provides a high degree of coherence between the macroscopic and microscopic terms through the unifying Skyrme force underlying both parts. A logical step towards improvements obviously consists in considering now the HF method as such. It was demonstrated very recently [7,8] that HF calculations in which a Skyrme force is fitted to essentially all the mass data are not only feasible, but can also compete with the most accurate droplet-like formulas available nowadays.

The force used in the latest HFBCS mass calculation of [8] is a conventional 10-parameter Skyrme force, along with a 4-parameter δ-function pairing force (pairing correlations are introduced in the framework of the BCS method). The Skyrme and pairing parameters are determined by fitting to the full data set of 1719 $A \geq 36$ masses, both spherical and deformed. The best fit is obtained for an effective nucleon mass larger than one ($M^* = 1.05\ M$) and a symmetry coefficient $J = 28$ MeV. In order to describe the $|N - Z| \leq 1$ nuclei, a phenomenological Wigner correction term is added to the total HFBCS binding energy. The resulting rms deviation from the 1772 masses with $A \geq 36$ amounts to 0.683 MeV. The MSk7 force has been used to estimate the complete mass table HFBCS1 made of 9200 nuclei, including all those lying between the drip lines over the range $Z, N \geq 8$ and $Z \leq 120$. The rms error to the 1888 nuclei in this range for which measured masses are given in the 1995 Audi-Wapstra compilation [9] is 0.738 MeV.

The quality of the mass models available is traditionally estimated by the rms error obtained in the fit to experimental data and the number of free parameters. However, this overall accuracy does not imply a reliable extrapolation far away from the experimentally known region in view of the possible shortcomings linked to the physics theory underlying the model. The reliability of the mass extrapolation is a criterion of first importance when dealing with astrophysics applications. The quality of a mass model can also be tested by its ability to estimate other quantities than masses, such as deformations, charge radii or fission barriers. ETFSI2 and HFBCS1 have proved their ability to reproduce such quantities. The ETFSI model has also been extended to the calculation of the fission barriers [10].In addition, since the calculation of reaction and β-decay rates requires the knowledge of various ground-state properties, mass models which do not provide all the quantities of relevance are consequently of reduced applicability for astrophysics. Coherent nucleosynthesis calculations demand the different nuclear quantities to be taken from one unique model.

In summary, the most popular mass models used nowadays for astrophysics applications are FRDM, ETFSI2 and HFBCS1. Despite the rough similarity in the quality of the data fits of all three mass formulae, the overall quality of the models are not identical. In particular, FRDM suffers from major shortcomings, such as the incoherent link between the macroscopic part and the microscopic correction, the instability of the mass prediction to different parameter sets, or the instability of the shell correction). ETFSI and HFBCS approaches are still subject to unsatisfactory treatment of the pairing force. At this stage, these three models should be regarded as providing a lower limit to the remaining uncertainties in the mass extrapolations. Unknown physical effects affecting exotic nuclei might not be described by neither of these models, and higher uncertainties can consequently be foreseen. For example, the existence of the shell quenching at the neutron drip line remains an open question that will be resolved by future developments of microscopic mass models constrained by new measurements. At the moment, striking differences emerge on extrapolating far from the experimental data, especially between HFBCS1 and FRDM predictions (HFBCS1 and ETFSI2 masses are relatively similar) [11].

B Nuclear level densities

In a similar way as for the determination of the nuclear ground state properties, until recently, only classical analytical models of NLD were used for practical applications. Although reliable microscopic models (in the statistical and combinatorial approaches) have been developed in the recent years, the back-shifted Fermi gas model (BSFG) approximation–or some variant of it– remains the most popular approach to estimate the spin-dependent NLD, particularly in view of its ability to provide a simple analytical formula. However, it is often forgotten that the BSFG model essentially introduces phenomenological improvements to the original analytical formulation of Bethe, and consequently none of the important shell, pairing and deformation effects are properly accounted for in such a description. Drastic approximations are usually made in deriving analytical formulae and often their shortcomings in matching experimental data are overcome by empirical parameter adjustments. It is well accepted that the shell correction to the NLD cannot be introduced by neither an energy shift, nor a simple energy-dependent level density parameter, and that the complex BCS pairing effect cannot be reduced to an odd-even energy back-shift (e.g [12]). A more sophisticated formulation of NLD than the one used in the BSFG approach is required if one pretends to describe the excitation spectrum of a nucleus analytically, especially because of the very high sensitivity of NLD to the different empirical parameters. For these reasons, large uncertainties are expected in the BSFG prediction of NLD, especially when extrapolating to very low (a few MeV) or high energies ($U \gtrsim 15\text{MeV}$) and/or to nuclei far from the valley of β-stability.

Several approximations used to obtain the NLD expressions in an analytical form can be avoided by quantitatively taking into account the discrete structure of the single-particle spectra associated with realistic average potentials (e.g. [13,14]). This approach has the advantage of treating in a natural way shell, pairing and deformation effects on all the thermodynamic quantities. The computation of the

NLD by this technique corresponds to the exact result that the analytical approximation tries to reproduce, and remains by far the most reliable method for estimating NLD (despite some inherent problems related to the choice of the single-particle configuration and pairing strength). A NLD formula based on the ETFSI ground state properties (single-particle level scheme and pairing strength) has already been proposed by [12]. Though it represents the first global microscopic formula which could decently reproduce the experimental neutron resonance spacings, some large deviations, for example in the Sn region, are found.

A new NLD formula within the microscopic statistical approach and based on the above-described HFBCS ground-state description has been constructed recently [15] to cure the ETFSI discrepancies. The HFBCS model provides in a consistent way the single-particle level scheme, pairing strength, as well as the deformation parameter and energy. The difficulty to describe the NLD of deformed nuclei has been resolved by introducing a phenomenological deformation damping function which takes two specific effects into account. First, an energy-dependent factor describes the transition from deformed to spherical shapes at excitation energies exceeding the deformation energy $E_{def} = E_{sph} - E_{eq}$ (where E_{eq} is the energy at the equilibrium deformation and E_{sph} the energy in the spherical configuration). No shape barriers is assumed in this simple picture. Second, the slightly deformed nuclei are described by including in the damping function a smooth deformation-dependent transition. The spherical approximation to the NLD is estimated with the use of a spherical single-particle level scheme, while the deformed NLD is derived from the deformed scheme at the equilibrium deformation. To avoid the unphysical divergence at low temperatures in the $T \to 0$ limit, the traditional formula is corrected by the asymptotic limit given by [16].

The constant-G strength is obtained by imposing that the pairing energy calculated with the MSk7 δ-pairing force and a cut-off energy of $\epsilon_\Lambda = \hbar\omega_0$ be the same as in the constant G-approximation with the constant cut-off energy $\epsilon_\Lambda = 20$ MeV. This increase of the cut-off energy from $\hbar\omega_0$ to 20 MeV leads to a global decrease of the pairing energy in the NLD application compared with the value derived in the HFBCS mass predictions. This inconsistent treatment of the pairing strength is found necessary to ensure an accurate fit to the experimental data on s-neutron resonance spacings, as shown in Fig. 2. The quality of the NLD formula is traditionally described by the rms deviation factor defined as

$$f_{rms} = \exp\left[\frac{1}{N_e}\sum_{i=1}^{N_e}\ln^2\frac{D_{th}^i}{D_{exp}^i}\right]^{1/2}, \qquad (1)$$

where $D_{th}(D_{exp})$ is the theoretical (experimental) resonance spacing and N_e is the number of nuclei in the compilation. For the microscopic HFBCS formula, $f_{rms} = 2.17$ on the 281 experimental data [17] which is comparable with the value of $f_{rms} = 1.97$ obtained with the phenomenological BSFG formula [18] on the same data set. The microscopic NLD formula also gives reliable extrapolation at low energies where experimental data on the cumulative number of levels is available

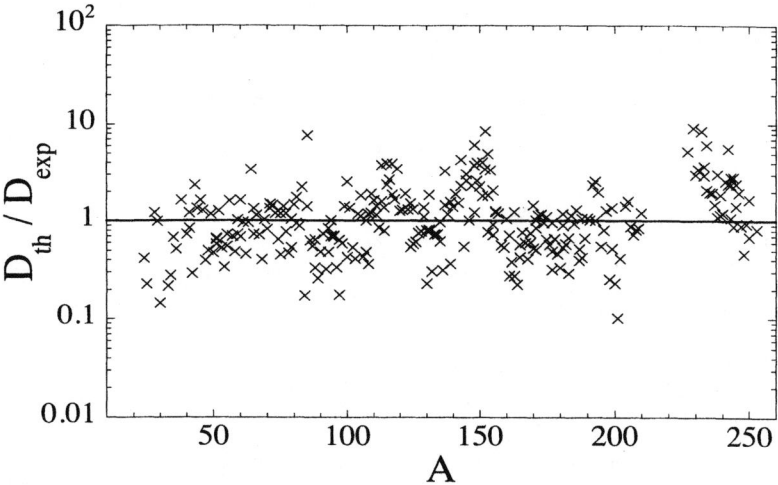

FIGURE 2. Ratio of theoretical HFBCS (D_{th}) to experimental (D_{exp}) s-neutron resonance spacings.

[15].

The NLD formula described in the present paper has been applied to the calculation of the spin-dependent NLD for more than 8000 nuclei ranging from Z=8 to Z=110 and tabulated in an energy and spin grid (U = 0.25 to 100 MeV and the lowest 15 spins). The corresponding table can be found at the website *http://www-astro.ulb.ac.be*.

C Optical potential

Most of the astrophysical applications make use nowadays of the nucleon-nucleus optical potential [19] derived from a Reid's hard core nucleon–nucleon interaction by applying the Brückner–Hartree–Fock approximation. An updated version was recently proposed by [20]. This semi-microscopic potential gives satisfactory results, though some improvements might be required in the low-energy description of the potential and the treatment of deformed nuclei.

Regarding the α-nucleus optical potential, the situation is less optimistic. The very-low energies of relevance in astrophysical environments (far below the Coulomb barrier) make the extrapolation of global potentials quite hazardous as shown by the new results in the ^{144}Sm$(\alpha,\gamma)^{148}$Gd experiment [21]. For these reasons, new global parametrizations of α-optical potential were recently proposed in order to take into account the strong energy dependence and nuclear structure effects affecting the imaginary part of the potential at low energies ($E \lesssim 20$MeV) [22,23]. However, experimental data at low energies [scattering data, α-capture or (n,α) cross sections] are scarce making the predictive power of the new parametrizations still uncertain. Cross section predicted with different potentials can differ by one order of magnitude. Much theoretical and experimental work remains to be done in this area.

D γ-ray strength function

The total photon transmission coefficient from a compound nucleus excited state is one of the key ingredients for statistical cross section evaluation. It strongly depends on the low-energy tail of the giant dipole resonance (GDR). In addition to the generalized Lorentzian model [24], new expressions of the γ-ray strength function were proposed recently [27,28]. These models include an improved description of the $E1$-strength function at energies below the neutron separation energy as derived from the theory of finite Fermi systems [25], and in particular, its non-zero $\varepsilon_\gamma \to 0$ limit. Dipole transitions to bound states have recently been investigated by means of the nuclear resonance fluorescence [26]. These experiments suggest the systematic existence of a so-called pygmy $E1$-resonance at energies below the neutron separation energy. Pygmy resonances have been observed in fp-shell nuclei as well as in heavy spherical nuclei near closed shells (Zr, Mo, Ba, Ce, Sn and Pb). The pygmy resonance is sometimes associated with oscillation of a small portion of nuclear matter relative to the rest of the nucleus. In particular, the existence of a neutron skin and its out-of-phase motion with respect to the neutron-proton core could generate soft vibrational [29] modes and enhance the neutron capture rates of exotic neutron-rich nuclei [27]. The total $E1$ strength in the pygmy resonance is small (around a few percent of the total GDR strength), but if located well below the neutron separation energy, it can significantly increase the radiative neutron capture cross-section. The quasi-particle phonon model has been successful in explaining the fine structure and fragmentation of the $E1$ strength and the presence of the pygmy resonance [26]. Nevertheless, much work remains to be done to estimate its systematic impact, in particular on the neutron capture rates by exotic neutron-rich nuclei.

III β^--DECAY RATES

The knowledge of β^--decay properties is crucial for understanding the r-process nucleosynthesis. For this application, the approximation of the Gamow-Teller (GT) transition is often accurate enough. Different nuclear models have been proposed for practical applications. In the gross theory [30], statistical arguments are used to estimate the smooth energy dependence of the GT and first forbidden strength function. Recent improvements in the so-called semi-gross theory [31] accounts for the inclusion of shell effects in the parent nucleus and the dependence of the one-particle strength function with the spin and parity of the decaying nucleon. The limitations of the method are essentially related to the neglect of coherent effects due to the effective NN-interaction.

A second simple, though microscopic, approach to nuclear β-decay properties is based on the proton-neutron QPRA. A complete model hamiltonian includes single-particle and pairing components, as well as schematic separable NN-interaction in the particle-hole (ph) and particle-particle (pp) channels. Two versions of the model have been applied to a global calculation of the $\beta^{\pm}$-decay half-lives throughout the nuclear chart. The first one [32] uses the deformed Nilsson+BCS quasi-particle formalism with a constant pairing and mass-dependent GT residual interaction in the ph and pp channels. The second version of the pnQRPA model [33] uses the finite-range droplet model (FRDM) with a folded Yukawa single-particle potential, a constant BCS pairing and a separable GT interaction in the ph channel. The simple pnQRPA has revealed its high efficiency in microscopic predictions of the β-decay properties. The main advantage is a physically sound description of the low-lying structures of the β-strength function. The limitations of the method are mainly due to the approximations made in such a scheme, i.e the use of the first order QRPA, an empirical one-body single-particle potential and a simple separable spin-isospin effective NN-interaction. In particular, the latter does not provide a universal treatment of the spin-isospin excitations of different multipolarities.

A fully self-consistent Hartree-Fock-Bogolyubov plus QRPA approach applicable to large-scale calculations of the ground state and β-decay properties has not been achieved yet (mainly for computational limitations). A practical step in this direction is found in the approximate microscopic approach ETFSI plus continuum QRPA (cQRPA) [34]. Based on the ETFSI ground state description, the strength function of the charge-exchange excitations and the resulting β-decay rate is calculated within the spherical cQRPA with the exact account for the single-particle continuum in the ph channel. The spin-isospin interaction in the ph channel is described by a renormalized one-pion exchange term plus a contact δ-term with a mass-independent Landau-Migdal constant g'_0=1.94 obtained by a fit to the experimental position of the GT resonance in ^{208}Pb. It should be mentioned that the effective spin-isospin strength used in the cQRPA calculations corresponds to a renormalized Landau–Migdal constant g'_0 with respect to the one constrained by the ETFSI Skyrme parameters. In the pp channel a form of the effective NN-interaction similar to that of the pairing is used with the strength constant derived from the experimental β^+-decay data. In this approach, the mass-independent finite-range effective NN-interaction ensures a universal description of the spin-isospin excitations of arbitrary multipolarity.

Despite this important effort, the β-decay half-lives of neutron-rich nuclei predicted by the existing models still differ significantly (Fig. 3). This has a non-negligible impact on r-process applications [34]. Fully self-consistent models for spherical and deformed nuclei need to be developed. The impossibility to describe at the moment the ground-state and the spin-isospin excitation with the same value of the Landau-Migdal constant (as extracted from experimental data) has to be investigated in deeper details. In addition, the influence of forbidden transitions and high-order QRPA effects on the β-decay rates still remain to be studied systematically.

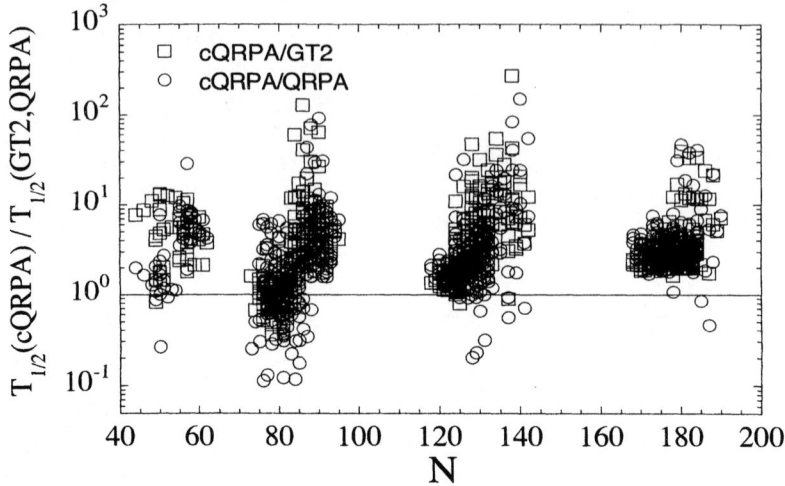

FIGURE 3. Comparison between the ETFSI+cQRPA [34], gross theory GT2 [30] and FRDM+QRPA [33] predictions of the β–decay half-lives of neutron rich nuclei near the closed shells.

IV CONCLUSIONS

Many astrophysics applications involve a large number of unstable nuclei and therefore require the use of global approaches. The extrapolation to exotic nuclei or energy ranges far away from experimentally known regions constrains the use of nuclear models to the most reliable ones, even if empirical approaches sometime present a better ability to reproduce experimental data. A subtle compromise between the reliability, accuracy and applicability of the different theories available has to be found according to the specific application considered. A continued effort to improve our predictions of the reaction and β-decay rates is obviously required. Priority should be given to a better description of the ground-state properties, nuclear level density and the α-nucleus optical potential, as well as a better understanding of given nuclear effects affecting exotic neutron-rich nuclei, such as the soft dipole modes. Fully self-consistent models for the description of β-decay properties of spherical and deformed nuclei still need to be developed. This continued effort to improve the microscopic nuclear predictions is concomitant with new measurements of masses and β-decay half-lives far away from stability, but also reaction cross sections on stable targets.

REFERENCES

1. Arnould, M., Takahashi, K., *Rep. Prog. Phys.* **99**, 1 (1998).
2. Goriely, S., in: Proc. of the 5th International Conf. on Nuclei in the Cosmos (Editions Frontières, eds. N. Prantzos, S. Harissopoulos), 1998, pp. 314.

3. Goriely, S., *Astron. Astrophys.* **325**, 414 (1997).
4. Möller, P., Nix, J. R., Myers, W. D., and Swiatecki, W. J., *At. Data Nucl. Data Tables* **59**, 185 (1995).
5. Aboussir, Y., Pearson, J. M., Dutta, A. K., and Tondeur, F., *At. Data Nucl. Data Tables* **61**, 127 (1995).
6. Goriely, S., in Proc. of the 10th Int. Symposium on Capture Gamma-Ray Spectroscopy and Related Topics (AIP, ed. S. Wender), 2000, pp. 287.
7. Tondeur, F., Goriely, S., Pearson, J. M., and Onsi, M., *Phys. Rev.* **C62**, 024308.
8. Goriely, S., Tondeur, F., and Pearson, J. M., submitted to *At. Data Nucl. Data Tables* (2000).
9. Audi, G., and Wapstra, A. H., *Nucl. Phys.* **A595**, 409 (1995).
10. Mamdouh, A., Pearson, J.M., et al., *Nucl. Phys.* **A644**, 389 (1998).
11. Goriely, S., in Proc. of the 2d Euroconference on Atomic Physics at Accelerators (2000) to be published in *Hyp. Int.*
12. Goriely, S., *Nuc. Phys.* **A605**, 28 (1996).
13. Decowski, P., Grochulski, W., et al., *Nuc. Phys.* **A110**, 129 (1968).
14. Arnould, M., Tondeur, F., Proc. Conf. on nuclei far from stability, Helsingør, CERN 81-09, vol1, p.229 (1981).
15. Demetriou, P., Goriely, S., in: Proc. of the 6th International Conf. on Nuclei in the Cosmos (2000) to be published in *Nucl. Phys. A*.
16. Grossjean, M.K., Feldmeier, H., *Nuc. Phys.* **A444**, 113 (1985).
17. Reference Input Parameter Library, IAEA-Tecdoc-1034 (1998) (also available at *http://iaeand.iaea.or.at/ripl*).
18. Rauscher, T., Thielemann, F.-K., and Kratz, K.-L., *Phys. Rev. C* **56**, 1613 (1997).
19. Jeukenne, J.P., Lejeune, A., and Mahaux, C., *Phys. Rev.* **C16**, 80 (1977).
20. Bauge, E., Delaroche, J.P., and Girod, M., *Phys. Rev.* **C58**, 1118 (1998).
21. Somorjai, E., Fülöp, Zs., et al., *Astron. Astrophys.* **333**, 1112 (1998).
22. Grama, C., Goriely, S., in: Proc. of the 5th International Conf. on Nuclei in the Cosmos (Editions Frontières, eds. Prantzos, N., Harissopulos, S.), pp. 463 (1998).
23. Rauscher, T., in: *Nuclear Astrophysics*, GSI, Darmstadt: Buballa, M. et al. (eds), pp. 288 (1998).
24. Kopecky, J., and Chrien, R.E., *Nucl. Phys.* **A468**, 285 (1987).
25. Kadmenskii, S.G., Markushev, V.P., and Furmann, V.I., *Sov. J. Nucl. Phys.* **37**, 165 (1983).
26. Govaert, K., Bauwens, F. et al., *Phys. Rev.* **C57**, 2229 (1998).
27. Goriely, S., *Phys. Lett.* **B436**, 10 (1998).
28. Mughabghab, S.F., and Dunford, C.L., *Phys. Lett.* **B487**, 155 (2000).
29. Van Isacker, P., Nagarajan, M.A., and Warne,r D.D., *Phys. Rev.* **C45**, R13 (1992).
30. Tachibana, T., Yamada, M., and Yoshida, N., *Prog. Theor. Phys.* **84**, 641 (1990).
31. Nakata, H., Tachibana, T., and Yamada, M., *Nucl. Phys.* **A625**, 521 (1997).
32. Hirsch, M., Staudt, A., and Klapdor-Kleingrothaus, H.-V., *At. Data Nucl. Data Tables* **51**, 244 (1992).
33. Möller, P., Nix, J.R., and Kratz, K.-L., *At. Data Nucl. Data Tables* **66**, 131 (1997).
34. Borzov, I., and Goriely, S., *Phys. Rev.* **C62**, 035501 (2000).

Cosmology, Cosmic Chemical Evolution, and the Age Problem

Grant J. Mathews

Center for Astrophysics
Department of Physics, University of Notre Dame
Notre Dame, IN 46635   gmathews@nd.edu

Abstract. This is an overview of the current efforts to clarify the age of the universe from nucleocosmochronology, globular clusters, local stars and white dwarf cooling, and the Hubble expansion rate. A particular emphasis is placed upon areas where nuclear physics continues to play a crucial role. There have been numerous recent breakthroughs in cosmology, nucleosynthesis and astrometry. These have significantly refined our understanding of cosmic ages. Although in the past these various methods seemed to imply different ages (the age problem), there is now mounting evidence that all of these methods point toward a concordant age of 13 ± 2 Gyr.

I   THE AGE PROBLEM IN THE PRE-*HST/HIPPARCHOS* ERA

The determination of the age of the universe has been a challenging problem for nuclear astrophysics since Rutherford [1] first analyzed radioactive ages based upon ^{232}Th and ^{235}U in terrestrial rocks. Over the years essentially four independent means have been developed to determine the cosmic age [2]. One of the first methods to be employed historically was the analysis of the nuclear radio chronometers. The available naturally occurring radioactive isotopes include ^{232}Th ($\tau = 14.1$ Gyr), ^{238}U ($\tau = 4.5$ Gyr), ^{187}Re ($\tau = 44$ Gyr), and ^{87}Rb ($\tau = 48$ Gyr). There are two basic problems associated with using this method of dating the time since the onset of nucleosynthesis. For one, finite temperature effects can alter the radioactive decay rates and nuclear reaction rates of the isotopes involved. The second problem concerns the fact that there is an unvoidable uncertainty in this method due to the unknown cosmic chemical evolution from the time of the first star formation to the formation of the solar system.

Historically, the next method to be applied [3] in the determination of the cosmic age was the Hubble expansion time. The idea is simple enough. For example, in an inflation-motivated, flat Einstein-DeSitter universe dominated by nonrelativistic matter, the cosmic expansion age is just $t_H = (2/3)H_0^{-1}$, where H_0 is the present value of the Hubble constant. This method, however, has been troubled by an uncertainty in the cosmic geometry as well as an ambiguity in the value for the Hubble constant H_0 itself.

The third method, which also falls within the venue of nuclear astrophysics is the age of the oldest globular clusters. Since these objects are comprised of a large sample of stars at a single distance, the H-R diagram for such systems can be contrasted with stellar evolutionary tracks. Identifiable phases of stellar evolution (most notably the main-sequence turn-off) can then be used to identify the age of the cluster. The problems with this method, however, include both the miriad of uncertainties associated with the standard one-dimensional non-rotating stellar evolutionary models, and the difficulty in precisely determining the distances to the oldest globular clusters.

The fourth method involves the identification of the oldest stars in the solar neighborhood. One can study, for example, the color magnitude diagrams for old open clusters, or the local F & G stars with well measured distances. Perhaps the most promising such method utilizes the luminosity function for near-by cool white dwarfs [4]. White dwarfs are after all believed to be well understood electron-degenerate objects which gradually cool, primarily by crystallization. For a more-or-less uniform star formation rate, their numbers should increase roughly exponentially with diminishing luminosity. This implies the existence of a large background of cool white dwarfs (CWDs). Indeed, recent scrutiny of the Hubble deep field survey has revealed that such a population exists.

More importantly, the observed sudden drop in the white-dwarf luminosity function for CWDs with $\log(L/L_\odot) \lesssim 4.5$ inplies a clear signature for the galactic age. In the pre-*Hipparchos/HST* era, the age problem generally referred to that fact that in addition to the large uncertainties of the various age determinants, cosmic ages inferred from the different methods did not necessarily agree to within their stated uncertainties.

For example, the nuclear chronometric ratios, ^{232}Th/^{238}U and ^{238}U/^{238}U, implied a nucleosynthesis age of 12 ± 2 Gyr, under the assumption of an exponentially decreasing star formation rate after an initial prompt enrichment. On the other hand, the expansion age involved an ambiguity in the Hubble constant itself. The Type Ia supernovae [5] tended to imply a value of $H_0 \approx 50 \pm 10$ km sec^{-1} Mpc^{-1}, for which $t_H = (2/3)H_0^{-1} \approx 13 \pm 2$ Gyr, while methods [6] which utilized various properties of external galaxies (e.g. Tully-Fischer, fundamental plane, velocity dispersions, etc.) as standard candles tended to favor $H_0 = 80 \pm 10$ km sec^{-1} Mpc^{-1} implying a very low expansion age, $t_H \approx 8 \pm 1$ Gyr. This is to be contrasted with the ages from the oldest globular clusters which tended to have inferred ages of $t_{GC} \approx 15 \pm 2$ Gyr [7] and the age from the white-dwarf luminosity function which was only $t_{WD} \approx 9.5 \pm 1$ Gyr.

These age disparities in fact led to a number of papers in which it was proposed (e.g. [8]) that an initial burst of star formation which produced the oldest GC's may have proceeded star formation and nucleosynthesis in the disk by as much as several Gyr.

II COSMOCHRONOLOGY IN THE POST-*HST/HIPPARCHOS* ERA

With this background in mind, it is interesting to look at the important developments of the past few years. First, let us consider the exciting advancements in cosmology. To clarify the dependence of the Hubble time on various cosmological parameters, let us begin with the usual Friedmann equation for the change of the Hubble expansion rate with time,

$$H^2 \equiv \left(\frac{\dot{R}}{R}\right)^2 = \frac{8}{3}\pi G\left(\rho_\gamma + \rho_m\right) + \frac{k}{R^2} + \frac{\Lambda}{3} , \qquad (1)$$

where ρ_γ is the mass energy deinsity in relativistic particles, (e.g. photons, light neutrinos, quarks, etc.) and ρ_m is the energy density in nonrelativistic matter (e.g. baryons, cold dark matter, etc.). One can next introduce a change of variables from mass energy to present contributions to the closure density, $\Omega_\gamma = \frac{8}{3}\frac{\pi G \rho_\gamma}{H_0^2}$, $\Omega_k = \frac{8}{3}\frac{\pi G k}{H_0^2}$, etc. and integrate the Friedman equation to find the cosmic age,

$$t_H = \frac{1}{H_0}\int \frac{dy}{y[\Omega_\gamma y^4 + \Omega_m y^3 + \Omega_k y^2 + \Omega_\Lambda]^{1/2}} , \qquad (2)$$

where the integration variable is related to the cosmic red shift $y = 1/R = (1+z)$, and by definition the present closure constants must sum to unity; $\Omega_\gamma + \Omega_m + \Omega_k + \Omega_\Lambda = 1$.

In recent times a much better quantification of the cosmological parameters in equation (2) has been obtained compared to what was available only a few years age. Let us start with the Hubble constant H_0.

A Determinations of H_0

One of the most important developments in cosmology in the last few years is that the long standing ambiguity in different methods to determine the Hubble constant has been largely removed. The determination of this parameter has particularly benefited from the availability of the Hubble Space Telescope (HST) which has for the first time identified Cepheid and RR-Lyrae variable stars in galaxies as distant as the Virgo cluster. This an exceedingly valuable calibration for the various distance indicators. Table 1 summarizes results from the Hubble Key project [9]. There is now universal agreement of an expansion rate of $H_0 = 71 \pm 3$ (Stat) $\pm$ 7 (sys).

Perhaps an even more exciting recent development in cosmology is the independent determinations of the cosmic equation of state from surveys of Type Ia supernovae (SNeIa) at high red shift. Assuming that the uncertainties in the SNeIa nuclear luminosities have been correctly calibrated for the low-metallicity progenitors

TABLE 1. Summary of HST Key Project Results for H_0

Method	H_0
Local Cepheid galaxies	$73 \pm 7 \pm 9$
Surface Brightness Fluctuation	$69 \pm 4 \pm 6$
Tully-Fisher clusters	$71 \pm 4 \pm 7$
Fundamental Plane	$78 \pm 7 \pm 8$
SNe Ia	$68 \pm 2 \pm 5$
SNe II	$73 \pm 7 \pm 7$
Combined	$71 \pm 3 \pm 7$

at high red shift, there is strong evidence that the universe is accelerating in recent history. This implies a cosmological constant (or perhaps cosmic "quintessence" or "k-essence" (cf. [10]) dominated universe at the present time. These results are complemented by advances in the measurement of the power spectrum of the cosmic microwave background (CMB) anisotropies on small ($< 1^\circ$) angular scales. Together, these two data support a flat $\Omega_k = 0$, $\Omega_m + \Omega_\Lambda = 1$ universe. The combined data sets imply [11] $\Omega_\Lambda = 0.7 \pm 0.1 (1\sigma)$ and $\Omega_m = 0.3 \pm 0.1 (1\sigma)$. This value for Ω_m is also independently confirmed by the observed X-ray luminosity of rich galactic clusters [12].

We can now insert these data into equation 2 to obtain a new expansion age of $t_H = 13.5 \pm 2.3$. With this result we see that the previous ambiguity has been removed. The cosmological age is no longer too low.

B Globular Clusters

The determined ages for the oldest globular clusters have been significantly impacted by the *Hipparchos* results. Although *Hipparchos* has not actually observed globular clusters, it has allowed accurate determinations of parallaxes (to 0.002") and proper motions (to 0.002"/yr) for $\sim 100,000$ stars in the Solar neighborhood. Using nearby main sequence stars to calibrate the luminosity of main sequence globular cluster stars implies a globular cluster distance which is significantly greater than previously estimated. This implies brighter luminosities and therefore significantly shorter main sequence lifetimes for stars in globular clusters. This has been analyzed by a number of authors [13,14]. When corrected for the presence of known binaries in the main sequence luminosity calibration there is universal agreement on a globular cluster age of 12.5±1.3 Gyr. This eliminates the previous problem of a globular cluster age which was too large.

The various uncertainties in the GC age has been analyzed by Monte-Carlo methods [14]. Of particular interest to nuclear physicists is that there remain significant uncertainties from the nuclear reaction rates. The most important reactions and their uncertainties can be summarized as follows:

$$p + p \rightarrow\ ^2H + e^+ + 2p \quad (0.2\%)$$

$$^3He +\ ^3He \rightarrow\ ^4He + 2p + \gamma \quad (6\%)$$

$$^3He +\ ^4He \rightarrow\ ^7Be + \gamma \quad (3,2\%)$$

$$^{12}C + p \rightarrow\ ^{14}N + \gamma \quad (15\%)$$

$$^{14}N + p \rightarrow\ ^{15}O + \gamma \quad (12\%)$$

$$^{16}N + p \rightarrow\ ^{17}F + \gamma \quad (16\%)$$

In principle, refinement of these reaction rates could reduce the GC age uncertainty by ±0.3 Gyr. However, the uncertainty is still dominated by the unknown abundances of α-elements (mostly oxygen) which contribute an uncertainty of ±2 Gyr. Presumably this uncertainty will diminish with modern light gathering power together with high resolution spectroscopy and improved model atmospheres.

C Old Local Stars

Considerable progress has been made [15] in dating the ages of the oldest local stars based upon the *Hipparchos* distances. The color magnitude diagrams for old open clusters, the total *Hipparchos* catalog, or the local F & G stars measured by *Hipparchos* have significantly reduced the implied ages of the oldest local stars.

There have also been several important recent developments in the determination of white-dwarf ages. In particular, it is now appreciated that white dwarfs are more complex than the simple uniform composition models [16] which have been often employed (e.g. [4]). For example, Hernanz et al. [17] have studied the stratification of abundances in white dwarfs. In particular, the presence of a Ne layer in Ne/O/Mg WDs implies significantly slower cooling and therefore longer WD ages.

At the same time there are other indications that WD's are more complicated objects than previously assumed. In [18] it is clearly shown that the mass-radius relation for several of the best determined white-dwarf systems exhibit significantly compacted radii compared to the standard models [16]. Indeed, the authors speculated that the three best measured radii are only consistent with white dwarf models if the cores are predominantly composed of iron. This poses a real challenge to nuclear astrophysics. We know that the only stars which form an iron core are massive ($m \geq 10$ M$_\odot$). Such stars unavoidable march toward Type II supernovae once they initiate the buildup of the iron core during quasiequilibrium silicon burning. The process of forming the iron core only takes a few days and inevitably leads to core collapse. There is, therefore, no opportunity to eject the outer layers and leave a Fe white dwarf before the Chandrasehkar mass is exceeded and the core collapses to leave behind a proto-neutron star. In other work [19] it has been shown that magnetic fields can also significantly affect white-dwarf structure, but not in a way to explain the observed compaction. Clearly, WD's are more complicated objects than is commonly assumed.

The observational data have also improved. The most recent survey [20] consists of 58 CWDs in what is the first sample to be both volume and luminosity complete. Fitting these data with the cooling curves of [17] and [21] leads to WD ages of 10±3 Gyr which is similar to the age deduced from open clusters and local F & G stars [15]. Allowing for $\sim 1 \pm 1$ Gyr for nucleosynthesis to begin in the disk [23] implies an age for the universe from white dwarfs and old stars of $t_{Stars} = 11 \pm 4$.

D Nucleocosmochronology

Finally, let us consider what has been learned from the nuclear chronometers. As we have heard at this conference there have been some significant improvements in the nuclear physics parameters of the Re/Os chronometer in particular. Regarding the uncertainties from chemical evolution there are some significant constraints on cosmic chemical evolution from HST. For example, there is now a clear picture of the cosmic star formation rate as a function of redshift [22]. Also, galactic chemo-dynamical models (cf. [23]) are beginning to give insight into the n-body hydrodynamical evolution of stars and gas along with the associated nucleosynthesis in the disk and halo. One of the most significant advances has been the completion of detailed surveys of heavy-element abundances in metal-poor halo stars [24]. These now constrain [24,25] the early chemical enrichment of the Galaxy.

Perhaps the most exciting recent development, however, is from high resolution HST measurements [26,27] of radioactive Th (undoubtedly ^{232}Th, $t_{1/2} = 14$ Gyr) on the surface of low-metallicity halo stars. These imply an age for the oldest halo stars of $t_{Nuc} = 15 \pm 4$ Gyr. The exciting aspect of this determination is that the surviving ^{232}Th was probably produced in a single supernova and there was almost no time for radioactive decay prior to being incorporated into the star. Thus, the uncertainty in the galactic chemical evolution has been removed.

The problem, however, is that unlike the solar system material where one can take ratios of thorium to uranium, on this star one must determine the initial abundance of thorium relative to other stable r-process nuclei produced. Since thorium and uranium are both produced by long chains of α-decay progenitors, much of the uncertainty due to the unknown initial production ratios can be removed by averaging of yields along the separate α-decay chains. Here, however, one must know the initial production of thorium relative to the stable r-process elements precisely. This poses a problem. Many r-process models can adequately produce the observed solar-system abundance peaks and yet produce little or no thorium (cf. [28]). The major uncertainty then comes from the uncertainties in models for the r-process. This is a problem since we are still trying to convince ourselves of where in Nature the r-process occurs [29]. A key reason for uncertainties in the r-process models remains the lack of reliable nuclear input data for the neutron-rich nuclei along the r-process path. It remains a challenge to nuclear physics to devise ways of measuring and/or reliably predicting those nuclear properties.

III CONCLUSIONS

As we have seen, great strides have been made toward determining the cosmic age and resolving the age problem. Summarizing the four ages discussed in this text, we find: $t_H = 13.5 \pm 2.3$; $t_{GC} = 12.5 \pm 1.3$; $t_{Stars} = 11 \pm 4$; and $t_{Nuc} = 15 \pm 5$. Taken together, these imply a weighted average of 13 ± 2 Gyr consistent with other recent determinations (cf. [30,31]). The methods described herein all now agree to within their putative uncertainties. Hence, it is now perhaps time to officially declare [32] that the age problem, as originally formulated, no longer exists.

Nevertheless, we still do have significant uncertainties in the various means to determine the age of the universe. These uncertainties are due at least in part to uncertainties in some important input nuclear data. Only by continued refinement of cosmological parameters, stellar models, and input nuclear data will an accurate cosmic age be ultimately determined.

ACKNOWLEDGMENTS

Work supported by DoE Nuclear Theory grant number DE-FG02-95ER40934 at the University of Notre Dame.

REFERENCES

1. Rutherford, E. Nature, 123, 313 (1929).
2. Cowan, J.J., Thielemann, F.-K. & Truran, J. W., ARA&A 447 (1991).
3. Hubble, E.P., 1936, ApJ, 84, 158, (1936); ApJ, 84, 270, (1936).
4. Winget, D. et al., ApJ, 315, L77 (1987).
5. Branch, D. ARA&A, 36, 17, (1998).
6. de Vaucouleurs et al., ApJ, 248, 408 (1981).
7. Lee, Y.-W., Demarque, P. & Zinn, ApJ, 350, 155 (1990).
8. Mathews, G. J. & Schramm, D. N. ApJ, 404, 468 (1993).
9. Freeman, W. L. Phys. Rep., 333, 13 (2000).
10. Brax, P. et al. PRD, 62, 103505 (2000).
11. Garnavich, P. et al., ApJ, 509, 74 (1998).
12. Bahcall, N., Phys.Scripta T85 32, (2000).
13. Reid, I.N. A.J., 114, 161 (1997); Gratton, R.G. et al., ApJ, 491, 749 (1997); D'Antons, A. et al., ApJ, 519, 534 (1997); Salaris, M., ApJ, 479, 665 (1997).
14. Krauss, L., Phys. Rep., 333, 33 (2000); B. Chaboyer et al., ApJ, 494, 96 (1998).
15. Carraro, G., in "The chemical evolution of the Milky Way : Stars vs Clusters, Vulcano (Italy), 20-24 September 1999, astro-ph/9911382 (1999).
16. Hamada, T. & Salpeter, E. E., ApJ, 134, 683 (1961).
17. Hernanz, M. et al., Astrophys. J., **434**, 652 (1994).

18. Provencal, J.L. et al. ApJ, 494, 759 (1998).
19. Suh, I. & Mathews, G.J. ApJ, 530, 949 (2000).
20. Knox, R.A., et al., Mon. Not. R. Astron. Soc. 306, 736 (1999).
21. Wood, M.A., ApJ, 386, 539 (1992).
22. Madau, P. et al. ApJ, 498, 106 (1998).
23. Burkert, A. et al. ApJ, 391, 651 (1992).
24. Ryan, S.G. et al., ApJ, 471, 254 (1996).
25. Chiappini, et al., ApJ, 515, 226 (1999).
26. Cowan, J.J., et al., Astrophys. J., **521**, 194 (1999).
27. Sneden, C., et al., Astrophys. J. Lett., **533**, L139 (2000).
28. Woosley, S. et al., Astrophys. J., **433**, 229 (1994).
29. Mathews, G.J. & Wilson, J.R. in "International Symposium on Origin of Matter and Evolution of Galaxies 97", Atami, Japan, 5-7 November 1997. Editors, S. Kubono, T. Kajino, K. I. Nomoto, and I. Tanihata. Singapore ; New Jersey : World Scientific, 1998., p.335; Meyer, B.S., ARA&A, 32, 153 (1994).
30. Chaboyer, B. Phys. Rep., 307 23 (1998).
31. Lineweaver, C.H., Science, 284, 1503, (1999).
32. Krauss, L., ApJ, 501, 461 (1998).

The Re/Os clock: open questions regarding the neutron cross sections

A. Mengoni[1,2] and F. Käppeler[3]

[1] *ENEA, Applied Physics Division, Via Don Fiammelli 2, I-40129 Bologna, Italy*
[2] *INFN, Bologna Section, Via Irnerio 46, I-40126 Bologna, Italy*
[3] *Institut für Kerphysik, FzK, Postfach 3640, D-76021 Karlsruhe, Germany*

Abstract. The role of neutron cross section data on the analysis and interpretation of the Re/Os clock is described in some detail. Proposals for reducing some of the remaining uncertainties on the nuclear physics aspects of the clock are discussed.

INTRODUCTION

The Re/Os clock was proposed by D. Clayton in 1964 [1]. The various uncertainties playing a role in the analysis of this clock were investigated by Yokoi and collaborators [2]. The effect of the temperature dependent β-decay rate of ^{187}Re and of the e^--capture rate of ^{187}Os during astration were pointed out. These aspects received recently a neat clarification [3] as a consequence of the experimental determination of the β-decay half-life of fully-stripped ^{187}Re ions [4]. Hence, the main source of uncertainty, as far as the nuclear physics aspects of the clock are concerned, are presently related to the neutron capture cross sections which determine the s-process yields. Here we address the problem of the influence of the neutron cross section data on the analysis of the clock, in particular on the ^{186,187}Os capture rates, the principal players of this game.

NEUTRON CROSS SECTIONS DATA NEEDS

From the very beginning [1] it clearly appeared that the neutron capture cross sections or, more precisely, the neutron capture cross section ratio of ^{186}Os to ^{187}Os, was a crucial quantity, necessary to make this clock useful. A simple direct estimate of the impact of the capture cross section ratio $R_\sigma = \sigma(186)/\sigma(187)$ on the age determination can be obtained using a simple exponential model for the ^{187}Re r-process enrichment rate. The ratio between the comsmoradiogenic abundance of

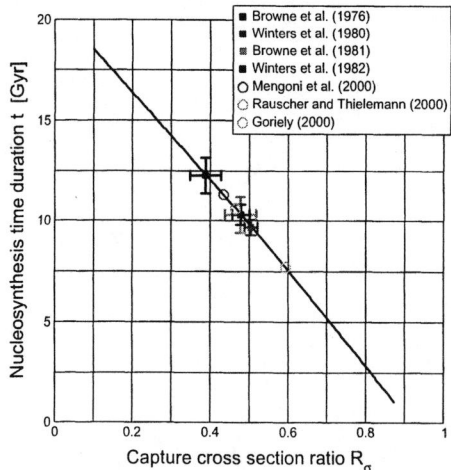

FIGURE 1. Time-duration of the nucleosynthesis related to the neutron capture cross section ratio R_σ. The experimental data are from Winters *et al.* [5] and Browne *et al.* [6]. The calculated values are from [7].

^{187}Os to the abundance of ^{187}Re is then given by

$$\frac{[^{187}\text{Os}]_c}{[^{187}\text{Re}]} = \frac{\Lambda - \lambda_{187}}{\Lambda} \frac{1 - \exp(-\Lambda t)}{\exp(-\lambda_{187} t) - \exp(-\Lambda t)} - 1.$$

Here, $\Lambda = (0.43t)^{-1} \text{Gr}^{-1}$ is the enrichment rate and $\lambda_{187} = 0.0164 \text{Gr}^{-1}$ the ^{187}Re β-decay rate (laboratory rate, for the present purpose). This ratio is directly related to R_σ by

$$\frac{[^{187}\text{Os}]_c}{[^{187}\text{Re}]} = \frac{[^{187}\text{Os}]/[\text{Os}] - F_\sigma R_\sigma [^{186}\text{Os}]/[\text{Os}]}{[^{187}\text{Re}]/[\text{Re}]} \frac{[\text{Os}]}{[\text{Re}]},$$

derived imposing the s-process condition $\sigma(A)[A] \approx const$. R_σ must evaluated with the capture cross sections averaged over a Maxwell-Boltzmann distribution of neutron energies at a given temperature and corrected for the effect of the thermal population of target states (incorporated into the F_σ factor in the relation above).

As can be seen in Figure 1, the time duration of the nucleosynthesis is within the range $7.5 \text{ Gyr} \leq t \leq 12.5 \text{ Gyr}$ with the presently available capture cross sections data. The slope of the curve shown in Figure 1 is $|\delta t/\delta R_\sigma| = 22.5$ Gyr. Even excluding the extreme values for the cross sections, an uncertainty of ≈ 0.1 has to be assigned to R_σ. This reflects into an age uncertainty of 2.3 Gyr. Obviously, the uncertainty in R_σ has to be reduced down to 1-2% if one wants to reduce the age uncertainty of the clock to the level of the uncertainties deriving from other sources, such as the galactic chemical evolution modeling.

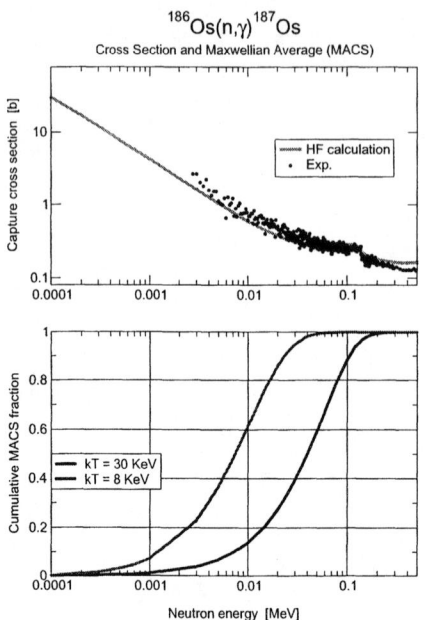

FIGURE 2. Neutron capture cross sections for ^{186}Os. The experimental data are from measurements made at ORELA. A comparison is shown with theoretical calculations based on the Hauser-Feshbach statistical model theory. The cumulative contribution to the Maxwellian averaged cross section (MACS) *vs* neutron energy is shown in the lower panel. Note the large contribution of the energy range below 10 keV to the $kT = 8$ keV situation, which dominates the stellar *s*-process environment. In this region data are very uncertain or even missing.

As an example, we show in Figure 2 the capture cross section of ^{186}Os as a function of the neutron energy. In the same figure, the cumulative fraction of the maxwellian averaged capture cross section (MACS) shows that, the lower energy side (say, $E_n \leq 10$ KeV) may influence a large fraction of the total rate (up to 60% for low temperatures).

STELLAR RATES

In general, the rate in a stellar environment may be influenced by the presence of low-lying excited states in the target nucleus. This is precisely the case in the ^{187}Os$(n,\gamma)^{188}$Os. Here, ^{187}Os has an excited state at 9.8 keV and several other states below 100 keV (see Figure 3). These states are populated in stellar environment for the temperatures of interest for the s-process ($kT \simeq 8\,\mathrm{keV}$ to $kT \simeq 30\,\mathrm{keV}$) and the capture cross section from these states are to be evaluated in order to provide the stellar rate.

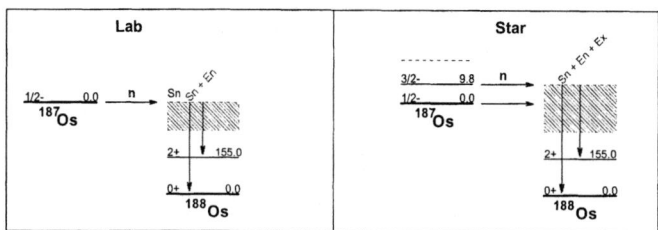

FIGURE 3. A scheme of the $n + {}^{187}\mathrm{Os}$ interaction processes in laboratory (left) and in a stellar environment (right) with some reaction channels shown.

An important role in these kind calculations is played by the presence of a super-elastic scattering channel, whenever the target nucleus is in an excited state. While in previous calculations of stellar enhancement factors the super-elastic scattering has been considered only by rough approximations, this effect can be treated in a self-consistent way once the inelastic scattering cross sections of $^{187}\mathrm{Os}$ will be available.

Because $F_\sigma \approx 0.85$ at $kT = 30$ keV, the effect of the stellar enhancement factor on the age determination gives an uncertainty of ≈ 2 Gyr. There is a plan to reduced this uncertainty with a neutron cross section measurement campaign, recently initiated at the CERN n_TOF facility [8].

REFERENCES

1. D. D. Clayton, *Ap.J.* **139** (1964) 637.
2. K. Yokoi, K. Takahashi, and M. Arnould, *Ap.j.* **117** (1983) 65-82.
3. K. Takahashi, in *Proc. of the Tours Symposium on Nuclear Physics III*, AIP Conference Proceedings 425, 1998, pag. 109.
4. F. Bosch, et al., *Phys. Rev. Lett.* **77** (1996) 5190.
5. R. R. Winters, R. L. Macklin, and J. Halperin, *Phys. Rev.* C **21** (1980) 563; R. R. Winters, and R. L. Macklin, *Phys. Rev.* C **25** (1982) 208.
6. J. Browne, J. P. Lamaze, and I. G. Schroder, *Phys. Rev.* C **14** (1976) 1287; J. Browne, and B. Berman, *Phys. Rev.* C **23** (1981) 1434.
7. T. Rauscher and F.-K. Thielemann, *Atomic Data and Nuclear Data Tables* **75** (2000), pag. 1; A. Mengoni et al., in preparation (2000); S. Goriely, in *Nuclear Data for Science and Technology*, (Italian Physical Society: Reffo et al.), 1997, pp. 811. Data downloaded from: http://www-astro.ulb.ac.be.
8. R. L. Aguiar et al., CERN/INTC 2000-040 (2000), unpublisheds.

Theoretical Nuclear Astrophysics in Tours: a Summary

Marcel Arnould* and Kohji Takahashi[†]

*Institut d'Astronomie et d'Astrophysique Université Libre de Bruxelles
Campus Plaine - CP 226, B-1050 Brussels, Belgium
†Max-Planck-Institut für Astronomie
Königstuhl 17, D-69117 Heidelberg, Germany

BACKGROUND

A century after the discovery of radioactivity and four decades since the famed article by Burbidge, Burbidge, Fowler and Hoyle, 'nuclear astrophysics' has by now claimed in modern science a position of its own. While this reflects the impressive progress having been made in the field, many problems remain and will certainly emerge. Along with many remarkable innovations in related nuclear-physics experiments being made, a deeper theoretical understanding of nuclear properties is still vigorously sought after, particularly for the purpose of extrapolations of known physics into the variety of peculiar astrophysical environments. [For starters: M. Arnould & K. Takahashi, Rept. Progr. Phys. **62** (1999) 395-464.]

That recent years have indeed witnessed a number of conferences that are wholly devoted to the subject is a delightful contrast to the situation some time ago when nuclear astrophysics was often no more than an entertainment immersed in either nuclear or astrophysics gatherings. An apparent side effect of the 'independence' is that it seems to have alienated some, particularly those who are working on 'proper' theoretical nuclear physics. [The importance of close cooperation with astronomers should be stressed as well.]

Considering the danger that some may grow weary of too many and similar conferences, and given the wide range of interests of the expected participants to this symposium (nuclear-, atomic-, astro-physics), we thought of a somewhat unusual format for its nuclear astrophysics sessions: Give ample time for a few formal talks and as much for informal 'round table discussions'. We hoped that the sessions would not be dull for 'outsiders' in the first instance. We were also convinced, perhaps naively, that free discussions (and even some skirmish) *in public* involving 'experts' and 'pedestrians' would highlight controversial issues more clearly, if not solving them. We were much aware of the difficulties of such a practice, like i) a nicely balanced selection of topics and speakers for the formal talks, and ii) coaxing the 'egos' of the other participants not assigned to formal presentations ("Why should I travel just for discussions ?"). Nevertheless, we went ahead with the idea in order to weigh its merits and demerits, considering this medium-sized meeting as an good test site.

The opening day of the symposium was devoted to theoretical nuclear astrophysics, during which we heard seven formal talks. A few 'short contributions' were added in order to stimulate the round-table discussions. Nuclear astrophysics sessions on the second day dealt with experimental efforts for the acquisition of thermonuclear reactions in quest (see the separate summary by H. Utsunomiya).

FORMAL PRESENTATIONS

Stellar Structure and Evolution: Nobert Langer (Utrecht) opened the symposium with a review on the tantalizing question of mixing mechanisms in stars. Adam Burrows (Arizona) gave a grand overview of the current issues of supernova physics. On the nuclear physics front, Masatomi Yasuhira's (Kyoto) comprehensive review dealt with phase transitions (e.g. kaon condensation) at very high densities in relation to neutron stars.

Handling Nuclear Reaction Networks: Konstantinos Kifonidis (Garching) described in detail the techniques and associated problems of implementing a nuclear reaction network into multi-dimensional hydrodynamical simulations. On a different vein, a network code with adaptive sizes at successive burning stages of nucleosynthesis in massive stars was introduced by Alexander Heger (Santa Cruz).

Nuclear Physics and Cosmochronology: Stephane Goriely (Brussels) made a thorough evaluation of existing theoretical models providing nuclear data for heavy-element nucleosynthesis computations. The current status regarding the different techniques of determining the age of the universe (hence the 'age problem') was tersely reviewed by Grant Mathews (Notre Dame).

QUESTIONS, ANSWERS, DISCUSSIONS

In what follows, we summarize some *issues* that were debated (more or less hotly, that is) during the symposium. Before going onto specific topics, however, let us summarize what we consider as challenges that nuclear astrophysics is facing in general, theoretical or experimental, which we had little time to discuss in detail:

- *Improve one-dimensional models for stars and stellar nuclear combustions of various kinds*

  During non-explosive evolutionary phases, much obviously remains to be done in order to device a reliable approximation for transport mechanisms, like 'convection' (see the contribution by N. Langer). This is essential indeed in order to predict the distribution of nuclides inside stars and at their surface.

In this respect, rotation in 'pseudo' one-dimensional models may be a key feature which has started to be investigated by various groups, but which clearly calls for more work. The question of the possible intrusion of chaos in stellar evolution modelling was also raised by the round table moderators. What is indeed the sensitivity of the models to initial conditions or to selected parameters ? Are the constructed models robust or fragile to changes in these quantities ? An attempt to answer these questions might be a fruitful route of investigation for the next millenium !

During certain explosive regimes (like those encountered in classical novae, Type Ia supernovae, X-ray bursts), one major problem has to do with the proper description of 'hydrodynamic combustions' (detonations or deflagrations), and on the coupling of the associated fluid motions and nuclear reaction networks (in this respect, see the contribution by K. Kifonidis);

- *Go beyond one-dimensional models*

The only way to describe properly fluid motions inside stars and their possible coupling with nuclear reactions is to construct multi-dimensional (multi-D) models. Several attempts have already been made, especially in the 2-D approximation. They concern namely the late evolution of deep layers of pre-supernova massive stars or several scenarios involving hydrodynamic burnings. A few heroic 3-D simulations have been conducted as well. The construction of reliable multi-D models certainly represents another major challenge for the future;

- *Model one-D or multi-D explosions of massive stars*

The explosion of a massive star leading to a Type II or a Type Ib/Ic supernova is not due to a hydrodynamic burning, but instead to the generation of an outward-moving shock front from the forming central residue (neutron star or even black hole) to the stellar surface. A suite of very difficult physical and numerical problems remain to be solved in order to provide self-consistent 1-D simulations of such events, and to predict from 'first principles' the ability for the shock wave to impart enough energy to the shocked material to be ejected with the observed characteristics of a supernova.

Further difficulties have to be faced. In particular, it has been made clear by the earlier than expected emergence of some γ-ray lines from the SN1987A supernova event that material synthesized in the deep interior of an exploding massive star could be transported quite efficiently into the outer layers. Various 2-D simulations seem to be able to account for such an efficient mixing, but much remains to be done. In addition, there is now mounting observational evidence that supernova ejecta are in many instances more or less strongly aspherical. At this point, only some very preliminary toy models have attempted to take this very important effect into account;

- *Get reliable experimental data and improve model predictions for nuclear reaction cross sections at energies of astrophysical interest*

  A key nuclear astrophysics challenge for non-explosive conditions relates to the necessity to know charged-particle induced reaction cross sections far below the Coulomb barrier. This dictates the need to explore the 'world of almost no event' (cross sections in the nb to pb range). This is an exceptionally difficult experimental task, even if one has mostly to deal with stable targets (see also the companion summary by H. Utsunomiya). As a consequence, many, and sometimes large, uncertainties remain in the rates of reactions involved in the non-explosive burnings of H, He [a famed example being the rate of $^{12}C(\alpha,\gamma)^{16}O$], C, Ne-O and Si. [See Angulo et al. (NACRE Collaboration), Nucl. Phys. **A656** (1999) 3-183 for a compilation of rates of H- or He-burning reactions.]

  The nature of the problem is different when dealing with explosive conditions. The energies of astrophysical interest are higher than in the non-explosive case due to higher temperatures. However, the corresponding larger cross sections have to be known on unstable targets in many instances. One thus replaces the problem of the exploration of the world of almost no event by the adventurous exploration of the world of exotism. The price to pay in order to enter this world is very high indeed ! (see a separate summary by H. Utsunomiya for an account of the experimental efforts in the field);

- *Identify really astrophysically important reactions*

  The selection for experimental and theoretical scrutiny of reactions which could have an important impact on astrophysical models and on their predictions regarding key observables certainly appears to be an obvious necessity to everybody. In particular, experimental nuclear physicists are always asking for a shopping list of reactions which are most recommended for study.

  In reality, the question is more easily asked than answered ! Apart from some clear cases involving in particular stable targets, the rest of the list is much less obvious. This concerns in particular reactions on unstable nuclei which might develop in explosive situations. The problem here is that many important facets of the modelling of stellar explosions where such reactions could take place are not under control yet. On top of that, the answer of a (non-exploding or exploding) stellar system to some change in the input physics is in general highly non-linear, so that naive expectations are in general of no value (see below). Among these naive expectations, one certainly has to put the conclusions often drawn from the solution of nuclear reaction networks in some a priori selected, and of course highly simplified, physical conditions. The frequent confusion made between solving a reaction network and constructing an astrophysics model may lead to *highly unsecure* conclusions concerning in particular the identification of reactions the rates of which are warmly recommended for, often very difficult, measurements;

- *Get reliable experimental data and/or improve model predictions for masses, β-decay rates, and for a variety of other nuclear physics-related questions of astrophysical relevance*

    Nuclear masses are clearly an essential ingredient entering the calculations of the whole nuclear physics input for astrophysical modellings. Much progress has been made recently in the measurements of masses of nuclei far from the valley of stability. In addition, models which come closer and closer to a microscopic description of the nucleus have been developed. The same is basically true for β-decay rate experiments or estimates. It remains to be demonstrated that more reliable (microscopic) theories are accurate enough for astrophysical purposes, and in particular for the modelling of processes involving highly exotic neutron-deficient or neutron-rich nuclei (see below).

    Other questions of substantial interest concern the establishment of an equation of state for supernovae or neutron stars, or many facets of the physics of the neutrinos.

– about Principles versus Pragmatism:

"How is it possible that a more 'reliable' theory is less 'accurate'?", demanded a few in the audience during S. Goriely's talk on models in use for predicting unknown nuclear data. This question may deserve some consideration as it illuminates a key factor apparently differentiating the viewpoint of 'standard' nuclear physicists from that of nuclear astrophysicists, and sometimes leading even to their mutual mistrust.

First, one has to recall that most nuclear-physics problems cannot be solved from first principles. Therefore, the best nuclear theorists hope to do is to set up models as close to those principles as possible. Highly reliable though it may be with regard to qualitative dependencies on physical variables, such a model often cannot reproduce the existing data with a level of accuracy that is required for astrophysical applications. In addition, 'most reliable' models could be too complex and/or too time-consuming to be handily applied to a large number of nuclides in a wide range of nuclear/astrophysical variable space. A resort is thus found in models that are more flexible (i.e., more adjustable parameters) with regard to the fit. In return, the best 'accuracy' (read: 'fit') does not necessarily mean the highest 'reliability' (read: 'predictive power') in extrapolation. The preferred compromise would be to adopt a model with phenomenological terms that are inspired by the more fundamental approaches.

– about the R-process Nucleosynthesis:

In his (not so) short contribution, K. Sumiyoshi (Numazu) asserted that a full r-process developed in the Type-II supernova 'hot bubble' (the term being, points out A. Burrows, "a misnomer and the term 'post-explosion wind' is more appropriate for the high entropy flow that emerges after the supernova".)

The apparent success was seemingly due to a very short expansion time scale, coupled with the assumed massive (2 $M_\odot$) and compact (10 km radius) nascent neutron star, which may not exist (at least as abundant as to make up the solar r-process material !). How is it if more realistic values, say 1.4 - 1.5 $M_\odot$, are adopted ? K. S. replies that 1.5 $M_\odot$ "might be still difficult". 1.6 - 1.8 $M_\odot$ may lead to entropy high enough to "produce up to the 2nd peak (of the solar r-process abundance curve) in the case of short expansion time", which "can be realized even for less massive neutron stars" (than assumed). If the expansion is indeed very fast, the nuclear network flow after the $\alpha-$ (capture) process dies out would go "through near the neutron-drip line by neutron captures already at $Z < 10$", K. S. stresses. A. Mengoni (Bologna) called attention to an experimental capability of measuring the neutron-capture cross sections of very neutron-rich light (C in particular) isotopes.

A.B. concludes: "We don't yet know whether the r-process occurs in the wind that emerges from the protoneutron star following the supernova explosion". Will the situation change in the 3-D case ? He snaps: "We don't know whether 3-D effects in the wind play a role in the yields of heavy nuclei". – Thank you very much: Still a lot to do !

– about *Equation of State for Supernovae:*

In his another (not so) short contribution, K. Sumiyoshi talked about a new table of nuclear equation of state (EOS) for supernova simulations. The underlying method relies on the so-called relativistic mean-field theory, with parameter values globally fitted to static properties (mass, radius) of known nuclei.

"How can you constrain from nuclear data the EOS of neutron-rich matter ?" asks H. Flocard (Orsay). K.S. replies: "From the neutron skin thickness. There is a clear correlation between the saturation density of asymmetric matter and the derivatives of pressure for neutron matter at normal nuclear matter density. The saturation density mainly determines the amount of neutrons outside proton distribution". How reliable is the EOS at very high densities ? The answer: "We take into account the behavior of the scalar and vector potentials in the relativistic Bruckner-Hartree-Fock theory by adding the non-linear terms in the Lagrangian" so as to reproduce the results of the latter (more 'reliable') theory.

We consider the above as a case exemplifying a good compromise between the 'reliability' and 'accuracy' (including the 'feasibility' for practical applications) of models for the nuclear many-body problem. In this respect, one can also consider the case of neutrino reaction rates at very high densities. A. Burrows states: "For completeness and accuracy, one must include the full energy and angle redistribution formalism in the neutrino transport. Indeed, this is time consuming and better, more efficient, techniques are required to properly explore their influence on collapse and explosion phenomenology".

– about *Nucleosynthesis in multi-dimensional Hydrodynamics:*

K. Kifonidis made a convincing case that, "for multi-dimensional nucleosynthesis calculations Eulerian methods are to be preferred over current Lagrangian approaches".

"How about numerical diffusion ?", asks N. Langer, which is inherent to the Eulerian scheme. "In spite of that", K. K. replies, "grid-based Eulerian schemes provide an accuracy which is hard to achieve with the popular (gridless) Lagrangian SPH scheme. The latter suffers from its Monte Carlo nature and its rather large intrinsic level of noise which artificially and severely enhance the convective mixing". What is the accuracy of the nucleosynthesis yields in the current 2-D Eulerian calculations ? K. K. notes that it is "strongly isotope-dependent" by taking the example of ^{44}Ti, which is formed by α-rich freeze-out in the layers of complete Si burning: "Numerical diffusion of α-particles into the layers of incomplete Si burning and the subsequent ^{40}Ca$(\alpha, \gamma)^{44}$Ti reaction can yield an erroneous ^{44}Ti contribution of the same order. In this specific case and using a low-order advective scheme for nuclear species with a spatial resolution of about 50 km, the yield is usually off by a factor of about 3", whereas by the "order of a few 10% for other isotopes". A numerical accuracy of a few % in all yields is "currently difficult to attain, even with the most accurate schemes and very fine zoning".

If nucleosynthesis calculations could be performed somehow 'off line', it would appeal to many nuclear astrophysicists who want to see the effects of improved nuclear reaction rates or of additional channels. Is it possible ? No, says K. K.: "Since we have to stick to Eulerian schemes in 2-D/3-D, we need to do the nucleosynthesis on-line. Advecting marker particles in the course of an Eulerian calculation yields the same problem in the case of strong shear and vortex flow as the Lagrangian approach: no direct control of the accuracy of the yields is possible because the distances between particles may become too large to resolve the temperature distribution accurately".

Concerning the 2-D simulation of SN 1987A by K. K., A. Burrows asks how the velocity distribution of ^{56}Ni evolves, and why the observed velocities cannot be reproduced. K. K. explains: "The maximum ^{56}Ni velocities about 1 s after core bounce are extremely high ($\sim$ 16000 km/s), but they decrease with time to about 3500 km/s around 30 min later when the clumps are about to enter a dense shell that has formed at the He/H interface of the star. In this shell, they drop to approximately 1500 km/s at about 10^4 s after the core bounce, when most of the ^{56}Ni mass is expanding with velocities around 700 km/s. The clumps are propagating with Mach numbers around unity through the shell medium, and thus dissipate a significant fraction of their energy in strong acoustic or even non-linear waves". Responding to A. B.'s another question about "why no strong mixing occurred at the He/H interface", K. K. says: "Because no seed perturbations were added to those layers", stressing that the aim of the study was "to demonstrate the impact of early neutrino driven convection on the Rayleigh-Taylor instabilities in the envelope".

Then, there is the canonical question if 2-D results could be misleading. On this, A. B. comments: "Indeed, 2-D is different from 3-D and will misrepresent the smaller-scale structures and the high-frequency power in the 'turbulent' flow. However, we think that 2-D hydrodynamics captures the large-scale structures, their velocities and scales. Nevertheless, 3-D calculations must be done to determine the true differences between 2-D and 3-D, and we must await the outcome of such simulations before passing judgement".

A. Heger wonders if 3-D effects would lead to higher ^{56}Ni velocities in SN 1987A than those of the 2-D model above, in better agreement with observation. K. K. observes: "There is definitely a difference between 2-D and 3-D clump propagation. Because of their different surface-to-volume ratios, genuine 3-D 'mushrooms' experience a smaller drag while propagating through their environment when compared with those found in 2-D, the latter being in fact tori". But, to see that effect precisely has to await "high-resolution 3-D calculations, which are still lacking".

Futhermore, K. K. adds: "Our 2-D supernova models indicate that convection between shock and gain radius (hence multi-D hydro) affects mainly 'complete' Si burning". Therefore, he expects "(significant ?) differences between 2-D and 3-D yields of the burning products", whereas O-, Ne- and C-burning "seems to be much less sensitive to dimensionality issues".

– about Nucleo-cosmochronology:

After the review by G. Mathews on different techniques of age determination and the current status of the results ("Are they conflicting each other ?"), a natural question asked was about how reliable nucleo-cosmochronology really was. Expectedly, S. Goriely takes exception to the uncertainties claimed in recent determinations of the age of the Galaxy with the use of the r-process element abundances in very metal poor stars. We concur with him in that a *reliable* 'theoretical' estimate of the needed Th yield by the r-process (in comparison with those of lighter elements) is next to impossible to obtain in lack of 'the' model of the r-process. [A bothering proof for this is that, historically, there appears a strong correlation between the ages derived from nucleo-cosmochronology and those from other techniques.]

One may find the ^{187}Re - ^{187}Os chronology as an alternative because of its (near) independence of r-process models. A. Mengoni presented a short contribution, summarizing the *nuclear* problems associated with this chronometry and depicting a planned experiment toward removing the major remaining uncertainty, i.e. neutron-capture cross sections of ^{187}Os in s-process environments. The same topic was dealt with in the following informal talk by S. G., who discussed about the theoretical uncertainties in evaluating the s-process contribution to ^{187}Os. One may hope that all the necessary nuclear physics data will soon become available with desired accuracy. This chronometry, relying on a relative abundance in the solar system, suffers from large uncertainties in modelling of the Galactic chemical evolution, however.

Assessment

It is highly fashionable these days to talk about scientific interdisciplinarity. Its practice is, however, always very challenging, even if nicely exciting. This certainly applies to nuclear astrophysics, and to meetings bringing together nuclear physicists and astrophysicists. Was Tours2000 successful in this respect ? Clearly, the nuclear astrophysics sessions reached high standards thanks to the quality of the speakers and the intrinsic interest of the topics covered. It is also our opinion that round-table discussions brought a special touch, and that it would be desirable to maintain, and even develop, them in the future.

Of course, the organization of these discussions was not perfect. In particular, our wishful thinking of involving 'standard' nuclear physicists into active discussions was hardly attained. It might well be that they were lost in the 'expert' jargons. To our relief, on the other hand, the speaker exclusively invited from that community (M. Y.) succeeded in explaining to us nuclear physics in a plain manner [e.g., the issue of kaon condensation even without writing down the (chiral) Lagrangians involved, which seem to be a must for nuclear theorists, but are definitely not what (nuclear) astrophysicists would have enjoyed]. The shortness in time of the informal discussions was also a matter of frustration for some. We have to keep these points in mind at the next occasion (if it ever comes, that is).

Another special touch was brought, quite unexpectedly, into one of the nuclear astrophysics round tables. We indeed had the pleasure to receive a statement from T. Kajino (Mitaka) by fax [see the summary by H. Utsunomiya]. Maybe it is time to give a serious thought to such a way of communication among participants at the conference site and those at remote locations (or even at short distances from the meeting place; we note indeed some 'no-shows' of expected contributors from quite nearby institutes).

Finally, we admit that this summary has also turned out to be unusual. The somewhat lengthy and unbalanced 'reproduction' of the discussions reflects our wishes to convey to the readers the enthusiasm displayed by some, whom we thank very much, on the chosen topics during the formal and informal sessions, and even in the post-conference period.

It remains to be hoped that the next millenium Tours2003 vintage will even meet better the challenges raised by nuclear astrophysics !

Advances in cross section measurements at low energies

M. Junker (LUNA-Collaboration))

Laboratori Nazionali del Gran Sasso
S.S.17bis km 18+910, 67010 Assergi(AQ), Italy

Abstract. Precise knowledge of nuclear cross sections at stellar energies are of importance for cosmology as well as for particle astrophysics. Measuring these cross sections at the relevant energies is quite challenging not only for the extremely low cross sections involved but also because a number of parameters like stopping powers, electron screening and angle straggling are not well known at low energies. The LUNA–Collaboration has installed two accelerators in the underground laboratories of the Laboratori Nazionali del Gran Sasso (LNGS) in Italy. Because of the reduced background from cosmic ray interactions in the detectors this facility provides a unique opportunity to study nuclear reactions with astrophysical relevance at energies at or close to stellar energies. Also the questions of electron screening and stopping power can be addressed.

I INTRODUCTION

Accurate knowledge of thermonuclear reaction rates is important in understanding the generation of energy, the luminosity of neutrinos, and the synthesis of elements in stars. Due to the Coulomb barrier (height E_c) of the entrance channel, the reaction cross section $\sigma(E)$ drops nearly exponentially with decreasing energy E. Thus it becomes increasingly difficult to measure $\sigma(E)$ and to deduce the astrophysical $S(E)$ factor defined by the equation

$$S(E) = \sigma(E)E\exp(2\pi\eta), \tag{1}$$

where $2\pi\eta = 31.29 Z_1 Z_2 (m/E)^{1/2}$ is the Sommerfeld parameter. The quantities Z_1 and Z_2 are the nuclear charges of the interacting particles in the entrance channel, m is the reduced mass in units of amu and E is the center-of-mass energy in units of keV. Although experimental techniques have improved significantly over the years to extend $\sigma(E)$ measurements to lower energies, for a long time it has not been possible to measure $S(E)$ within the thermal energy region in stars called the Gamow energy window. Its energy depends on the stellar temperature and lies far below the height of the Coulomb barrier. Instead, the observed $\sigma(E)$ data at higher energies had to be extrapolated to thermal energies. As always in physics, such an extrapolation into the unknown can lead to considerable uncertainties [1].

The low-energy studies of thermonuclear reactions in a laboratory at the Earth's surface are hampered predominantly by the effects of cosmic rays in the detectors. Passive shielding provides a reduction of γ's and neutrons from the environment, but it produces at the same time an increase of γ's and neutrons due to the cosmic-ray interactions in the shield itself. A 4π active shielding can only partially reduce the problem of cosmic-ray background. An excellent solution is to install an accelerator facility in a laboratory deep underground. As a pilot project, a 50 kV accelerator facility has been installed in the Laboratori Nazionali del Gran Sasso (LNGS), where the flux of cosmic-ray muons is reduced by a factor 10^6. The Laboratory for Underground Nuclear Astrophysics (LUNA) pilot project was designed primarily for a renewed study of the ^{3}He(^{3}He,2p)^{4}He reaction (Q=12.86 MeV) in the energy range of the solar Gamow peak which has its maximaum at 21.9 keV for a central star temperature of $T = 15.5 \cdot 10^6$ K. Based on the success of this project a 400 kV machine has been mounted underground at LNGS. The facility will provide the possibility to measure several key reactions of the pp-chain, the CNO-cycle and the MgAl-cycle at extremely low energies [2].

II TECHNICAL ASPECTS

The LUNA underground accelerator laboratory consists of two accelerators. Basic design features for both machines are high intensity proton and helium beams with a precise knowledge of the beam energy. While the first characteristic results from the fact that cross sections are extremely low at sub-coulomb energies the latter reflects the exponential energy dependents of the cross section. High efficiency and good background recognition characterize the detector setup used. Due to the long measuring times caused by the low count rates and the uncomfortable working conditions in the underground laboratory the system must be as reliable as possible and a remote control is favorable.

A The 50 kV accelerator at LNGS

The 50 kV accelerator facility consists of a duoplasmatron ion source, an extraction and acceleration system and a double-focusing 90° analyzing magnet with adjustable pole faces. The energy spread of the ion source is less than 20 eV, the plasma potential energy deviates by less than 10 eV from the voltage applied to the anode, and the emittance of the source is 2 cm rad eV$^{1/2}$. The ion source provides a stable beam current of about 1 mA over periods of up to 3 weeks. The high voltage of the accelerator is provided by a power supply with high stability. The high voltage is determined with a resistor chain contained in an airtight plexiglas tube and a digital multimeter. The resistor chain is built as a voltage divider, with fifty 20 MV resistors and one 100 kV resistor. This HV-measuring device was calibrated at the PTB in Braunschweig (Germany) to a precision of $5 \cdot 10^{-5}$. Within this precision no instability has ever been observed during normal operation. The LUNA facility is equipped with an interlock system, which allows the system to run without an operator on site.

B The 400 kV accelerator at LNGS

In 1998 INFN has approved the funding for installing a new 400 kV accelerator in the underground laboratories of LNGS. The accelerator has been constructed High Voltage Engineering Europe B.V. (The Netherlands) and is actually being commissioned. The terminal is located in a pressure vessel filled with 20 bar of N_2/CO_2 mixture. The high voltage is generated by an inline Cockroft Walton power supply. The RF ion source is mounted directly on the accelerator tube. After the accelerator tube, the only ion optical element is a 45° analyzing magnet, keeping the number of optical components to a minimum. Steering the beam properly on the target is done by this magnet and a magnetic Y-steerer. The system is as short as possible and only magnetic optical elements are used in order to maintain the important space charge compensation that is needed in order to prevent brightness degradation. The use of a shortening rod guaranties a dynamical range of the machine which reaches from 400 kV to accceleration voltages lower than 50 kV. The high voltage ripple measured with a pickup plate is as low as low as ± 4 V at 200 kV. The machine provides proton beams of 100 (500) μA at 30 (400) keV which pass through a 4.3 mm and a 10 mm aperture placed at 1200 and 2200 mm after the exit of the analyzing magnet. The stability of the beam energy will also be check making use of a narrow resonance at 338 keV in the reaction $^{26}Mg(p,\gamma)^{27}Al$.

C Target Technology

Aiming to measure low energy cross sections requires special attention to the target technology used. Important design aims are high target density, good stability and cleaness. High target stability is important not only in order to get a reasonable count rate but also in order to deduce $S(E)$ reliably. Clean targets help to reduce the influence of beam induced background. In particular all kinds of passive layers in front of the the target must be avoided, as they alter the enegy of the impinging beam and spread up its energy.

Generally speaking there are two target technologies available: Windowless gas target systems and solid state targets.

1 Windowless Gas Target Systems

In windowless gas target systems the target is formed by a gas cell. The beam enters this cell passing a small aperture which limits the flux of the target gas into the beam line and asures a homogenious pressure distribution in the target cell. An efficient pumping system upstream the beam line guarentees to reach a pressure of about $1 \cdot 10^{-5}$ mbar in the accelerator.

Because of the continuos gas flux through the target cell windowless gas targets offer a high stability of target stoechometry and density. The target density can easily be determind by measuring the pressure in the target cell while the stoechometry and the homogenity can be checked making use of nuclear resonances. Windowless gas targets allow for easy changing of the target density. However, gas target systems are limited to targets which are available as gaseous materials. This limitation may be overcome sometimes by studying reactions in inverse kinematics.

Another drawback of gas target systems is the extended target area limiting the possibility of determining angular distributions and inducing doppler broadening to the spectra of the ejectiles. Gas jet targets provide a solution to this problem but it is more difficult determine their target density. In addition gas targets and jets in particular are quite complex system which require accurate planning and maintenance in order to guarantee a good performance.

2 Solid State Target

Solid state targets are poduced by evapolarting the target material on a backing or by implanting the target into the backing. Solid state targets have a high target density. This is an advantage in experiments which aim to measure angular distributions. When using solid state targets much attention must be given to the deposition of dead layers on the target surface and to changes of target stoechometry due to sputtering, diffusion, etc. Also target inhomogeneities can be a problem. These problems can be reduced by using liquid nitrogen cooling traps, efficient water cooling and wobbling of the beam over the target surface. The effects of the beam impinging on the target can also be studied by techniques like recoil spectroscopy or nuclear reaction analysis.

D Detectors for γ-spectroscopy

The detectors used with the underground accelerators must provide high efficiency and low intrinsic activity. In addition they must provide the possibility of background recognition. This goal can be achieved by generating triggers based on coincidence conditions or on pulse shape analysis. A good resolution of the detector reduces dramatically the background in the detector.

One concept to design a detection system is to locate the target inside the borehole of a massive BGO–detector. In this configuration a very efficient detector material covers a solid angle of 4π. As the reactions take place inside the detector and thus all gammas generated in a reaction are detected in coincidence they are summed up intrinsically in the detector. The signature of the reaction is a peak in the spectrum located at the Q-value of the reaction plus the beam energy. The detector used by the LUNA-collaboration has a length of 20 cm a diameter of 21 cm and a borehole diameter of 6 cm. The absolute efficiency is about 80%. The detector consists of six optically seperated segments, each observed by two photomultipliers at either side. By generating the trigger on a fast coincidence of the two photomultipliers which observe the same segment, electronic noise can be reduced efficiently. The segmentation aims to collect information on angular distributions and branching ratios. It may also help to reduce the effect of the intrinsic background in the detector. Unfortunately BGO-detectors do not provide a good energy resolution and in addition they are often polluted with ^{207}Bi generated in atmospheric nuclear weapon tests.

Germanium detectors would provide the possibility of getting a much lower background and more detailed information on nuclear physics parameters. Though the efficiency of Germanium is low compared to BGO, due to the good resolution of the detectors the signal to noise ratio is better. The best solution would be to construct a modular germanium detector with allows covering a solid angle of almost 4π. This detector would give a clear signature of the reaction like in the case of the BGO-detector but with a much better signal to noise ratio while the single elements can give clues on the nuclear properties of the reaction studied.

III PHYSICS CASES

A The Reaction ^{3}He(^{3}He,2p)^{4}He

The reaction ^{3}He(^{3}He,2p)^{3}He is a member of the hydrogen burning proton-proton chain, which is predominantly responsible for the energy generation and neutrino luminosity of the sun. The ^{3}He(^{3}He,2p)^{3}He reaction represents in the exit channel a three-body breakup: if the breakup is direct, one should observe a continuous energy distribution of the ejectiles described by phase-space considerations; if the breakup follows a sequential process, the energies of the ejectiles are described by two-body kinematics. Before LUNA, the ^{3}He(^{3}He,2p)^{3}He cross section measurements stopped at the center of mass energy of 24.5 keV, just at the upper edge of the thermal energy region of the sun [4]. The LUNA-facility has allowed to study this important reaction over the full range of the solar Gamow peak ranging from 20 keV down to 16 keV. The cross section is as low as 8 pbarn at E_{cm}=25 keV and about 20 fbarn at E_{cm}=16.5 keV corresponding to a count rate of about 2 events/month, rather low even for the "silent" experiments underground. The energy dependence of the astrophysical $S(E)$ factor determined with this machine is consistent with the predictions based on an extrapolation from higher energies [6].

In addition it is possible to obtain conclusions on the effect of the tail of a hypothetical narrow resonance lying in the not measured low-energy region. Upper limits for the strength of such a resonance have been calculated. A detailed description of the calculations is given in [5]. At energies $E_R=$ 9 keV one concludes that the presence of a resonance does not alter the solar neutrino spectrum significantly [6,7]. The same conclusion applies in the energy region between 9 and 18 keV for resonance widths ≤0.5 keV. Resonances with width below 0.05 keV do not contribute significantly [5,6]

B Studies of Electron Screening and Energy Loss

In order to interpret cross sections measurements at low energy correctly and to construct realistic stellar models, good understanding of the electron screening effect is needed. For this reason the LUNA-collaboration has studied the reaction D(^{3}He,p)^{4}He at the 50 kV accelerator at LNGS. The experiment has been carried out reducing sources of systematic errors like energy loss and angle straggling as much as possible.

The 50 kV LUNA accelerator at LNGS delivered a ^{3}He beam with energies E_{ext} between 25 keV and 12 keV in the laboratory system. The beam impinged on a windowless gas target. Different target pressures p where chosen for each energy E_{ext}. Large area silicon detectors were mounted in the gas target chamber. As the beam lost part of its energy while passing the as target, the effective beam energy E in the center of mass system increases with decreasing target pressure at each energy E_{ext} and thus $E = E(p)$. Because of the exponential energy dependence of the cross section $\sigma(E(p))$ this pressure dependence is of particular importance for deriving the astrophysical S-factor $S(E)$ from the cross section $\sigma(E)$ determined experimentally. The lost energy has been determined usually applying the stopping power tables of Ziegler at al. However, the energy loss at low energies quoted in these tables rely on extrapolation and thus induce an additional uncertainty in the data analysis as pointed out by Langanke et al [7]. For this reason the experiment was carried out the way that the stopping power could be extracted from the data used to determine the electron screening. Details of the data reduction and there interpretation are given in [9]. The stopping powers obtain with the measurements using a ^{3}He beam on a D$_2$ gas target are consistent with the data of [8]. The screening potential U_e obtained is 132 eV, consistent with dynamic Hartree Fock calculations [10]. In order to get the more information on stopping power and screening, the system ^{3}He+d is being exploited at the moment by bombarding a ^{3}He target with a D$^+$ beam using the identical technique. As in the previous case the target is in the form of an atomic gas this experiment may provide important clues on energy loss and electron screening at low energies.

C Future Prospects

The first measurements to be carried out at the new accelerator will continue the research on the nuclear reaction relevant in the sun. At first the reaction ^{14}N(p,γ)^{15}O will be studied, which determines the velocity of the CNO–cycle. Though the CNO cycle gives no significant contribution to the energy generation in the sun, it is important to know the neutrino flux produced by the decays of ^{15}O and ^{13}N in view of upcoming solar neutrino experiments like BOREXINO.

At solar energies the cross section of ^{14}N(p,γ)^{15}O is dominated by a subthreshold resonance at -504 keV. The effect of this resonance has not been observed directly up to now but only by fitting high energy resonances. The measurements of the LUNA–collaboration will employ a windowless gas target and the BGO–summing detector in oder to obtain the overall cross section down to E_{cm}=80 keV. In a second step, measurements using high purity germanium detectors are planned in order to get more information on the reaction mechanism, which will allow to put the extrapolations on a more solid ground.

Another reaction to be studied is ^{3}He(^{4}He,γ)^{7}Be. It determines the first branchching in the pp-chain together with ^{3}He(^{3}He,2p)^{4}He. Because the low Q-value of 1.6 MeV this reaction must be studies with Germanium detectors shielded against the environmental activity. With this technique the cross section can be determined down to E_{cm}=60 keV, while the lowest data point measured up to now is at E_{cm}=187 keV

REFERENCES

1. C. Rolfs and W.S. Rodney, Cauldron in the Cosmos (Uiniversity of Chicago Press, 1988)
2. G. Fiorentini, R.W. Kavanagh and C. Rolfs, Z. Phys. A350 (1995) 289
3. U. Greife et al., Nucl. Instr. Meth. A350 (1994) 327
4. A. Krauss et at., Nucl. Phys. 467 (1987), 273
5. M. Junker et al, Phys. Rev. C 57 (1998), 2700
6. R. Bonetti et al, Phys. Rev. Lett. 82 (1999), 5205
7. K. Langanke et al, Phys. Lett. B 369 (1996), 211
8. P. Prati et al., Z. Phys A350 (1994), 171
9. H. Costantini et al., Phys. Lett. B 482 (2000), 43
10. T.D. Shoppa et al., Nuc. Phys. A 605 (1996), 387

Nuclear Astrophysics with a Recoil Mass Separator

Frank Strieder

Institut für Experimentalphysik III[1]
Ruhr-Universität Bochum, D-44780 Bochum, Germany

Abstract. Radiative capture reactions, like (α,γ)- and (p,γ)-reactions, are of great importance for the understanding of the different burning phases in stars. In most cases laboratory studies of some key reactions are very difficult due to the low cross section at the relevant Gamow energy where the stellar burning occurs. A new approach to measure these capture cross sections involves a two-sided differentially pumped gas target, a recoil mass separator, and a ΔE-E detector telescope (allowing for particle identification) as detection system. This combination allows a direct measurement of the produced recoils in inverse kinematics. The direct observation of the recoils requires an efficient recoil mass separator to filter out the incident beam particles from the recoils. The recoil separator must not only have a high filtering power but also a high transmission of the recoils (for the selected charge state) between the gas target chamber and the ΔE-E telescope. The feasibility of the separation of projectiles and recoils with a mass difference of 1 amu to 1 part in 10^{11} or more has been demonstrated in various experiments. A few of these experiments are discussed in this paper.

INTRODUCTION

The quest of nuclear astrophysics is to understand how chemical elements that make up our world are formed in the cosmos. Nucleosynthesis started in the Big Bang, and continues in the various stages of stars. Chemical elements are released into the Universe following explosive collapse of a star. While a great deal is now known due to the pioneering efforts of Hans Bethe, Carl F. von Weizsäcker, and William Fowler, nevertheless major questions are still present [1].

One of the important open questions is still the solar neutrino puzzle. The observed solar neutrino fluxes on the earth provide no unique picture of the microscopic processes in the sun [2]. Neutrino oscillations have been invoked to explain the discrepancy between observation and model predictions, but nuclear inputs to solar models play still an important role. In particular, the astrophysical S factor at the Gamow energy $E_0 = 18$ keV of the capture reaction ^{7}Be$(p,\gamma)^8$B influences

---

[1] The reported projects are supported by the Deutsche Forschungsgemeinschaft (Ro 429/35-1, Gr 1577/3-1) and the Istituto Nazionale di Fisica Nucleare.

sensitively the calculated flux of high-energy solar neutrinos and must therefore be known with adequate precision. An experiment at the Accelerator Mass Spectrometry facility of the University of Naples, Italy, aiming in the direct measurement of this capture reaction served as a pilot project for a design of a sophisticated recoil mass separator for other reactions involving stable nuclei.

This separator is being installed at the 4 MV Dynamitron tandem accelerator of the Ruhr-Universität Bochum, Germany, and designed mainly for an improved measurement of the key reaction $^{12}C(\alpha,\gamma)^{16}O$ in inverse kinematics. The capture reaction $^{12}C(\alpha,\gamma)^{16}O$ takes place in the helium burning of Red Giants and determines not only the nucleosynthesis of elements up to the iron region but also the subsequent evolution of massive stars, the dynamics of a supernova, and the kind of remnant after a supernova explosion. For these reasons, the cross section should be known with a precision of at least 10 %. In spite of tremendous experimental efforts in measuring the cross section with standard techniques over nearly 30 years, one is still far from this goal.

At the higher temperatures of the explosive events, it has been proposed [3] that radioactive species play a role through various sub-Coulomb barrier nuclear reactions with hydrogen and helium in the stars. Processes such as radiative alpha and proton capture can occur with radioactive species, opening up the hot CNO cycle, and leading eventually to the rp (rapid proton capture) process. These processes would involve proton rich species in the vicinity of the $Z = N$ line and

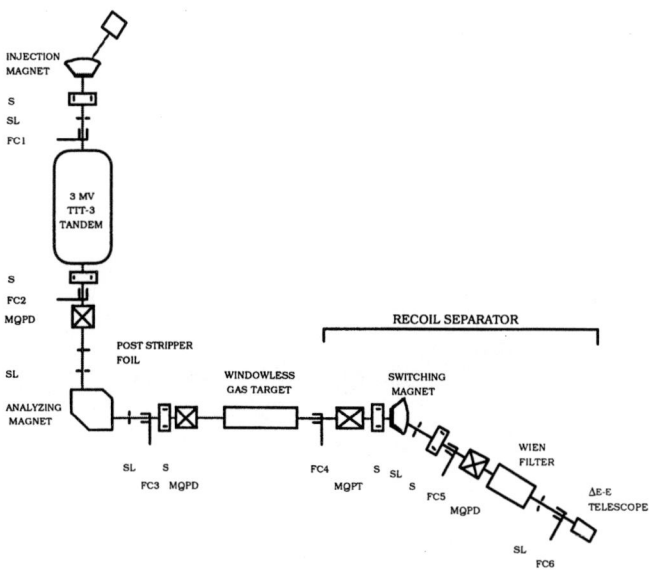

FIGURE 1. Schematic diagramm of the 3 MV TTT-3 tandem accelerator at Naples, the beam transport system, the windowless gas target, and the recoil separator.

proceed at least into the medium mass regime. While some measurements have been performed over the last 10 years many still remain to be studied in detail. In general such studies can only be performed using an intense accelerated radioactive ion beam and inverse kinematics. A new facility for radioactive ion beams is ISAC (Isotope Separation and Acceleration), situated on the TRIUMF cyclotron in Vancouver, Canada. The radioactive beam with A < 30 will be accelerated to an energy ranging from 0.15 to 1.5 MeV/amu, a range optimal to study such radiative capture reactions.

ABSOLUTE CROSS SECTION OF $^1\text{H}(^7\text{Be},\gamma)^8\text{B}$

The aim of a project at the 3 MV TTT-3 tandem accelerator in Naples (Fig. 1) was to provide an improved cross section value of the reaction $^7\text{Be}(p,\gamma)^8\text{B}$ in the nonresonant region, i.e. at E_{cm} = 992 keV. The reaction was studied in inverse kinematics, $p(^7\text{Be},\gamma)^8\text{B}$, i.e. a radioactive ^7Be ion beam was guided into a windowless gas target system filled with H_2 gas (pressure $p(H_2)$ = 5 mbar), thus avoiding the problems of target stoichiometry. As a novel technique, the ^8B residuals nuclides were detected directly in an efficient recoil separator.

At the given energy the ^8B recoils emerge from the gas target with an energy E_{8B} = 7 MeV and several charge states. The fully stripped $^8\text{B}^{5+}$ recoils were selected and tuned in the separator, since for this charge state the intensity of the

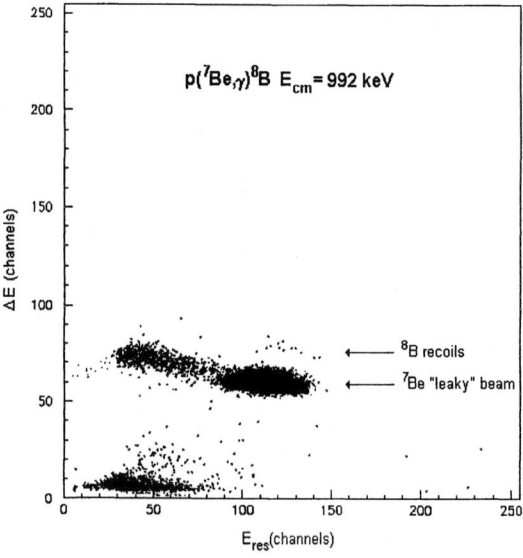

FIGURE 2. Two dimensional density plot of the ΔE-E telescope with the recoil separator tuned to the $^8\text{B}^{5+}$ nuclides from $^1\text{H}(^7\text{Be},\gamma)^8\text{B}$ reaction. The observed structures are identified.

leaky ^{7}Be beams should be minimised [4]. The charge state probability of boron ions in the H$_2$ gas was measured using a ^{11}B^{4+} beam with the same energy per amu. The mean value of the relevant 5$^+$ charge state is 65 %.

The recoil separator (Fig. 1) includes a magnetic quadrupole triplet, a 30° switching magnet, a magnetic quadrupole doublet, a Wien filter, and a conventional ionisation chamber. For the tuning of the separator, a pilot ^{7}Li^{3+} beam with the same rigidity as the ^{8}B^{5+} recoils was used. After this tuning, hydrogen gas was filled into the gas target system and the ^{7}Li^{3+} beam was retuned through the separator (to take into account the energy loss of the ^{7}Li beam in the gas) leading to a set of optimum values for the magnetic field of the switching magnet, the Wien filter, and the magnetic quadrupoles. The resulting values were then corrected for differences (1 %) in the energy loss for the pilot ^{7}Li beam, the actual ^{7}Be beam and ^{8}B recoils. Finally, the magnetic field of the Wien filter was scaled down to account for the different velocities between ^{7}Li^{3+} and ^{8}B^{5+}. The same procedure was repeated using other pilot beams leading to the same final optimum values for the transport parameters within the acceptance of the separator.

The resulting identification matrix of the ΔE-E telescope is shown in Fig. 2. The ^{8}B events, N = 13 ± 4, are well resolved from a band of counts due to leaky ^{7}Be beams, which correspond to a suppression factor of about 1×10^{-10} for the incident ^{7}Be projectiles. This result leads to $\sigma(E) = 0.41 \pm 0.11 \mu$b for the absolute cross section of p(^{7}Be,γ)^{8}B at E$_{cm}$ = 992 keV. Within the present statistical uncertainty the result is consistant with the value recommended recently [5]. The details of this measurement are reported in [6] and references therein.

This experiment demonstrates the feasibility of the techniques used in the study of this reaction. An improvement of the statistical uncertainty attainable with the present technique could be archived - besides the possibility of using a prohibitive amount of activity of the order of 1 TBq - by increasing the accelerator transmission and/or by using a more probable charge state for the accelerated ^{7}Be ions, which would require a terminal voltage not accessible to the Naples accelerator.

THE ERNA-PROJECT: ^{12}C(α,γ)^{16}O

The capture reaction ^{12}C(α,γ)^{16}O is a key reaction of nuclear astrophysics due to its influence on several areas of stellar evolution. All previous direct measurements have focused on the observation of the capture γ-rays, including one experiment [7] that combined γ-detection with coincident detection of the ^{16}O recoils produced in the reaction. Due to the low cross section and various background problems depending on the exact nature of the experiments, γ-ray data with still inadequate precision were limited to center-of-mass energies between 1.2 and 3.2 MeV. On the basis of the experiences of the Naples experiment and to improve the experimental situation concerning the reaction ^{12}C(α,γ)^{16}O, a new experimental approach is in preparation at the 4 MV Dynamitron tandem accelerator in Bochum, called ERNA (European Recoil separator for Nuclear Astrophysics). In this approach, the reaction is initiated in inverted kinematics, $^4\text{He}(^{12}\text{C},\gamma)^{16}\text{O}$, i.e. a ^{12}C beam is guided into a windowless ^4He jet gas target and the kinematically forward-directed ^{16}O recoils are detected downstream on the beam line. The direct observation of the ^{16}O recoils requires an efficient recoil separator to filter out the intense ^{12}C beam particles from the ^{16}O recoils: the number of ^{16}O recoils per incident ^{12}C projectile is 1×10^{-18} for a cross section $\sigma = 1$ pb and a target density $n(^4\text{He}) = 1 \times 10^{18}$ atoms/cm^2. The recoil separator must also filter out beam contaminants, small-angle elastic scattering products, and background events from multiple scattering processes leading to a degraded tail of the projectiles. If the filtering of the separator is sufficiently effective, the ^{16}O recoils can be counted directly in a ΔE-E telescope placed in the beam line at the end of the recoil separator, where the telescope allows for particle identification. It is expected that the high detection efficiency of the ^{16}O recoils and the negligible contribution of cosmic-ray events in the ΔE-E coincidences allows a measurement of the $^4\text{He}(^{12}\text{C},\gamma)^{16}\text{O}$ cross section to as low as $\text{E}_{cm} = 0.7$ MeV ($\sigma \approx$ 1pb).

Beam contaminants

Since the ^{12}C projectiles and the ^{16}O recoils have essentially the same momentum and since the ^{12}C ion beam emerging from the accelerator passes a momentum filter (analysing magnet), a nearly complete elimination of any ^{16}O beam contaminant in the ^{12}C ion beam incident on the ^4He gas target is of utmost importance for the new approach: the ^{16}O beam contaminant and the ^{16}O recoils cannot be distinguished in the recoil separator, since both have the same momentum.

In a first ERNA report [8], a Wien filter and a ΔE-E telescope were used to investigate the ^{16}O beam contamination accompanying a momentum-filtered ^{12}C beam. The resulting identification matrix (Fig. 3) shows indeed the presence of a contaminant ^{16}O beam. The energy of the contaminant ^{16}O beam is about 3/4 the energy of the ^{12}C incident beam, as expected from the momentum filter for equal charge states. Furthermore, the leaky ^{12}C beam has about the same velocity as that of the contaminant ^{16}O beam, as expected from the action of the Wien filter (velocity) filter, and represents a velocity-filtered section of the degraded tail of the ^{12}C beam. The intensity ratio of ^{16}O to ^{12}C was found to be $P_0 = 6 \times 10^{-10}$, i.e. much higher than the intensity ratio 1×10^{-18} between the ^{16}O recoils and ^{12}C projectiles at $\text{E}_{cm} = 0.7$ MeV, or even 5×10^{-14} at $\text{E}_{cm} = 2.45$ MeV with maximum radiative capture cross section $\sigma \approx 50$ nb. If the injection magnet after the ion source of the accelerator is set at mass 16 (rather than at mass 12 for the ^{12}C ion beam), the telescope shows the dominant presence of ^{16}O ions, at the same point in the matrix. Thus, the main source of the ^{16}O beam contamination lies in the ion source setup arising from the finite mass resolution of the injection magnet.

The data (Fig. 3) indicate that the ^{12}C ion beam intensity can be suppressed by a single Wien filter to about 4×10^{-8}. Since the studies of $^4\text{He}(^{12}\text{C},\gamma)^{16}\text{O}$ using the recoil separator ERNA include the combination of a Wien filter, a momen-

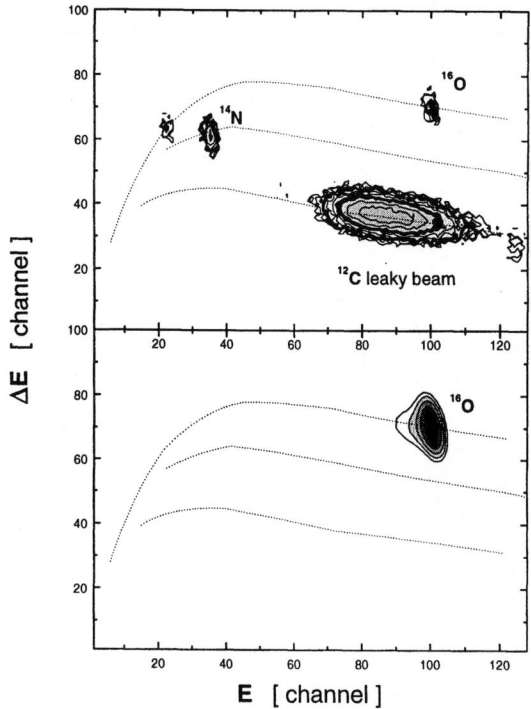

FIGURE 3. The ΔE-E identification matrix for a $^{12}C^{3+}$ ion beam is shown with the injection magnet of the accelerator (Fig. 4) set at mass 12 (upper part) and at mass 16 (lower part): the contaminant ^{16}O beam appears at the same point in the matrix.

tum filter, and another Wien filter (see below), the above results suggest that the needed suppression factor of 1×10^{-15} can be achieved with ERNA. The ^{16}O beam purification of a momentum-filtered $^{12}C^{3+}$ ion beam (from the 4 MV Dynamitron tandem) using a single Wien filter is about $6 \times 10^{-10} \times 4 \times 10^{-8} = 2 \times 10^{-17}$, where we have assumed an ^{16}O degraded tail identical to that of the ^{12}C ion beam. This purification is not quite sufficient for a measurement with $\sigma \approx 1pb$ using a single Wien filter. For this reason, one Wien filter will be installed - in the final setup of the ERNA project - before the analysing magnet and a second Wien filter will be placed between the analysing magnet and the jet gas target (Fig. 4), where this setup should provide a sufficient ^{16}O beam purification for the ERNA aims.

Ion beam optics

Although the ^{16}O recoils - produced in the ^{4}He jet gas target via the reaction ^{4}He(^{12}C,γ)^{16}O - are kinematically forward-directed, the emission of the capture γ-rays (energy E_γ) leads to an emission cone of half-angle $\Theta = \arctan(E_\gamma/pc)$, where p is the momentum of the ^{16}O recoils and c the velocity of light. Associated with the γ-ray emission there is also a spread $\Delta p/p$ in momentum. For example, at $E_{cm} = 0.7$ MeV ($E_\gamma = 7.9$ MeV) one finds $\Theta = 1.8°$ and $\Delta p/p = 6.2$ %. At $E_{cm} = 0.7$ MeV, the cone has reached a diameter of 3.1 cm at a 0.5 m distance from the jet gas target. Thus, shortly after the jet gas target there must be a focusing element followed by filter-elements and other focusing elements up to the site of the telescope, where all elements must have an angle acceptance of at least $\Theta = 1.8°$ and a momentum acceptance of at least $\Delta p/p = 6.2$ %, in order to transport the ^{16}O recoils with 100 % transmission to the telescope. This requirement demand a compact design of the jet gas target system involving several pumping stages, where the present technical plan involves an extension of 35 cm on both sides of the jet gas target.

The calculations of ion beam optics for the recoil separator ERNA were performed up to third order using the program COSY INFINITY [10]. They showed that the above requirements can be fulfilled with the setup indicated in Fig. 4. After the jet gas target, the separator will consist sequentially of the following elements: (i) a magnetic quadrupole triplet (MQT; length 105 cm, inner diameter 10.6 cm), (ii) a Wien filter (WF3, DANFYSIK) containing two electrostatic plates (12 cm width, 50 cm effective length, 7 cm gap), the maximum electric field is ±40 kV with an magnetic field of 0.135 T, (iii) a magnetic quadrupole singlet (MQS1, length 5.2

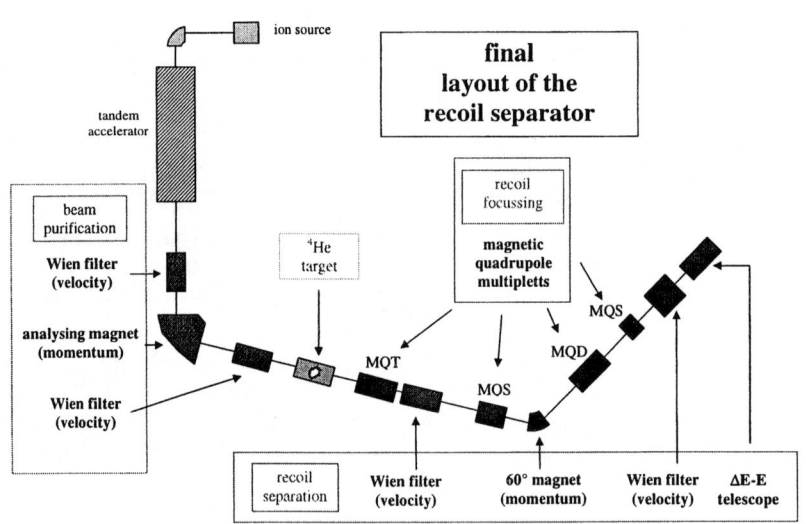

FIGURE 4. Schematic diagram of the complete ERNA setup (MQT = magnetic quadrupole triplet, MQD = magnetic quadrupole doublet, MQS = magnetic quadrupole singlet)

cm, inner diameter 10.6 cm), (iv) a 60° dipole magnet (radius of curvature: 40 cm, 7.6 cm gap, max. B-field: 1.3 T), (v) a magnetic quadrupole doublet (MQD, length 50.8 cm, inner diameter 10.6 cm), (vi) a second Wien filter (WF4, electrical plates: 6.3 cm width, 57.8 cm effective length, 7 cm gap), (vii) a magnetic quadrupole singlet (MQS2, length 5.2 cm, inner diameter 7.7 cm), and (viii) the ΔE-E telescope [8]. The calculations were performed assuming a parallel ^{12}C ion beam of 3 mm diameter and a pointlike ^{4}He jet gas target with a density of 1×10^{18} atoms/cm^2. To calculate the emittance of the ^{16}O recoils, the energy loss of the ^{12}C projectiles and the ^{16}O recoils in the gas target as well as the recoil from the capture γ-rays to the ^{16}O ground state were taken into account; the effects of angle and energy straggling in the target have also been included. The calculations were performed over the planned energy region, E_{cm} = 0.7 to 5.0 MeV, where at each energy the most probable charge state was selected for the ^{16}O recoils. The trajectories emitted within the entire emission cone in the jet gas target are 100 % transmitted to the telescope, where they enter the telescope within a diameter of about 30 mm and are nearly parallel.

Conclusions

Since the setup used for the measurement of the beam suppression factor [9] is nearly identical to that in the final ERNA layout, a successful direct measurement of the cross section of the reaction ^{12}C$(\alpha,\gamma)^{16}$O with a recoil separator seems to be feasible. First results of this experiment are expected for summer 2001.

THE DRAGON-FACILITY AT TRIUMF-ISAC

The DRAGON (Detector of Recoils And Gammas Of Nuclear reactions) facility is being installed at the new ISAC accelerated radioactive beam facility at TRIUMF in Vancouver, Canada. This detector system is designed to measure the rates

TABLE 1. Specifications for the DRAGON Electromagnetic Separator

Optical Path Length: 20.4 m	Mass Dispersion:	0.46 cm/% 1st stage
Acceptance: angular ($\pm$20 hor. & $\pm$25 vertical)		2 cm/% 2st stage
velocity ($\pm$2 %)	Mass Resolution:	90 after 1st stage
		150 after 2st stage
	Spot Size:	5 mm after 1st stage
Components		
First sector dipole magnet: 1 m, 10 cm gap, 50°, 0.22 T		
Second dipole magnet: 0.81 m, 12 cm gap, 75°, 0.27 T		
First electric dipole: 2 m, 20°, 10 cm gap, $\pm$200kV		
Second electric dipole: 2.5 m, 35°, 10 cm gap, $\pm$160kV		
Q1: 10.8 cm, Q2-Q8: 15.9 cm, Q9-Q10: 15 cm		
Sext 1-4: 16 cm		

of radiative capture reactions involving exotic nuclei. Such rates are important in elucidating the processes which play a role in various explosive nucleosynthesis phenomena such as novae, supernovae, and x-ray bursts and therefore are important for the production of most of the stable elements. In particular most bottle-neck reactions that determine the breakout to higher-order burning processes need further study. The first study involving a radioactive beam will be ^{21}Na(p,γ)^{22}Mg which leads the burning in novae from the cold to the hot NeNa cycle, further studies are aiming in the measurement of the reactions: ^{15}O(α,γ)^{19}Ne, ^{19}Ne(p,γ)^{20}Na, ^{23}Mg(p,γ)^{24}Al, and ^{25}Al(p,γ)^{26}Si. At this new detection system it is planned to measure reaction rates in inverse kinematics in the energy range between 0.15 and 1.5 MeV/amu.

The DRAGON facility is composed of an extended (10 cm) gas target, possibly surrounded by a BGO based gamma array of high geometric efficiency. The ions enter the separator with the first magnetic dipole used to separate one charge

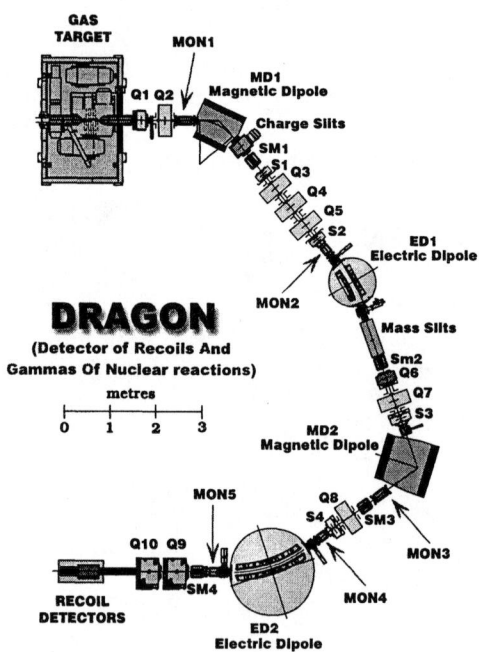

FIGURE 5. Detailed layout plan for DRAGON (Q = magnetic quadrupole lens)

state of both the beam and reaction product. The second main component of the separator is an electrostatic dipole which operates at up to 40 kV/cm. This device is used to separate the beam and recoil based upon the different energies. A second sequence of magnetic and electrostatic dipoles are used to provide further separation. It is expected that a beam suppression factor of about 10^{12} will be achieved in this stage. The specifications of the separator components are given in Tab. 1 while Fig. 5 shows the actually layout plan for DRAGON.

A local time-of-flight measurement over a distance of 50 cm in front of the end detector will be used to improve the ion identification properties of the detection system. A time-of-flight start signal provided by a MCP/C foil electron detection system [11] coupled with a stop signal from an energy detector will provide the primary measurement of the mass of the resultant ions from the separator. It is expected that an additional beam suppression factor of up to 100 will be realized with these approaches. Finally, a gamma array based upon BGO crystals [12] is planned, situated around the gas target, to provide a coincident signal from the prompt reaction gammas.

REFERENCES

1. C.E. Rolfs, W.S. Rodney, Cauldrons in the Cosmos (University of Chicago Press, 1988).
2. J.N. Bahcall and M.H. Pinsonneault, *Rev. Mod. Phys.* **64**, 885 (1992).
3. R.E. Wallace and S.E. Woosley, *Ap. J. Suppl.* **45**, 398 (1981).
4. L. Gialanella et al., *Nucl. Instr. Meth.* **A376**, 174 (1996).
5. C. Angulo et al., *Nucl. Phys.* **A656**, 3 (1999).
6. L. Gialanella et al., *Eur. Phys. J.* **A7**, 303 (2000).
7. R.M. Kremer et al., *Phys. Rev. Lett.* **60**, 1475 (1988).
8. D. Rogalla et al., *Nucl. Instr. Meth.* **A437**, 266 (1999).
9. D. Rogalla et al., *Eur. Phys. J.* **A6**, 471 (1999).
10. M. Berz: Computational aspects of design and simulation (COSY INFINITY), *Nucl. Instr. Meth.* **A298**, 473 (1990).
11. P. Boccaccio et al., *Nucl. Instr. Meth.* **A243**, 599 (1973).
12. J. Rogers et al., "A Gamma Ray Detector for DRAGON at ISAC", NSERC Grant Application 1998, P.I.=J. D'Auria.

Radioactive Beams in Nuclear Astrophysics

Michael S. Smith*

*Physics Division, Oak Ridge National Laboratory [1]
Oak Ridge, TN 37830

Abstract. Reactions on unstable nuclei play an important role in a diverse range of astrophysical phenomena including stellar explosions such as novae, X-ray bursts, and supernovae. They also play a role in the evolution of our own Sun, red giant stars, supermassive stars, the early universe, and other astrophysical environments. Beams of radioactive nuclei are now offering a direct probe of the nuclear physics occurring in these sites. The astrophysical motivation for radioactive beam development and the techniques to produce these beams are discussed. Some important radioactive beam measurements and plans for future radioactive beam facilities are also presented.

ASTROPHYSICAL MOTIVATION FOR RADIOACTIVE BEAMS

Reactions on unstable nuclei play an important role in astrophysical phenomena ranging from the early universe to the late stages of stellar evolution, from quiescent nuclear burning to explosive events, and from our own sun to the outer reaches of the Universe. Nuclear reactions on unstable nuclei influence and in some cases determine the energy generation, element synthesis, and other features of these astrophysical environments.

Astrophysical Environments

Novae Nova explosions are accretion-driven explosions caused by the transfer of mass from one star to a white dwarf companion star [1]. The mass transfer and subsequent rise in temperature and pressure can initiate a violent runaway thermonuclear explosion ($\sim 10^{38}$ - 10^{45} ergs released), resulting in the synthesis of heavy elements (up to mass ~ 40) and their subsequent ejection into space. The explosion also influences the evolution of binary star system. These catastrophic

---

[1] Managed by UT-Battelle, LLC, for the U.S. Department of Energy under contract DE-AC05-00OR22725.

stellar events are characterized by extremely high temperatures and densities - greater than 10^8 K and 10^3 g/cm^3, respectively. Under such conditions, (p,γ) and (α,p) reactions can rapidly (on timescales of ns to minutes) produce unstable nuclei on the proton-rich side of the valley of stability. Any such nuclei (decaying via e$^+$-emission) produced with half-lives longer than, or comparable to, the mean time between nuclear reactions can potentially undergo subsequent nuclear processing.

Reactions on proton-rich radioactive nuclei are crucial in these explosions [2,3], producing abundances which are very different than those from the hydrogen burning occurring in non-explosive environments [4,5] and generating energy up to 100 times faster than in quiescent stars. Some radioactive nuclei (those with lifetimes $\tau \geq 100$ s) synthesized in explosions may be carried by convection to the top of the envelope before they decay (and make a small contribution towards powering the expansion [6]). Observations of the γ-ray lines (especially the 511- keV emission of ^{18}F) resulting from such radioactive decays in the envelope may provide stringent tests of nova models [7–9]. The γ-ray emissions depend sensitively on the amount of radionuclides synthesized by nuclear reactions in the explosion, which in turn depends on the rates of nuclear reactions on radioactive isotopes [10,11]. Examples of important reactions include ^{17}F(p,γ)^{18}Ne, ^{18}F(p,α)^{15}O, ^{19}Ne(p,γ)^{20}Na, ^{20}Na(p,γ)^{21}Mg, ^{21}Na(p,γ)^{22}Mg, ^{22}Na(p,γ)^{23}Mg, and ^{25}Al(p,γ)^{26}Si [2,12]. These can lead to hydrogen burning through the rapid proton capture process (rp-process), involving (p,γ) reactions near the proton drip line competing with e$^+$ - decay and reaction cycles (e.g., the Ne-Na and Mg-Al cycles). The rates of such reactions on unstable isotopes are needed to understand the nova phenomenon.

X-ray Bursts and X-ray Pulsars Other accretion-driven phenomena important in astrophysics include X-ray bursts and X-ray pulsars. These can occur when material is accreted onto the surface of a neutron star, where temperatures and densities can reach over 10^9 K and 10^6 g/cm^3, respectively [13,14]. The ensuing explosive hydrogen burning can synthesize isotopes with masses up to 80 - 100 or beyond [14–16] via reactions in the αp- and rp-processes. Recent studies of nucleosynthesis in these violent explosions suggest that their X-ray luminosity and neutron star crust composition are influenced by the nuclear reactions (most involving proton-rich radioactive isotopes) used in the model [17]. For example, the rates of reactions in the sequence ^{12}C(p,γ)^{13}N(p,γ)^{14}O(α,p)^{17}F (p,γ)^{18}Ne(α,p)^{21}Na(p,γ)^{22}Mg are crucial because they give the maximum flux of X-rays and serve as the gateway to the synthesis of heavier nuclei [14,18]. Properties of nuclei along the proton dripline are important, along with proton capture reaction rates, to understanding these violent explosions.

Supernova Explosions Approximately half of the isotopes heavier than iron are believed to be produced by the r-process, which may occur in violent core-collapse supernova explosions [19]. The high-entropy wind off the surface of the neutron star created in such explosions is believed to have sufficiently high neutron density (10^{20} cm^3) and temperature ($\geq 10^9$ K) along with the presence of heavy "seed" nuclei to enable the formation of heavy elements. The primary nuclear reactions involved in the r-process are (n,γ), (γ,n), and beta decays on neutron-rich unstable isotopes. At these temperatures and densities, the neutron capture reactions are so fast ($\sim$ seconds) that the sequence of reactions is determined primarily by the properties of the neutron-rich unstable nuclei rather than the neutron capture rates [20]. Information such as masses, neutron separation energies, beta-decay lifetimes, beta-delayed neutron decay probabilities, and level densities are needed for approximately 2000 isotopes on the neutron-rich side of stability to calculate the synthesis of nuclei in the r-process. Since the properties of all of these nuclides will never be measured, it is important to combine measurements of the properties of a few key nuclides with nuclear models to obtain accurate global predictions of nuclear properties relevant for the r-process [21].

Additionally, explosive H and He burning on proton-rich radioactive isotopes may occur in the outer envelopes of core-collapse supernovae [3]. The propagation of a supernova shock wave into the outer layers may heat and compress material sufficiently to ignite brief occurrences of explosive burning. The nuclear reactions would involve proton-rich radioactive isotopes and be similar to those occurring in very hot novae or X-ray bursts.

Supermassive Stars Super Massive Stars (SMSs) are potentially formed in the very early universe ("Population III" stars), are composed of only H and He (i.e., they have zero metallicity), and have enormous masses - more than 10^5 times the mass of the sun [22]. These stars form and quasi-statically contract to a point of general relativistic instability. They then possibly collapse into supermassive black holes, presenting one explanation of the observations of these objects at the active center of many galaxies (Active Galactic Nuclei [AGN]) [23]. The many new observations of black holes, the uncertain origin of AGNs, and the possible gravitational wave signatures of AGN formation have resulted in a revitalization of work on supermassive stars [24,25]. The evolution of SMSs depends upon a number of factors including the nuclear energy generated by rapid hydrogen burning during the collapse phase [26]. The Hot pp Chain [27,28] may initiate hydrogen burning through the Hot CNO cycle and even (in some cases) the rp-process in SMSs, significantly effecting the SMS evolution by causing the star to explode rather than collapse [26,28-30]. The relevant reactions include proton- and α-captures on proton-rich radioactive nuclei such as $^7\text{Be}(p,\gamma)^8\text{B}$, $^8\text{B}(\alpha,p)^{11}\text{C}$, and $^{11}\text{C}(p,\gamma)^{12}\text{N}$ [27,28]. Reactions in the Hot pp Chain may potentially occur in other astrophysical environments such as novae [31].

Inhomogeneous Big Bang Nucleosynthesis In the early universe, the phase transition from quarks and gluons to hadrons could possibly cause proton-rich and neutron-rich regions to form [32]. This could enable nuclear reactions on some light unstable isotopes to alter the abundances formed in the early universe compared to homogeneous Big Bang models. Examples of important reactions are $^8\text{Li}(\alpha,n)^{11}\text{B}$, $^8\text{Li}(n,\gamma)^9\text{Li}$, and $^8\text{Li}(d,n)^9\text{Li}$ [33]. Recent observations of the spectrum of the cosmic microwave background radiation power spectrum [34] suggest a universal mass density that is consistent with some inhomogeneous Big Bang models but outside the range normally quoted for standard Big Bang models [35].

Red Giant Stars Radioactive isotopes also play a role in the non-explosive nuclear burning occurring in the red giant stars - specifically, in the thermally-pulsing He-shells of low-mass Asymptotic Giant Branch stars [36–38]. Here, s-process nucleosynthesis occurs: a series of slow ($\sim$ minutes to years) neutron captures on stable nuclides interspersed with beta decays. The s-process produces $\sim$ 80 radioactive nuclides that are one neutron beyond stability. These long-lived "branch point" nuclei, such as ^{85}Kr, ^{134}Cs, and ^{154}Eu, can either beta decay or undergo neutron capture. The capture - to - decay branching ratio can be written as a function of the stellar temperature and neutron density; therefore, measurements of the neutron capture on these branch points can provide an important constraint on the conditions under which s-process burning occurs [38,39].

The Sun Energy is generated at the core of our Sun through the burning of hydrogen in the pp-chains. One of the side branches of these chains is the fusion of protons with radioactive ^{7}Be. Although unimportant for energy generation, the rate of this reaction strongly determines the number of high-energy neutrinos measured in most terrestrial solar neutrino detectors. The longstanding overabundance of neutrinos predicted by theory compared to the measured neutrino flux, the "Solar Neutrino Problem" [40], continues to attract intense international attention on many fronts. The largest nuclear physics uncertainty in the predictions of solar neutrino fluxes is due to the uncertainty in the ^{7}Be(p,γ)^{8}B reaction rate [41].

Other Possibilities There are other astrophysical sites [42] - e.g., the accretion disk around black holes [43] - where temperatures and densities may be sufficient for explosive hydrogen burning to occur. In these environments, reactions on proton-rich radioactive isotopes may play an important role, as they do in novae and X-ray bursts.

The Importance of Radioactive Beams

Critical comparisons of models of these astrophysical phenomena with observations require a knowledge of nuclear reactions on radioactive nuclei and the properties of the unstable nuclei themselves. Almost all of the nuclei of interest (with a few exceptions such as ^{7}Be, ^{44}Ti, and others) are short lived and unsuitable for targets. Experiments to directly obtain nuclear reaction and structure information on these nuclides have therefore been, until recently, impossible due to the lack of intense radioactive nuclear beams. Models currently employ reaction rate estimates based on systematic properties of nuclear states, on information from analog nuclei, on partial resonance information from stable beam transfer reaction studies, and on statistical model calculations. In some cases, these estimates can be incorrect by orders of magnitude (e.g., [3]). Because of these uncertainties, critical comparisons of theory to astrophysical observations to determine, for example, the temperatures, densities, and duration of stellar explosions or the origin of the elements in our world, are difficult to make. The recent availability of radioactive beams has initiated a new era in laboratory nuclear astrophysics - one in which previously unattainable nuclear reaction cross sections can be measured and subsequently incorporated into an emerging generation of sophisticated, computationally intensive models of astrophysical phenomena.

RADIOACTIVE BEAM PRODUCTION TECHNIQUES

Isotope Separator On Line Technique. One approach to radioactive beam production, pioneered at CERN, is the Isotope Separator On-Line (ISOL) technique [44], now used at Louvain-la-Neuve [45] and at Oak Ridge National Laboratory [46]. One accelerator bombards a thick target with a beam of stable nuclei to produce radioactive atoms. These atoms diffuse out of the target material and effuse through a transfer line to an ion source where they are ionized and extracted. The radioactive ions are then mass separated from other ions and accelerated to energies needed for experiments by a second accelerator. The ISOL technique can produce very high beam qualities and (in some cases) purities, with reasonable intensities [47]. The disadvantages: only a few radioactive beam species can be generated from each combination of production target and primary beam; the effectiveness of the technique is very dependent upon the chemistry of the element; and beams with short lifetimes (≤ 1 s) are difficult to produce because of the time required for the radioactive atoms to diffuse out of the target.

A related approach is the Ion Guide Isotope Separator On Line (IGISOL) technique [48]. A beam interacts with a thin target in a chamber containing a stopping gas. The reaction products exit the target, are stopped in the gas, are swept out of the chamber via electric fields (an ion guide) to an ion source, are ionized, and then sent to a mass separator. So far, radioactive species produced in this manner have not been reaccelerated, but the proposed Rare Isotope Accelerator [49] will use this technique to generate accelerated unstable beams.

Projectile Fragmentation A complementary radioactive beam production technique used at Michigan State University [50], RIKEN [51], GANIL [52], and GSI [53] employs the fragmentation that occurs when a high-energy beam of stable heavy ions passes through a thin target [54]. In this Projectile Fragmentation (PF) method, the desired (radioactive) fragments are then mass separated from other fragments and steered towards the experimental area. The PF technique can produce beams of very short lifetimes ($\sim \mu$s or less), and the same beam and target combination can be used to produce many different beam species. The disadvantages are that high beam quality, purity, and intensity are difficult to obtain, and astrophysical energies (typically less than ~ 2 MeV/u for direct measurements) cannot be reached.

Other Production Techniques Another technique for producing radioactive beams consists of forming the species of interest "in-flight" as the product of a reaction between primary beam particles with nuclei in a gas or thin solid target. The products pass through the target and are then focused on a secondary target to measure the reaction of interest. This has been used for radioactive beams at

Notre Dame University [55] with solid production targets and at Argonne National Laboratory [56] with gas production targets.

A final technique for producing radioactive beams has been used for long-lived species such as ^{7}Be, ^{18}F, ^{44}Ti, and ^{56}Ni. Material is bombarded by an accelerator beam for a duration up to several half-lives of the species of interest. The material containing the long-lived nuclei is in some cases chemically processed and then transported to a conventional ion source (e.g., a sputter ion source) where the radioactive atoms are ionized and extracted. Another approach avoids chemical processing by bombarding the material and then rotating it in front of an ion source. These "batch mode" production approaches have been used at Argonne National Laboratory [57] and Naples [58,59] and will soon be used at Oak Ridge National Laboratory [60] and other facilities in the future.

MEASUREMENTS WITH RADIOACTIVE ION BEAMS

The first experiment using an accelerated beam of radioactive ions was a measurement of the ^{13}N(p,γ)^{14}O resonant cross section at Louvain-la-Neuve in 1991 [61,62]. Since then, many more measurements have been carried out [63]. The majority of the experimental techniques used are those employed in traditional nuclear spectroscopy measurements, with an emphasis on techniques enabling measurements with beam intensities far less than the 10^{10} ions/s or greater typically obtained in stable beam experiments. Radioactive beam experiments also require the capability to handle undesirable beam characteristics such as high contamination, large emittance, and low energy resolution. Finally, many of the astrophysically relevant reactions involve the fusion of heavy ions with very light nuclides (e.g., p, n, α) and therefore require that a heavy beam bombard a light target. Such inverse kinematics measurements have both significant advantages and challenges. Some important radioactive beam measurements will be discussed in this Section along with a number of successful experimental techniques.

Direct Measurements of Astrophysical Reactions

The cross sections of radioactive heavy ions interacting with hydrogen or helium nuclei have been directly measured with radioactive beams for some reactions important for astrophysics. For explosive nucleosynthesis occurring in novae and X-ray bursts, the reactions ^{13}N(p,γ)^{14}O [61,62], ^{18}F(p,α)^{15}O [64–71], and ^{18}Ne(α,p)^{21}Na [72] have been measured, and upper limits have been placed on the ^{18}F(p,γ)^{19}Ne [73] and ^{19}Ne(p,γ)^{20}Na [74] resonance yields. For the explosive burning occurring in supernovae, ^{44}Ti(α,p)^{47}V has been measured [57], and ^{7}Be(p,γ)^{8}B [58,59] has been directly measured to help understand the burning at the solar core.

The techniques used in these measurements can be grouped into four categories. One technique involves bombarding a gas hydrogen or helium target with a radioactive heavy ion beam and counting the heavy products of the (p,γ) and (α,γ) capture reactions in a recoil separator. Recoil detection in inverse kinematics has many advantages over capture γ-ray detection techniques [75]. Most importantly, a high recoil detection efficiency is possible with a device of moderate acceptance due to the forward focusing of the recoils in the laboratory frame ($\theta_{lab} \leq 0.5°$). This enables 100 % of the recoils to enter the separator and up to 40 % to be detected at the focal plane (after a charge-state selection is made in the separator). Another advantage is a low background since the detectors at the focal plane are far from the high-background radiation area near the target. Recoil detection also offers a variety of detection schemes at the recoil separator focal plane - such as time-of-arrival, Z-identification, delayed-activity detection, and recoil-γ coincidence measurements. The recoil detection approach is, however, very challenging because: (1) the spatial distributions of the recoils and the unreacted projectiles overlap, so the separator must be located along the beam axis and therefore accept all of the unreacted beam particles; (2) the recoils are only 10^{-10} to 10^{-15} times as intense as the projectiles entering the separator due to the small reaction cross section; and (3) the projectiles and recoils have approximately the same momentum, and differ in mass and velocity by only a few percent. It is therefore necessary to optimize the recoil separator to collect one mass group (the recoils) at the final focus with the highest possible (10^{-10} to 10^{-15}) suppression of scattered projectiles. Additionally, the final focus of the separator must have detectors designed to distinguish between the scattered projectiles and recoils, which is quite challenging in view of the low energy (0.4 - 2 MeV/u), and often low Z, of the particles to be identified.

The recoil detection technique for proton capture reactions was first proven viable with a measurement of p(^{12}C,^{13}N)γ with a small, non-optimized recoil separator at Caltech [75]. A 10^{-10} suppression of scattered beam particles was obtained with this device. Since then, other separators have been used for radioactive beam experiments. The Fragment Mass Analyzer (FMA) at Argonne National Laboratory was used with a batch mode ^{18}F beam to set an upper limit on the strength of an important resonance in the ^{18}F(p,γ)^{19}Ne reaction [73]. The FMA was also used with a batch mode ^{44}Ti beam to measure recoils from the ^{44}Ti(α,p)^{47}V reaction, thereby determining the rate of this reaction during explosive nucleosynthesis in supernovae [57]. A separator at Naples was used to measure the ^{7}Be(p,γ)^{8}B cross section at 1 MeV in the center of mass [58,59]. Other separators which are being commissioned or constructed include the Daresbury Recoil Separator at ORNL [76], ARES at Louvain-la-Neuve [77], and DRAGON at TRIUMF-ISAC [78]. Intensities of roughly 10^7 s^{-1} are generally required for these direct measurements.

A second approach to detecting recoils from proton- or α-capture reactions is to directly detect the capture γ-rays. This technique has a low efficiency because there is no kinematic focusing of the γ-rays and the detectors have a low intrinsic efficiency. It also has high backgrounds from 511-keV γ-rays caused by the decay of scattered proton-rich radioactive beam particles in the chamber. The only case to date where these difficulties have been overcome is the measurement of the ^{13}N(p,γ)^{14}O reaction at Louvain-la-Neuve [61], using NaI detectors with specially modified photomultiplier tubes to handle the high background counting rate [79]. Recoil-γ coincidence measurements may be carried out in the future because of the potential for increasing the rejection of scattered beam events at the focal plane (at the expense of a significantly lower total detection efficiency).

A third direct measurement approach involves detecting non-prompt decay products from a capture reaction. This technique was used, for example, to investigate the ^{19}Ne(p,γ)^{20}Na reaction at Louvain-la-Neuve [74]. A CH$_2$ target was bombarded by a radioactive ^{19}Ne beam, and the resulting ^{20}Na nuclei were detected by their β-delayed α-decay to ^{16}O using Si strip detectors and track detectors. The betas from the ^{20}Na decay were detected in a separate experiment using a solenoid magnet backed with a stack of plastic detectors.

Reactions with two charged particles in the final state - such as (p,α) and (α,p) - can also be measured directly with a fourth technique. The light-ion products of these reactions can be detected in arrays of silicon strip detectors arranged to cover a large solid angle around the target. The high segmentation of modern arrays (with typically over 100 strips [80,81]), coupled with high-density fast electronics [82,83], enable high count rates in the entire array and excellent resolution in angle and energy. Typical detectors have strips of a few mm height and widths from a few mm to a few cm, with thicknesses ranging from 100 - 1000 μm. Solid angle coverage of such arrays can be as high as 25 % or more, corresponding to an even higher detection efficiency for some reactions because of the kinematic focusing of the reaction products. Detectors with a transmission geometry can be stacked to enable particle identification by measuring both energy loss and total energy [84,85]. Double Sided Strip Detectors (DSSDs) give two-dimensional position information by segmenting both the front and back of the detectors. The heavy ion products of the (p,α) and (α,p) reactions can be detected in additional Si strip arrays or by other detectors (e.g., gas ionization counters) placed downstream of the target. These techniques have been used for direct measurements of the ^{18}F(p,α)^{15}O reaction at Louvain-la-Neuve [64,67,71], Argonne National Lab [65,66], and Oak Ridge National Lab [68–70]. This reaction is the dominant destruction mechanism for the long-lived radioisotope ^{18}F in novae, a potential target of γ-ray astronomy. Although initial measurements of one of the dominant resonances in this reaction had serious discrepancies in the resonance width and energy [64–67], more recent measurements have resolved these discrepancies and obtained precise values for these parameters [68–71]. Si strip detectors have also been used for a direct measurement of the ^{18}Ne(α,p)^{21}Na [72] reaction at Louvain-la-Neuve. This reaction is part of the dominant sequence leading to the synthesis of heavy elements in X-ray bursts.

Indirect Techniques

Some experiments with radioactive beams do not involve direct measurement of the reaction occurring in astrophysical environments. Rather, experiments to determine nuclear structure information needed to calculate reaction rates can be extremely valuable for progress in the field.

Transfer Reactions Transfer (e.g., (d,p) [86], (^{3}He,^{6}He) [87], and many others) and charge-exchange (e.g., (^{3}He,t) [88]) reactions have been extensively used with stable beams to determine stellar reaction rates by populating near-threshold resonances dominating the reactions of interest and measuring their energies, spins, widths, spectroscopic factors, and other properties. By using radioactive beams, the number of reactions that can be investigated with this approach is greatly expanded. These measurements require energies typically greater than 5 - 10 MeV/u to ensure that the reaction is direct (rather than compound nuclear) and that the reaction products have sufficient energy to be detected. Such measurements typically employ a magnetic spectrograph or recoil separator to detect the heavy reaction products. Measurement of the light and heavy ions in coincidence enables the use of radioactive beams with stable isobar contaminants. One example is a measurement of the ^{56}Ni(d,p)^{57}Ni reaction at Argonne National Laboratory [89] to investigate the ^{56}Ni(p,γ)^{57}Cu reaction, important in X-ray burst nucleosynthesis. States in ^{57}Ni that are isobaric analogs to states near the ^{56}Ni + p threshold in ^{57}Cu were studied in this experiment. The (d,p) reaction can also be used to obtain level information and neutron spectroscopic factors needed to calculate neutron capture rates of neutron-rich unstable nuclei [90]. Such an approach is, for example, the only way to study (n,γ) reactions occurring in r-process nucleosynthesis. Such (d,p) measurements would require solid CD$_2$ targets or a D$_2$ gas jet target coupled to a magnetic spectrometer or recoil separator, and most likely utilize an array of charged-particle detectors near the target for proton detection. Beams of approximately 10^5 s^{-1} would give reasonable counting rates in such experiments.

Transfer reactions such as (^{14}N,^{13}C) can also be used to indirectly determine the non-resonant direct capture contribution of the reaction $A(p,\gamma)B$. The overlap of the tail of B's wavefunction with that of the A + p channel when A and p are separated at a large distance is determined. This overlap - the Asymptotic Normalization Coefficient (ANC) - is similar to a spectroscopic factor but has less dependence on the wave function in the nuclear interior [91]. This approach has been used to investigate the ^{7}Be(p,γ)^{8}B reaction for the solar neutrino problem [92] and the ^{11}C(p,γ)^{12}N reaction important for supermassive stars [93].

Scattering Another approach involves the scattering of radioactive heavy ions off of protons to discover resonances that may enhance proton capture reaction rates and to make precision measurements of resonant properties (energy, total and partial widths, spin, and parity). This technique works especially well for s-wave resonances that have proton widths greater than 1 keV. Low radioactive ion beam intensities (10^3 s^{-1}) can be used to search for resonances with the (p,p) reaction, whereas such searches are not feasible with (p,γ) reactions because the capture cross sections are orders of magnitude smaller. A scattering technique was used with a thin CH_2 target, for example, to give the first unambiguous evidence for an s-wave resonance in ^{18}Ne that dominates the ^{17}F(p,γ)^{18}Ne reaction in stellar explosions [81,94]. Whereas nine previous studies with stable beams had been unable to locate this resonance, a very low intensity radioactive ion beam (8000 ^{17}F ions/s) was sufficient to find and precisely measure the resonance energy and total width. Scattering reactions have also been used with thick targets to study the ^{13}N(p,γ)^{14}O [95], ^{18}F(p,α)^{15}O [64], and ^{19}Ne(p,γ)^{20}Na [74] reactions.

Nuclear structure measurements Mass measurements, lifetime measurements, and level density determinations are needed for nuclei near the proton dripline and neutron-rich isotopes which play a role in, for example, the p- and r-processes in supernovae and the rp-process in X-ray bursts. Although measuring these properties for all of the thousands of nuclei of involved in these processes is impossible, global calculations of these properties [21,97–99] can be significantly enhanced by making measurements on a few crucial nuclei.

Storage rings [102], ion traps [103,104], time-of-flight mass spectrometers [105], and other techniques [106,107] have all been successfully employed for measuring masses. Measurements with unaccelerated beams of radioactive, heavy, neutron-rich nuclei have been made at CERN ISOLDE [100] to learn about the properties of nuclei in the r-process path - such as ^{130}Cd [101]. Lifetime measurements for dripline nuclei important for astrophysics can be efficiently made with projectile fragmentation beams because many of the isotopes produced can be simultaneously detected at the focal plane of a separator. This technique, which enables the properties of many nuclei to be rapidly determined, has been used to investigate nuclei important to rp-process nucleosynthesis in X-ray bursts - especially those isotopes one proton richer than $N = Z$ waiting points such as ^{64}Ge and ^{68}Se. Determinations of which fragments are proton-bound or, at least, which have a lifetime longer than the $\sim$1 μs flight time through a separator, were made with very low beam intensities (as low as a few per day). Implanting these nuclides into Si detectors at the focal plane enabled lifetime measurements to be made. Specific, important examples include the GANIL measurement demonstrating that ^{69}Br is most likely proton unbound [108], and a measurement at MSU of the lifetime of ^{65}As showing that the decay is primarily via beta emission rather than by proton emission [109].

Inverse Techniques It is advantageous in some cases to measure the inverse of the reaction that occurs in astrophysical environments. This may be because of difficulty of producing the beam or target, or an unfavorable energy of the products, of the forward reaction. The Coulomb dissociation technique [110] is one inverse approach to direct capture measurements used successfully for some reactions - such as the dissociation of ^{14}O into ^{13}N + p [111,112], the inverse of the ^{13}N(p,γ)^{14}O reaction. The ^{7}Be(p,γ)^{8}B has also been studied via the dissociation of ^{8}B [113–116], and the time-inverse of the Hot pp-chain ^{11}C(p,γ)^{12}N reaction has been studied [31,117,118] using the Coulomb dissociation of a radioactive ^{12}N beam. There are, however, limitations of this technique: it only probes the ground state transition of the gamma strength; there are difficulties with gamma transitions of mixed multipolarities; and the final state products can be influenced by the intense coulomb field of the target.

Another inverse approach successfully used involves measuring the inverse (p,α) instead of the astrophysically important forward (α,p) reaction, thereby avoiding a Helium target or an unavailable beam. Important examples are the measurements of ^{17}F(p,α)^{14}O at Argonne National Laboratory [119] and Oak Ridge National Laboratory [85] to determine the ^{14}O(α,p)^{17}F rate. The contributions of excited states in the final nucleus (e.g., ^{14}O(α,p)^{17}F*) to the total rate cannot be neglected and must be determined separately, however. The ^{8}Li(α,n)^{11}B reaction occurring in inhomogeneous Big Bang nucleosynthesis is an example where excited state contributions were found to be a factor of five over contributions from the ground state [120].

FUTURE OUTLOOK AND SUMMARY

Current radioactive beam facilities have changed our understanding of a wide variety of astrophysical phenomena including novae, x-ray bursts, supermassive stars, supernovae, red giants, the Sun, the Big Bang, and others. We are, however, entering an exciting new era in science with radioactive beams: a number of new radioactive beam facilities are under construction, others are proposed, and a number of upgrades to existing facilities are planned or in progress. Facilities under construction include ISAC at TRIUMF [78] and the RIB Factory at RIKEN [121]. Upgrades in progress or planned include the coupled cyclotron upgrade at MSU [122], ARENAS3 at Louvain-la-Neuve [123], GSI [124], and REX-ISOLDE [125]. Proposals for new facilities include the E-arena at the Japan Hadron Project [126] and the Rare Isotope Accelerator (RIA) [49]. These facilities promise to provide a far greater variety of beams at higher intensities than ever before. In particular, RIA involves a new hybrid approach that may deliver ISOL-quality beams with the selectability, fast delivery time, and ease of development characteristic of PF facilities. Beams from these future facilities will enable even greater insights into a variety of fascinating astrophysical environments.

REFERENCES

1. Starrfield S. *Phys. Rept.* 311:371 (1999)
2. Wallace RK, Woosley SE. *Ap. J. Suppl.* 45:389 (1981)
3. Champagne AE, Wiescher M. *Ann. Rev. Nucl. Part. Sci.* 42:39 (1992)
4. Vanlandingham KM, et al. *Mon. Not. R. Aston. Soc.* 282:563 (1996)
5. Chin Y, Henkel C, Langer N, Mauersberger R. *Ap. J. Lett.* 512:L143 (1999)
6. Shore S, et al. *Ap. J.* 421:344 (1994)
7. Gomez-Gomar J, et al. *Mon. Not. R. Aston. Soc.* 296:913 (1998)
8. Harris MJ, et al. *Ap. J* 522:424 (1999)

9. Hernanz M, et al. *Ap. J.* 526:L97 (1999)
10. Wiescher M, et al. *Astron. Astrophys.* 160:563 (1986)
11. Jose J, Coc A, Hernanz M. *Ap. J.* 520:347 (1999)
12. Wiescher M, Görres J, Schatz H. *J. Phys.* G 25: R133 (1999)
13. Taam RE, Woosley SE, Weaver TA, Lamb DQ. *Ap. J.* 413:324 (1993)
14. Schatz H, Bildsten L, Cumming A, Wiescher M. *Ap. J.* 524:1014 (1999)
15. Schatz H, et al. *Phys. Rept.* 294:167 (1998)
16. Wiescher M, Schatz H, Champagne AE. *Philos. Trans. R. Soc. London A* 356: 2105 (1998)
17. Koike O, Hashimoto M, Arai K, Wanajo S. *Astron. Astrophys.* 342:464 (1999)
18. Arai K, Hashimoto M. *Astron. Astrophys.* 254: 191 (1992)
19. Meyer BS. *Ann. Rev. Astron. Astrophys.* 32:153 (1994)
20. Cowan JJ, Thielemann F-K, Truran JW. *Phys. Rept.* 208:267 (1991)
21. Kratz KL, et al. *Ap. J.* 403:216 (1993)
22. Fowler WA. *Rev. Mod. Phys.* 36:545 (1964)
23. Kormendy J, Richstone D. *Ann. Rev. Astron. Astrophys.* 33:581 (1995)
24. Fuller, G. priv. comm. (2000).
25. Mezzacappa, A. priv. comm. (2000).
26. Fuller G, et al. *Ap. J.* 307:675 (1986)
27. Arnould M, Norgaard, H. *Astron. Astrophys.* 42:55 (1975)
28. Wiescher M, et al. *Ap. J.* 343:352 (1989)
29. Mitalas R. *Ap. J.* 290:273 (1985)
30. Jorissen A, Arnould M, *Astron. Astrophys.* 221:161 (1989)
31. Lefebvre A, et al. *Nucl. Phys. A* 592:69 (1995)
32. Malaney RA, Mathews GJ. *Phys. Rept.* 229:145 (1993)
33. Malaney RA, Fowler WA. *Ap. J.* 333:14 (1988); 345:L5 (1989)
34. Jaffe AH, et al. http://xxx.lanl.gov/abs/astro-ph/0007333 (2000)
35. Kainulainen K, Kurki-Suonio H, Sihvola E. *Phys. Rev. D* 59:083505 (1999)
36. Gallino R, et al. *Ap. J.* 334:L45 (1988)
37. Straniero O, et al. *Ap. J.* 440:L85 (1995)
38. Kaeppeler F, Thielemann F-K, Wiescher M. *Ann. Rev. Nucl. Part. Sci.* 48:175 (1998)
39. Kaeppeler F, et al. *Ap. J.* 354:630 (1990)
40. Bahcall JN, Pinsonneault MH. *Rev. Mod. Phys.* 64:885 (1992)
41. Adelberger EG, et al. *Rev. Mod. Phys.* 70:1265 (1998)
42. Schatz H, et al. *Phys. Rept.* 294:167 (1998)
43. Arai K, Hashimoto M. *Astron. Astrophys.* 254:191 (1992)
44. Ravn HL. *Phys. Rept.* 54:201 (1979)
45. Darquennes D, et al. *Phys. Rev. C* 42:R804 (1990)
46. Alton GD, Beene JR. *J. Phys. G* 24:1347 (1998)
47. Welton RF, et al. *Nucl. Inst. Meth. B* 159:116 (1999)
48. Penttild H, et al. *Nucl. Instrum. Meth. B* 126:213 (1997)
49. http://srfsrv.JLab.org/ISOL/
50. Sherrill BM, et al. *Nucl. Inst. Meth. B* 56/57:1106 (1991)
51. Kubo T, et al. *Nucl. Inst. Meth. B* 70:309 (1992)

52. Gillibert A, et al. *Phys. Lett. B* 176:317 (1986)
53. Clerc H-G, et al. *Nucl. Instr. Meth. B* 70:265 (1992)
54. Morrissey DJ, Sherrill BM. *Phil. Trans. Royal Soc. A* 356:1985 (1998)
55. Lee MY, et al. *Nucl. Inst. Meth. A* 422:536 (1999)
56. Harss B, et al. *Phys. Rev. Lett.* 82:3964 (1999)
57. Sonzogni AA, et al. *Phys. Rev. Lett.* 84:1651 (2000)
58. Gialanella L, et al. *Nucl. Instr. Meth. A* 376:174 (1996)
59. Gialanella L, et al., *Eur. Phys. J. A* 7:181 (2000)
60. Liu Y, et al. *Bull. Am. Phys. Soc.* 45:83 (2000)
61. Decrock P, et al. *Phys. Rev. Lett.* 67:808 (1991)
62. Delbar Th., et al. *Phys. Rev. C* 48:3088 (1993)
63. Smith MS, Rehm KE. *Ann. Rev. Nucl. Part. Sci.* in preparation (2001)
64. Coszach R, et al. *Phys. Lett. B* 353:184 (1995)
65. Rehm KE, et al. *Phys. Rev. C* 52:R460 (1995)
66. Rehm KE, et al. *Phys. Rev. C* 53:1950 (1996)
67. Graulich JS, et al. *Nucl. Phys. A* 626:751 (1997)
68. Bardayan, DW, et al. *Phys. Rev. C* 62:042802(R) (2000)
69. Bardayan, DW, et al. In *Nuclei in the Cosmos VI*, ed. K. Langangke, submitted (2000)
70. Bardayan, DW, et al. *Phys. Rev. C* (2000) submitted.
71. Graulich JS, et al., *Phys. Rev. C* (2000) submitted.
72. Bradfield-Smith W, et al. In *Nuclei in the Cosmos V*, ed. N Prantzos, S Harissopulos, pp. 419. Gif-sur-Yvette, France:*Editions Frontières Conf. Proc.* (1998)
73. Rehm KE, et al. *Phys. Rev. C* 55:R566 (1997)
74. Page RD, et al., *Phys. Rev. Lett.* 73:3066 (1994)
75. Smith MS, Rolfs C, Barnes CA. *Nucl. Inst. Meth. A* 306:233 (1991)
76. Smith MS, et al. In *Stellar Evolution, Stellar Explosions and Galactic Chemical Evolution*, ed. A Mezzacappa, pp. 511. Bristol, UK:*IOP Conf. Proc.* (1998)
77. Graulich JS, et al. In *Nuclei in the Cosmos V*, ed. N Prantzos, S Harissopulos, pp. 471. Gif-sur-Yvette, France:*Editions Frontières Conf. Proc.* (1998)
78. D'Auria J. In *Nuclei in the Cosmos V*, ed. N Prantzos, S Harissopulos, pp. 435. Gif-sur-Yvette, France:*Editions Frontières Conf. Proc.* (1998)
79. Bonnet L, et al. *Nucl. Inst. Meth. A* 292:343 (1990)
80. Sellin PJ, et al. *Nucl. Inst. Meth. A* 311:217 (1992)
81. Bardayan DW, et al. *Phys. Rev. Lett.* 83:45 (1999)
82. Davinson T, et al. *Nucl. Inst. Meth. A* 288:245 (1990)
83. Davinson T, et al. *Nucl. Inst. Meth. A* 454:350 (2000)
84. Bradfield-Smith W, et al. *Nucl. Inst. Meth. A* 425:1 (1999)
85. Blackmon JC, et al. In *Nuclei in the Cosmos VI*, ed. K. Langangke, submitted (2000)
86. Champagne AE, et al. *Phys. Rev. C* 42:2730 (1990)
87. Kubono S, et al. *Phys. Rev C* 43:1821 (1991)
88. Smith MS, et al. *Nucl. Phys. A* 536:333 (1992)
89. Rehm KE, et al. *Phys. Rev. Lett.* 81:3341 (1998)
90. Kraus G, et al. *Z. Phys. A* 340:339 (1991)

91. Gagliardi CA, et al. *Phys. Rev C* 59:1149 (1999)
92. Azhari A, et al. *Phys. Rev. Lett.* 82:3960 (1999)
93. Tang X, et al. *Bull. Am. Phys. Soc.* 45:44 (2000)
94. Bardayan DW, et al. *Phys. Rev. C* 62:055804 (2000)
95. Delbar T, et al. *Nucl. Phys. A* 542:263 (1992)
96. Coszach R, et al. *Phys. Rev C* 50:1695 (1994)
97. Möller P, Nix JR, Kratz K-L. *At. Data Nucl. Data Tables* 66:131 (1997)
98. Rauscher T, Thielemann F-K, Kratz K-L. *Phys. Rev. C* 56:1613 (1997)
99. Goriely S. In *Proc. 10th Int. Symp. Capture Gamma-Ray Spectroscopy and Related Topics*, ed. S. Wender, pp. 287, New York: *AIP Press* (1999)
100. Kratz KL, et al., *Hyperfine Int.* 129:(in press)(2000)
101. Kratz KL, et al., *Z. Phys. A* 325:489 (1986)
102. Schlitt B, et al. *Nucl. Phys. A* 626:315C (1997)
103. Bollen G, et al. *Nucl. Inst. Meth. A* 368:675 (1996)
104. Savard G, et al. *Nucl. Phys. A* 626:353C (1997)
105. Siefert HL, et al. *Z. Phys. A* 349:25 (1994)
106. Mittig W, et al. *Ann. Rev. Nuc. Part. Sci.* 47:27 (1997)
107. Lunney D. In *Nuclei in the Cosmos V*, ed. N Prantzos, S Harissopulos, pp. 296. Gif-sur-Yvette, France:*Editions Frontières Conf. Proc.* (1998)
108. Blank B, et al. *Phys. Rev. Lett.* 74:4611 (1995)
109. Winger JA, et al. *Phys. Lett. B* 299:214 (1993)
110. Baur G, Rebel H. *Ann. Rev. Nucl. Part. Sci.* 46:321 (1996)
111. Motobayashi T, et al. *Phys. Lett. B* 264:259 (1991)
112. Kiener J, et al. *Nucl. Phys. A* 552:66 (1993)
113. Iwasa N, et al. *Phys. Rev. Lett.* 83:2910 (1999)
114. Motobayashi T, et al. *Phys. Rev. Lett.* 73:2680 (1994)
115. Kelley JH, et al. *Phys. Rev. Lett.* 77:5020 (1996)
116. von Schwarzenberg J, et al. *Phys. Rev. C* 53:R2598 (1996)
117. T. Motobayashi, in *ENAM98: Exotic Nuclei and Atomic Masses*, eds. B.M. Sherrill, D.J. Morrissey, C.N. Davids, *AIP Press*, New York (1998) 882.
118. T. Minemura *et al.*, RIKEN Accel. Prog. Rep. **33** (2000) 63.
119. Harss B, et al. *Phys. Rev. Lett.* 82:3964 (1999)
120. Boyd RN, et al. *Phys. Rev. Lett.* 68:1283 (1992)
121. http://www.rarf.riken.go.jp/ribf/index.html
122. http://www.nscl.msu.edu/departments/accelerator_rd/ccp/index.htm
123. http://www.fynu.ucl.ac.be/projets/rnb/
124. Henning W. priv. comm. (2000)
125. Kester O, et al. In *Proc. 15th Int. Conf. Applications of Accelerators in Research and Industry*, ed. by JL Duggan, IL Morgan, pp. 309. New York:*AIP Conf. Proc. 475* (1999); Van Duppen P, et al. *Nucl. Instrum. Meth. B* 139:128 (1998)
126. Nomura T. In *Radioactive Nuclear Beams,* ed. WD Myers, JM Nitschke, EB Norman, pp. 13. Singapore:*World Sci. Conf. Proc.* (1990); Kubono S. *Prog. Theor. Phys.* 96:275 (1996)

The Trojan Horse Method in Nuclear Astrophysics

M. Aliotta^{a1}, M. Lattuada^{b,c}, M.G. Pellegriti^{b,d}, R.G. Pizzone^{b,d},
C. Spitaleri^{b,d}, Dj. Miljanic^{e}, C. Rolfs^{a}, S. Typel^{f}, H. H. Wolter^{f}

^{a} Lehrstuhl für Experimentalphysik III, Universität Bochum - Bochum, Germany
^{b} Laboratori Nazionali del Sud - Catania, Italy
^{c} Dipartimento di Fisica e Astronomia, Università di Catania - Catania, Italy
^{d} Dipartimento di Metodologie Fisiche e Chimiche per l'Ingegneria,
Univeristà di Catania - Catania, Italy
^{e} Rudjer Boskovic Institute - Zagreb, Croatia
^{f} Sektion Physik, Ludwig-Maximilians Universität - München, Germany

Abstract.
Because of the Coulomb barrier, reaction cross sections in astrophysics cannot be accessed directly at the relevant Gamow energies, unless very favourable conditions are met (e.g. LUNA - underground experiments). Theoretical extrapolations of available data are then needed to derive the astrophysical S(0)-factor. Various indirect processes have been used in order to obtain additional information on the parameters entering these extrapolations.

The Trojan Horse Method is an indirect method which might help to bypass some of the problems typically encountered in direct measurements, namely the presence of the Coulomb barrier and the effect of the electron screening. However, a comparison with direct data in an appropriate energy region (e.g. around the Coulomb barrier) is crucial before extending the method to the relevant Gamow energy. Additionally, experimental and theoretical tests are needed to validate the assumptions underlying the method.

The application of the Trojan Horse Method to some cases of interest is discussed.

INTRODUCTION

In nuclear astrophysics, reaction cross sections between charged particles represent a key quantity for various astrophysical models.

Customarily, experimental data are given in the form of the astrophysical $S(E)$-factor $S(E) = E\sigma(E)\exp(2\pi\eta)$, where the exponential drop in the cross section due to the Coulomb barrier is explicitly taken out. Here $\sigma(E)$ represents the *bare* nuclear cross section, which is obtained from the laboratory measured cross section after

---

[1] Alexander von Humboldt Fellow

correction for the so-called electron screening effect [1]. Indeed nuclear reactions studied in the laboratory involve projectiles and targets which are usually in the form of ions and neutral atoms or molecules, respectively. The electron cloud shields the nuclear charge and leads to an enhancement in the measured cross section. In order to extract a reliable S(E)-factor at the relevant astrophysical energies, a good understanding of the electron screening effect is needed. However, the presence of the Coulomb barrier prevents a direct measurement of reaction cross sections at thermal (i.e astrophysical) energies. Alternatively, indirect methods can be used to study processes which are different from, but closely related to, those of astrophysical relevance so that additional information can be obtained.

THE METHOD AND ITS ASSUMPTIONS

The Trojan-Horse Method (THM) [2] is an indirect method based on the quasi-free breakup mechanism, involving a reaction with three particles in the final state. A given reaction of astrophysical interest, say $t(p,a)b$, can then be studied by choosing a suitable three-body reaction $A(p,ab)s$ under specific conditions and assumptions. These are the following:

a) the target nucleus A can be considered as being predominantly composed by two clusters of nucleons, i.e. $A = t + s$;

b) the interaction between projectile and target leads to the breakup of the latter into its constituent clusters;

c) cluster s does not participate in the p-A interaction and it is therefore referred to as the *spectator* (conversely, cluster t represents the *transferred* particle);

d) the relative energy in the entrance $A + p$ channel is chosen to overcome the Coulomb barrier, so as to avoid a reduction in the three-body cross section;

e) the reaction between t and p can be induced at relatively small energies, because of the Fermi motion of t inside A which compensates (at least partially) for the relative energy in the $A + p$ system.

These assumptions have two important consequences.

Firstly, both the breakup of the target and the p-t interaction can be regarded, to a good approximation, as being processes independent of one another. In a plane wave impulse approximation (PWIA) the three-body cross-section can then be factored into the product of a term describing the breakup of A and one describing the two-body p-t interaction [3]:

$$\frac{d^3\sigma}{dE_{ab}^f d\Omega_{ab} d\Omega_{Bs}} = \text{KF}\, |\Phi(\vec{p}_s)|^2 \left.\frac{d\sigma^N}{d\Omega}\right|_{pt \to ab} \qquad (1)$$

Here KF is a kinematic factor which can be calculated from experimental conditions (B standing for the ab system). The second term $|\Phi(\vec{p}_s)|^2$ is the ground state momentum distribution of nucleus A and it represents the probability of finding cluster s at momentum $\vec{p}_s$. The third term is the two-body cross section, related to that of astrophysical interest. In principle it is off-shell, but in the PWIA it is identified with the on-shell cross section.

Secondly, the fact that the transferred particle t is *hidden* inside nucleus A (hence the name of the method), together with the hypothesis that the breakup of A takes place in the nuclear interaction region, allows us to consider the process as being virtually free from the effect of the Coulomb interaction and, at the same time, unaffected by electron screening. In this respect it could be said that the derived two-body cross section represents the bare nuclear cross section, that is the key quantity needed for astrophysical applications.

A comparison with the usual cross section measured in a direct way can only be done after correcting for the effect of the reduced Coulomb barrier in the two-body reaction. In an heuristic approach based on the arguments above, this can be done by multiplying $d\sigma^N/d\Omega$ by the Coulomb penetrability. Because in the PWIA the nuclear cross section is obtained with an arbitrary normalization, direct data have to be available at least above the Coulomb barrier to carry out the normalization procedure. Indeed, the agreement between the two cross sections at energies above the Coulomb barrier (i.e. where also direct data become independent of the exponential decrease of cross section) represents a necessary condition for the application of the THM at the astrophysically relevant energies. Although the THM does not allow for absolute measurements of a reaction cross section (or alternatively of the associated $S(E)$-factor), it nevertheless provides its energy dependence, which is what is needed in the extrapolation to astrophysical energies.

The formalism of the Trojan-Horse Method has been developed in detail [4] from standard reaction theory. For the given breakup reaction the relevant T-matrix element T_{fi} is conveniently calculated in the post-form distorted wave Born approximation (DWBA). Applying the surface approximation [4], i.e. using the asymptotic form of the ab relative wave function, a relation of T_{fi} to the S-matrix elements of the two-body reaction of astrophysical interest can be established. In a further plane wave approximation for the relative motion in the initial $A + p$ and the final $B + s$ channel one obtains the cross section as

$$\frac{d^3\sigma}{dE_{ab}d\Omega_{ab}d\Omega_{Bs}} = KF \left|W(\vec{Q}_{Bs})\right|^2 \frac{16\pi^2}{(k_{pt}Q_{pA})^2} \frac{v_{ab}}{v_{pt}} \frac{d\sigma^{TH}}{d\Omega_{pt}}(ab \to pt), \qquad (2)$$

which resembles the PWIA, eq. (1), in structure. The momentum distribution of the nucleus A is replaced by $\left|W(\vec{Q}_{Bs})\right|^2$ which is the Fourier transform of the cluster ground-state wave function multiplied with the interaction V_{st} and thus directly related to $|\Phi|^2$. The argument is given by

$$\vec{Q}_{Bs} = \vec{k}_{Bs} - \frac{m_s}{m_s + m_t}\vec{k}_{pA} \qquad (3)$$

where $\hbar\vec{k}_{ij}$ denotes the relative momentum between nuclei i and j. Neglecting the binding energies of the nuclei we have $\vec{Q}_{Bs} \approx -\vec{k}_{ts}$ which is just $-\vec{k}_s$ in the target frame.

The two-body cross section

$$\frac{d\sigma^{TH}}{d\Omega_{pt}}(ab \to pt) = \frac{1}{4k_{ab}^2}\left|\sum_{lm}(2l+1)P_l(\hat{Q}_{pA}\cdot\hat{k}_{ab})\left[S_l J_l^{(+)} - \delta_{(ab)(pt)} J_l^{(-)}\right]\right|^2 \quad (4)$$

with the total (nuclear + Coulomb) S-matrix elements S_l for the reaction $a + b \to p + t$ has the form of a usual cross section except for the functions

$$J_l^{(\pm)} = k_{pt} Q_{pA} \int_R^\infty dr\, r\, j_l(Q_{pA}r)\, u_l^{(\pm)}(k_{pt}r) \quad (5)$$

which are a consequence the off-shell nature of the two-body process. In this expression spherical Bessel functions j_l and Coulomb wave functions $u_l^{(\pm)} = e^{\mp i\sigma_l}(G_l \pm i F_l)$ appear. Furthermore the momentum

$$\vec{Q}_{pA} = \vec{k}_{pA} - \frac{m_p}{m_p + m_t}\vec{k}_{Bs} \quad (6)$$

enters. Due to the surface approximation, the cutoff radius R is usually chosen as the sum of the radii of nuclei p and t. The argument of the Legendre polynomial P_l in eq. (4) is just the cosine of the cm scattering angle of the two-body reaction. From a low energy approximation of $J_l^\pm$ one can justify the heuristic approach discussed above to extract the direct cross section using the Coulomb penetrability in the case of an inelastic two-body reaction. The functions $J_l^\pm$ compensate for the Gamow factor $\exp(-2\pi\eta)$. Studying an elastic reaction with the THM requires the decomposition of the cross section $\frac{d\sigma^{TH}}{d\Omega_{pt}}$ into nuclear and Coulomb contributions in order to assess the reduction of the Coulomb scattering amplitude as compared to the free two-body scattering. This is qualitatively different from the inelastic case and simple approximations seem unavailable at the moment. Calculations for specific reactions with the more refined expression (2) are in progress.

SOME APPLICATIONS

The THM has already been applied in the PWIA formalism of eq. (1) in some cases in order to identify the quasi-free reaction mechanism, to test the method and to obtain the astrophysical $S(E)$-factor wherever the assumptions are fulfilled [5-8]. In the following sections we will report on the quasi-free ^{12}C-α elastic scattering (a test case) and on the ^{7}Li(p,α)^{4}He and ^{6}Li(d,α)^{4}He reactions (measurement of $S(E)$ and estimate of screening potential).

Typically, the experimental approach requires the coincidence measurement of the total energies of two of the three outgoing particles (say a and b) at specific angles θ_a and θ_b, so as to determine completely the kinematic properties of the final state. The choice of these angles depends on the reaction and reflects the kinematic conditions where the quasi-free process is favoured. They are called *quasi-free angles*.

Of course, other mechanisms can lead to the same final state of interest. It is therefore of primary importance to discriminate such processes, either theoretically or experimentally, from the quasi-free one, in which the reaction $p + t \to a + b$ is not affected by the presence of s.

Once the bf quasi-free mechanism has been identified and selected, a measurement of the three-body cross section allows for the extraction of the two-body cross section of interest by inverting eq.(1), provided the momentum distribution $|\Phi(\vec{p}_s)|^2$ is known.

The quasi-free ^{12}C-α elastic scattering

The ^{12}C$(\alpha,\gamma)^{16}$O reaction cross section at the Gamow energy $E_0 \sim 300$ keV is believed to be dominated by the contribution of two subthreshold states in ^{16}O located at -45 keV ($J^\pi=1^-$) and -245 keV ($J^\pi=2^+$) in the α-^{12}C center-of-mass system [1] (and references therein). An accurate knowledge of the reduced widths of these states could greatly help to constrain the extrapolation of the astrophysical $S(E)$-factor down to the Gamow peak [9]. It is worth remembering here that no direct data are available below about $E_{cm}=1$ MeV and consequently large discrepancies still remain on the extrapolated value of $S(E_0)$ [10] (and references therein). Attempts to measure these partial widths have been made (e.g. [11,12]) but the values obtained differ quite markedly from one another. Because of its features, the THM may be able to determine these important quantities. This can be done by studying, for example, the quasi-free elastic scattering ^{12}C(^{6}Li,α^{12}C)^{2}H, where the α particle participating in the scattering process is initially bound in the ^{6}Li nucleus.

As discussed in the previous section, however, the preliminary step requires a comparison between direct and indirect data around the Coulomb barrier. This work has been extensively reported in [8]. Here we shall briefly recall the main aspects which will serve as a guide to the application of the THM.

The choice of the ^{6}Li nucleus as a target has been made:

a) because of its large probability of being clustered into an α-d configuration, and

b) because the relative motion of its constituent clusters is $l = 0$, which, as we shall see, simplifies the application of the method.

Because the α-d relative motion is in an $l = 0$ state, the deuteron momentum distribution inside ^{6}Li is peaked around zero [13]. That is, $p_d = 0$ is the most probable value of momentum attained by the spectator deuteron. It is therefore natural to choose the quasi-free angles as those corresponding to $p_d = 0$ (and hence $E_d = 0$) in the laboratory system, which obviously simplifies the three-body kinematics calculations.

The angle of emission for the ^{12}C particle in the α-^{12}C center-of-mass system can be obtained by the relation [3]

$$\theta_{cm} = \arccos\frac{(\mathbf{v}_p - \mathbf{v}_t)\cdot(\mathbf{v}_C - \mathbf{v}_\alpha)}{|\mathbf{v}_p - \mathbf{v}_t|\,|\mathbf{v}_C - \mathbf{v}_\alpha|}, \qquad (7)$$

where the vectors $\mathbf{v}_p, \mathbf{v}_t, \mathbf{v}_C$ and $\mathbf{v}_\alpha$ are the velocities of the projectile, the transferred α-particle, and the two outgoing ^{12}C and α particles, respectively.

These quantities are calculated from their corresponding momenta in the lab-system. In particular, because of the quasi-free assumption, the momentum of the transferred particle is equal and opposite to that of the spectator [3]. Knowledge of the momentum and energy of the spectator in the final state provides information on the momentum and energy of the transferred particle at the time of interaction with the projectile.

In the experiment the range of spectator momenta, where the assumptions of a quasi-free breakup mechanism are justified, has been investigated in detail.

The ^{6}Li(^{12}C,α,^{12}C)^{2}H experiment was performed at the SMP Tandem Van de Graaff accelerator of the Laboratori Nazionali del Sud (LNS) - Catania, which provided a 18 MeV ^{12}C beam on a ^{6}Li-enriched LiF target. Details on the experimental setup can be found in [8]. Briefly, α's and C-particles were detected in coincidence by means of two $\Delta E - E$ telescopes, each consisting of an ionization chamber (IC) and a position sensitive silicon detector (PSD).

The first evidence of the presence of the quasi-free mechanism was found through a series of angular correlation analyses as reported in [8]. Thereafter, experimental data have been evaluated in two steps: first the experiment was described by means of a Monte-Carlo simulation using PWIA for the cross section under the hypotheses of the quasi-free mechanism (see below). This allowed the determination of the conditions where the process actually dominates. Under these conditions, the two-body excitation function was then extracted and compared with the results of a direct α–^{12}C elastic scattering measurement [14] in the energy range $E_{cm} = 2.5$–3.5 MeV.

The Monte-Carlo simulation of the experiment has been carried out under the assumption that the mechanism giving rise to the reaction is purely quasi free, so that the three-body cross section can be calculated according to eq. (1). The momentum distribution has been obtained from the Fourier transform of the ^{6}Li ground-state wave function, assuming for the α-d interaction a Woods-Saxon potential adjusted so as to reproduce the ^{6}Li binding energy. The two-body cross section entering eq. (1) was obtained from a multilevel R-matrix parametrization of direct elastic scattering data as reported in [15]. The geometrical efficiency of the experimental setup and the detection energy thresholds of both detectors have also been taken into account in the simulation. A calculation of the error in the relative α-^{12}C energy led to a value of about 70 keV.

The result of the simulation (histogram) together with the experimental data (points) are shown in Figure 1 for various values of the deuteron momentum and for a center-of-mass angle $\theta_{cm} = 120°$ ($\Delta\theta = 5°$).

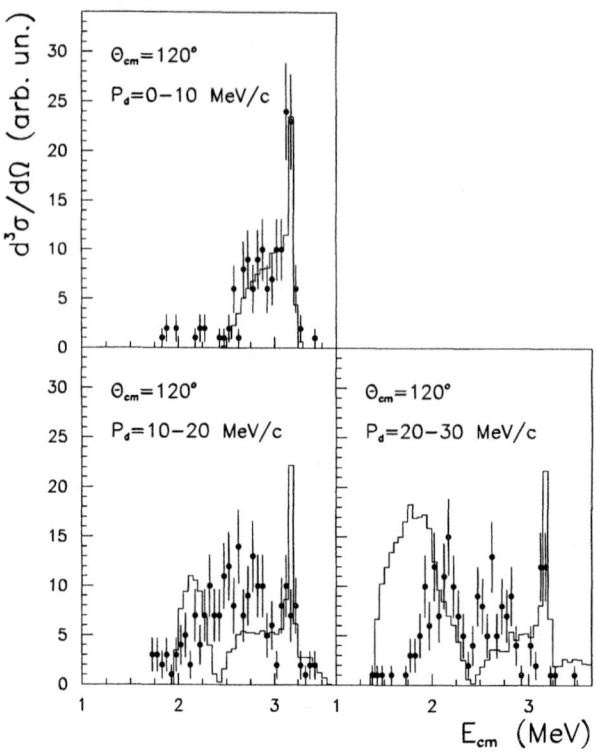

FIGURE 1. Monte-Carlo simulation of the three-body differential cross section for various ranges of the deuteron momentum

The comparison shows an excellent agreement at low spectator momenta (i.e. $p_d \leq 10$ MeV/c), suggesting that the approximations used to describe the process are plausible and give results in agreement with the experimental data. However, a substantial disagreement remains for higher p_d values. This seems to indicate the need for a better theoretical description in which distortion effects for larger p_d values are accounted for. Because of this, only events with $p_d \leq 10$ MeV/c have been considered for the extraction of the two-body cross section. Figure 2 shows the final comparison between the extracted two-body cross section and the direct data of Kettner et al. [14].

On the basis of the results obtained so far it is now possible to extend the application of the THM to lower energies and therefore explore a region much closer to that of astrophysical interest. Further work in this direction is still in progress.

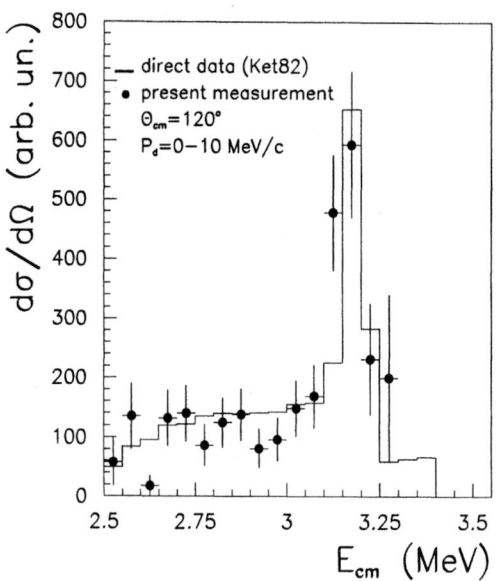

FIGURE 2. Comparison between direct (histogram) and indirect (points) excitation function for the two-body $\alpha-{}^{12}$C scattering.

The ^{7}Li(p,α)^{4}He reaction

The importance of the ^{7}Li(p,α)^{4}He reaction in astrophysics is related to the so-called Li problem. It is known that lithium (of which ^{7}Li is the most abundant isotope) is produced during the very early stages of the evolution of the universe, but that it is mainly destroyed during stellar evolution to an extent which, among other factors, depends on the rate of the ^{7}Li(p,α)^{4}He reaction (e.g. [16]). The reaction cross section measured so far, however, would lead to a lithium abundance in contrast with observations (e.g. [17]).

The THM has been used here through the ^{7}Li(d,2α)n reaction where the role of the Trojan horse is now played by the deuteron. Investigation of the quasi free process for this reaction at energies $E_{^7\text{Li}} = 28 - 48$ MeV is reported in [18]. In a more recent paper [7] the $S(E)$-factor has been derived in the energy range $E_{cm} = 0 - 0.3$ MeV at the SMP Tandem Van de Graaff accelerator of the Laboratori Nazionali del Sud - Catania.

In this case, no particle identification was required because the high Q-value (15.121 MeV) for the ^{2}H(^{7}Li,$\alpha\alpha$)n reaction makes it possible to distinguish the α's of interest from those arising from different reactions with impurities in the target.

Detection of the outgoing α particles has therefore been performed by using six silicon position sensitive detectors (PSD), three on each side of the beam axis. Calculations of quasi-free angles require that the two α's are emitted with a relative angle $\theta_{rel} \sim 90°$ almost independently of the beam energy. The detectors were centered at angles which would allow the coverage of the largest number of quasi-free angle pairs.

Coincidences were measured between PSDs on opposite sides of the beam. However, while coincidences between detectors 1 and 4, 2 and 5, 3 and 6 fulfilled the requirement $\theta_{rel} \sim 90°$, the remaining ones did not. This allowed us to cross-check the method, since in these cases no quasi-free contribution is expected.

The selected angular ranges corresponded to kinematical conditions under which the momentum p_s of the undetected spectator neutron ranges from ~ -80 MeV/c to ~ 80 MeV/c. This assures that the bulk of quasi-free contributions falls inside the investigated regions, since the FWHM of the neutron momentum distribution is about 110 MeV/c.

It was found that the ^{2}H(^{7}Li,$\alpha\alpha$)n reaction proceeded mainly through the formation of the 16.6 and 16.9 states in ^{8}Be, whose subsequent decay into two α's represented a sequential process which had to be separated from the quasi-free mechanism [7].

After subtraction of sequential-decay contributions from the experimental spectra, quasi-free data have been plotted as a function of the relative ^{7}Li-p energy E, defined in the so-called post-collision prescription as:

$$E = E_{\alpha_1,\alpha_2} - Q \qquad (8)$$

where Q (=17.346 MeV) is the Q-value for the two-body ^{7}Li(p,α)^{4}He reaction.

The two-body nuclear differential cross section was then obtained from expression of eq. (1), where $|\Phi(\vec{p}_s)|^2$ was calculated in terms of a Hulten wave function to describe the $n - p$ motion in ^{2}H [18].

The astrophysical $S(E)$-factor derived on the basis of the extracted cross section has been compared with direct data. Preliminary results show a good overall agreement at energies above about 100 keV, whereas a disagreement due to the electron screening effect is observed at lower energies. Further details on this work and on the final data analysis will be given in a forthcoming paper.

The ^{6}Li(d,α)^{4}He Reaction

It has been suggested that the abundance of ^{6}Li can provide hints as to the value of the cosmological parameter η, the baryon density Ω_B of the universe and also on the abundance of ^{7}Li. According to the inhomogeneous big bang model ^{6}Li can be destroyed through the ^{6}Li(d,α)^{4}He reaction in neutron-rich regions (e.g. [19]).

In this context, the THM was applied to the study of the reaction ^{6}Li(^{6}Li,$\alpha\alpha$)^{4}He to derive the $S(E)$ factor for the ^{6}Li(d,α)^{4}He reaction in the energy range $E_{cm} = 0 - 0.7$ MeV.

The experiment was perfomed using the EN Tandem Van de Graaf accelerator of the Institut Rudjer Boskovic in Zagreb where a ^{6}Li^{2+} beam was used to bombard an isotopically enriched ^{6}Li$_2$O target. The outgoing α-particles were detected by means of three position-sensitive silicon detectors placed on opposite sides of the beam axis at angles of −60°, 73° and 103°. These angles were selected by the three-body kinematics for the emission of the two α-particles on the assumption of a quasi-free breakup occurring either in the target or in the projectile.

After selecting data corresponding to the quasi-free mechanism with a procedure similar to that described in the previous sections, the two-body cross section was obtained on an event-by-event basis by dividing the three-body cross-section by the calculated kinematical factor and by the appropriate value of the momentum distribution as reported by [20].

The normalisation to the direct data [17] was carried out at $E_{cm} \geq 0.7$ MeV, where the barrier penetration effect is expected to be negligible and no screening effect is present. Also in this case a comparison between the $S(E)$-factor obtained from direct data and that derived with the THM proved to be very satisfactory. However, the analysis is still in progress and definite results will have to await for a forthcoming publication.

Conclusions

The work presented shows the possibility of measuring the astrophysical $S(E)$ factor at energies relevant to astrophysical applications by means of the THM. Because the extracted cross section is, to a first approximation, the nuclear part only, the effects of the Coulomb barrier can be neglected. The THM can therefore be regarded as an independent tool to estimate the effects of electron screening by comparing the direct cross section with the cross section for bare nuclei (from the THM). However, because of the numerous assumptions underlying the method, a deeper theoretical understanding of the physics is required to validate the method. Much work in this direction is currently underway.

REFERENCES

1. C.E. Rolfs and W.S. Rodney, *Cauldrons in the Cosmos*, The University of Chicago Press - Chicago, 1988.
2. G. Baur, *Phys. Lett. B* 178 (1986) 135.
3. A.K. Jain et al., *Nucl. Phys. A* 153 (1970) 49.
4. S. Typel and H.H. Wolter, *Few-Body Systems*, (2000) accepted for publication.
5. S. Cherubini et al., *Ap. J.* 457 (1996) 855.
6. G. Calvi et al. *Nucl. Phys. A* 621 (1997) 139.
7. C. Spitaleri et al., *Phys. Rev. C* 60 (1999) 055802.
8. C. Spitaleri et al., *Eur. Phys. J. A* 7 (2000) 181.
9. T.A. Tombrello et al., *Essays in Nuclear Astrophysics*, Cambridge University Press, 1982.
10. R.E. Azuma et al., *Phys. Rev. C* 50 (1994) 1194.
11. F.D. Becchetti et al., *Nucl. Phys. A* 305 (1997) 293 and 313.

12. F.D. Becchetti et al., *Nucl. Phys. A* 344 (1997) 336.
13. M. Lattuada et al., *Nuovo Cimento A* 83 (1984) 151.
14. K.U. Kettner et al., *Z. Phys. A* 308 (1982) 73.
15. R. Plaga et al., *Nucl. Phys. A* 465 (1987) 291.
16. M. Audoze, *Nucleosynthesis and Chemical evolution*, Obs. Publ. Geneva, p. 429, 1986.
17. S. Engstler et al., *Z. Phys. A* 342 (1992) 471.
18. M. Zadro et al., *Phys. Rev. C* 40 (1989) 181.
19. K.M. Nollet et al., *Phys. Rev. C* 56 (1989) 1114.
20. M. Lattuada et al., *Z. Phys. A* 330 (1988) 1848.

First Result of the Cross Sectional Measurement of ^{3}He-^{3}He Solar Reaction in OCEAN

T. Itahashi, N. Kudomi, K. Kume, K. Takahisa, S. Yoshida,
H. Ejiri, H. Toki, and Y. Nagai

Research Center for Nuclear Physics, Osaka Univ., Ibaraki, Osaka, Japan

M. Komori

Dept. of Phys. Facl. of Sci. Osaka Univ., Toyonaka, Japan

H. Ohsumi

Dept. of Culture and Education, Saga Univ., Saga, Japan

Abstract. The first result in OCEAN measurement of the fusion reactions ^{3}He(^{3}He,2p)α at the energy of 40 to 50 keV by means of a low-energy, high current accelerator are reported. The accelerator in this facility can produce an intense beam of ^{3}He^{1+} and ^{3}He^{2+} ions of more than 1mA. A detection efficiency for proposed detector assembly of ΔE-E counter telescope is simulated with GEANT program and it expects a detection efficiency about 10% for the two proton coincidence for ^{3}He+^{3}He→2p+α reaction. The accuracy of Monte Carlo program was checked by D(^{3}He,p)α reaction by replacing the target gas to deuterium.

INTRODUCTION

A high brightness ion source and a precise low energy beam accelerator are indispensable tools in the study of fusion reactions in nuclear astrophysics. When a gas target is combined with a powerful ion source, such as a proton, deuteron, or helium isotope ion source, the necessary conditions appropriate for the measurement of extremely low cross-section events in this region can be fulfilled. Although the fusion reaction, p+p→d+e+ν, is the operative reaction in the solar combustion of hydrogen, as well as the initial reaction in the chain-reaction for producing photons and neutrinos, the rate of p +p→d+e+ν is too slow ($\sim 10^{-52}$cm^2) to be measured experimentally at 6 keV, the energy range of the actual solar reaction. Of the

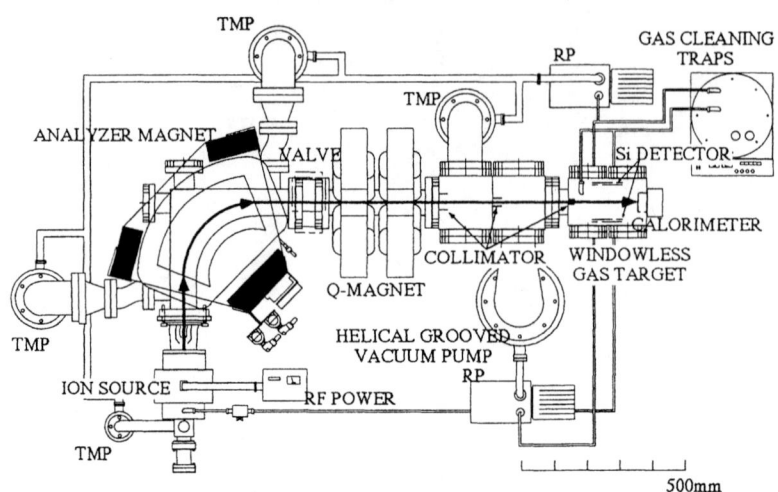

FIGURE 1. Total system of a compact accelerator, beam transport and gas target for experiments in nuclear astrophysics.

reactions that follows the basic fusion, such as d+p→^{3}He+γ, ^{3}He+^{3}He→2p+α, and ^{3}He+α→^{7}Be+γ, we have focused on the cross-section measurement of the ^{3}He+^{3}He reaction at the effective energy E_{cm}=17-27 keV. The reaction manifests the so-called neutrino problem in the sun and can be used to verify the standard solar model [1]. Currently the LUNA group in the Laboratori Nationali del Gran Sasso (LNGS) has presented data down to 20.7 keV center of mass energy [2]. The present paper describes the construction of a compact ion accelerator facility and results of the first measurement in the energy region of 40keV to 45keV.

EXPERIMENTAL APPARATUS

The experimental apparatus, OCEAN (**O**saka Univ. **C**osmological **E**xperimental **A**pparatus for **N**uclear Physics), consists of (1) a powerful ion source that provides an intense current of ^{3}He^{1+} or ^{3}He^{2+} more than 1 mA at incident energies of 30-50 keV, (2) a low-energy beam transport with good transmission, (3) a windowless gas target and a circulation/purification system, (4) a reliable calorimeter, (5) detectors for reaction identification, and (6) an electronics and data acquisition system. The layout of this accelerator is shown in Fig. 1

A Ion source and extraction electrodes

An intense ion source that can produce ^{3}He^{1+} and ^{3}He^{2+} ions is essential for the present study. The NANOGAN^{TM} was obtained from PANTECH, it confines high-temperature electrons produced by the electron cyclotron resonance (ECR) and in its assembled form becomes an ECR ion source at 50 kV potential with 10 GHz, 200 watt RF generator (CPI,VZX-6383G5). The design of multi-electrodes for ion extraction has been completed, as has a test model based on the design [4–6]. Recently, we obtained the analyzed current of ^{3}He^{1+}, with less than 10 watts for RF power. The total current obtained so far was 3010 μA at the source extraction and 1203 μA ^{3}He^{1+} at target position. We have further investigated the operational conditions in order to get a larger current($\geq$100 μA) of ^{3}He^{2+} ions. It is also helpful to study the reaction in a wider energy range [7]. The required energy range of ^{3}He ions should be between 30 to 50 keV, in which the astrophysical S-factor data for ^{3}He+^{3}He fusion reaction can be deduced.

B Low energy beam transport

Despite the fluctuations of the beam, a nearly invariant beam form could be achieved at the exit using the multi-electrodes extraction system. In addition, from a source to a target beam transport should be designed to maintain a high transport efficiency and other desirable beam qualities and thereby to allow precision measurement of the rare nuclear reaction ^{3}He+^{3}He→2p+α. To maintain the minimum collimator aperture, we calculated the dimension of the beam at the target position by varying the parameters of elements and drift lengths to create smaller dx and dy. The beam transmission efficiency from the ion source through the target is about 30%.

C Window-less gas target

The windowless gas target for a study of ^{3}He+^{3}He reaction consists of differential pumping and gas recirculation and purification system (Fig. 2). For ^{3}He gas target of less than the pressure of 1mbar in the chamber, the pumping system should be composed of the several stages between the target chamber and beam transport. Thus we prepared a helical grooved vacuum pump as a main pump for evacuating the gas flow not only at the viscose region but also at the higher vacuum region. The recirculation system consists of purification via a cryo-pump without charcoal adsorbent, which is similar system as a cryo-trap technology without a consumption of liquid nitrogen. From the experimental study, we realized that the pressure at the target chamber was stayed 3.1($\pm$0.1) Torr for about four days.

As pointed out by Rolfs et al. [2,9,7], the deuterium contamination both in target and in the beam resulting from the water vapour is a crucial problem for obtaining the low energy data because d+^{3}He reaction-cross section is 10^6 times higher than that of ^{3}He+^{3}He reaction. Therefore, we have measured the deuterium contamination in commercial ^{3}He gas by detecting HD$^+$ separately with Accelerator Mass Spectrometry(AMS). The experiment was carried out using the RCNP K=140 AVF

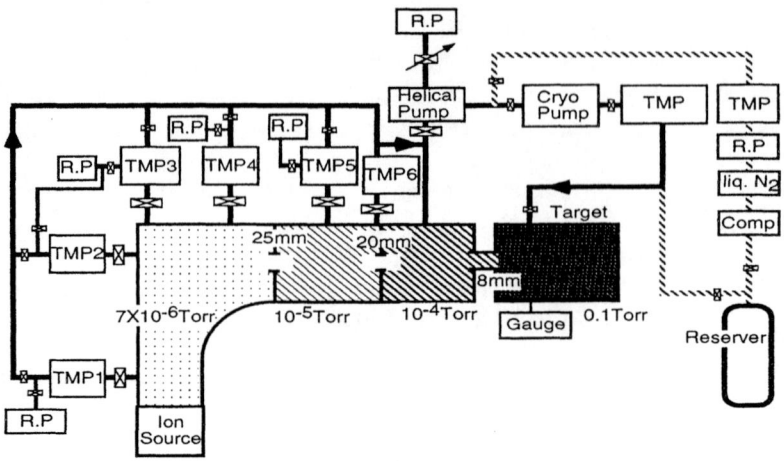

FIGURE 2. Total system of a window-less gas target evacuation and recirculation.

cyclotron. The cyclotron accelerator and the NEOMAFIOS ion source were operated only for the experiment on the beam injection line to the post accelerator (RCNP Ring cyclotron) [8].

On the other hand, the deuterium contamination in target was also estimated as follows. Total system was evacuated and the circulation and cryo-pump was operated. Then the pressure at target chamber was 1.2×10^{-2} Torr, and at helical pump was 7.6×10^{-7} Torr, respectively. The H_2O component in the residual gas was measured by quadrupole mass spectrometer, and this experiment indicates that the H_2O component was about 20 %. Assuming that the amounts of H_2O is the same and abundance of deuterium is same as natural abundance(0.014 %), even in the pressure of 0.3 Torr of ^{3}He gas, we can deduce the deuterium contamination(D_2O) is less then the order of ppm. This will be satisfactory for this measurement as will discussed below. In addition to this study, the amounts of deuterium contamination in the gas target was also evaluated by ^{3}He+^{3}He reaction experiment, by detecting the 14.7MeV proton, and was found to be about 0.1ppm.

D Calorimeter

The beam calorimeter was designed and fabricated so as to determine the number of incident particles. Charge integration with a usual Faraday cup and a commercial current integrator is difficult when a gas target is used, because of charge exchange effects in the ion beam. Under present experimental conditions, such as a gas target with a pressure of 0.3 Torr, the calorimetric method is crucial.

In the present calorimeter a heat flux sensor (OMEGA HFS-3) is used to measure the heat transfer from a hot part to a cold part. For precise measurements of the

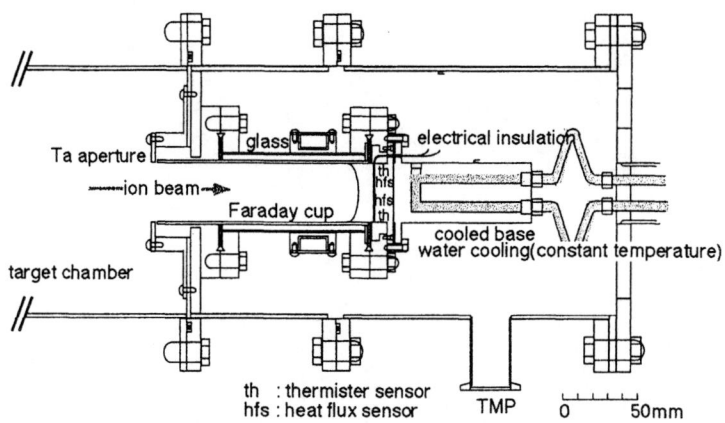

FIGURE 3. Cross sectional view of the calorimeter.

heat transfer, we made use of the following conditions: 1.)In vacuum, the current of the ion beam can be measured by standard charge integration where the calorimeter cup serves as a Faraday cup. 2.)To reduce the conduction and convection losses of heat, the Faraday cup and heat sink are installed in a vacuum chamber. 3.)The heat capacity of the calorimeter should be reduced as much as possible in order to get better time response. With present heat flux sensor, temperature differences of less than 0.001 °C can be detected easily. Therefore the temperature of the heat sink of the calorimeter should be stabilized to better than 0.1 °C. As shown in Fig. 3, the calorimeter consists of a solid copper heat sink (100mm length, 38mm diameter) with water channels, and a Faraday cup(140 mm, 38 mm diameter, 1.5 mm wall) in front of the heat sink. Between the copper base and the thin plate, two heat flux sensor are sandwiched with thermister temperature sensors. These are originally insulated electrically. As shown in Fig. 3, thermister temperature sensors are also located to measure the temperature of an ambient or Faraday cup base and the cooled heat sink. These are installed in a stainless steel pipe (40 mm length, 10.5 cm diameter), which can be evacuated by a small turbo-molecular pump.

A test was made by ^{3}He^{2+} beam of 40kV. The relation between beam current(I) and heat flux(H) can be written as,

$$I \cdot \delta t = k_1 \cdot H \cdot \delta t + C \cdot \delta T, \tag{1}$$

where T is the temperature of the calorimeter and C is the heat capacitance. The term $C \cdot \delta T$ shows the temperature of the calorimeter depends on the incident beam current. Thus, if the intensity I is changed, the converted heat is both used to heat the calorimeterand is transfered to the cold base. The temperature difference between front and cold base is larger, the transferred heat may also be large, thus the second term $C \cdot \delta T$ can be rewritten as $k_2 \cdot \delta H$. Thus the equation (1) can be written as,

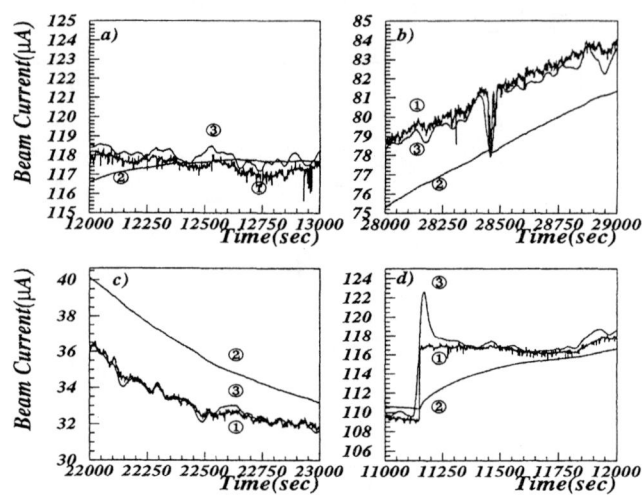

FIGURE 4. Beam current, measured by ① Faraday cup, ②, calculated from $k_1 \cdot H$ and ③ $k_1 \cdot H + k_2 \cdot \delta H/\delta t$ as function of time. Measurements were made for a)stable beam current b)increasing beam current c)decreasing. beam current d)for beam current step.

$$I = k_1 \cdot H + k_2 \frac{\delta H}{\delta t}. \tag{2}$$

In this test, the beam current was measured by a Faraday cup, as a gas pressure of 10^{-6} Torr. The beam current was calculated from the HFS output by comparison to Faraday cup value. The parameter k_1 and k_2 can be determined as follows.

(1)*Parameter k_1* : If the system is stable, i.e., incident beam I is and T is stable, the second term of equation (2) can be ignored. In this condition, the parameter k_1 can be derived. (2)*Parameter k_2:* Parameter k_2 can be also derived once parameter k_1 is derived as in (1). Here, the term $\delta H/\delta t$ was estimated for averaged time scales of 3, 7, 15 and 30 sec, It was found that the scales were setting 30 sec. (3)*Comparison with beam current* : Fig. 4 shows the beam current as a function of time measured by those methods using a Faraday cup, $k_1 \cdot H$, and $k_1 \cdot H + k_2 \frac{dT}{dt}$ with different conditions. If the beam current is stable, slowly increased or decreased, calculated currents measured by HFS well reproduce the one measured by the Faraday cup as shown in Fig. 4 a), b) and c)). On the other hand, if the beam current is suddenly changed(Fig. 4 d), calculated currents over-sing of the one measured by HFS. Further improvements of this system are necessary. Fig 5 shows the accuracy of the calculated beam current and shows the accuracy of beam current of HFS is about 2%.

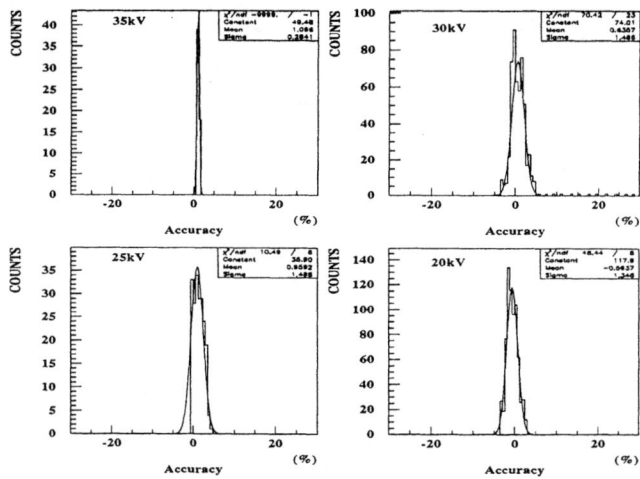

FIGURE 5. The accuracy in the form of $(I(HFS) - I(FC))/I(FC)$, for the calculated beam current where I(xx) denotes the beam current measured by HFS and Faraday cup(FC).

E Detector Configuration and Simulation

In order to detect the reaction particles by ^{3}He+^{3}He reaction, where Q-value is 12.86 MeV, we should install the counter telescopes which surrounds the gas target. Four ΔE-E telescopes placed around the beam axis are planned for the two proton coincidence measurement. The ΔE and E detectors in each telescopes have an active area 2500 mm^2, the ΔE detector has a thickness of 140 μm and the E detector has a thickness of 1500 μm. We are employing the Monte Carlo calculation with GEANT3 program code to find the optimum detector set up for an efficient and background free measurement. The expected ultra rare reaction rate is around a few events per day or less, and the typical single background rate of the silicon detectors is one event per hour or more. To remove such accidental events, two proton coincidence is required for the identification of the present reaction. The estimated counting rates for true and fake signal are summarized in Table 1. Here the fake events in the condition of pp-coin are contributed from accidental coincidence of two protons from two events of ^{3}He(D,p)^{4}He reaction. Here, the accuracy of Monte Carlo calculation was checked by ^{3}He+D reaction by reserving the deuteron as a target gas. This reaction is very useful to check the experimental situation and simulation code as the followings. 1.) The generated proton from ^{3}He+d reaction has energy of 14.7 MeV, thus the response of the detector is simple. 2.) The energy of proton from this reaction is in the same range of ^{3}He+^{3}He reaction. 3.) The cross section of ^{3}He+d reaction is six orders of magnitude larger than the ^{3}He+^{3}He reaction, thus the larger statistics is available. From this test

TABLE 1. The expected counting rates for $^3\text{He}(^3\text{He},2p)^4\text{He}$ and $^3\text{He}(D,p)^4\text{He}$ reaction. p-single indicates that at least single proton is detected in detector and pp-coin. indicates that the two proton coincidence is required in off-line analysis, respectively.

	$^3\text{He}+^3\text{He}$		$^3\text{He}+\text{D}$	
E_{cm}	p-single	pp-coin	p-single	pp-coin
$100\mu\text{A}$ $^3\text{He}^{2+}$				
50	1.2×10^4	4.0×10^3	4.3	4.3×10^{-5}
40	1.2×10^3	4.0×10^2	1.9	1.9×10^{-5}
30	3.7×10	1.2×10	0.5	0.5×10^{-5}
$1000\mu\text{A}$ $^3\text{He}^+$				
25	6.0×10	20	4.5×10^{-1}	4.5×10^{-6}
22	9.0	3.0	2.1×10^{-1}	2.1×10^{-6}
20	2.0	7.0×10^{-1}	1.2×10^{-1}	1.2×10^{-6}

experiment, the systematic error from this part is about 3%.

MEASUREMENT OF THE REACTION AND RESULTS

The several conditions for the measurement of $^3\text{He}+^3\text{He}$ reaction are summarized in Table 2. Fig. 6(Left) shows the obtained energy distribution between ΔE and E counter, and fig. 6(Right) shows the accepted region of $^3\text{He}+^3\text{He}$ reaction. This region was estimated as the followings. 1)The real $^3\text{He}+^3\text{He}$ reaction was generated by Monte Carlo simulation. 2)The Background contribution from $^3\text{He}+\text{D}$ reaction

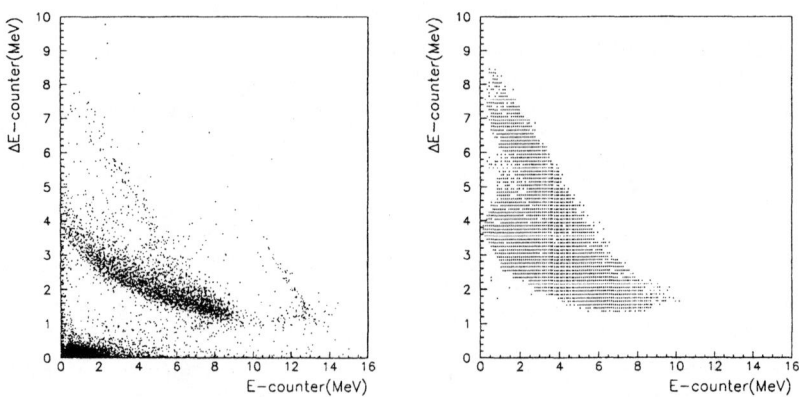

FIGURE 6. (Left)Energy distribution of $^3\text{He}+^3\text{He}$ reaction for E and ΔE-counter. (Right)Estimated effective area of E-ΔE distribution, for $^3\text{He}+^3\text{He}$ reaction.

TABLE 2. Conditions for the measurement of ^{3}He+^{3}He reaction.

E_{cm}(keV)	45.3	43.3	41.3	39.3
real time(sec)	94910	80404	80081	84944
live time(sec)	94857	80357	80034	84892
beam current(μA)	103	90.1	98.7	86.4
target press.(Torr)	7.51×10^{-2}	6.79×10^{-2}	6.79×10^{-2}	7.31×10^{-2}
target temp.(°C)	27.1	27.3	27.1	27.0
^{3}He+^{3}He(counts)	2426	1001	693	408
^{3}He+d(counts)	151	56	59	42
BG ^{3}He+d(counts)	1.42	0.53	0.55	0.39
BG other(counts)	14.4	12.2	12.1	12.9
cross section(barn)	1.43×10^{-8}	9.03×10^{-9}	5.59×10^{-9}	3.39×10^{-9}
s-factor(MeV·b)	5.02	5.14	5.32	5.61
Total systematic error = 3.6%				

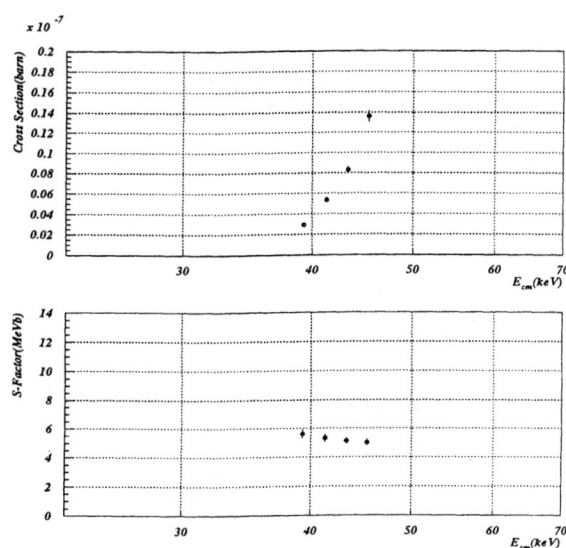

FIGURE 7. Cross section and S-Factor from this measurement.

was generated by Monte Carlo simulation, and those of cosmic rays and so on, were obtained by the measurement of experimental background run. 3)ΔE and E distribution was divided into 16000 points, and signal to noise ratio were evaluated by taking into account above three component in each point. From this procedure, cross sections for this reaction and S-factors were obtained. Fig. 7 shows the obtained cross sections and S-factors.

ACKNOWLEDGMENT

This work was supported by the Grant-in-Aid of Scientific Research, Ministry of Education, Science, Culture and Sports.

REFERENCES

1. J. N. Bahcall, et al., Rev. Mod. Phys. 54 (1982) 767.
2. M. Junker, et al., Phys. Rev. C57 (1998) 2700 ; C. Arpesella et al., Phys. Lett. B389 (1998) 452 ; U. Greife et al., Nucl. Inst. Meth. A350 (1994) 327.
3. H. Ohsumi, et al., RCNP Annual Report, RCNP (1995), p. 175.
4. T. Itahashi, et al., Rev. Sci. Inst. 69 (1998) 1032 ; T. Itahashi, et al., The 11th Symp. on Acc. Sci. and Tech., Hyogo, Japan, p. 45.; T. Itahashi, et al., Rev. Sci. Inst. 71 (2000) 1075.
5. P. Sortais, et. al., Proc. of the 12th Intern. Workshop on ECR ion sources, RIKEN, Japan (1995) p. 45.
6. K. Takahisa et al., 24th INS Symposium on ECR Ion sources and their Applications, April 25-27, RIKEN (1995) p.115, and *RCNP Annual Report*, RCNP (1994) p.190.
7. G. Fiorentini, et al., Z. Phys. A350, (1995) 289.
8. K. Langanke, et al., Phys. Lett. B369 (1996) 211.
9. U. Greife, et al., Z. Phys. A351, (1995) 107.
10. Proc. of the 19th INS Symp. Cooler Rings and Their Applications, Tokyo, Japan (1990)
11. CH. Thomann and J.E. Benn, *Nucl. Instr. and Meth.* **138**, 293(1976); A.E. Vlieks, et al., *Nucl. Instr. and Meth.* **213**, 291(1983).
12. J.M. Nitschke, *Nucl. Instr. and Meth.* **206**, 355(1983).

New measurements of astrophysical (γ,n) reaction rates using a quasi-thermal photon bath from bremsstrahlung

K. Vogt, M. Babilon, W. Bayer, K. Denefleh, J. Enders[1],
D. Galaviz, T. Hartmann, C. Hutter, P. Mohr, S. Volz, A. Zilges

*Institut für Kernphysik, Technische Universität Darmstadt,
Schlossgartenstrasse 9, D-64289 Darmstadt, Germany
email: vogt@ikp.tu-darmstadt.de*

Abstract. The astrophysical reaction rates for the (γ,n) photodisintegration reactions of ^{204}Pb, ^{196,198,204}Hg, and ^{197}Au will be determined using a quasi-thermal photon bath which is produced from bremsstrahlung. Here we present the first part of this experiment. The reaction ^{197}Au(γ,n)^{196}Au is discussed in detail because this reaction was measured in several experiments using different experimental techniques.

I INTRODUCTION

The bulk of heavy nuclei with masses above $A \approx 60$ has been synthesized by neutron capture reactions in the astrophysical s- and r-processes. However, there are about 30 neutron-deficient stable nuclei in this mass region which cannot be produced by neutron capture. These so-called p-nuclei have typically low natural abundances of about 0.01 to 1 % [1].

The p-nuclei have been synthesized by photon-induced reactions in the astrophysical γ-process. The oxygen- and neon-rich layers of type II supernovae are good candidates for the astrophysical site of the γ-process. Typical parameters are temperatures of $(2-3) \times 10^9$ K, densities of $\rho \approx 10^6$ g/cm^3, and time scales τ in the order of one second [2,3]. The γ-process starts from seed nuclei synthesized earlier in the s- and r-processes. The most important reactions for the γ-process are (γ,n) and (γ,α) reactions. (γ,p) reactions are suppressed by the Coulomb barrier; note that the proton binding energy is typically much higher than the α binding energy for heavy neutron-deficient nuclei.

Up to now practically no experimental data exist for the extended reaction network of the γ-process [3]. The huge amount of existing (γ,n) data [4,5] has mainly

---

[1] present address: National Superconducting Cyclotron Laboratory, Michigan State University, South Shaw Lane, East Lansing, MI 48824 - 1321, U.S.A.

been measured around the giant dipole resonance (GDR) and does not cover the astrophysically relevant energy range close to the reaction threshold [6], and no data have been measured for the p-nuclei because of their low natural abundances. The nucleosynthesis calculations which use huge reaction networks depend on theoretical cross sections which are calculated using the statistical model.

The energy distribution of a thermal photon bath at a temperature T is given by the Planck distribution

$$n_\gamma(E,T) = \left(\frac{1}{\pi}\right)^2 \left(\frac{1}{\hbar c}\right)^3 \frac{E^2}{\exp(E/kT) - 1} \qquad (1)$$

where $n_\gamma(E,T)$ is the number of γ-rays at energy E per unit of volume and energy interval. In a photon-induced reaction B(γ,x)A the distribution leads to a temperature dependent decay rate $\lambda(T)$ of the initial nucleus B

$$\lambda(T) = \int_0^\infty c\, n_\gamma(E,T)\, \sigma_{(\gamma,x)}(E)\, dE \qquad (2)$$

with the speed of light c and the cross section of the γ-induced reaction $\sigma_{(\gamma,x)}(E)$. Obviously, λ is also the production rate of the residual nucleus A.

The integrand of Eq. (2) is given by the product of the steeply decreasing photon density $n_\gamma(E,T)$ and by the increasing cross section $\sigma_{(\gamma,x)}(E)$. This leads to an integrand which has its main contributions in a very narrow energy window around $kT/2$ above the neutron threshold with a typical width of about 1 MeV. This fact

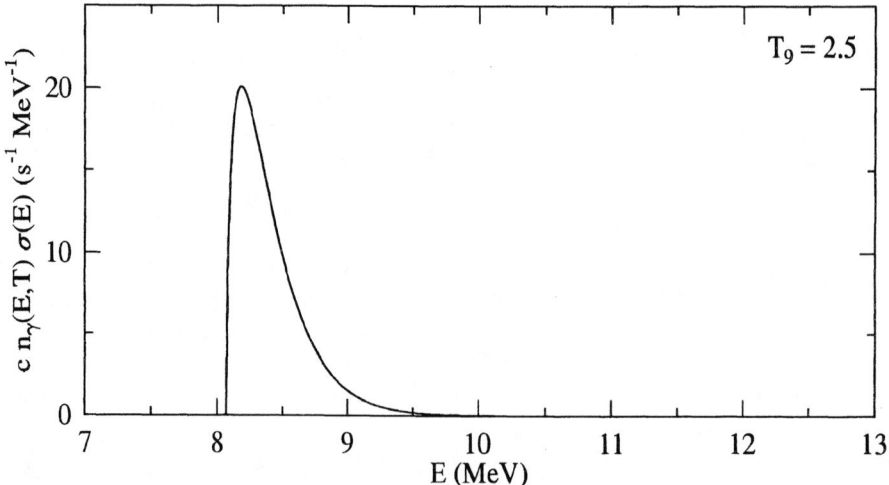

FIGURE 1. Gamow-like window for the ^{197}Au(γ,n)^{196}Au reaction. The integrand of Eq. (2) is shown for a temperature $T_9 = 2.5$. The cross section is taken from the Lorentzian GDR parametrization of [7] with an additional threshold behavior $\sigma \sim \sqrt{(E - E_{\text{thr}})}$ which was matched to the GDR parametrization 1 MeV above the threshold E_{thr}. The reaction rate λ is given by the integral over the shown curve: $\lambda(T_9 = 2.5) = 9.4\,\text{s}^{-1}$.

is shown in Fig. 1 for the ^{197}Au$(\gamma,n)^{196}$Au reaction at a temperature of $T_9 = 2.5$. Further details about this Gamow-like window for (γ,n) reactions can be found in Ref. [6].

II EXPERIMENTAL PROCEDURE

In a recent paper [8] we have shown that the thermal photon bath at temperatures $T_9 = 2 - 3$ can be simulated by a careful superposition of bremsstrahlung spectra with different endpoint energies. This quasi-thermal spectrum was used to determine astrophysical (γ,n) reaction rates for several platinum isotopes by the photoactivation technique.

The experiments are performed at the superconducting electron accelerator S-DALINAC installed at Technische Universität Darmstadt [9]. The high current electron beam of the injector provides electron beams with energies up to about 10 MeV and currents up to 50 μA. The electrons are stopped in a massive copper target. The bremsstrahlung passes a massive copper collimator and hits the target roughly 1.5 m behind the bremstarget. Photon fluxes up to $10^5/(\text{keV s cm}^2)$ can be obtained at the target position. The incoming photon flux is monitored by measuring well-known excitations in the ^{11}B(γ,γ') reaction simultaneously with the activation. The photon scattering setup is described in [10]. After the activation the samples are mounted in front of a high-purity germanium (HPGe) detector to determine the number of (γ,n) reactions from the γ-ray activity of the sample.

In the present experiment we will measure the (γ,n) reaction rates of ^{204}Pb, ^{196,198,204}Hg, and ^{197}Au. Reaction rates of lead, mercury, and platinum are the basic ingredients for the calculation of the production factors of the two p-nuclei ^{190}Pt and ^{196}Hg. The ratio of these two production factors varies within a factor of 30 from $r \approx 1.5 - 3.0$ [2] to $r \approx 0.10 - 0.15$ [11] in different nucleosynthesis calculations.

As first part of the experiment three samples consisting of 149 mg Au, 2616 mg HgS, and 12.91 g Pb were activated simultaneously for roughly 12 hours at an endpoint energy $E_0 = 10$ MeV and a beam current of about 25 μA. A typical γ-ray spectrum from the activated gold sample shows decay lines of ^{196}Au from the ^{197}Au$(\gamma,n)^{196}$Au reaction (Fig. 2). Decay lines from the nuclei ^{196}Au, ^{203}Pb, ^{195,197,203}Hg, and from isomers in ^{204}Pb and ^{199}Hg could be clearly identified in the γ-ray spectra from the different samples. The half-lives have been measured and show good agreement with the adopted values.

In this first part of the experiment the samples were mounted directly behind the bremstarget at a distance of about 6 cm. At this position the photon flux is roughly a factor of 300 higher than at the usual target position behind the collimator. This ratio has been verified experimentally by activating one platinum sample at this position and comparing the yield to the usual target position. Photon spectra in front of and behind the collimator are shown in Fig. 3. The spectra were calculated using the GEANT simulation code [12]. It has to be pointed out that discrepancies

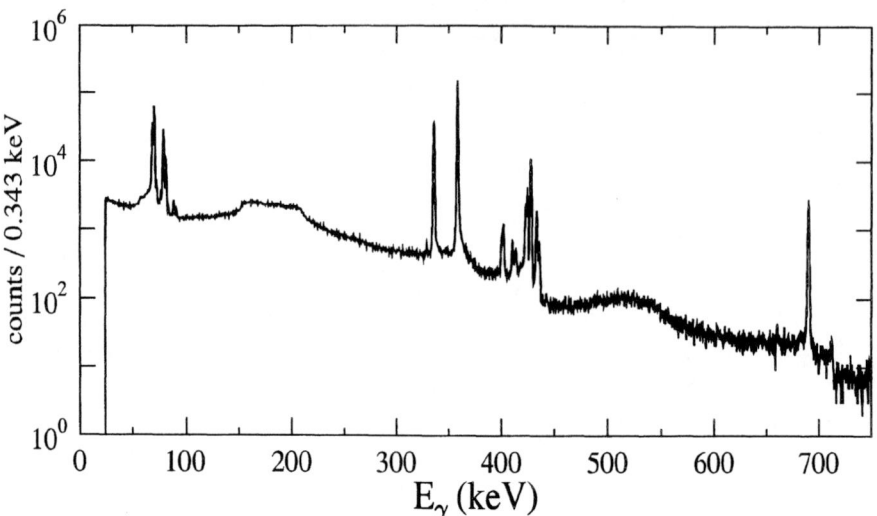

FIGURE 2. Photon spectrum from the decay of ^{196}Au, activated in the ^{197}Au$(\gamma,n)^{196}$Au reaction. All shown lines come from the ^{196}Au decay. Because of the small distance between the activated sample and the HPGe detector one can also see several sum lines from the coincident emission of γ-rays (356 keV, 333 keV) and X-rays ($\approx$ 70 – 80 keV).

between the GEANT simulations and the experimentally determined photon flux have been found especially close to the endpoint energy of the spectra [13].

The activation experiments in front of the collimator have one serious drawback: there is no possibility to normalize the incoming photon flux. The possibility of relative measurements will be discussed in Sect. III. Nevertheless, such an experiment gives valuable information. Because of the higher photon flux, a large number of (γ,n) reactions occurs in the target, and the higher activity can be used to measure the relevant half-lives with high accuracy. Especially for the residual nuclei of the (γ,n) reactions from the rare p-nuclei the half-lives are only known with limited precision [14]. The necessary amount of target material at the usual target position can be estimated from the signal-to-background ratio in the activation spectrum. The purity of the target can be checked by the presence or absence of background lines. Finally, one gets a rough estimate on the photodisintegration cross section.

III THE REACTION ^{197}Au$(\gamma,n)^{196}$Au

The reaction ^{197}Au$(\gamma,n)^{196}$Au has been chosen for several reasons. ^{197}Au is the only stable gold isotope, and therefore large amounts of isotopically pure target material are available. The ^{197}Au$(\gamma,n)^{196}$Au cross section has been measured around the GDR by Refs. [7,15,16] using quasi-monoenergetic photons from positron annihilation. The data are shown in Fig. 4 together with the result of a conventional bremsstrahlung experiment [17].

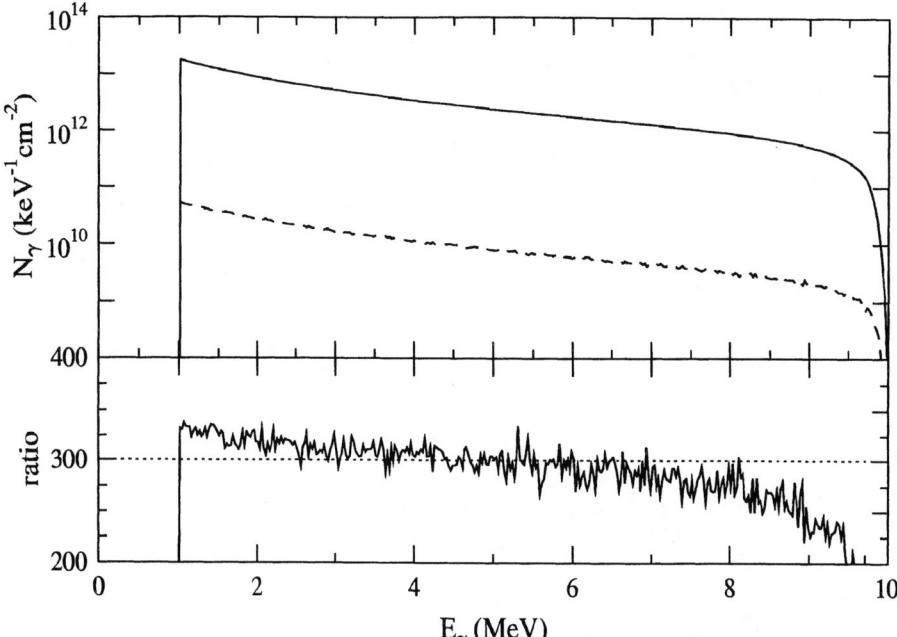

FIGURE 3. Upper: bremsstrahlung spectra calculated from GEANT simulations in front of (full line) and behind the collimator (dashed line). Lower: ratio between the photon flux in front of the collimator and at the usual target position behind the collimator.

Recently, this reaction has been measured using new types of photon sources. In a gold sample which was hit by a petawatt laser decay lines of ^{196}Au from the ^{197}Au(γ,n)^{196}Au reaction could be detected [18], and the ^{197}Au(γ,n)^{196}Au reaction was measured directly using a monochromatic laser-induced Compton backscattering γ-ray source [19]. This new technique is also described in [20].

In our activation experiment with bremsstrahlung photons the experimental yield is given by

$$Y \sim \int N_\gamma(E)\, \sigma(E)\, dE \qquad (3)$$

with the photon flux N_γ and the (γ,n) cross section σ. If one assumes a typical threshold behavior of the cross section

$$\sigma(E) = \sigma_0 \times \sqrt{(E - E_{\text{thr}})/E_{\text{thr}}} \qquad (4)$$

one can deduce the constant σ_0 directly from the measured yield. As pointed out above, there are relatively large uncertainties in the determination of the photon flux N_γ, and it is not clear whether the assumption in Eq. (4) holds in the measured energy range from $E_{\text{thr}} = 8.071$ MeV up to the endpoint energy $E = 10$ MeV. Additional uncertainties come from the self-absorption of γ-rays in the relatively

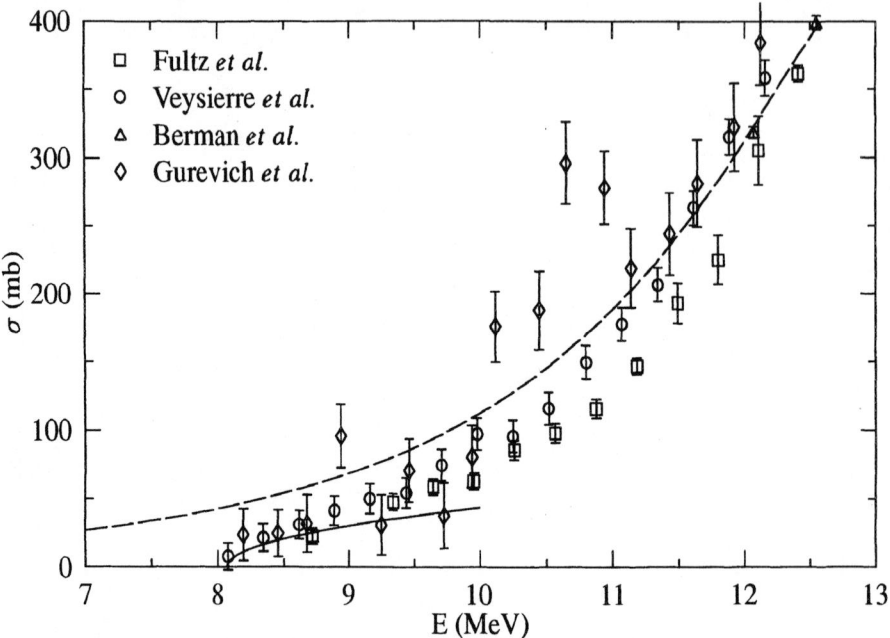

FIGURE 4. Cross section of the reaction ^{197}Au$(\gamma,n)^{196}$Au between the neutron threshold ($E_{\text{thr}} = 8.071$ MeV) and the GDR ($E_{\text{GDR}} = 13.73$ MeV) [7,15–17]. Berman et al. [7], Veyssiere et al. [16], and Fultz et al. [15] used quasi-monochromatic photons from positron annihilation in their experiments; Gurevich et al. [17] used electron bremsstrahlung. The Lorentzian parametrization of the GDR from [7] is shown as dashed line. The Lorentzian parametrization overestimates the cross section close to the reaction threshold. Additionally, the result of the present experiment is shown (full line from threshold to 10 MeV). Here we present the original data of [7,15–17]; note that [7] recommends to reduce the data of [16] by 8% and not to use the old data from [15]. The numerical data were taken from the compilation [5].

thick samples. From our activation experiment we deduce $\sigma_0 = 90 \pm 20$ mb in good agreement with the (γ,n) data measured close above the threshold (see Fig. 4). All measured σ_0 values and reaction rates λ are listed in Table 1. The systematic uncertainties will be reduced in the second part of the experiment where thin samples will be activated at the usual target position, and the photon flux will be monitored (see Sect. II).

The ^{197}Au$(\gamma,n)^{196}$Au reaction is one of the best analyzed photodisintegration reactions; it was measured independently by very different experimental techniques. Therefore, this reaction is well-suited as a standard for (γ,n) experiments with monoenergetic photons. However, close to the reaction threshold the quality of the data is not sufficient because the experimental uncertainties in the first MeV above the reaction threshold (which is the astrophysically relevant energy region, see Fig. 1) are in the order of 20% to 100%. These huge uncertainties should be reduced by a combination of the final results of this experiment and Ref. [19].

TABLE 1. Preliminary results for the constant σ_0 and reaction rates λ at $T_9 = 2.5$ for the (γ,n) reaction of the nuclei ^{204}Pb, ^{196,198,204}Hg, and ^{197}Au. The reaction rates λ are calculated from the σ_0 values with the assumption that the cross section follows Eq. (4) in the relevant energy region. Note the smaller reaction rate of the ^{197}Au$(\gamma,n)^{196}$Au reaction compared to the number given in the caption of Fig. 1 because the measured (γ,n) cross section is significantly smaller than the extrapolation of the Lorentzian curve in Fig. 1.

nucleus	^{204}Pb	^{196}Hg	^{198}Hg	^{204}Hg	^{197}Au
E_{thr} (keV)	8394	8339	8484	7495	8071
σ_0 (mb)	140 ± 30	135 ± 30	130 ± 30	80 ± 20	90 ± 20
λ (s^{-1})	1.5	1.9	0.9	48.7	4.2

The usage of a standard for (γ,n) reactions with bremsstrahlung is complicated by the following fact. The yield in bremsstrahlung experiments is given by the integral in Eq. (3). Because of the decreasing photon flux at higher energies and the increasing cross section the integrand is again peaked at energies relatively close above the reaction threshold (comparable to Fig. 1). The ^{197}Au$(\gamma,n)^{196}$Au reaction can be used as standard for (γ,n) reactions only for reactions with thresholds very close to the neutron threshold in ^{197}Au. For (γ,n) reactions with different neutron thresholds the shape of the bremsstrahlung spectrum has to be known very accurately; otherwise systematic errors appear in the data analysis because the standard ^{197}Au monitors the photon flux close above the ^{197}Au threshold whereas the target yield is determined by the photon flux close above the target threshold. Because of the known uncertainties in the shape of the photon flux from bremsstrahlung [13] the usage of gold as standard in bremsstrahlung experiments leads to non-negligible systematic uncertainties.

IV CONCLUSIONS AND OUTLOOK

A quasi-thermal photon spectrum can be produced by the superposition of bremsstrahlung spectra with different endpoint energies. In this work we have presented the first part of the new experiment on the nuclei ^{204}Pb, ^{196,198,204}Hg, and ^{197}Au. The preliminary results show that an experimental determination of astrophysical reaction rates is possible for all these nuclei. Future work on these nuclei is underway along the lines of our pioneering experiment performed on the platinum isotopes ^{190,192,198}Pt [8].

Furthermore, we have discussed the ^{197}Au$(\gamma,n)^{196}$Au reaction in detail. This reaction has been measured in many experiments by different experimental techniques and is one of the best-known (γ,n) reactions. However, there are still large uncertainties at energies close to the reaction threshold which is the astrophysically relevant energy region. These uncertainties will be reduced in the near future by different experiments.

ACKNOWLEDGMENTS

This work was supported by Deutsche Forschungsgemeinschaft DFG (contracts Zi 510/2-1 and FOR 272/2-1).

REFERENCES

1. D. L. Lambert, Astron. Astrophys. Rev. **3** (1992) 201.
2. S. E. Woosley and W. M. Howard, Astrophys. J. Suppl. **36** (1978) 285.
3. M. Rayet, N. Prantzos, and M. Arnould, Astron. Astrophys. **227** (1990) 271.
4. S. S. Dietrich and B. L. Berman, At. Data Nucl. Data Tables **38** (1988) 199.
5. I. N. Boboshin et al., The Centre for Photonuclear Experiments Data (CDFE) nuclear data bases, *http://depni.npi.msu.su/cdfe*.
6. P. Mohr et al., Proc. *Nuclei in the Cosmos 2000*, Nucl. Phys. A (2000).
7. B. L. Berman et al., Phys. Rev. C **36** (1987) 1286.
8. P. Mohr et al., Phys. Lett. B, in press (2000).
9. A. Richter, Proc. 5th *EPAC*, Barcelona 1996, ed. S. Myers, IOP Publ., Bristol, p. 110.
10. P. Mohr et al., Nucl. Inst. Meth. Phys. Res. A **423** (1999) 480.
11. V. Costa et al., Astron. Astrophys. **358** (2000) L67.
12. R. Brun and F. Carminati, GEANT Detector Description and Simulation Tool, CERN Program Library Long Writeup W5013 edition, CERN, Geneva (1993).
13. K. Vogt et al., to be published.
14. P. Mohr et al., Europ. Phys. J. A **7** (2000) 45.
15. S. C. Fultz et al., Phys. Rev. **127** (1962) 1273.
16. A. Veyssiere et al., Nucl. Phys. **A159** (1970) 561.
17. G. M. Gurevich et al., Nucl. Phys. **A351** (1981) 257.
18. T. E. Cowan et al., Phys. Rev. Lett. **84** (2000) 903.
19. H. Utsunomiya, private communication (2000).
20. H. Utsunomiya et al., Phys. Rev. C, in press (2000).

Nuclear reaction in the static burning of stars

C. Rolfs

Institut für Physik mit Ionenstrahlen, Ruhr-Universität Bochum, Germany

1 INTRODUCTION

Investigations during the last 60 years have shown [1-3] that we are connected with distant space and time not only by our imaginations but also through a common cosmic heritage: the chemical elements that make up our bodies. These elements were created by nuclear reactions in the hot interiors of remote and long-vanished stars over many billions of years. Their nuclear fuels finally spent, these giant stars met death in cataclysmic explosions, scattering far and wide the atoms of heavy elements synthesized deep within their cores. Eventually this material, as well as material lost by smaller stars during red-giant stages, collected into clouds of gas in interstellar space; these, in turn, slowly collapsed giving birth to new generations of stars, thus leading to a cyclic evolution that is still going on. The present picture is that all elements from carbon to uranium have been produced entirely within stars during their fiery lifetimes and explosive deaths. A few of the lightest elements were formed before the stars even existed, during the birth of the universe itself. In addition, a few of the most reactive light elements appear to have been synthesized in intergalactic space by cosmic rays. Thus, theories of nucleosynthesis have identified the most important sites of element formation and also the diverse nuclear processes involved in their production. The detailed understanding of our cosmic heritage combines astrophysics and nuclear physics, and forms what is called nuclear astrophysics. Nuclear reactions are at the heart of nuclear astrophysics: they influence sensitively the nucleosynthesis of the elements in the earliest stages of the universe and in all the objects formed thereafter, and control the associated energy generation, neutrino luminosity, and evolution of stars. A good knowledge of the rates of these reactions is essential to understanding this broad picture.

In hot stellar matter the energies of the moving nuclei can be described by a Maxwell-Boltzmann distribution, $\Phi(E) \propto E \exp(-E/kT)$, where T is the local temperature and k the Boltzmann constant. Folding $\sigma(E)$ with this energy (or velocity) distribution leads to the nuclear reaction rate per pair of nuclei [3]:

$$<\sigma v> = (8/\pi\mu)^{1/2} (kT)^{-3/2} \int_0^\infty \sigma(E) \exp(-E/kT) \, dE, \qquad (1)$$

where v is the relative velocity of the pair of nuclei, E is the center-of-mass energy, and $\mu = m_1 m_2/(m_1+m_2)$ is the reduced mass of the entrance channel. In order to cover the different evolution phases of stars, i.e. from main-sequence stars (T ≈ 10^7 K) to supernovae (T ≈ 10^9 K), one must know the reaction rates over a wide range of temperatures, which in turn requires the availability of σ(E) data over a wide range of energies. It is the challenge to the experimentalist to make precise σ(E) measurements over a wide range of energies, as our fragmented knowledge of nuclear physics prevents us from predicting σ(E) on purely theoretical grounds.

For the important class of charged-particle-induced reactions, there is a repulsive Coulomb barrier in the entrance channel of height $E_c = Z_1 Z_2 e^2/r$, where Z_1 and Z_2 are the integral nuclear charges of the interacting particles, e is the unit of electric charge, and r is the interaction radius. Due to the tunneling effect through the barrier, σ(E) drops nearly exponentially with decreasing energy (Fig. 1):

$$\sigma(E) = S(E) \, E^{-1} \exp(-2\pi\eta(E)), \qquad (2)$$

where $\eta(E) = 2\pi Z_1 Z_2 e^2/hv$ is the Sommerfeld parameter (h = Planck constant). The function S(E) contains all the strictly nuclear effects, and is usually referred to as the nuclear or astrophysical S-factor. If (2) is inserted in (1), one obtains

$$<\sigma v> = (8/\pi\mu)^{1/2} \, (kT)^{-3/2} \int_0^\infty S(E) \exp(-E/kT - b/E^{1/2}) \, dE, \qquad (3)$$

with $b = 2(2\mu)^{1/2} \pi^2 e^2 Z_1 Z_2 / h$. Since for nonresonant reactions S(E) varies slowly with

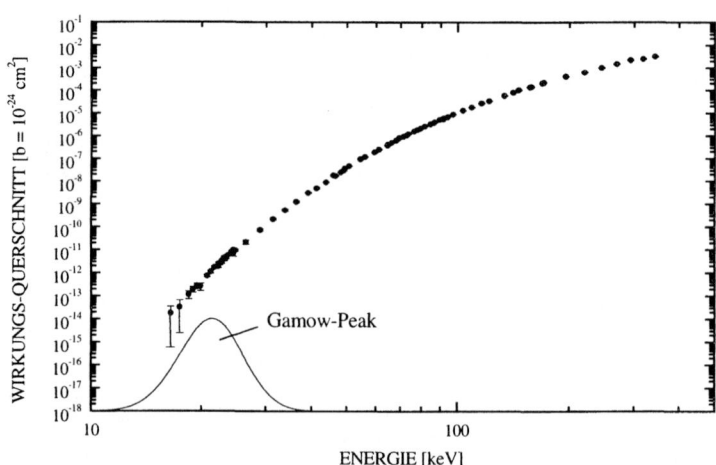

FIGURE 1. Energy dependence of the cross section σ(E) for the ^{3}He(^{3}He,2p)^{4}He reaction. The Gamow peak is at E_0 = 21 keV for solar conditions (central temperature T ≈ 15×10^6 K). The low-energy data have been obtained by LUNA [9].

energy, the steep energy dependence of the integrand in (3) is governed primarily by the exponential term, which is characterized by a peak near an energy E_0 that is usually much larger than kT. The peak is frequently referred to as the Gamow peak; for a constant S(E) value over the energy region of the peak, one finds $E_0 = (bkT/2)^{2/3}$. This is the effective mean energy for a given reaction at a given temperature. Approximating the peak by a Gaussian function, one finds an effective width $\Delta = 4(E_0 kT)^{1/2}/3^{1/2}$. Thus, nuclear burning takes place predominantly over the energy window $E_0 \pm \Delta/2$, the stellar thermal energy range for which information on $\sigma(E)$ must be obtained.

Due to the steep drop of $\sigma(E)$ at subcoulomb energies, it becomes increasingly difficult to measure $\sigma(E)$ as E is lowered. Although experimental techniques have improved significantly over the years [3], extending $\sigma(E)$ measurements to lower energies (with a low-energy limit at E_L, corresponding to a reaction yield in a detector of roughly 1 event per hour), it has not been possible to measure $\sigma(E)$ at stellar thermal energies, as $E_0/E_c \approx 1/100$ for hydrogen-burning reactions (i.e. pp-chain and CNO-cycles) in main-sequence stars such as our sun. Instead, the measured energy dependence of $\sigma(E)$ at higher energies ($E > E_L$) must be extrapolated to stellar energies, using the S(E) factor defined in (2). Such an "extrapolation into the unknown" can lead to considerable uncertainty. At energies lower than E_L there might be a change of reaction mechanism, or of the centrifugal barrier, or there might be a contribution of narrow or subthreshold resonances to $\sigma(E)$ at stellar energies. The danger of such extrapolations was strikingly demonstrated in the case of $^2H(d,\gamma)^4He$ [4], for example, where new low-energy data increased the extrapolated values by a factor of 1000, mainly due to a change in the centrifugal barrier (d- to s-waves). Thus, new experimental approaches are needed to reduce the uncertainties of the $\sigma(E)$ extrapolations, at least for the key reactions in hydrogen-, helium-, and carbon-burning [1-3].

2 THE LUNA PROJECT

The observed solar neutrino fluxes in the existing neutrino detectors are not consistent with the current "standard" picture of the microscopic processes in the sun [5]. A possible solution for this solar neutrino problem may be found in one or more of the areas of neutrino physics (e.g., the recent experimental evidence for neutrino oscillations at Super-Kamiokande), solar physics (models), or nuclear physics. In view of the important and fundamental conclusions regarding non-standard physics, which may be deduced from the results of present and future solar neutrino observations, it is of utmost importance to place all of the predictions on a solid basis. The nuclear physics component of this problem involves the cross sections of the hydrogen-burning reactions extrapolated to the relevant solar energies. Clearly, new experimental approaches are needed to reduce the uncertainties of these extrapolations.

Low-energy studies of thermonuclear reactions in a laboratory at the earth's surface are hampered predominantly by background effects of cosmic rays in the detectors, leading typically to more than 10 background-events per hour in common detectors. Conventional passive or active shielding around the detectors can only partially reduce the problem of cosmic ray background. The best solution is to install an accelerator facility in a laboratory deep underground. As a pilot project, a 50 kV accelerator facility has been installed in the Laboratori Nazionali del Gran Sasso (LNGS), where the flux of cosmic-ray muons is reduced by a factor of 10^6 compared with the flux at the surface [6]. This unique project, called LUNA (Laboratory for Underground Nuclear Astrophysics), was designed [7] primarily for a renewed study of ^{3}He(^{3}He,2p)^{4}He at low energies, aiming to reach the solar Gamow peak at $E_o \pm \Delta/2 = 21 \pm 5$ keV. This goal has been reached [8,9] with a detected reaction yield of about 1 event per month at the lowest energy, $E = 16$ keV, with $\sigma \approx 20$ fb or 2×10^{-38} cm^2. Thus, the cross section of an important reaction of the pp-chain has been directly measured for the first time at solar thermal energies (Fig. 1); in principle, extrapolation is no longer needed in this reaction. The work demonstrated the research potential of LUNA and that all of the experimental requirements in such low-rate, time-consuming experiments can be fulfilled. Installing larger facilities at LNGS, such as a 400 kV high-current accelerator at the present time, will open the possibility of improving our knowledge of other key reactions by shifting their measured low-energy limit E_L significantly closer to E_0 or even to within the Gamow peak.

3 ELECTRON SCREENING

In the extrapolation of $\sigma(E)$ using (2), it is assumed that the Coulomb potential of the target nucleus and projectile is that resulting from bare nuclei. However, for nuclear reactions studied in the laboratory, the target nuclei and the projectiles are usually in the form of neutral atoms or molecules and ions, respectively. The electron clouds surrounding the interacting nuclides act as a screening potential: the projectile effectively sees a reduced Coulomb barrier. This in turn leads to a higher cross section, $\sigma_s(E)$, than would be the case for bare nuclei, $\sigma_b(E)$. There is an enhancement factor, $f_{lab}(E) = \sigma_s(E)/\sigma_b(E) \approx \exp(\pi\eta U_e/E)$ [10], where U_e is the electron-screening potential energy (e.g. $U_e \approx Z_1Z_2e^2/R_a$, with R_a an atomic radius). Note that $f_{lab}(E)$ increases exponentially with decreasing energy. For ratios $E/U_e > 1000$, shielding effects are negligible, and laboratory experiments can be regarded as essentially measuring $\sigma_b(E)$. However, for $E/U_e < 100$, shielding effects begin to become important for understanding and extrapolating low-energy data. Relatively small enhancements arising from electron screening at $E/U_e \approx 100$ can cause significant errors in the extrapolation of cross sections to lower energies, if the curve of the cross section is forced to follow the trend of the enhanced cross sections, without correction for the screening. Note that for a stellar plasma, the value $\sigma_b(E)$ must be known because the screening in the plasma will be quite different from that in the laboratory nuclear-reaction studies, i.e. $\sigma_{plasma}(E) = f_{plasma}(E)\ \sigma_b(E)$, and $f_{plasma}(E)$ must be explicitly included for each situation. A good understanding of electron-screening effects in the laboratory is needed to arrive at reliable $\sigma_b(E)$ data at low energies. Experimental studies of reactions involving light nuclides [9, 11-16] have shown the expected exponential enhancement of the cross section at low energies. However, the observed enhancements were in all cases significantly larger than could be accounted for from available atomic-physics models. This situation is disturbing because if the effects of electron screening are not understood under laboratory conditions, they are most likely to be not fully understood in a stellar plasma. A solution to the laboratory puzzle might be found in one (or all) of the following areas: the assumed energy-loss predictions from stopping-power codes at low energies (section 5), the assumed nuclear-reaction models at energies far below the Coulomb barrier (section 6), and the assumed atomic-physics models. All of these areas require additional experimental and theoretical efforts. An improved understanding of laboratory electron screening may also help eventually to improve the corresponding understanding of electron screening in stellar plasmas. It is in the nature of astrophysics that many of the processes and most of the objects one tries to understand are physically inaccessible. Thus, it is important that those aspects that can be studied in the laboratory be rather well understood. The electron-screening project addresses one such aspect.

4 ENERGY LOSS OF DEUTERONS IN ^{3}He GAS

For absolute $\sigma(E)$ measurements at subcoulomb energies, an accurate knowledge of the effective beam energy associated with the observed reaction yield is as important as the yield measurements themselves. In the analysis of such data, the effective energy in the target involves always energy-loss corrections, which are extracted from a standard compilation [17]. The compilation is based on experimental data down to energies around the Bragg peak, while at lower energies - relevant to nuclear astrophysics – the experimental data are extrapolated with theoretical guidance. In recent studies of the $d(^3He,p)^4He$ reaction (Q = 18.4 MeV) at the LUNA facility [18], the observed energy loss of ^{3}He ions in D_2 molecular gas was in good agreement with the extrapolated values of the compilation. For studies of the inverted reaction, i.e. $^3He(d,p)^4He$, energy-loss data are needed for deuterons in ^{3}He gas. These energy-loss measurements were carried out at the 100 kV accelerator of the Ruhr-Universität Bochum [19] involving the LUNA setup [7-9]. Briefly, the beam entered the target chamber of a differentially pumped gas-target system through apertures of high gas-flow impedance and was stopped in a calorimeter. The ^{3}He gas pressure in the target chamber, $P \leq 1.00$ mbar, was measured with a Baratron capacitance manometer. The main pressure drop occurred across the entrance aperture to the target chamber, while the extended target zone between the entrance aperture and the calorimeter (length d = 30.1 cm) was characterized by a constant gas pressure. For each run, the average power deposited by the beam on the calorimeter was deduced from the difference between transistor powers needed to keep the beam dump at the same temperature, with the beam off and on. The detector setup consisted of four, 1 mm thick Si detectors of 5x5 cm^2 area (each) placed around the beam axis: they formed a 5 cm long parallelepiped in the target chamber. In going through the gas of the target chamber, the beam experienced an energy loss to the middle of the detector setup, at a distance z = 14.1 cm from the middle of the entrance aperture.

At a given incident energy E_d, the reaction yield $Y(E_d,P) = N/WP$ was obtained as a function of gas pressure P, where N is the number of observed protons from the reaction ^{3}He(d,p)^{4}He (in the detector setup) and W is the integrated beam power (deduced from the calorimeter). The yield is related to the cross section $\sigma(E_d,P)$, for which one arrives - from (2) and the yield definition - to the expression

$$\alpha Y(E_d,P) = (1 + \varepsilon(E_d) \, \rho_o \, d \, P / P_o \, E_d) \, \sigma(E_d,P) / \sigma(E_d, P \to 0)$$
$$= (1 + \varepsilon(E_d) \, \rho_o \, d \, P / P_o \, E_d) \exp(-2\pi\eta(E-\Delta E))(E-\Delta E)^{-1} / \exp(-2\pi\eta(E))E^{-1}$$
$$= (1 + \varepsilon(E_d) \, \rho_o \, d \, P / P_o \, E_d)(1 - \varepsilon(E_d) \, \rho_o \, z \, (\pi \, \eta(E) - 1) \, P / P_o \, E_d) + ..., \quad (4)$$

where α is a normalisation constant, E is the center-of-mass energy of the incident beam, ΔE is the center-of-mass energy loss of the incident beam over the distance z, $\varepsilon(E_d)$ is the stopping power of the deuterons in the ^{3}He target gas, ρ_o and P_o are the density and pressure of the gas at STP, respectively, and the symbol P→0 indicates the

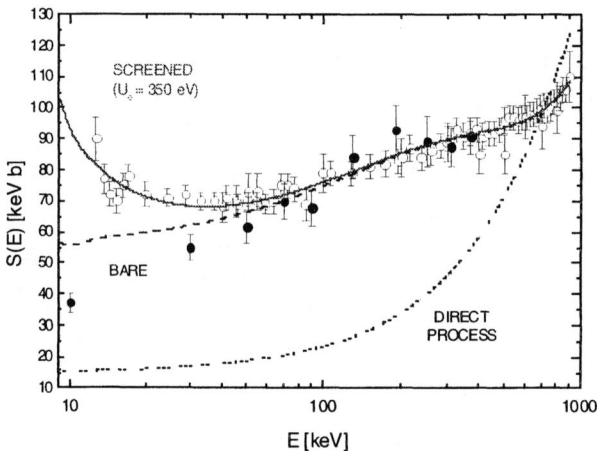

FIGURE 2. Astrophysical S-factor for the reaction ^{7}Li(p,α)α as obtained by direct measurements (open circles). The dotted curve represents the S-factor for a p-wave direct process alone. The dashed curve is the result of a fit to the direct data at E $\geq$ 100 keV using the direct process and two subthreshold resonances, and the solid curve is the S-factor for screened nuclei derived from equation 2 with U_e = 350 eV. The data obtained by the Trojan-horse-method (filled circles) agree fairly well with the dashed curve.

limit of zero pressure. The first term corrects for the energy loss of the beam arriving at the calorimeter. Since relative values of the cross section are involved here, only statistical errors have to be included in the analysis. The deduced energy-loss values show (e.g. Fig. 2 in [20]) a threshold behavior near $E_d = 18$ keV, where the values reach the domain of nuclear stopping power. The threshold in the electronic stopping power arises from the minimum energy transfer $E_{e,min}$ in the 1s→2s electron excitation of the He target atoms, $E_{e,min} = 19.8$ eV, which translates into a minimum deuteron energy $E_{d,min} = (m_d/4m_e)(1+m_e/m_d)^2 E_{e,min} = 18.2$ keV (m_d = deuteron mass, m_e = electron mass). Below this energy, the electron cloud of the He atom cannot be excited via an ion-electron interaction and thus the electronic energy loss vanishes: the He atoms become transparent for the deuterons. Below the threshold energy, solely the nuclear stopping power is left. To our knowledge, a sharp threshold effect in the energy loss behavior has not been predicted. The observed threshold behavior is a quantum effect and may be compared in a way with superconductivity.

In principle, the observed threshold behavior should occur in many ion-target combinations. For the case of ^{3}He ions in D_2 molecular gas, the D_2 molecule can be dissociated involving an energy of about 0.6 eV with a corresponding threshold energy near $E_{He} = 0.8$ keV; above this energy one may expect no significant deviation from the compilation, as observed [18]. In metallic targets, there should be no threshold effect at all since the electrons can be excited continuously within overlapping or partially filled energy bands. However, in insulators or semiconductors with separated band gaps, e.g. about 10 eV for diamond, the threshold effect should exist and is - for Al implantation in diamond - at $E_{Al} = 120$ keV. Indeed, the range of low-energy ion implantation has been observed to be larger than expected and was suggested as arising from channeling effects; the present work offers an alternative explanation. Below a given threshold energy, one has the possibility to measure directly – and to our knowledge for the first time - the nuclear stopping power, which may help to improve the quantitative understanding of this process. In turn, the process is the basis of many ion-beam applications in materials science such as sputtering and angle-straggling effects in implantation. The observed threshold behavior may have important consequences also in nuclear astrophysics, i.e. the properties of a stellar plasma, where the chemical elements are present in form of atoms and assumed to be nearly fully ionized, with thermal energies of $kT = 1.3$ keV at the solar center. However, rigorous consequences have to await the results of detailed calculations.

With the energy loss information just described, the ^{3}He(d,p)^{4}He reaction was restudied also at the 100 kV accelerator in Bochum. The results led to an electron screening potential energy $U_e = 210$ eV consistent with previous work [11,12] but about a factor larger than the adiabatic limit from atomic physics, $U_{ad} = 102$ eV. The puzzle is now reduced to either nuclear or atomic physics. With regard to nuclear physics a new method (section 7) may provide the needed information.

5 THE TROJAN HORSE METHOD

The Trojan horse method (THM) allows one to measure the energy dependence of the bare astrophysical S-factor, $S_b(E)$, down to the relevant low energies, free of the effects of the Coulomb barrier and electron screening. The principle of the THM has been discussed previously ([21] and references therein). Briefly, a particle a strikes a nucleus A, where A is described by a wave function with a large amplitude for a s-b cluster configuration. Under appropriate kinematic conditions the particle a then interacts only with the part b of the target nucleus A, while the other part s behaves as a spectator to the process $a+b\ (+s) \rightarrow c+d\ (+s)$. In order to completely determine the kinematic properties of the spectator s, the energies E_c and E_d of the two particles c and d must be measured in coincidence at specific angles θ_c and θ_d, respectively. In the plane-wave impulse approximation the three-body cross section may be expressed as

$$d^3\sigma/dE_c d\Omega_c d\Omega_d \propto (KF)\ |\Phi(p_s)|^2\ d\sigma^N/d\Omega, \tag{5}$$

where KF is a kinematic factor containing the final-state phase-space factor, $|\Phi(p_s)|^2$ is the momentum distribution of the spectator s inside the nucleus A, and $d\sigma^N/d\Omega$ is the differential nuclear cross section for the reaction $a+b \rightarrow c+d$. With the known terms KF and $|\Phi(p_s)|^2$ one can derive $d\sigma^N/d\Omega$ from a measurement of $d^3\sigma/dE_c d\Omega_c d\Omega_d$, whereby integration over the solid angle leads finally to the total cross section $\sigma^N(E)$. Furthermore, if the bombarding energy is chosen to be above the Coulomb barrier in the incident channel of the reaction $a+A \rightarrow c+d+s$, the particle b can be brought into the nuclear interaction zone to induce the reaction $a+b \rightarrow c+d$. If the Fermi motion of particle b inside A compensates at least in part for the initial projectile velocity v_a, the reaction $a+b \rightarrow c+d$ can be induced at a low relative energy between a and b, relevant to nuclear astrophysics. Note that the deduced reaction cross section $\sigma^N(E)$ is the nuclear part alone, since the Coulomb barrier has already been overcome in the entrance channel. The corresponding astrophysical $S^N(E)$ factor is then derived from the relation $S^N(E) = E\ \sigma^N(E)$ [21], where $S^N(E)$ represents the S-factor for bare nuclides, since the projectile energy is above the height of the Coulomb barrier for the entrance channel. Since $S(E)$ as defined by (2) contains the term $\exp(2\pi\eta)$ representing approximately the tunneling through the Coulomb barrier (for s-waves), a comparison of $S^N(E)$ with the $S(E)$ factor data from direct measurements requires the introduction of the term $\exp(2\pi\eta)$ and the actual transmission factor $T_{l=0}(E)$,

$$S_b(E) = S^N(E)\ T_{l=0}(E)\ \exp(2\pi\eta). \tag{6}$$

The absolute scale for $S_b(E)$ is obtained by normalisation of the THM data to the direct data at energies where the effects of electron screening are negligible. Thus, the energy dependence of $S_b(E)$ should be identical to that derived by the direct measurements, except at low energies, where the two data sets should differ due to the effects of electron screening. In turn, the value of U_e can then be obtained by comparing the two data sets.

The ^{7}Li(p,α)α reaction was studied directly ([12,24] and references therein), where the resulting data are displayed in Fig. 2. A theoretical analysis of the data is missing, although it was suggested [25] that such an analysis may have to include a direct process as well as two $J^\pi = 2^+$ subthreshold resonances at $E_{R1} = -0.62$ MeV and $E_{R2} = -0.33$ MeV with total widths $\Gamma_{R1} = 108$ keV and $\Gamma_{R2} = 74$ keV (^{8}Be states at 16.63 and 16.92 MeV excitation energy). We have followed this suggestion in a simple parametrisation of the data. Firstly, the reaction can only proceed via p-waves in the entrance channel. For a direct process (DP) alone, the p-wave centrifugal barrier leads to an S(E) factor falling much steeper than the data (dotted curve in Fig. 2). We included thus the two p-wave subthreshold resonances and the associated interference terms:

$$S_b(E) = S_{DP}(E) + S_{R1}(E) + S_{R2}(E) \pm 2(S_{R1}(E)S_{R2}(E))^{1/2}\cos\Phi_{R1,R2} \pm 2(I_R S_{DP}(E) S_{R1}(E))^{1/2}\cos\Phi_{DP,R1}$$
$$\pm 2(I_R S_{DP}(E) S_{R2}(E))^{1/2}\cos\Phi_{DP,R2} \quad (7)$$

where the direct process is described by $S_{DP}(E) = N\, P_{l=1}(E)/P_{l=0}(E)$ ($P_l(E)$ = penetrability for s- and p-waves; N = free parameter) and the high-energy tail of each subthreshold resonance is parametrised by a Breit-Wigner expression [3], with the reduced proton width $\theta_{l=1}^2$ taken as free parameter. The resonance phase is described by $\Phi_{Ri}(E) = \arctan(0.5\Gamma_{Ri}(E)/(E-E_{Ri}))$, where the energy dependence of the total width includes the contributions of both the proton- and alpha-channels, leading to $\Phi_{DP,Ri} = \Phi_{Ri}$ and $\Phi_{R1,R2} = \Phi_{R1} - \Phi_{R2}$. Finally, the factor $I_R = 5/16$ represents a statistical factor between the $J^\pi = 2^+$ resonances and the possible angular momenta of the direct process ($J^\pi = 0^+$ to 3^+). A fit to the data at energies $E \geq 100$ keV (i.e. no significant effects of electron screening expected) leads to the dashed curve in Fig. 2, for the parameters $N = 0.55$, $\theta_{l=1}^2(R1) = 1.04$, $\theta_{l=1}^2(R2) = 0.13$, and the sign-combination --+ of the 3 interference terms. At zero energy one finds $S_b(0) \approx 40$ keV b, which may be compared with the adopted value $S_b(0) = 59$ keV [24,25]. An improved analysis is highly desirable since the present low $S_b(0)$ value leads to a higher calculated ^{7}Li abundance in big-bang nucleosynthesis [1-3] influencing the conclusions on the universal baryon density. Assuming the validity of the calculated $S_b(E)$ curve, a comparison with the direct data at $E \leq 100$ keV leads to $U_e = 350$ eV (solid curve in Fig. 2). This value is substantially higher than the adiabatic limit (175 eV) and is not understood at present. Since the direct measurements have also been carried out in inverted kinematics with nearly identical results [12], it is unlikely that the direct data are heavily influenced by incorrect energy-loss values.

The ^{7}Li(p,α)α reaction was also studied with THM, i.e. using the reaction ^{2}H(^{7}Li,αα)n ($A \equiv p$-n cluster, $s \equiv n$ = spectator) [21-23]. The THM data (Fig. 2) have been normalized to the direct data in the energy region $E = 100$ to 370 keV. At energies above $E = 100$ keV the direct and THM data agree within errors (after normalising), supporting the validity of the THM, while at lower energies the THM data appear to confirm the calculated $S_b(E)$ curve for bare nuclides.

Although it is not yet clear that all relevant components of the THM have been included in current analyses, the present work demonstrates how well the THM and direct measurements complement one another in determining both low-energy cross sections and associated electron-screening effects. Further work on the theoretical aspects of the THM and on comparisons with low-energy data, as well as improved experimental data on low-energy energy-loss, are urgently needed before one can have complete confidence in applying the THM to the problem of determining astrophysically important nuclear reaction cross sections.

6 THE ERNA PROJECT

The capture reaction $^{12}C(\alpha,\gamma)^{16}O$ (Q = 7.16 MeV) takes place during the helium burning stage of red giants [1-3] and is a key reaction of nuclear astrophysics. The cross section at the relevant Gamow-energy, $E_o \approx 0.3$ MeV, determines not only the nucleosynthesis of elements up to the iron region, but also the subsequent evolution of massive stars, the dynamics of supernovae, and the kind of remnants after supernova explosions. For these reasons, the cross section $\sigma(E_o)$ should be known with a precision of at least 10%. In spite of tremendous experimental efforts over nearly 30 years [26-33], one is still far from this goal. All previous efforts have focused on the observation of the capture γ-rays, including one experiment that combined γ-detection with coincident detection of the ^{16}O recoils produced in the reaction. Due to the low radiative capture cross section and various backgrounds depending on the exact nature

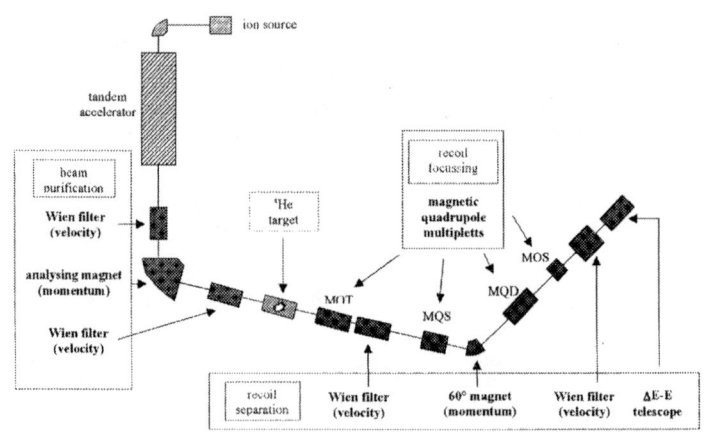

FIGURE 3. Schematic diagram of the final technical layout of ERNA (see text).

of the experiments, γ-ray data with useful, but still inadequate, precision were limited to center-of-mass energies 1.2 MeV ≤ E ≤ 3.2 MeV.

To improve the situation, a new experimental approach is in preparation at the 4 MV Dynamitron tandem accelerator in Bochum, called ERNA (**E**uropean **R**ecoil separator for **N**uclear **A**strophysics, Fig. 3). In this approach, the reaction is initiated in inverted kinematics, ^{4}He(^{12}C,γ)^{16}O, i.e. a ^{12}C ion beam is guided into a windowless ^{4}He jet gas- target and the kinematically forward-directed ^{16}O recoils are detected downstream on the beam line. The direct observation of the ^{16}O recoils requires an efficient recoil separator to filter out the intense ^{12}C beam particles from the ^{16}O recoils: the number of ^{16}O recoils per incident ^{12}C projectile is 1×10^{-18} for σ = 1 pb and a target density $n(^4\text{He}) = 1 \times 10^{18}$ atoms/cm^2. The recoil separator must also filter out beam contaminants, small-angle elastic scattering products, and background events from multiple scattering processes leading to a degraded tail of the projectiles. If the filtering of the separator is sufficiently effective (with a beam suppression factor of the order $R_{rec} = 1 \times 10^{-14}$ at E = 0.7 MeV), the ^{16}O recoils can be counted directly in a ΔE-E telescope placed in the beam line at the end of the recoil separator, where the telescope allows for particle identification. Previous measurements [34] have shown that a suppression factor for the telescope alone of $R_{tel} = 1 \times 10^{-4}$ can be achieved leading to a combined suppression factor of $R_{tot} = R_{rec} R_{tel} = 1 \times 10^{-18}$ at E = 0.7 MeV for the planned separator.

Since the ^{12}C projectiles and the ^{16}O recoils have essentially the same momentum and since the ^{12}C ion beam emerging from the accelerator passes a momentum filter (analysing magnet), a nearly complete elimination of any ^{16}O beam contaminant in the ^{12}C ion beam incident on the ^{4}He gas target is of utmost importance for the new approach: the ^{16}O beam contaminant and the ^{16}O recoils cannot be distinguished in the recoil separator, since both have the same momentum. In a first ERNA report [35], a Wien filter and a ΔE-E telescope were used to investigate the ^{16}O beam contamination accompanying a momentum-filtered ^{12}C beam: the intensity ratio of ^{16}O to ^{12}C was found to be about $P_0 = 6 \times 10^{-10}$, i.e. much higher than the intensity ratio 1×10^{-18} between the ^{16}O recoils and the ^{12}C projectiles at E = 0.7 MeV, or even 5×10^{-14} at E = 2.45 MeV with maximum radiative capture cross section σ ≈ 50 nb.

In recoil separators that include momentum analysis, it is necessary to make a charge state selection of the recoils, causing a reduction in the number of recoils transmitted through the separator. However, since there is usually – in the equilibrium charge state distribution - a charge state representing about 50% of the total recoils produced, this reduction is not too serious. Since the ^{16}O recoils are produced in the ^{4}He jet gas-target, their charge state distribution depends however on the geometric location within the target: those ^{16}O recoils produced in the upstream part of the target will most likely reach an equilibrium charge state distribution in the passage of the remaining target length, while those ^{16}O recoils produced near the downstream end of the target will not. Thus, not all ^{16}O recoils produced will be characterized by an equilibrium charge state distribution and this feature can lead to significant uncertainties in the cross section determination. It was found previously [34], that the charge state distribution of the recoils also depends on the incident charge state of the projectiles, called a memory effect. A possible solution for these problems lies in the installation of an additional gas stripper (e.g. Ar gas) shortly after the jet gas-target, which is presently in the stage of technical development.

Although the ^{16}O recoils - produced in the ^{4}He jet gas-target via the reaction ^{4}He(^{12}C,γ)^{16}O - are kinematically forward-directed, the emission of the capture γ-rays (energy E_γ) leads to an emission cone of half-angle $\theta = \arctan(E_\gamma/pc)$, where p is the momentum of the ^{16}O recoils and c is the velocity of light. Associated with the γ-ray emission there is also a spread $\Delta p/p$ in momentum. For example, at E = 0.7 MeV (E_γ = 7.9 MeV) one finds θ = 1.8° and $\Delta p/p$ = 6.2%, and at E = 5.0 MeV (E_γ = 12.2 MeV) one finds θ = 1.0° and $\Delta p/p$ = 3.6%. At E = 0.7 MeV, the cone has reached a diameter of 3.1 cm at a 0.5 m distance from the jet gas-target. Thus, shortly after the jet gas-target there must be a focusing element followed by filter elements and other focusing elements up to the site of the telescope, where all elements must have an angle acceptance of at least θ = 1.8° and a momentum acceptance of at least $\Delta p/p$ = 6.2%, in order to transport the ^{16}O recoils with 100% transmission to the telescope. This requirement demands a compact design of the jet gas-target system involving several pumping stages, where the present technical plan involves an extension of 35 cm on both sides of the jet gas-target.

The high detection efficiency of the ^{16}O recoils and the negligible contribution of cosmic-ray events in the ΔE-E coincidences of the telescope probably allow a measurement of the ^{4}He(^{12}C,γ)^{16}O cross section to as low as E = 0.7 MeV ($\sigma \approx$ 1 pb), if the following requirements can be fulfilled: (i) a ^{12}C beam suppression R_{rec} = 1×10^{-14} in the separator, (ii) a free selection of the charge state for the ^{16}O recoils in the separator, (iii) a relative ^{16}O beam contamination of $P_{tot} \leq 1\times10^{-20}$ in the incident ^{12}C beam, (iv) a 100% beam-energy-independent separator detection efficiency of all ^{16}O recoils of the selected charge state emerging from the jet gas-target, and (v) an equilibrium charge state distribution of the ^{16}O recoils after the jet gas-target. In the second ERNA report [36], calculations of ion beam optics for the filtering and focusing elements of ERNA were reported fulfilling the requirement (iv). For the requirement (iii) the ^{16}O beam contamination relative to the incident momentum-filtered ^{12}C ion beam (with E_{lab} = 10 MeV) could be reduced from P_0 = 1×10^{-11} to $P_{tot} < 2\times10^{-29}$ (and most likely to P_{tot} = 1.4×10^{-35}) by the installation of Wien filters both before and after the analysing magnet (Fig. 3); this purification level is far better than the level required by ERNA. The setup used for the measurement of the beam suppression factor is nearly identical to that in the final ERNA layout (Fig. 3), where a compact jet gas-target system will be installed; between the first Wien filter and the dipole magnet there will be a sizable side port used as a ^{12}C beam dump, whereby the intense ^{12}C beam – deflected by the Wien filter - will be removed effectively from the beam axis. This side port was simulated in the preliminary studies by the installation of a 30 cm-long pipe (Φ = 17 mm, otherwise completely closing the beam pipe) in the 4" beam pipe at a 1.5 m distance downstream from the Wien filter, where the deflected ^{12}C beam was hitting the 4" beam pipe at a point downstream of the entrance of the 30 cm-long pipe. This setup led to a beam suppression factor of $R_{rec} = 1 \times 10^{-13}$; the count rate was 0.25/s, while the telescope can handle a count rate of about 5 kHz, thus $R_{tel} = 5 \times 10^{-5}$ and $R_{tot} = R_{rec} R_{tel} = 5 \times 10^{-18}$, a value close to the ERNA requirement (ii). The experiments indicate that a free choice of the charge state for the ^{16}O recoils is possible (here: $q_C = q_O = 3^+$). The exception is the charge state combination $q_C = 3^+$ and $q_O = 4^+$, for which the momenta and velocities of ^{12}C and ^{16}O are identical (due to their mass ratio ¾) and thus no filtering is possible with ERNA. However, this represents no serious problem: at high energies (say $E_{lab} \geq$ 5 MeV) one may choose $q_C = 3^+$ and $q_O = 5^+$ (or 6^+) or $q_C = 4^+$ and $q_O = 5^+$ (or 6^+), while at low energies one can choose $q_C = 2^+$ and $q_O = 3^+$ (or 4^+). In all these cases, the charge state probability for the ^{16}O recoils is of the order of 50%: e.g. at $E_{lab}(q_C=4^+) = 20$ MeV one finds a probability $\varphi(q_O=6^+) = 49\%$, and at $E_{lab}(q_C=2^+) = 4.0$ MeV one finds $\varphi(q_O=4^+) = 45\%$.

7 SUMMARY

Impressive progress has been achieved in the knowledge of nuclear reaction rates. However, there remains much critical work to be done in the future to arrive at reliable data for many key reactions and processes. New techniques continue to be developed for this purpose, for nuclear reaction rates involving both stable and radioactive nuclei, and research in nuclear astrophysics will remain an exciting pursuit for many years to come.

REFERENCES

1. E.M.Burbidge, G.R.Burbidge, W.A.Fowler, F.Hoyle: Rev.Mod.Phys. 29(1957)547
2. W.A.Fowler: Rev.Mod.Phys. 56(1984)149
3. C.Rolfs, W.S.Rodney: Cauldrons in the Cosmos (University of Chicago Press, 1988)
4. C.A.Barnes et al.: Phys.Lett. 197(1987)315
5. J.N.Bahcall, M.H.Pinsonneault: Rev.Mod.Phys. 64 (1992) 885
6. G.Fiorentini, R.W.Kavanagh, C.Rolfs: Z.Phys. A350(1995)289
7. C.Arpesella et al.: Nucl.Instr.Meth. A360(1995)607
8. M.Junker et al.: Phys.Rev. C57(1998)2700
9. R.Bonetti et al.: Phys.Rev.Lett. 82(1999)5205
10. H.J.Assenbaum, K.Langanke, C.Rolfs: Z.Phys. A327(1987)461
11. S.Engstler et al.: Phys.Lett. B202(1988)179
12. S.Engstler et al.: Z.Phys. A342(1992)471
13. C.Angulo et al.: Z.Phys. A345(1993)231
14. P.Prati et al.: Z.Phys. A350(1994)171
15. U.Greife et al.: Z.Phys. A351(1995)107

16. D.Zahnow et al.: Z.Phys. A359(1997)211
17. H.Andersen, J.F.Ziegler: The Stopping and Ranges of Ions in Matter (Pergamon, New York, 1977) and SRIM-2000
18. H.Costantini et al.: Phys.Lett. B482(2000)43
19. A.Krauss et al.: Nucl.Phys. A465(1987)150
20. A.Formicola et al.: Eur.Phys.J. A8(2000)443
21. C.Spitaleri et al.: Phys.Rev. C60(1999)55802
22. M.Zadro et al.: Phys.Rev. C40(1989)181
23. G.Calvi et al.: Nucl.Phys. A621(1997)139
24. C.Angulo et al.: Nucl.Phys. A656(1999)3
25. C.Rolfs, R.W.Kavanagh: Nucl.Phys. A455(1986)179
26. P.Dyer, C.A.Barnes: Nucl.Phys. A233(1974)495
27. K.U.Kettner et al.: Z.Phys. A308(1982)73
28. A.Redder et al.: Nucl.Phys. A462(1987)385
29. R.M.Kremer et al.: Phys.Rev.Lett. 60(1988)1475
30. J.M.L.Ouellet et al.: Phys.Rev. C54(1996)1982
31. G.Roters, C.Rolfs, F.Strieder, H.P.Trautvetter: Eur.Phys.J. A6(1999)451
32. D.Rogalla, Diplomarbeit, Ruhr-Universität Bochum (1997) (to be submitted)
33. L.Gialanella, Thesis, Ruhr-Universität Bochum (1999) (to be published)
34. L.Gialanella et al.: Nucl.Instr.Meth. A376(1996)174
35. D.Rogalla et al.: Nucl.Instr.Meth. A437(1999)266
36. D.Rogalla et al.: Eur.Phys.J. A6(1999)471

Nuclear Reactions in Hot and Explosive Burning: A Summary

Hiroaki Utsunomiya[†], Michael S. Smith[*], Toshitaka Kajino[‡]

[†]*Department of Physics, Konan University,
Okamoto 8-9-1, Higashinada, Kobe 658-8501, Japan*

[*]*Physics Division, Oak Ridge National Laboratory* [1] *Oak Ridge, TN 37830, USA*

[‡]*Division of Theoretical Astrophysics, National Astronomical Observatory, 2-21-1 Osawa, Mitaka, Tokyo 181-8588, Japan*

Abstract. We give a summary of the round table discussions on nuclear reactions in hot and explosive burning with emphasis on four topics: primordial nucleosynthesis in relation to the recent Boomerang and Maxima-1 observations, the $^{15}O(\alpha,\gamma)^{19}Ne$ reaction in the hot CNO cycle, the p- process, and the Re/Os cosmochronometer.

Key Reactions in Explosive Primordial Nucleosynthesis

Cosmological Quest: The recent observation of the power spectrum of the cosmic microwave background (CMB) by Boomerang [1,2] and Maxima-1 [3,4] has exhibited that the flat-universe model ($\Omega_M + \Omega_\Lambda = 1$) is most plausible with high Ω_b values: $\Omega_b = 0.05$ and $H_o = 70$ km/s/Mpc (Boomerang); and $\Omega_b = 0.07$ and $H_o = 60$ km/s/Mpc (Maxima-1). This implies that the standard big-bang model [5–7] for primordial nucleosynthesis, which predicts $\Omega_b \sim 0.02$ ($H_o = 70$ km/s/Mpc), may be in serious trouble.
Astrophysical Quest: The CMB power spectrum is in favor of the flat universe model $\Omega_M + \Omega_\Lambda = 1$, as required by inflationary cosmology [8–10]. The higher baryon density $\Omega_b = 0.05 \sim 0.07$ should be explained in better big-bang models for primordial nucleosynthesis.
Nuclear Astrophysical Quest: The baryon inhomogeneous big-bang model [11–14] allows Ω_b as large as 0.08, predicting 3 – 4 orders of magnitude larger

---

[1]) Managed by UT-Battelle, LLC, for the U.S. Department of Energy under contract DE-AC05-00OR22725.

primordial abundance of ^{9}Be and ^{11}B. Since (n), ^{3}H, ^{8}Li, ^{8}B play significant roles in producing A > 8 elements in an inhomogeneously segregated neutron/proton number density distribution, many reactions of lights nuclides including ^{3}H as well as ^{8}Li should be studied. Several key reactions are ^{4}He(^{3}H,γ)^{7}Li, ^{8}Li(^{4}He,n)^{11}B*, and ^{7}Li(^{3}H,n)^{9}Be, whose reaction cross sections are poorly known.

The neutrino degenerate universe models allow $0.015 < \Omega_b < 1$ in the inhomogeneous [15] and even homogeneous [16] Big-Bang models that indicate strong dependence on Ω_b of the primordial abundances of ^{11}B and the heavier elements. Since ^{11}B is produced in the ^{4}He(^{3}H,γ)^{7}Li(^{4}He,γ)^{11}B and ^{4}He(^{3}He,γ)^{7}Be(^{4}He,γ)^{11}C($\beta\nu$)^{11}B reactions, these reactions and also the competing destruction reactions ^{11}B(x,y) and ^{11}C(x,y) (with x = p, ^{3}H, ^{4}He, and y = γ, p, n) should be studied.

The ^{15}O(α,γ)^{19}Ne Reaction

The ^{15}O(α,γ)^{19}Ne reaction is important in stellar explosions such as novae and X-ray bursts [17]. At sufficiently high stellar temperatures and densities, this reaction bypasses a waiting point in the Hot CNO cycle - the slow beta-decay of ^{15}O. It triggers the reaction sequence ^{15}O(α,γ)^{19}Ne(p,γ)^{20}Na(p,γ)^{21}Mg ... which is thought to be one of two processes by which CNO "seed" nuclei may be processed to masses greater than 20. Even a small "leakage" through this reaction cycle can influence the abundances synthesized and energy generated during stellar explosions.

The rate of the ^{15}O(α,γ)^{19}Ne reaction is dominated by the contribution from a resonance at E_{cm} = 504 keV, corresponding to the 4033-keV state in ^{19}Ne [18]. The strength of this resonance is determined by the alpha width, and is estimated to be extremely small - on the order of μeV [19]. A direct measurement with an ^{15}O radioactive beam will be difficult: with a current of 10^{10} s^{-1}, a gas target of density 10^{18} cm^{-2}, and a recoil detection efficiency of 40 %, the count rate is expected to be less than 1 per hour. Such a measurement is planned at the TRIUMF ISAC facility with the DRAGON recoil separator [20]. An alternative approach with an ^{15}O radioactive beam is to measure an alpha-transfer reaction such as ^{15}O(^{6}Li,d)^{19}Ne to determine spectroscopic factors of the resonance of interest; this technique can be calibrated by making both direct (α,γ) and indirect (^{6}Li,d) measurements on higher lying states which have a higher cross section [21].

A number of experiments have attempted to determine the strength of this resonance via measurements of the decay properties (i.e., the alpha-branching ratio). This approach enables the resonant contribution to the reaction rate to be calculated. The ^{19}F(^{3}He,t)^{19}Ne*(α)^{15}O reaction was used to measure alpha-branching ratios states with energies greater than the 4033-keV level [18]. The sensitivity of this measurement for alpha-branching ratios was approximately 1 %, due to small alpha particle detector solid angle and a strong background from the ^{19}F(^{3}He,d)^{20}Ne reaction. Therefore, this experiment was not sensitive to the 504-keV level which is expected to have a branching ratio of 10^{-4}. A recent updated version of this measurement [22] with larger alpha detectors failed to give an improvement over the earlier study. At Louvain-la-Neuve, a novel approach using a radioactive ^{18}Ne beam and the ^{18}Ne(d,p)^{19}Ne*(α)^{15}O reaction attempted to determine this branching ratio via a triple coincidence measurement [23]. However, the beam intensity was too low to get a value for the 504-keV resonance alpha-branching ratio.

Progress in obtaining this branching ratio will require populating the resonant state of interest at the highest rate possible and measuring the very weak α-decay branch with the highest efficiency. There is still opportunity to use stable beam transfer reactions such as ^{19}F(^{3}He,t)^{19}Ne*(α)^{15}O to get the alpha-branching ratio. Such a measurement would require a high-efficiency spectrometer with a focal plane detector to measure the tritons, giving the rate of population of the 4033-keV state. The tritons would be measured in coincidence with a modern array of silicon strip detectors in the target chamber covering a very large solid angle to get the alpha particles from the decay of this level. Such an approach, essentially an updated version of the setup used in [18], could potentially determine the branching ratio of this important 4033-keV ^{19}Ne level and hence its contribution to the ^{15}O(α,γ)^{19}Ne reaction rate.

The p-process

p-nuclei: Approximately 30 stable nuclides heavier than iron (the p-nuclei) have a different nucleosynthetic origin from the s- and r-process nuclei produced by neutron capture [24]. They are present on the proton-rich side of the valley of the beta stability with small solar system abundance (0.01 - 1%). Possible scenarios of synthesizing the p-nuclei are photodisintegration of s- and r-process nuclei [(γ,n), (γ,α), and (γ,p) reactions] in balance with inverse capture reactions in the O/Ne-rich layers in Type II supernovae (SN II) [25–28], and/or the rp-process in accreting neutron stars or black holes [29]. The recent calculation based on a realistic SN II model [30] shows that the former scenario can overcome the notorious underproduction of Mo and Ru when taking into account the uncertainties allowed in the NACRE compilation [31] in the rate of the ^{22}Ne(α,n)^{25}Mg reaction.

Destruction: As discussed in [32], the rate of the (γ,n) photodisintegration is determined by the energy integral of the product of two factors: the energy distribution of photons in the stellar photon bath (Planck distribution) and photoneutron cross sections. The product defines a narrow energy window for the (γ,n) reactions similar to the Gamow window for the charged-particle-induced reactions. The effective energies lie immediately above the neutron threshold with a typical width $\sim$ 1 MeV. A pioneering work of measuring the (γ,n) reaction rate on platinum and mercury isotopes has been undertaken at Darmstadt using the superposition of bremsstrahlung spectra with different endpoint energies [32,33]. The activation technique employed in this work has a great sensitivity to (γ,n) destruction reactions on p-nuclei with small abundance.

Construction: On the other hand, the construction of the p-nuclei proceeds through (γ,n) reactions on abundant parent nuclei. It seems that the (γ,n) reactions need cross sectional reinvestigation despite the elaborate efforts of measuring (γ,n) cross sections for E1 giant resonances in the past [34,35]. The two major groups at Livermore and Saclay employed the same quasi-monochromatic γ-ray source of positron annihilation in flight but different neutron detectors: BF_3 counters embedded in paraffin or polyethylene moderators; and Gd-doped liquid scintillator, respectively. It is rather well known that the absolute cross sections of the two groups are often inconsistent with each other. In particular, cross sections near neutron thresholds need to be measured with caution from astrophysical viewpoint where the conventional Lorentzian fit considerably overestimates the data.

LIC γ source: A new γ-ray source called laser inverse-Compton (LIC) γ rays has become available at the Electrotechnical Laboratory (ETL) [36]. Quasi-monochromatic γ rays of 1 - 40 MeV are produced by inverse Compton scattering of Nd:YLF laser photons incident on relativistic electrons in the accumulator ring TERAS. The γ-ray energy is varied by changing the electron energy. The pencil-like beam has a flux of $\sim 10^4$ photons/sec/mm^2 and energy spread of a few % in FWHM below 9 MeV [36]. The advantage of the LIC γ rays is that they form an intense peak at energies of astrophysical relevance compared with the continuum bremsstrahlung near the endpoint energy and that they are purely quasi-monochromatic compared with the positron annihilation source which includes a bremsstrahlung component. The LIC γ beams were used to measure (γ,n) cross sections on ^{9}Be [37]. This γ-ray source enables one to measure photoneutron cross sections by direct neutron counting for the production of the p-nuclei. An interesting case is the ^{139}La(γ,n)^{138}La reaction for an odd-odd p-nucleus ^{138}La which faces a serious underproduction in the p-process calculation [28].

A test: The (γ,n) cross section for ^{197}Au is one of best-documented cross sections measured with different γ sources and neutron-counting techniques, but highly controversial results came out (see [33]). As a test case, the cross section was re-measured near the neutron threshold with both the LIC photons [38] and the bremsstrahlung [33].

^{180}**Ta:** In the construction process, not only total (γ,n) cross sections in the low-energy tail of the E1 giant resonance but also the branching to final states in the p-nuclei are often important to measure. For example, the rarest solar nuclide ^{180}Ta is present not as the ground state ($T_{1/2}$ = 8.1 hr) but as an isomeric 9$^-$ state ($T_{1/2} \geq 10^{15}$ yr). Although the s-process [39,40] and the ν-process [41] may contribute to the synthesis of the rarest nuclide, we essentially lack experimental knowledge of the ^{181}Ta(γ,n) cross sections leading to the 9$^-$ state in ^{180}Ta. This cross section $\sigma(9^-)$ can be determined from $\sigma(9^-) = \sigma(total) - \sigma(gs)$, where $\sigma(total)$ is the total (γ,n) cross section on ^{181}Ta and $\sigma(gs)$ is the ^{181}Ta(γ,n) cross section leading to the ground state of ^{180}Ta. The total cross section can be determined by means of the direct neutron counting, while the latter with the activation technique. In the activation technique, one can measure 93 keV-γ transitions in ^{180}Hf or converted K X-rays which is produced by the electron capture of ^{180}Ta(gs) [24 % branching].

The Re/Os Cosmochronometer

Probing final excited states in photonuclear reactions may help to make the Re/Os cosmochronometer [42] more reliable. The s-process contribution to the abundance of ^{187}Os through the 9.8 keV state populated in the stellar photon bath is an important issue to the cosmochronometer [43]. Beside the inelastic neutron scattering measurement on ^{187}Os [44], insight into the ^{187}Os(9.8)(n,γ)^{188}Os reaction may be obtained in the inverse process, i.e., photoexcitation of ^{188}Os. After photoexciting ^{188}Os to the low-energy tail of the E1 giant resonance, the branching to the excited states in ^{187}Os including the 9.8 keV state may be probed by using the technique of γ-ray spectroscopy.

Acknowledgement

H.U. and M.S. are grateful to T.K., who made unusual participation into the round table discussion in Tours2000 by fax, for sharing the authorship of this article. H.U. is also grateful to C. Rolfs, who constructed the main scheme of the round table discussion on experimental nuclear astrophysics but regrettably failed to appear in Tours, for writing a separate summary for key reactions in static burning in stars. Due to the divided roles of writing the summary, it turned out that the Itahashi's result with OCEAN found no place to be accommodated. Herein we congratulate on his first (*not last*) result of ^{3}He(^{3}He,2p)^{4}He cross sections in the energy range of 40 - 50 keV with expectation of *subsequent* results. H.U. regrets that the round table discussion suffered from a tight schedule because it was sandwiched between the extended lunch time and the rigorous-departure time for excursion. Nevertheless, we shared such feeling that this lively style of discussions can be a uniqueness of Tours Symposium.

REFERENCES

1. Bernardls P *et al.*, *Nature* 404: 955 (2000)
2. Lange A *et al.*, preprint *(astro-ph/0005004)* (2000)
3. Hanany S *et al.*, submitted to *Astrophys. J. Lett*, and preprint *(astro-ph/0005123)* (2000)
4. Balbi A *et al.*, submitted to *Astrophys. J. Lett*, and preprint *(astro-ph/0005124)* (2000)
5. Wagoner RV, Fowler WA, and Hoyle F, *Astrophys. J.* 148:3 1967
6. Yang J *et al.*, *Astrophys. J.* 281:493 1984
7. Walker TP *et al.*, *Astrophys. J.* 376:51 1991
8. Guth AH, *Phys. Rev. D* 23:347 1981
9. Linde AD, *Phys. Lett. B* 108:389 1982
10. Albrecht A, and Steinhardt PJ, *Phys. Rev. Lett.* 48:1220 1982
11. Alcock CR, Fuller GM, and Mathews GJ, *Astrophys. J.* 320:439 1987

12. Applegate JH, Hogan C, and Scherrer RJ, *Astrophys. J.* 329:572 1988
13. Kajino T, and Boyd RN, *Astrophys. J.* 359:267 1990
14. Orito M, Kajino T, Boyd RN, and Mathews GM, *Astrophys. J.* 488:515 1990
15. Kajino T, and Orito M, *Nucl. Phys. A* 629:538c (1998); Orito M, and Kajino T, (2000), in preparation
16. Orito M, Kajino T, Mathews GJ, and Boyd RN, *submitted to Astrophys. J., and preprint (astro-ph/0005446)* (2000)
17. Champagne AE, Wiescher M. *Ann. Rev. Nucl. Part. Sci.* 42:39 (1992)
18. Magnus PV *et al.*, *Nucl. Phys. A* 506:332 (1990)
19. Mao ZQ *et al.*, *Phys. Rev. Lett.* 74:3760 (1995)
20. D'Auria J. In *Nuclei in the Cosmos V*, ed. N Prantzos, S Harissopulos, pp. 435. Gif-sur-Yvette, France:*Editions Frontières Conf. Proc.* (1998)
21. Parker PD. Private Communication (1995)
22. Kubono S *et al.*, In *Nuclei in the Cosmos V*, ed. N Prantzos, S Harissopulos, pp. 409. Gif-sur-Yvette, France:*Editions Frontières Conf. Proc.* (1998)
23. Laird A *et al.*, In *Nuclei in the Cosmos V*, ed. N Prantzos, S Harissopulos, pp. 415. Gif-sur-Yvette, France:*Editions Frontières Conf. Proc.* (1998)
24. Lambert D.L., *Astron. Astrophys. Rev.* 3:201 (1992)
25. Woosley S.E., Howard W.M., *Astrophys. J. Suppl.* 36:285 (1978)
26. Rayet M., Prantzos N., Arnould M., *Astron. Astrophys.* 227:271 (1990)
27. Prantzos N., Hashimoto M., Rayet M., Arnould M., *Astron. Astrophys.* 238:455 (1990)
28. Rayet M. *et al.*, *Astron. Astrophys.* 354:740 (1995)
29. Schatz H. *et al.*, *Phys. Rep.* 294:167 (1998)
30. Costa V., Rayet M., Zappala R.A., Arnould M., *Astron. Astrophys.* 358:L67 (2000).
31. Angulo E. *et al.*,, *Nucl. Phys. A* 656:3 (1999)
32. Morh P. *et al.*, Phys. Lett. B, in press
33. Vogt K. *et al.*, *contributed paper in this symposium*
34. Dietrich S.S., Berman B.L., *At. Data Nucl. Data Tables* 38:199 (1988)
35. Boboshin I.N. *et al.*, *The center for photonuclear experiments data (CDFE) nucler data bases*, http://depni.npi.msu.su/cdfe
36. Ohgaki H. *et al.*, *IEEE Trans. Nucl. Sci.* 38:386 (1991)
37. Utsunomiya H. *et al.*, *Phys. Rev. C*, in press
38. Utsunomiya H. *et al.*, *unpublished*
39. Yokoi K., Takahashi K., *Nature* 305:198 (1983)
40. Némes Zs., Käppeler F., Reffo G., *Astrophys. J.* 392:277 (1991)
41. Woosley S.E. *et al.*, *Astrophys. J.* 356:272 (1990)
42. Clayton D.D., *Astrophys. J.* 139:637 (1964)
43. Yokoi K., Takahashi K., Arnould M., *Astrophys. J.* 117:65 (1983)
44. Mengoni A., *contributed paper in this symposium*

ATOMIC CLUSTER PHYSICS (ACP)

Atomic clusters and atomic nuclei

Claude Guet

*Département de Recherche Fondamentale sur la Matière Condensée, CEA-Grenoble,
17, rue des Martyrs, F-38054 Grenoble CEDEX 9, France*

Abstract. In simple metals the conduction electrons are approximatively independent and free. This is also the case for clusters of simple metal atoms. Nucleons in nuclei also behave as delocalized and independent fermions. This generic behavior invites us to bring out analogies between metal clusters and nuclei. The analogies manifest themselves in the elementary modes of excitation of finite quantum systems. However there are also major differences that arise from the presence of ions in metal clusters. Fission of nuclei and clusters exemplifies these differences.

INTRODUCTION

In the last decade there has not been any important conference in cluster physics without contributions from some nuclear physicists and in return there has been many nuclear physics meetings that explicitely devoted at least one session to atomic clusters. Several workshops have specifically focused their theme on analogies and differences between atomic clusters and nuclei. Why are atomic clusters so interesting for nuclear physics and vice-versa?

Clusters are aggregates of atoms or molecules with a well-defined size varying from a few constituents to several tens of thousands. Cluster physics lies between atomic and molecular physics on the one hand and condensed matter physics on the other. It is the paradigm for understanding how matter builds up from its elementary constituents. Thus studying clusters may shed new light on bulk matter and theoretical tools suited to it. Moreoever the finite nature of the number of constituents leads to novel structural and thermodynamic properties with no equivalent in the bulk. As a prime fact, clusters are distinguished from bulk matter in so far as their properties are strongly affected by the existence of a surface involving a large fraction of the constituents; for example in a cluster of 55 atoms of argon more than 30 are on its surface.

The nature of forces binding the atoms inside atomic clusters rules their classification. In clusters of metal atoms the bonding is provided by the delocalized valence electrons. This implies that ions embedded in the electron gas are subjected to genuine many-body forces. In metal clusters the spectrum of electronic eigenstates at low temperature consists of discrete, separated levels. The level separation can be so large that this structure, which acts in addition to the geometric shell structure, should modify considerably the optical, magnetic, and superconducting properties compared to those of the bulk metal. Experimental measurements carried out in recent years on free, unsupported clusters confirm these special properties due to the spatial confinement of the electrons [1–4].

Undoubtedly the physics of metal clusters owes much to nuclear physics [5–8]. Over more than 50 years nuclear theory has developed concepts and powerful methods to elucidate the complex many-body problem for finite fermionic systems. Their extension to metal clusters which often was carried by nuclear theorists provided basic models for understanding the abundant set of data from cluster beam experiments. By a favourable twist of fate nuclear physics has gained a noticeable impetus from cluster physics. This is the case for the investigation of pairing and odd-even staggering, shells and supershells which we shall discuss later.

A Common features in clusters and nuclei

Low energy nuclear physics contemplates atomic nuclei as an ensemble of N neutrons and Z protons held together by an effective short range nuclear force. The effective force which implicitely contains all underlying physics arising from the mesic fields or quark contents results fromthe many-body screening of the in medium strong force. It is characterized by a set of Landau parameters for the Fermi liquid which nuclear matter is. This is similar to the effective electron-electron interaction in an electron liquid. As to the structure and dynamics of both nuclei and metal clusters not only the effective interaction plays a role but the determining factor is the number of constituents.

Consider n interacting fermions occupying a volume R^3 with a Fermi energy, ϵ_F. For a metal cluster, saturation implies that $R = r_S n^{1/3}$ where r_S is the Wigner Seitz radius of the corresponding bulk metal, and $\epsilon_F \simeq 2/r_S^2$ (atomic units ,$\hbar = m = e = 1$, are used). Near the Fermi surface the average energy-level spacing is of the order of

$$\delta\epsilon \approx \epsilon_F/n \approx \frac{\hbar v_F}{R} n^{-1/3}, \qquad (1)$$

where v_F is the Fermi velocity. For $n \approx 100$ and $\epsilon_F \simeq 3eV$ one sees that the electronic level splitting will not be washed out by cristal field effects which are less than $0.01eV$. This explains why the quantum properties of metal clusters are expected to be overwhelmingly dominated by the finite system of delocalized electrons. The first strong experimental evidences came with the pioneering beam experiments that W.D. Knight and his group carried at Berkeley in the mid eighties [9]. Sodium clusters were produced by expansion of sodium vapor into a vacuum through a supersonic nozzle. The experiment consisted in measuring the occurence frequency of different sizes. The abundance spectrum shows pronounced peaks for clusters having 8, 20, 40 atoms. These numbers can be associated with the closure of shells for fermions in a three-dimensional harmonic oscillator. Spherical shell closure leads to an enhanced stability which explains the experimental observation as there is one delocalized electron per atom.

The observation of magic numbers associated to electrons immediately triggered the interest of nuclear physicists who looked at metal clusters as an exciting challenge for theory: hopefully one can get a deeper insight into properties of finite fermionic systems by extending the study to systems of electrons under various conditions. Unlike nuclei whose size is limited by the Coulomb repulsion between protons ($A = N + Z \leq 300$) neutral clusters have sizes which can vary as much as one likes. There is no real experimental limitation to the number of atoms inside the cluster as far as the electronic shell structure still dominates the competing structure due to the geometrical arrangement of ions (also leading to magic numbers) without being washed out by too a large temperature. Over the years nuclear theory has provided a wealth of predictions of how specific features of nuclear properties (collective excitations for example) ought to depend upon the size A. As these A-dependences often are weak real nuclei just probe the low-A tail. The most spectacular example is the following. The theoretical prediction of fermionic supershells in a cavity [10] could never be confirmed in nuclei because the size period of oscillations is large. Twenty years later in a dramatic manner the new availability of metal clusters with sizes up to several thousands atoms provided the missing confirmation [11].

Theoretical models and experimental facts have shown that small alkali-metal clusters of simple metal atoms have shapes of their lowest energy configuration which most often are far from spherical. This is also the case for nuclear shapes. This is a manifestation of the Jahn-Teller effect well known in molecular physics. A fermion system with a degenerate partially occupied last shell spontaneously deform in order to lift the degeneracy. This spontaneous symmetry breaking lowers the total energy of the system. In molecules there is a rearrangement of the atoms whereas in nuclei it is the self-induced shape that deforms. For metallic clusters a reasonable approximation is to neglect all details of the ionic structure and to replace it by a uniform positive charge density distribution having the bulk density, and correspondingly a given r_s parameter. This is known as the jellium approximation. A further approximation, called the ultimate jellium, gets rid of this density parameter and assumes that this positive distribution exactly adjusts to the electronic density ensuring local neutralization [12]. This is just the extension to finite systems of the electron liquid whose saturation density and energy are unique. The lowest-energy shapes of such electron liquid clusters are strikingly similar to those of symmetric nuclei with the same number of fermionic constituents [12]. This clearly shows that to lowest order it is not the interaction that governs the shape but the number of fermions in a self-generated system.

Pairing correlations between nucleons lead to differences between odd and even nuclei that are observed on many nuclear properties such as level densities near the ground state, binding energies, moments of inertia, alpha decay and fission probabilities. Pairing correlations are intimately related to superconductivity (or superfluidity) in macroscopic systems. In superconductors there is a condensate of bosons. These identical bosons are the pairs of bound electrons near the Fermi surface. The coherence length, ξ, which is of the order of the spatial extent of the correlated Cooper pairs, is of the order of

$$\xi = \frac{\hbar v_F}{\Delta} \qquad (2)$$

where Δ is the energy associated with the binding of the pairs. Conventional superconductors have small Δ (less than 1 mev) and thus coherence lengths larger by several orders of magnitude than the typical lengths of the material. This leads to striking phenomena such as the Meissner effect and flux quantization. For a supposedly superconductor small particle of radius R the coherence length in units of R is, according to Eq.(1 and Eq.(2):

$$\frac{\xi}{R} = \frac{\delta}{\Delta} n^{2/3} \ . \qquad (3)$$

For nuclei ($\Delta \approx 1 MeV$, v_F is about 0.3 the light velocity) the coherence length is of the order of $50 fm$ that is about 5 times larger than the radius of the heaviest nuclei. Thus the spectacular macroscopic features directly show neither in the nuclei nor in the metallic particles of nanometric sizes. The question that immediately arises is that of a critical size for superconductivity. This is a tricky question. As we can infer from nuclear physics there ought to be observables that are sensitive to pairing correlations even for sizes well below the coherence length.

Indeed experiments on aluminum grains of radius less than a hundredth of the coherence length have revealed the existence of a superconducting excitation gap which is driven to zero by an applied magnetic field [13]. Moreover the experiment can tell whether the number of electrons ($\sim 10^6$) is odd or even. As in nuclei the number of fermions in a cluster is constant. This strict constraint requires a theoretical treatment beyond the standard BCS theory in order to restore the symmetry. In this respect powerful methods are now available [14]. In addition one should address the question of the origin and magnitude of the attractive interaction in small metallic clusters namely what is the role of phonons in small clusters?

In addition to the analogies based on finite Fermi liquids, there are other fruitful analogies between nuclei and atomic clusters either metallic or not. As finite systems both nuclei and atomic clusters can undergo fragmentation. Fission and fragmentation are phenomena entirely specific to finite systems. They yield much information on density fluctuations and on the effects of long-range interactions. Their study requires first of all a thorough understanding of the thermodynamics of small systems. As to the latter, methods for exploring the multidimensional potential energy surface of a complex system of n particles (rich in minima and saddle points) as well as classical and quantal techniques for determining thermodynamic properties of clusters of various types have all undergone rapid development in recent years, whether these techniques are analytical or numerical [15,16]. The study of phase changes in clusters has revealed to us the full richness (and complexity) of finite systems [17]. One observes rapid variations of thermodynamic properties but no discontinuity, and bands of phase coexistence rather than simple curves. Nuclear physicists have been much concerned with the nuclear equation of state. In order to approach it one has no alternative to the study of phases and phase changes in finite nuclei: normal-superfluid, liquid-gas, quark-gluon confinement-deconfinement. It is a privilege of cluster physics to have the thermodynamics properties of corresponding bulk matter a priori available. Because we can study small systems in detail and follow how they behave as they approach the large number limit we can use them to get new insight into the phases and phase transitions of bulk matter. Moreover the lessons we would have learnt from cluster physics will surely be valuable for the extrapolation procedures that are unavoidable in the nuclear case.

B Differences between clusters and nuclei

So far we have emphasized the analogies between clusters and nuclei. We deliberately have had a simplistic approach aimed at pointing out some generic features of finite quantum liquids and finite size effects. However the realm of clusters extends far beyond. The major (but trivial) difference that immediately suggests itself is that nuclear matter is unique (it has a single equation of state just like the electron liquid) whereas all elements around us have their own equation of state. As for bulk matter, it is the constituent element in a cluster that will determine its density and its binding energy. In the ultimate jellium which was mentioned earlier there is only one type of cluster which happens to be close to sodium.

Powerful specific methods to understand the physics and chemistry of clusters have been steadily improved. One has to cope with a much more complex N-body problem than encountered in nuclear physics. The complexity arises from the different energy scales and the low level of symmetry of disordered structures. Let us consider our favourite metal clusters. Beside the delocalized electrons in the few eV range and with a typical femtosecond dynamics there are ions. The core electrons are in the energy range of the order of tens of eV. It implies that a detailed description of the electron-ion interaction which should take into account all core electrons will preferentially leave the place to pseudo-potential methods. A pseudo-potential has its first eigenstates and eigenvalues matching the valence states of the atom with frozen core ion; most often this is a non-local potential. Ions have masses of the order of several tens of thousands of electron masses. They move at a much slower pace than the electrons within a time scale of the order of picoseconds. One immediately realizes that any accurate study of dynamical processes will face a huge obstacle due to these very different timescales [18]. As an example consider the decay of some collective excitation. In the nuclear case a dynamical theory based on real time like the Time Dependent Hartree-Fock method would cope with this problem without too much difficulties because nucleonic motion and collective motion (surface modes, compressional modes,..) are on the same scale. In the metal cluster case an electronic excitation can decay not only into particle-hole excitations but also into vibrational excitations of the ionic structure through the electron-phonon coupling. This coupling is weak and requires to carry a coupled electron ion dynamics calculation over a picosecond timescale. But the time step must be very small, less than a tenth of femtosecond in order to preserve a reasonable description of the electron dynamics [18].

Both nuclei and charged metallic clusters fission whenever the electrostatic pressure can no longer be overcome by the surface pressure. While the later tends to preserve the spherical shape, the former tends to deform and thus to destabilize the drop. Lord Rayleigh investigated the stability of surface-charged droplets against infinitesimal multipole fluctuations. The first mode to be destabilized with increasing charge is the quadrupole. It is actually in nuclear physics that the richness of the phenomenology associated with charged liquid drops has been most revealed. Here the charge is carried by the protons and the isospin part of the strong nuclear interaction forces the electrical charge to be homogeneously distributed over the whole volume instead of being surface distributed as in conducting droplets. This is a dramatic difference. Fission of nuclei leads to a quasi-symmetric binary division. As Rayleigh had expected, it is indeed the quadrupole mode that triggers the large amplitude deformation which develops in a multidimensional energy surface and brings the nucleus to some saddle-point. Beyond this saddle the nucleus irreversibly goes towards scission. In charged metal clusters the net electric charge is highly mobile on the surface. So to speak there is a decoupling of charges and masses. As a consequence Coulomb thermally unstable metallic clusters decay essentially by the emission of small singly charged fragments [19] and so far there has not been any clear evidence for symmetric fission.

REFERENCES

1. de Heer W., *Rev. Mod. Phys.* **65**, 611 (1993).
2. Nishina Y. and Sugano S. (Eds), *Small Particles and Inorganic Clusters, ISSPIC7*, 1994, Kobe, Japan, *Surf. Rev. Lett.* **3** (1996).
3. Andersen H. H. (Ed), *Small Particles and Inorganic Clusters, ISSPIC8*, 1996, Copenhagen, Denmark, *Z. Phys. D* **40** (1997).
4. Bonard J. M. and Châtelain A. (Eds), *Small Particles and Inorganic Clusters, ISSPIC9*, 1998, Lausanne, Switzerland, *Eur.Phys.J. D* **9** (1999).
5. Brack M. ,*Rev. Mod. Phys.* **65**, 677 (1993).
6. Abe Y.*et al* (Eds), *Similarities and Differences between Atomic Nuclei and Clusters*, July 1997, Tsukuba, Japan, *AIP Conference Proceedings* **416** (1998)
7. Bertsch G. F. and Broglia R. A.*Oscillations in Finite Quantum Systems*. Cambridge University Press (1994)
8. Guet C. *et al* (Eds), *Atomic Clusters and Nanoparticles, Les Houches Lectures Session LXXIII* Springer Verlag (2001)
9. Knight W D. 1984 *et al*, *Phys. Rev. Lett.* **52**, 2141 (1984).
10. Balian R., and Bloch. C.,*Ann. Phys.* **69**, 76 (1971).

11. Pedersen J. *et al* ,*Nature* **353**, 733 (1991).
12. Häkkinen H. *et al*, *Phys. Rev. Lett* **78**, 1034 (1997).
13. Black C. T. , Ralph D. C., and Tinkham M., *et al* , *Phys. Rev. Lett* **76**, 688(1996).
14. Balian R., Flocard. H., and Vénéroni M., *Phys. Rep.* **317**, 251(1999).
15. Wales D. J. in *Atomic Clusters and Nanoparticles, Les Houches Lectures Session LXXIII Springer Verlag* (2001)
16. Chekmarev S., in *Atomic Clusters and Nanoparticles, Les Houches Lectures Session LXXIII Springer Verlag* (2001)
17. Berry R. S. in *Theory of Atomic and Molecular Clusters, Ed.: Jellinek J.*, Springer , 1999, pp. 1-26.
18. Blaise P., Blundell S. A., and Guet C. , *Phys. Rev. B* **55,** 15856 (1997).
19. Näher U. *et al*, *Physics Reports* **275**, 245 (1997).

Shell Effects on Fission of Metal Clusters

Masato Nakamura

*College of Science and Technology, NIHON University,
7-24-1, Narashino-dai, Funabashi 274-8501 Japan*

Abstract. Shell effects play important roles in the fragmentation of multiply charged metal clusters. They affect the competition between fission and evaporation near the critical size. It also determines the product size distribution of fission fragments. Due to shell effects, decay to energetically most favorite channel is observed when the parent cluster is relatively cold. In the case where parent clusters are hot, the product size distribution shows even-odd oscillation. The spheroid model qualitatively explains these phenomenon.

INTRODUCTION

When alkali- or noble- metal atoms make up a cluster, a valence electron moves relatively flat and smooth mean potential since positive background charge is screened by the other electrons. Because of this fact, the electronic shell structure similar to that in a nucleus is observed in metal clusters. In this article, we will discuss how the shell effects manifest themselves in the fragmentation of metal clusters.

In this decade fragmentation (fission and evaporation) of multiply charged metal clusters has been experimentally observed. [1] In the fragmentation of multiply charged clusters, fission and evaporation are competitive processes unless a parent cluster is extremely cold. They are similar to nuclear fission and neutron evaporation in heavy nuclei, respectively. Thus the liquid drop model (LDM) which has been used in the study of nuclear reactions is valid to expresses the energetics of these processes. If we follow the model, the fissility parameter, which is the half the ratio of the Coulomb energy to the surface energy determines the stability of a multiply-charged cluster. This parameter is proportional to z^2/n, where z is the charge of a parent cluster and n the number of atoms in the cluster, while the energy required for monomer evaporation is almost independent of the size. Thus if we restrict ourselves to fragmentation from doubly charged clusters, larger clusters tend to evaporate while small clusters undergo fission. Such a tendency has been observed in fragmentation of alkali- and noble-metal clusters [2-4]. The LDM well explains this trend [2]. As for the product size distribution of fission fragments, however, we see that the model cannot account for. For example, experiments by Mainz group shows that emissions of trimer ions are predominantly observed among various fission channels from doubly charged silver clusters.[4] In the early experiments by Katakuse and coworkers[5], on the other hand, the fragment size distribution from doubly charged silver cluster shows somewhat different profile. It has an even-odd oscillation as a function of product size. In both cases, the shell effects seem to govern the distribution.

The trimer ion is extremely stable for its shell-closing structure. Even-odd oscillation is due to doubly occupancy of a single level. To shed light on the shell effects in fission of metal clusters, we propose the following model. As a typical example, the model is applied to the fragmentation of doubly charged silver clusters.

MODEL

The Touching Spheroid Method

The energy diagram for the fission ($Ag_n^{2+} \rightarrow Ag_p^+ + Ag_{n-p}^+$) and the monomer evaporation ($Ag_n^{2+} \rightarrow Ag_{n-1}^{2+} + Ag$) of doubly charged silver clusters is schematically shown in Fig. 1. The most important physical parameters to describe these processes are the fission barrier and energy required for monomer evaporation, respectively. Once we get these values, we can obtain product size distribution as well as lifetime using a statistical model. The calculation of the evaporation energy is relatively easier because it is given by the energy difference. It is not easy, however, to determine the barrier height from a first principle calculation. For it is difficult to determine the geometric configuration of atoms in a cluster at the transition state. To bypass this difficulty, the touching sphere model [1] is often used. The model assumes that two spherical fragment clusters are contact at the transition state.

Let us denote E_n^{m+} for the energy of an isolated Ag_n^{m+}, the fission barrier height $B_{n,p}^{2+}$ is written as

$$B_{n,p}^{2+} = E_{contact} - Q_{n,p}^{2+}, \qquad (1)$$

with

$$E_{contact} = \frac{e^2}{r_p + r_{n-p} + 2\delta R}, \qquad (2)$$

where $r_n = r_s n^{1/3}$ is the radius of a daughter cluster, δR the surface diffuseness and the heat of reaction $Q_{n,p}^{2+}$ is written as

$$Q_{n,p}^{2+} = E_p^+ + E_{n-p}^+ - E_n^{2+}. \qquad (3)$$

Since the calculation of $E_{contact}$ is trivial, we have only to calculate the heat of reaction.[1] The energy required for monomer evaporation $D_{n,1}^{2+}$ is written as

$$D_{n,1}^{2+} = E_n^{2+} - E_{n-1}^{2+} - E_1^0. \qquad (4)$$

In this way, both fission barrier and evaporation energy can be obtained from the energy of an isolated cluster. So far, in most treatments, both parent and daughter clusters are assumed to be spherical. Here we are going to replace these spheres by spheroids. In the *touching spheroid model*, we assume that two spheroidal clusters are in contact with keeping an axial symmetry. The shape of the spheroid is optimized

---
[1] Actually the barrier height will be lowered due to the polarization effects

so that the energy of each isolated cluster becomes the minimum value while taking the volume as constant. The energy of the cluster is calculated by using the shell correction method. In what follow, we briefly show the method.

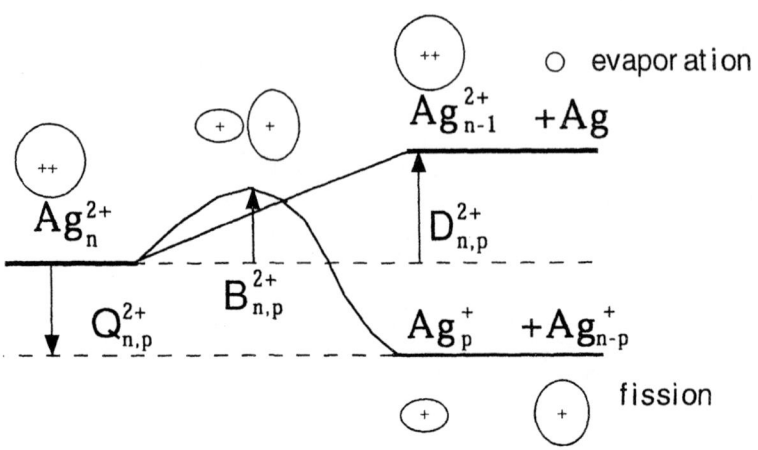

FIGURE 1. Energy diagram for fission and evaporation

The Shell Correction Method

The shell correction method was originally used in the study of nuclear fission by Strutinsky and coworkers. The theory is explained in detail elsewhere[6]. In the model, the total energy is expressed as a sum of two parts; the liquid droplet part E_{LD} and the shell correction ΔE_{shell} namely,

$$E_{tot} = E_{LD} + \Delta E_{shell}. \tag{5}$$

A merit for doing this is that the liquid-drop energy can be calculated from the information on bulk properties of the system. The liquid-drop part is composed of the volume term E_{vol}, the surface term E_{sur}, and the Coulomb energy $E_{Coulomb}$,

$$E_{LD} = E_{vol} + E_{surface} + E_{Coulomb} \tag{6}$$

The volume of a cluster V is assumed to be proportional to the number of atoms in a cluster n,

$$V = \frac{4}{3}\pi r_s^3 n, \tag{7}$$

where r_s is the Wigner-Seiz radius. The Coulomb energy is calculated by assuming that the charge distributes the surface of the cluster.

The shell correction term is written as

$$\Delta E_{shell} = \int_0^{E_H} \varepsilon G(\varepsilon)d\varepsilon - \int_0^{E_F} \varepsilon g(\varepsilon)d\varepsilon, \tag{8}$$

where $E_H(E_F)$ is the highest-occupied (Fermi) level, and $G(g)$ is the quantized (smoothed) level density of the system. In calculating these quantities, we assume that a valence electron is confirmed to the cluster by the square-well potential with an infinity wall. The smoothed level density for such a case was obtained by Balian and Bloch [7]. In the case of spheroid, it is written in an analytical form.

If a cluster is spherical, the calculation is trivial. However, a spherical cluster is often unstable toward deformations in the case where the ground state level is degenerate. In fact, Clemenger [8] showed that clusters are stabilized by deformations in such a case. This situation is nearly the same as the way that Nielsen showed as for nuclei. Here we assume that a cluster is spheroidal. The deformation is optimized so that the total energy of the cluster takes the minimum value with keeping the volume as constant. In Fig. 2, we show schematically an optimized shape of Ag_n^+ for $1 \le n \le 20$. [9] We see that they are spherical for magic numbers but not spherical for non-magic numbers. In Fig. 3, we show the total energies of singly charged silver cluster as a function of n. The volume term is dropped for it is simply proportional to n. The LDM part of the energy shows a smooth variation as a function of n. If we add the shell correction term for a spherical cluster, the total energy has a downward casps at the magic numbers. If we assume that a cluster is spheroidal and optimize deformation, the total energy increases step by step with an interval of either 2 or 4. This is due is the splitting of degenerate levels due to the spheroidal deformation.

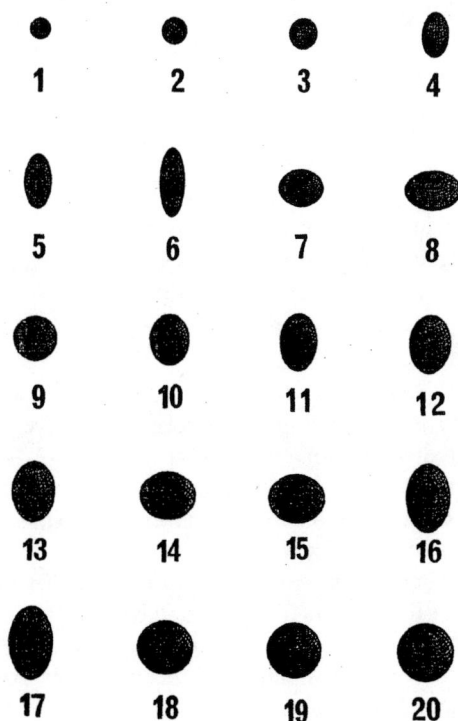

FIGURE 2 An energetically optimized shape of Ag_n^+

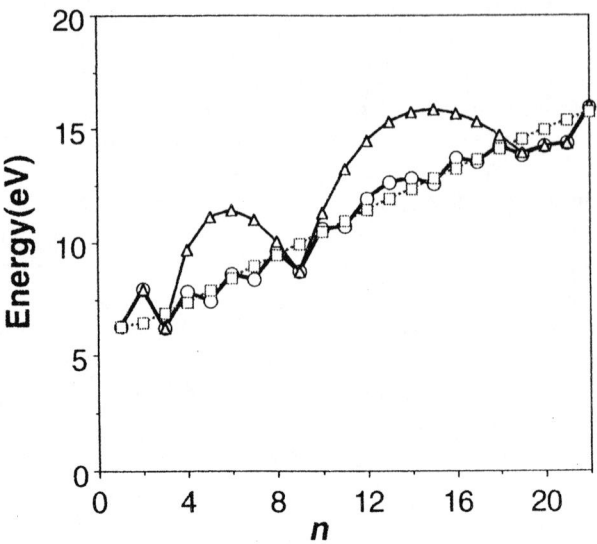

FIGURE 3 The total energies of Ag_n^+ : □ by the LDM, △:the shell correction term is added while cluster is assumed to be spherical and ○ : the shell correction term is added and cluster is energetically optimized spheroidal

COMPARISON WITH EXPERIMENTS

Using the method mentioned above, we calculate the fission barrier and the evaporation energy. We are going to make a comparison with the experiments on fragmentation of doubly-charged silver clusters. One is recent experiment by S. Krückeberg et al. [4d] and the other is early experiment by Katakuse et al.[5]

Comparison with Mainz group results

Krückeberg et al. have observed the collision-induced-dissociation of doubly (and also singly) charged silver clusters. Furthermore they have also recently determined the energy required for fission/evaporation from doubly-charged silver clusters. [4e] In their case, temperature of parent cluster seems to be relatively low since net internal energy in a cluster is estimated as several eV. Thus, it is assumed that only energetically most favorite process takes place. They showed that only fission with a trimer ion and evaporation with a neutral monomer are possible decay channels. Fission is observed for the clusters of $n \leq 16$, and evaporation takes place for n≥17. However, an evaporation is observed also for n=11 and 15. According to the LDM, there exist a critical size n_c above which a cluster undergoes evaporation and below which fission takes place. In this case, the n_c is roughly estimated as 14. Because of shell effects, the evaporation energy for even-electrons system is larger than that for odd-electrons system. Thus, the fission barrier can be lower than the evaporation energy above critical size and *vice versa*.

To probe, we calculate both the fission barrier and evaporation energy from doubly charged silver clusters with using the method mentioned above. Among several fission channels, the energetically most favorite one is chosen as a candidate to compete with the evaporation. Table 1 lists which one is lower than the other between fission and evaporation. In the table, F denotes fission, and E for evaporation. Our calculation reproduces the tendency that smaller clusters tend to undergo fission and larger to evaporate. Furthermore, the calculation also reproduces the experimental result that the evaporation energy is lower than fission barrier for $n=11$. As for n=15, 18, 20, we cannot reproduce the experimental result. For these cases, the fission barrier and the evaporation energy are close each other.[2]

Another aspect of the shell effects consists in the fact that only an emission of trimer ion is experimentally observed among various fission channels. In our calculation, an emission of trimer ion and that of 9-mer ion always compete. They are quite stable for their electronic shell closing structure. However an emission of 9-mer is not at all observed in their experiment except for $n=12$. We do not have a clear explanation why only trimer ion is observed in their case. It should be noted that, the energy barrier is lowest for symmetric fission in the case of $n=18$ and two 9-mer ions are produced. But this fission cannot be detected in the time-of-flight experiments.

TABLE 1. Competitions between fission and evaporation

n	9	10	11	12	13	14	15	16	17	18	19	20	21	22	23	24
Experiment	F	F	E	F	F	F	E	F	E	E	E	E	E	E	E	E
Model	F	F	E	F	F	F	F	F	E	F	E	F	E	E	E	E

Comparison with Katakuse's group results

Katakuse and coworkers observed decays of doubly charged silver clusters in the size range of $12 \leq n \leq 22$ [5]. They produced doubly (and also singly) charged silver clusters by impacting Xe ions with the kinetic energy of 7 keV onto silver surface. They produced clusters in such a way that the parent clusters would have much higher temperature than that in the other experiment [4]. Thus, a variety of fission channels have been observed in their experiment while fragmentation into a few channels was seen in the others. The distribution as a function of fragment size p often shows an even-odd oscillation. Note that only the larger fragment has been directly detected and that the smaller fragment is guessed from the conservation of mass.

If we assume the Arrehnius law, the distribution $f_n(p)$ should be proportional to $\exp(-B_{n,p}^{2+}/k_B T)$, where T is the temperature of the parent cluster and k_B is the Boltzmann constant. Then,

$$\ln[f_n(p)] \propto -B_{n,p}^{2+} . \tag{9}$$

In Fig. 4, we plot the fission-barrier height $B_{n,p}^{2+}$ as a function of fragment size p calculated in the present model together with the logarithmic distribution $\ln[f_n(p)]$ in

---

[2] They have perfectly reproduced experimental results by using empirical values for shell correction energies.[4e]

an arbitrary unit. The calculation of barrier height $B_{n,p}^{2+}$ is made by using the touching spheroid model. For comparison, the energy barrier obtained by the touching sphere model as well as that by the LDM are plotted in the same figure.

Here we will comment for three typical cases. As for other cases, see Ref. 9.

(1) decay from Ag_{16}^{2+}

According to the LDM, the barrier height shows a smooth variation as a function of fragment size p. It is contrary to the experiment. If we add the shell correction term, the touching sphere model reproduces peaks at the magic numbers. But it does not reproduce an even-odd oscillation seen in the experiment. By taking into account of the deformation of clusters, we can reproduce the even-odd oscillation observed in the experiment. By fitting experimental data, we can roughly estimate the temperature as 0.8 eV.[3] From this comparison, it is not liquid drop part of the energy but the shell correction term which determines the product size distribution. Touching spheroid model well explains the experiment. The accordance between the touching spheroid model and the experiment is surprisingly well for n=12-16 and 18.

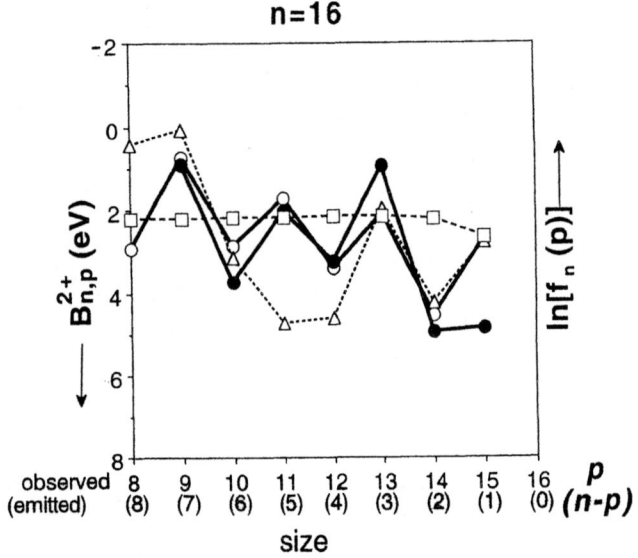

FIGURE 4A Calculated energy barriers for fission from doubly charged silver clusters as functions of product size p, together with the experimentally measured logarithmic distribution $\ln[f_n(p)]$ in an arbitrary unit for n =16 , ●: the experimental data (Katakuse *et al.* [5]) ○ : the touching spheroid model, △: the touching sphere model and □:the LDM

---

[3] This value may be somewhat too high. The shell correction method overestimates the shell correction term by factor of two or three. Then the actual temperature may be much lower than this value

(2) decay from Ag_{17}^{2+}

For this case, the present model cannot explain experimental data. We must notice, however, that the evaporation energy is lower than fission barrier for $n=17$. Then let us consider the following two successive reactions that evaporation takes place before fission occurs, namely,

$$Ag_{17}^{2+} \rightarrow Ag_{16}^{2+} + Ag$$
$$Ag_{16}^{2+} \rightarrow Ag_p^+ + Ag_{16-p}^+ \quad (10)$$

We plot the fission barriers for the second process, namely, barrier height of Ag_{16}^{2+} as a function of p in the same figure. It shows nearly the same trend as the experiment. This fact implies us that monomer evaporation takes place before fission.

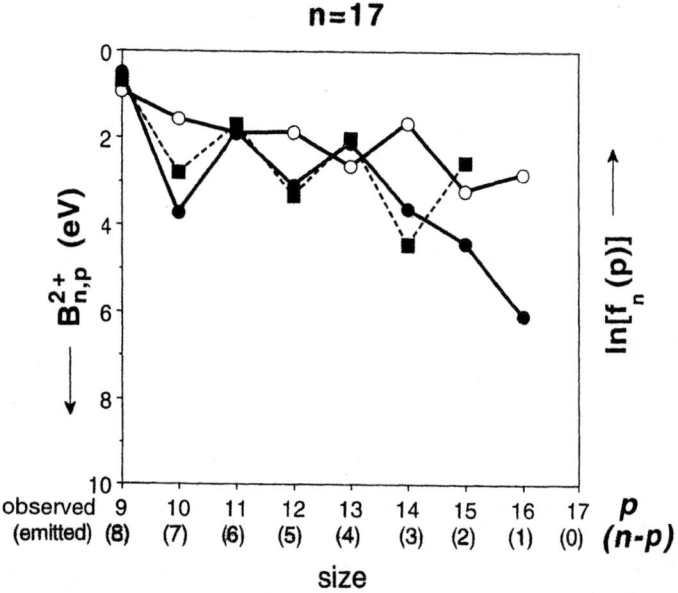

FIGURE 4B The same figure as Figure 4A for $n=17$, ■: the fission barrier for $n=16$ (fission after evaporation)

(3) Decay from Ag_{20}^{2+}

In this case, there is a little discrepancy between our touching spheroid model and the experiment. Our model reproduces an even-odd oscillation seen in the experiment. The experiment shows, however, a larger intensity for quasi-symmetric fission ($p \approx n-p$) than that for strongly-asymmetric fission ($p \gg n-p$) while our model does not give such a tendency. Furthermore, fragment size distribution from $n=20$ and that from $n=22$ show nearly the same trend. We do not yet know the reason for this discrepancy. A candidate to explain this discrepancy is successive decay model which will be discussed in the next section.

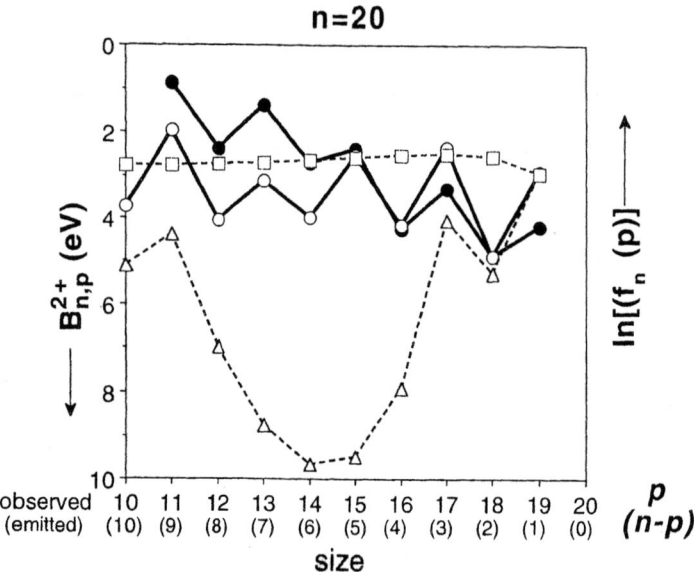

FIGURE 4C The same figure as Figure 4A and 4B, for $n=20$

DISCUSSION

Krückeberg et al. speculated that the product size distribution observed in Katakuse et al.'s experiment is not for direct fission, but as a result of successive decays. They assume that every fragment from doubly charged clusters is produced through emissions of monomers and trimer ions. They suppose in such a way because they observed only an emission of trimer ion or neutral monomer. Of course, we cannot deny the possibility of such chain reactions in Katakuse et al.'s case. However, when the parent cluster is hot, many decay channels other than monomer / trimer emission are open. A direct fission process is more energetically favorite than successive processes. Furthermore, the present model well explains the even-odd oscillation seen in the experiment. Thus it is natural to assume that the direct fission will be a main process for smaller clusters. For larger clusters, the net internal energy of the cluster becomes so large that it is plausible that successive decays take place. A reason for discrepancy between theory and experiment for the case of $n = 20 - 22$ may be due to the effects of successive decays.

CONCLUSION AND FUTURE PROBLEMS

Shell effects perturb the competition between fission and evaporation near the critical size. It also governs the product size distribution of fission fragment. These findings are revealed through the comparison between the theory and experiment.

At low temperature, only the decay to the energetically favorite channel is observed. At high temperature, the decay will be opened a to variety of channels. Thus the size distribution shows oscillation due to shell effects. The present model is valid for a qualitative feature of the size distribution of fission fragments. Both Mainz group's result and Osaka group's result can be explained in the same framework.

Although the present model explains an overall profile in the dynamics of fission and evaporation, the theory must be improved to evaluate the accurate value of fission barrier and evaporation energy. It is still under discussion whether fragment distribution in Katakuse *et al.*'s case is due to direct fission or as a result of successive decays. To solve this problem, the experiment should be 'perfect' so that every fragments can be detected. We hope that future study will solve the problems.

ACKNOWLEDGMENTS

The author expresses sincere thanks to I. Katakuse and H. Itoh for useful discussion. He is grateful for A. Ichimura for reading this manuscript.

REFERENCES

1. U. Näher, S. Bjøronholm, S. Frauendorf, F. Garcias, and C. Guet; Phys. Rep. **285**, 245 (1997)
2. (a) W. A. Saunders; Phys. Rev. Lett. **64**, 3046 (1990) (b) W. A. Saunders; Phys. Rev. **A46**, 7028 (1992)
3. (a) C. Bréchignac, Ph. Cahuzac, F. Carlier, and J. Leygnier; Phys. Rev. Lett. **63**, 1368 (1989) (b) C. Bréchignac, Ph. Cahuzac, F. Carlier and J. Leygnier, and A. Safari; Phys. Rev. **B44**, 11386 (1989) (c) C. Bréchignac, Ph. Cahuzac, F. Carlier, and M. de Frutos, Phys. Rev. Lett. **64**, 2893 (1990) (d) C. Bréchignac, Ph. Cahuzac, F. Carlier and M. de Frutos, R. N. Barnett, and U. Landman; Phys. Rev. Lett. **72**, 1636 (1994) (e) C. Bréchignac, Ph. Cahuzac, F. Carlier, and M. de Frutos; Phys. Rev. **B49**, 2825 (1994)
4. (a) S. Krückeberg, G, Dietrich, K. Lützenkirchen, L. Schweikhard, C. Walther, and J. Ziegler; Hyperfine Interactions **108**, 107 (1997) (b) L. Schweikhard, P. Beiersdorfer, W. Bell, G, Dietrich, S. Krückeberg, K. Lützenkirchen, B. Obst, and J. Ziegler; Hyperfine Interactions **99**, 97 (1996) (c) S. Krückeberg, G. Dietrich, K. Lützenkirchen, L. Schweikhard, C. Walther, and J. Ziegler; Z. Phys. **D40**, 341 (1997) (d) L. Schweikhard, G. Dietrich, S. Krückeberg, K. Lützenkirchen, C. Walther, and J. Ziegler; Rapid Communications in Mass Spectrom. **11**, 1592 (1997) (e) S. Krückeberg, G. Dietrich, K. Lützenkirchen, L. Schweikhard, C. Walther, and J. Ziegler; Eur. Phys. J. **D9**, 145 (1999)
5. I. Katakuse, H. Itoh, and T. Ichihara; Int. J. Mass Spectrom. and Ion Processes **97**, 47 (1990)
6. M. Brack, J. Damgaard, S. A. Jensen, H. C. Pauli, V. M. Strutinsky and C. Y. Wong; Rev. Mod. Phys. **44**, 320 (1972)
7. R. Balian and C. Bloch; Ann. Phys. (N.Y.) **60**, 401 (1970)
8. K. Clemenger; Phys. Rev. **B32**, 1359 (1985)
9. M. Nakamura, Phys. Rev. **A60**, 2222 (1999)

Evaporation of finite systems: the case of atomic clusters

C. Bréchignac, Ph. Cahuzac, B. Concina, J. Leygnier, B. Villard

*Laboratoire Aimé Cotton, C.N.R.S. UPR 3321,
Bât. 505 Université Paris Sud, 91405 Orsay Cedex, France*

abstract: A finite system dissipates an excess of internal energy by fragmentation or evaporation of part of its constituents. We have studied experimentally the evaporation of small atomic cluster ions Li_n^+, $n=$ 5-41, by time-of-flight mass spectrometry. We observed the dissociation pathways and measured the evaporation ratio and kinetic energy release for clusters produced in well controlled conditions. We used statistical models, and deduced from measurements the cluster dissociation energies, internal energies, kinetic energy release distributions and temperatures.

pacs: 36.40 Qv Stability and Fragmentation of clusters, 82.20 db Statistical theories

I INTRODUCTION.

Fragmentation is commonly used to probe the physical properties of finite systems at non zero temperature. It has been observed for systems on extremely different size scales, from nuclei, molecules, atomic or molecular clusters, to macroscopic droplets. This paper is devoted to the evaporation of atomic clusters, $M_n^+ \longrightarrow M_{n-p}^+ + M_p$ for $n = 5\text{-}41$ and $p = 1\text{-}2$.

Measuring the evaporation of a finite system brings first of all information on its stability [1–3]. A deepest analysis also probes the behavior of internal energy in a microcanonical system, and the characteristic time at which an excess of internal energy is dissipated by evaporation. The lifetime of a finite system having a given internal energy is a key parameter.

Other data accessible from experiment are the dissociation pathways. The way a system fragmentates gives much information on its internal structure. Dissociation pathways depend primarily on the energetic of the different accessible evaporation channels [2,3]. The evaporation of an isolated system is limited by the dynamics and the dissociation pathways also depend on the volume of the phase space accessible for each channel, that means on the reaction entropy change [4].

A third observable is the kinetic energy released in the decomposition process (KER). The kinetic energy of the ejected particle brings an information on the kinetic energy of the particles inside the system, i.e. on its temperature. For microcanonical systems containing a well defined energy, measuring directly the temperature is of fundamental interest in order to test the models used for the thermodynamics of finite systems. We measure independently the internal energy and the temperature of small clusters. Such observations may lead to a better understanding of phase transitions, as evaporation or fragmentation can also be considered as phase changes in an isolated system [5,6].

This paper is organized as follows : in part II, we present experimental techniques that we used to determine the evaporation rates and the dissociation pathways of atomic clusters, as well as the kinetic energy released in the process. In part III, we mention statistical models that we used to relate the lifetime of a cluster to the internal energy it contains. In Part IV is presented a quantitative interpretation of evaporation experiments that led to the measurement of the cluster dissociation energies. We give in section V data on the kinetic energy released in the evaporation of atomic clusters, with a discussion of the energy and KER distributions in those particles.

Keeping in mind the analogy with nuclear matter, we will focus on clusters that can be described as a system of fermions in a potential well, *i.e.* metal clusters. We will describe experiments concerning the metal clusters having the simplest electronic structure, the lithium atom clusters.

II EXPERIMENTAL INVESTIGATION OF EVAPORATION

The physical properties of free atomic clusters are generally investigated using time-of-flight (TOF) mass spectrometry. The essential elements of a typical apparatus are shown Fig 1.

A metal sodium or lithium vapor diffuses in cold helium gas and condenses in neutral droplets of 300-500 atoms. Clusters are transported in the gas stream in a beam which enters a high vacuum chamber. They are ionized by a pulsed laser ($h\nu = 3.50\ eV$) between the first two electrodes of a multiplate accelerating system. The charged aggregates M_n^+ are accelerated to a final kinetic energy 9 keV, and enter a field free drift tube where they resolve into mass separated packets that hit the detector at an arrival time $t_n \propto \sqrt{n}$.

We have to detail the ionization-acceleration sequence because it defines the internal properties of the clusters. The laser fluency is high enough to induce the absorption of many photons in each cluster, which is heated up. It dissipates its excess of internal energy by the sequential evaporation of many atoms, and shrinks down to low masses, typically in the size range 1-50 atoms in our experimental conditions. The effective mass dispersion by acceleration takes place between the plates 2 and 3 of the system ($V_0 - V_1 \ll V_1$), between time t_1 and t_1'. The time is

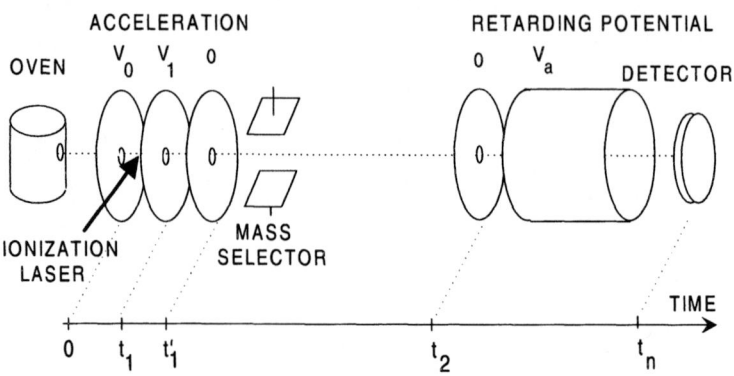

FIGURE 1. Schematic diagram showing the essential elements of the apparatus. The characteristic duration t_1, t'_1 and t_2 are typically in the range of 0.5, 1 and 10 μs respectively.

referred to the laser pulse $t = 0$. All the cluster ions having the same number of atoms n at time t'_1 form a packet that propagates freely to the detector. A pulsed electrostatic gate can deflects all the other mass packets from the beam.

The dynamics of the evaporation from M_n^+ clusters can be analyzed by a second mass dispersion at a time t_2 larger than t'_1. A static potential applied on an electrostatic plate creates a decelerating field seen by the cluster at a time t_2. M_{n-p}^+ fragments formed between t'_1 and t_2 reach the detector later than their precursor. Their masses are determined by their arrival time.

The metastable decay (unimolecular decay) of Li_9^+ clusters is displayed on Fig. 2 which presents time-of-flight mass spectra obtained in two situations : (a) the cluster ions are formed by evaporation of larger particles just after the laser pulse (t=0). Li_9^+ clusters are mass dispersed by acceleration (time interval $[t_1, t'_1]$) and reach the detector in a single peak corresponding to Li_9^+ particles having a lifetime larger than t'_1 (all the other mass peaks have been deflected from the detector axis by the mass gate). (b) a new mass analysis is set at time t_2. The Li_9^+ particles having a lifetime lower than t_2 evaporate p atoms, are decelerated as Li_{9-p}^+ clusters and reach the detector later than Li_9^+ at a time which identifies Li_8^+ and Li_7^+ products. This interpretation in term of lifetimes will be quantitatively justified in section IV. Denoting S_9, S_8 and S_7 the surfaces of the peaks labeled Li_9^+, Li_8^+ and Li_7^+ on Fig. 2 (b), we observe that:

- a fraction $F_n = \dfrac{S_8 + S_7}{S_9 + S_8 + S_7} \simeq 7\ \%$ has evaporated between t'_1 and t_2. That means that 7% of the initially selected Li_9^+ are metastable with lifetimes shorter than or in the range of t_2. This ratio is called 'evaporation ratio'. We showed [3] that it increases from 3 % to 50 % when n varies from 5 to 41.

- the dissociation pathways are

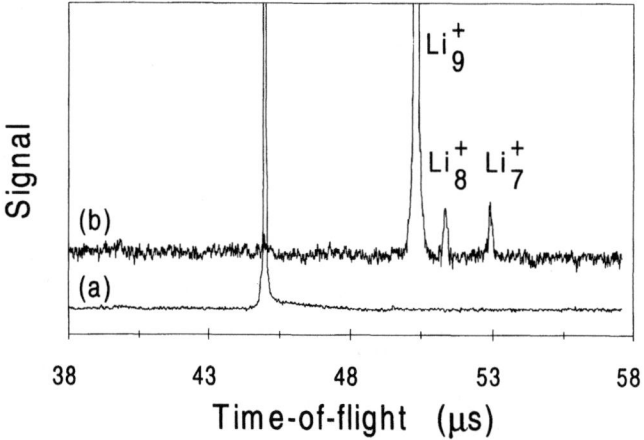

FIGURE 2. TOF spectra for (a) the parent clusters Li_9^+ mass dispersed at t'_1. (b) the cluster ions Li_9^+, Li_8^+ and Li_7^+ obtained after the metastable decay of the parent clusters on a time interval $[t'_1, t_2]$.

$$Li_9^+ \longrightarrow Li_8^+ + Li \qquad \text{ratio} \qquad F_{n,1} = \frac{S_8}{S_9 + S_8 + S_7} \qquad (1)$$

$$Li_9^+ \longrightarrow Li_7^+ + Li_2 \qquad \text{ratio} \qquad F_{n,2} = \frac{S_7}{S_9 + S_8 + S_7} \qquad (2)$$

The characterization of a dimer loss, instead of two sequential monomer evaporation steps was made by reionization of the neutral fragment and mass spectrometry. Moreover, the internal energy of Li_9^+ is such (section IV) that Li_8^+ produced by the evaporation of a monomer possesses an internal energy lower than its dissociation energy and can not evaporate again. The dissociation pathways for Li_n^+ clusters (n=5-41) were published earlier [3].

- the neutral fragments Li and Li_2 are detected in a weak peak Fig 2 (b) at the same arrival time as Li_9^+ on Fig 2 (a). They have been measured in a separate experiment with a better signal to noise ratio. The TOF peaks corresponding respectively to the Li_9^+ parents and to the light neutral fragments are plotted on Fig 3. The peak formed by the metastable decay light fragments is broader. This demonstrates that some kinetic energy is released in the decomposition process. The method used to extract this quantity is described in section V.

We have illustrated from experimental results that some of the clusters produced in those conditions possess lifetime in the range of $t_2 \simeq 10\mu s$. We mention now different simple models taken from the literature that can, starting from the cluster lifetime, estimate the value of the internal energy it contains.

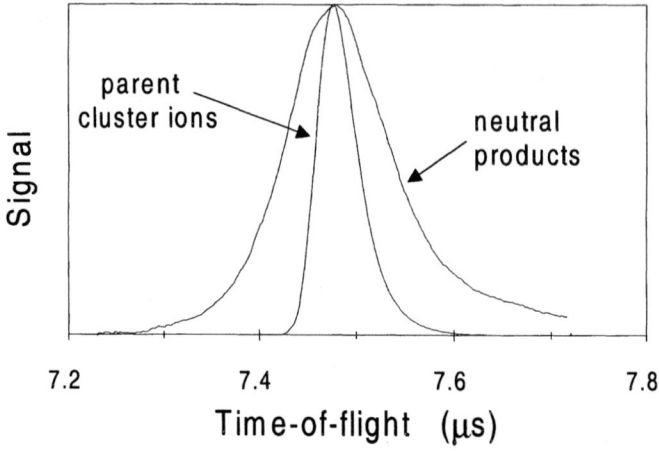

FIGURE 3. Normalized TOF mass peak profiles for (a) Li_9^+ parent clusters (b) neutral fragments produced by evaporation of Li_9^+. The peak broadening is the signature of the Kinetic Energy Release.

III INTERNAL ENERGY AND LIFETIME

A n-atom cluster containing an internal energy E^* evaporates a monomer with the rate constant $k_n(E^*)$, inverse of the lifetime $\tau_n(E^*)$. The statistical models estimate $k_n(E^*)$ from the comparison of the number of states accessible in the phase space for the system in its initial state $[M_n^+(E^*)]$ or in its final state $[M_{n-1}(E^* - D_n - \epsilon) + M + \epsilon]$. D_n is the dissociation energy of the parent cluster, ϵ is the kinetic energy released in the process.

Within the Weisskopf theory based on the microreversibility principle, the lifetime is given by [7]

$$\frac{1}{\tau_n(E^*)} = 8\pi g R^2 m \nu^3 \frac{3n-6}{E^*}\left(1 - \frac{D_n}{E^*}\right)^{3n-8} \tag{3}$$

where g is the electronic degeneracy of the emitted monomer ($g = 2$ for sodium or lithium monomers), m is the mass of the monomer, R is the radius of the parent cluster. This formula is obtained when taking for the cluster an harmonic density of states [8] using the bulk parameters.

Hervieux and Gross proposed to take into account the anharmonic effects and to estimate the density of state from the entropy of an atom in the bulk [9]. They also considered the surface contribution to the density of states [10].

Another possibility is to estimate the cluster lifetime by analogy with the statistical description of atom evaporation from a surface. In the standard kinetic theory, the evaporation time is given by [11]

$$\frac{1}{\tau} = \frac{A}{\sqrt{2\pi m T}} P(T) \tag{4}$$

where A is the surface area and P(T) the vapor pressure in equilibrium with the bulk surface at temperature T. Taking for $P(T)$ the expression given by the kinetic theory in the harmonic description of the bulk matter and considering a sphere of radius R, one obtains [12,13]

$$\frac{1}{\tau} = 8\pi g R^2 m \nu^3 \frac{1}{kT} \exp{-\frac{D}{kT}} \quad (5)$$

There is asymptotic equivalence between Weisskopf and kinetic gas theory for large clusters and high excitation energies (in the harmonic limit) as demonstrated by Brink and Stringari [12,13].

As the expression coming from a microscopic analysis (Weisskopf formula) or from the kinetic theory are consistent, we propose to use a semi-empirical approach combining the kinetic approach and a parametrization by the bulk entropy parameters. In Eqn. (4), we use for P the Clausius Clapeyron expression

$$\ln(P) = \frac{\Delta S}{R} - \frac{\Delta H}{RT} \quad (6)$$

where we take for ΔS the entropy measured for the evaporation of one atom from a bulk surface ('bulk approximation' [9]). We replace ΔH by the parent cluster dissociation energy, and introduce the finite size character by changing the term $\exp{-\frac{D}{kT}}$ by $(1 - \frac{D_n}{E^*})^{3n-8}$, by analogy with Weisskopf's formula and get finally

$$\frac{1}{\tau_n(E^*)} = 10^5 \frac{4\pi R^2}{\sqrt{2\pi m}} \sqrt{\frac{3n-6}{E^*}} \exp{\frac{\Delta S}{R}} \left(1 - \frac{D_n}{E^*}\right)^{3n-8} \quad (7)$$

The advantage of this model is twofold : first, it treats the anharmonicity in the density of state from bulk tabulated measurements, within the approximations considered by Gross and Hervieux [9]. This formula is analytically consistent with Weisskopf's formula in the harmonic limit. Second, it can be applied for clusters evaporating a dimer or a larger particle M_p, provided that the partial pressure of M_p molecules in equilibrium with the bulk is known. We use this model in the next section to calculate evaporation rates for alkali clusters evaporating dimers, taking the data for bulk lithium evaporation from reference [14].

IV QUANTITATIVE INTERPRETATION OF ATOMIC CLUSTERS EVAPORATION

In section II, we have interpreted the metastable decay of atomic clusters in term of lifetimes. Using the relations between the lifetime, the internal energy and the dissociation energy (section III), we extract from the evaporation experiments the values of the dissociation energies of the clusters, as well as their internal energy.

We refer again to the conditions in which the clusters are prepared. The neutral clusters containing 300 to 500 atoms are warmed up and ionized by the laser at

$t = 0$, and cool down by evaporation of a few tenth of atoms until the time t_1 of dispersion by acceleration (see Fig. 1). As in an evaporation step a fraction of internal energy typically equal to the dissociation energy is lost by the cluster, each evaporation step is slower than the previous one. Consequently, we can consider that a cluster Li_n^+ existing at t_1 was produced by evaporation of Li_{n+1}^+ between $t = 0$ and t_1, and neglect the duration of the previous evaporation steps. These experimental conditions are those identified by Klots as an 'evaporative ensemble'. The evaporation of such particles present very general characteristics described by the 'Evaporative Ensemble Model' (EEM) [15].

We consider a precursor Li_{n+1}^+ [dissociation energy D_{n+1}, internal energy E_{n+1}^*, evaporation rate $k_{n+1}(E_{n+1}^*)$] existing at $t = 0$. A cluster Li_n^+ [D_n, $E^* = E_{n+1}^* - D_{n+1}$, $k_n(E^*)$] is formed from its precursor and not yet evaporated at a time t_1 with a probability $P(n, t_1, E^*)$ solution of the rate equations

$$dP(n+1, t, E_{n+1}^*) = -k_{n+1}(E_{n+1}^*)P(n+1, t, E_{n+1}^*)dt \qquad (8)$$
$$dP(n, t, E^*) = -k_n(E^*)P(n, t, E^*)dt + k_{n+1}(E_{n+1}^*)P(n+1, t, E_{n+1}^*)dt \qquad (9)$$

We get

$$P(n, t1, E^*) = \frac{k_{n+1}(E_{n+1}^*)}{k_{n+1}(E_{n+1}^*) - k_n(E^*)} \left[e^{-k_n(E^*)t_1} - e^{-k_{n+1}(E_{n+1}^*)t_1} \right] \qquad (10)$$

Experimental conditions are such that the internal energy of the precursor at $t = 0$ can be considered as randomly distributed. We then observe by mass dispersion at t_1 an ensemble of Li_n^+ clusters for which the internal energy is described by the distribution $P(n, t_1, E^*)$.

The distribution of internal energies for an ensemble of Li_9^+ clusters produced in those conditions is plotted Fig. 4. $k_{n+1}(E_{n+1}^*)$ and $k_n(E^*)$ are calculated for each E^* by the semi-empirical gas kinetic model (formula 7).

The first observation is that the internal energies Li_9^+ can contain are bracketed between two extreme values. The 'Evaporative Ensemble Model' shows that the distribution width is approximately the cluster dissociation energy. We report on Fig. 4 the lifetimes calculated for some energies by Eqn. (7). As they vary very fast with E^* for a given dissociation energy, they are distributed over many orders of magnitude. For this reason, we considered in section II that clusters having an internal energy associated to a lifetime lower than t_2 all have evaporated before t_2.

We go on describing the evaporation as it takes place during the experiment. We consider Li_n^+ clusters existing at t_1 with a given energy E^* and the corresponding evaporation rate $k_n(E^*)$. They survive the acceleration duration $[t_1, t_1']$ with the probability $\exp -k_n(E^*)(t_1' - t_1)$ and enter the time of flight spectrometer at t_1'. Particles evaporating during the acceleration are not detected in the n-atom mass peak. We note $k_{n,1}(E^*)$ and $k_{n,2}(E^*) = k_n(E^*) - k_{n,1}(E^*)$ the evaporation rates for the evaporation of a monomer, a dimer respectively, evaluated by Eqn. (7). There is evaporation of a monomer (of a dimer) before the second mass analysis

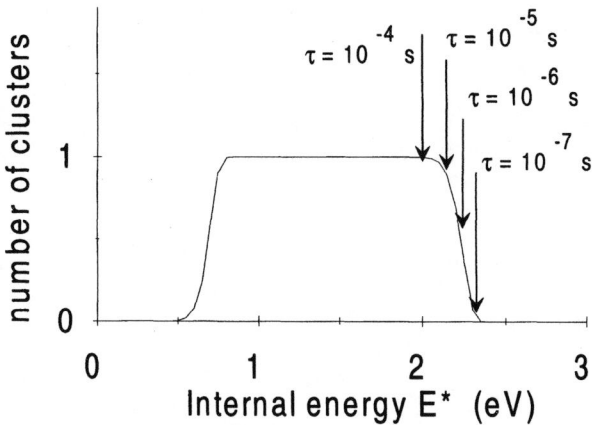

FIGURE 4. Distribution of the internal energy for an ensemble of Li_9^+ clusters formed experimentally in the 'evaporative ensemble conditions'. Mean lifetimes calculated for some values of the internal energy (formula 7) with $D_{9,1} = 1.60\ eV$ and $D_{9,2} = 1.64\ eV$

(time t_2) with the probabilities $\frac{k_{n,i}}{k_n}[1 - \exp -k_n(t_2 - t_1')]$ for $i = 1, 2$ respectively. The evaporation fractions measured experimentally and defined in section II are obtained by integration over all the parent cluster internal energies :

$$F_{n,i} = \frac{\int P(n, t_1, E^*) \exp -k_n(t_1' - t_1) \frac{k_{n,i}}{k_n}[1 - \exp -k_n(t_2 - t_1')] dE^*}{\int P(n, t_1, E^*) \exp -k_n(t_1' - t_1)\, dE^*} \qquad (11)$$

The dissociation energies $D_{n,1}$ and $D_{n,2}$ for the evaporation of a monomer (a dimer) from Li_n^+ are used as fitting parameters so that the calculated evaporation fractions $F_{n,1}$ and $F_{n,2}$ reproduce the measured values. They also have to satisfy the energy constraint

$$D_{n,2}^+ - D_{n,1}^+ = D_{n-1,1}^+ - D_2 \qquad (12)$$

where D_2 is the neutral dimer dissociation energy, resulting from the Born-Haber cycle

$$M_n^+ \begin{array}{c} \nearrow M_{n-1}^+ + M \\ \searrow M_{n-2}^+ + M_2 \end{array} \nearrow M_{n-2}^+ + M + M \qquad (13)$$

We used for this work the evaporation rates given by the semi-empirical kinetic model described in section III for monomer and dimer evaporation. We give in

n	$D_{n,1}^+(eV)$	$D_{n,2}^+(eV)$	n	$D_{n,1}^+(eV)$	$D_{n,2}^+(eV)$	n	$D_{n,1}^+(eV)$	$D_{n,2}^+(eV)$
5		1.40	18	1.19	1.35	31	1.22	1.37
6	1.14		19	1.31	1.44	32	1.20	1.36
7	1.34	1.42	20	1.20	1.45	33	1.24	1.38
8	1.10	1.38	21	1.31	1.45	34	1.22	1.40
9	1.60	1.64	22	1.12	1.37	35	1.24	1.40
10	0.83	1.37	23	1.31	1.27	36	1.22	1.40
11	1.30	1.07	24	1.17	1.32	37	1.23	1.39
12	1.11	1.35	25	1.20	1.31	38	1.23	1.40
13	1.20	1.25	26	1.19	1.33	39	1.24	1.41
14	1.15	1.29	27	1.21	1.34	40	1.22	1.40
15	1.23	1.32	28	1.20	1.35	41	1.24	1.40
16	1.14	1.31	29	1.23	1.37			
17	1.22	1.30	30	1.21	1.38			

TABLE 1. Dissociation energies of lithium cluster ions, determined through the Evaporative Ensemble Model when using the evaporation rates given by the semi-empirical gas kinetic model.

Tab. 1 the set of dissociation energies deduced from the fitting procedure. They reflect the well known saw-tooth behavior predicted for the metal clusters by the Jellium model, with particularly high dissociation energies (stability) for the so-called 'magic numbers' (Li_9^+, Li_{21}^+ Li_{41}^+, $\cdots$) having the same physical interpretation as in nuclear physics [16]. These more refined values lie within the error bars of our previous results [3] using Weisskopf's formula either for monomer and dimer evaporation. Calculation with various methods show a good agreement with these measurements [17–21].

V KINETIC ENERGY RELEASE

We showed in section II that the release of kinetic energy in the process

$$Li_n^+(E^*) \longrightarrow Li_{n-1}^+(E_{frag}^* = E^* - D_n - \epsilon) + Li + \epsilon \qquad (14)$$

can be experimentally investigated through the time-of-flight mass peak profiles.

The parent cluster has a velocity $v_n = \sqrt{\frac{2q}{nm}\frac{V_0+V_1}{2}}$ where n, q, $\frac{V_0+V_1}{2}$ are the number of atoms, charge and birth potential of the parent ion, m is the mass of the lithium monomer. The kinetic energy ϵ is shared between the large and the light fragment. The latest acquires the velocity $v_1 = \sqrt{\frac{2(n-1)}{nm}\epsilon}$ in the center of mass frame, with an isotropic angular distribution. The arrival time of this fragment depends on (a) the projection of its velocity on the detector axis, (b) the parent velocity dispersion correlated to the parent mass peak width (apparatus function), (c) the distribution of the kinetic energy released by a n-atom parent having an internal energy E^*, characterized by its profile and its mean value $<\epsilon>$. Taking into account these three points, we simulate the time-of-flight mass peak profile for the neutral fragments and adjust $<\epsilon>$ to give the best fit with the measured peak profile. The detailed procedure will be published elsewhere [23]. An example is given Fig. 5.

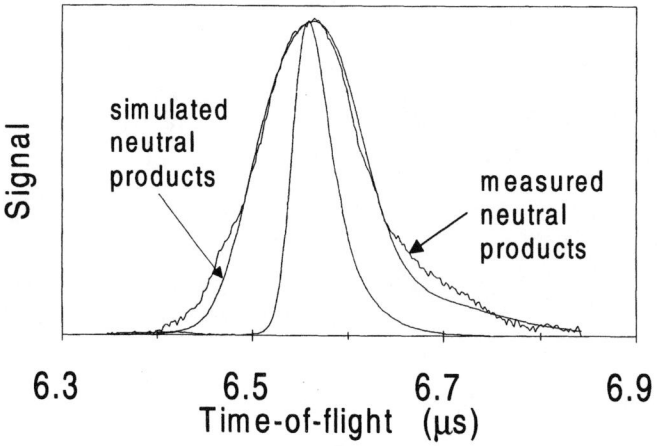

FIGURE 5. TOF mass peak profiles for the decomposition $Li_8^+ \longrightarrow Li_7^+ + Li$. Narrow peak : parent ion mass peak. Broader peak : measured neutral fragments mass peak. Full line : simulated neutral fragment mass peak for a mean kinetic energy release $<\epsilon> = 0.032\ eV$.

The kinetic energy distribution we use is the n-dependent simple expression given by Engelking [22]. In a modified QET-RRK model using an harmonic density of states, he showed that M_n^+ having an internal energy E^* and a dissociation energy D_n releases by evaporation a kinetic energy ϵ with the distribution

$$f_{<\epsilon>}(\epsilon_t) \propto \epsilon \left(\frac{3n-7}{2} <\epsilon> -\epsilon\right)^{3n-10} \qquad (15)$$

Within this harmonic description, the mean value of ϵ can be related to the fragment clusters temperature by $<\epsilon> = 2\frac{E^*-D_n}{3n-7} = 2\frac{E^*_{frag}}{3(n-1)-6} = 2kT_{frag}$ when using the evaporation energy balance $E^*_{frag} = E^* - D_n - \epsilon$. The thermalization between the light neutral and heavy ionic fragments is discussed from molecular dynamic calculations in ref. [24]. The kinetic energy of the ejected atom presents large fluctuations. For the evaporation $Li_8^+ \longrightarrow Li_7^+ + Li$, we find $<\epsilon> = 0.032 \pm 0.003\ eV$ and a temperature $T_{frag} = 180\ K$. The kinetic energy distribution (formula 15) is plotted on Fig. 6 (a). The vertical line represents the mean value $<\epsilon>$.

After this determination of the fragment cluster temperature, we now estimate independently the internal energy this cluster contains. An ensemble of cluster parents Li_n^+ is formed just after $t = 0$ and evaporate between t_1' and t_2. Their internal energy is deduced from Eqn.10 and corresponds to the distribution

$$P_{evap}(n, E^*) \propto \left[e^{-k_n t_1'} - e^{-k_n t_2}\right] \quad (16)$$

From the energy balance $E_{frag}^* = E^* - D_n - \epsilon$, we deduce the distribution of the fragment cluster energies. The quantity of interest is the mean internal energy per mode in the fragment cluster produced in our experiment, $E_{frag}^*/(3(n-1)-6)$. It is illustrated on Fig. 6 (b) for $Li_8^+ \longrightarrow Li_7^+ + Li$ reaction. We note that the internal energy per mode distribution is narrow as compared to the kinetic energy distribution. A factor of two for Fig. 6 (a) and Fig. 6 (b) horizontal scales allows a

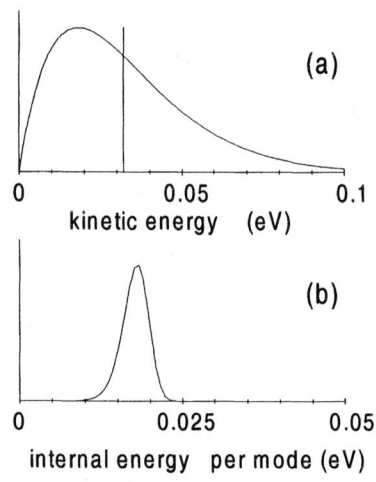

FIGURE 6. $Li_8^+ \longrightarrow Li_7^+ + Li$ evaporation. (a) distribution of the kinetic energy released in the evaporation (formula 15). (b) : distribution of the internal energy per mode for an ensemble of fragment cluster ion Li_7^+ for the mean value $<\epsilon> = 0.032$ eV obtained from measurements.

direct graphic comparison between the fragment cluster temperature $kT_{frag} = \frac{<\epsilon>}{2}$ and mean internal energy per mode $\frac{E_{frag}^*}{3(n-1)-6}$. We obtain a rather good agreement, as expected for harmonic systems.

VI CONCLUSION

The experimental investigation of atomic cluster evaporation was briefly reviewed. Evaporation reveals to be a powerful tool for measuring the physical properties of a finite system : we observed the dissociation pathways and measured evaporation ratio and kinetic energy release, from which we deduced dissociation energies, internal energies and temperatures of alkali atom clusters.

Internal energy and temperature are determined independently. We have checked on one example that the temperature (Li_7^+, 180 K) actually coincides with the mean internal energy per mode, as predicted for a solid phase cluster by harmonic statistical models. This demonstrates the good consistency of the statistical models applied to the description of small atomic clusters.

REFERENCES

1. 'Clusters of Atoms and Molecules I', Ed. H.Haberland, Springer (1991)
2. C. Bréchignac, Ph. Cahuzac, J. Leygnier, J. Weiner, J. Chem. Phys **90** 1492 (1989)
3. C. Bréchignac, H. Bush, Ph. Cahuzac, J. Leygnier, J. Chem. Phys **102** 6692 (1994)
4. C. Bréchignac, Ph. Cahuzac, N. Kébaïli, J. Leygnier Phys. Rev. Lett. **81** 4612 (1998)
5. D.H.E. Gross, M.E. Madjet, Z. Phys B **104** 541 (97)
6. T.L. Beck, J. Jellineck, R.S. Berry, J. Chem Phys. **87** 545 (1987)
7. V. Weisskopf Phys. Rev. **52** 295 (1937)
8. P. Fröbrich, Physics Letters A **202** 99 (1995)
9. D.H.E. Gross, P.A. Hervieux Z. Phys. D **35** 27 (1995)
10. P.A. Hervieux, D.H.E. Gross, Z. Phys. D **33** 295 (1995)
11. L.D. Landau, E.M. Lifshitz, Statistical Physics. Oxford. Pergamon Press 1969
12. D.M. Brink, S. Stringari, Z. Phys. D **15** 257 (1990)
13. G. F. Bertsch, N. Oberhofer, S. Stringari, Z. Phys. D **20** 123 (1991)
14. Thermochemical properties of inorganic substances, O. Knacke, O. Kubaschewski, K. Hesselmann Editors, Springer Verlag 2nd Ed. (1991)
15. C.E. Klots, Z. Phys. D **5** 83 (1987)
16. W.A. de Heer, W.D. Knight, M.Y. Chou, M.L. Cohen, 'Electronic shell structures and metal clusters' in Solid State Physics, edited by H. Ehrenreich and D. Turnbull (Academic, New York), vol 40 p 93
17. P Fantucci, V. Bonačić-Koutecký, J. Koutecký, Z. Phys D **12** 307 (1989)
18. F. Spiegelmann, D. Pavolini, J. Chem Phys **89** 4954 (1988)
19. D.M. Lindsay, Y. Wang, T.F. George, J. Chem. Phys **86** 3500 (1987)
20. F. Spiegelmann, R. Poteau, Comments in Atom. and Mol. Phys. **31** 395 (1995)
21. V. Akulin, C. Bréchignac, A. Sarfati, Phys. Rev. Lett. **75** 220 (1995)
22. P.C. Engelking, J. Chem Phys **87** 936 (1987)
23. C. Bréchignac, Ph. Bréchignac, Ph. Cahuzac, B. Concina, J. Leygnier, P. Parneix, B. Villard, submitted.
24. A. Strachan, C.O. Dorso, Phys. Rev. C **59** 285 (1999)

Time-dependent mean-field method in electronic systems

Kazuhiro Yabana

Institute of Physics, University of Tsukuba
Tsukuba 305-8571, Japan

Abstract. As in nuclear physics, the time-dependent mean-field theory offers a useful basis to describe excitations and collision phenomena in electronic systems. We have developed a real-time, real-space simulation method for electronic dynamics with the first-principle Hamiltonian. We report some results of our recent studies including linear response calculations of atomic clusters and molecules, and calculations of charge transfer reaction with highly charged ion.

INTRODUCTION

Common concepts and methods are often useful both in electronic and nuclear systems since they are the condensed systems of many-fermions. In theoretical side, it is remarkable that the one-body approximation derived with a variational principle from the energy functional succeeds in describing quantitatively the ground state properties of electronic and nuclear systems; the density-functional theory (DFT) in electronic systems, and either the Skyrme Hartree-Fock or relativistic mean-field theories in nuclear systems.

The DFT is now recognized as one of the most successful first-principle methods to calculate the ground state properties of electronic systems. An extension of the DFT to electronic dynamics under time-dependent external field is now developing rapidly. It is known as the time-dependent density-functional theory (TDDFT) [1]. Like the time-dependent Hartree-Fock theory in nuclear physics, the TDDFT has been applied to linear response properties, such as the optical responses of atoms [2], molecules [3,4], atomic clusters [5], and solids [6,7]. It has also been applied for collision problems of short-pulse laser [1] and highly charged ion [8].

More than twenty years ago, a real-time, real-space method to solve the time-dependent mean-field equation was invented in nuclear theory [9]. We have recently implemented the real-space, real-time method for electronic systems with first-principle Hamiltonian [10]. The method has turned out to be quite efficient to calculate optical responses of various materials and to describe collision phenomena.

In this report, I would like to show our recent results with the method. I also mention the subjects currently under progress.

OPTICAL RESPONSE OF FINITE SYSTEMS

The basic equation describing the electronic dynamics is the time-dependent Kohn-Sham equation:

$$\left\{-\frac{\hbar^2}{2m}\nabla^2 + V_{ion}(\mathbf{r}) + e^2 \int d\mathbf{r}' \frac{n(\mathbf{r}',t)}{|\mathbf{r}-\mathbf{r}'|} + \mu_{xc}(n(\mathbf{r},t)) + V_{ext}(\mathbf{r},t)\right\}\psi_i(\mathbf{r},t)$$
$$= i\hbar\frac{\partial}{\partial t}\psi_i(\mathbf{r},t), \quad (1)$$

where V_{ion} is the electron-ion interaction which we use the norm-conserving pseudopotential with separable approximation. The electron density is given by $n(\mathbf{r},t) = \sum_i |\psi_i(\mathbf{r},t)|^2$. μ_{xc} is the so-called exchange-correlation potential. We assume an adiabatic approximation, employing the same function as that in the static calculation. $V_{ext}(t)$ is an external field applied to the system.

We solve the equation discretizing the three-dimensional Cartesian coordinate in uniform mesh and discretizing the time variable as well. As in nuclear theory, the higher-order finite-difference approximation for the differential operator gives accurate results with rather course mesh spacing [11]. A Taylor expansion of the short-time evolution operator up to a finite order gives stable time-evolution [9].

The optical response is calculated making use of the time-frequency Fourier transformation [10,12]. We first solve the static Kohn-Sham equation to obtain the ground state. At time $t = 0$, we apply an instantaneous external field. For dipole response, our external field is given by $V_{ext} = kz\delta(t)$. The ground state Kohn-Sham orbitals $\phi_i(\mathbf{r})$ get a phase factor immediately after the instantaneous field. The wave function at $t = 0_+$ is given by $\psi_i(\mathbf{r},t=0_+) = e^{ikz}\phi_i(\mathbf{r})$. We then calculate time evolution without any external field, and obtain dipole polarization as a function of time, $p(t) = e^2 \int d\mathbf{r} z n(\mathbf{r},t)$. The dynamical polarizability can be obtained by the time-frequency Fourier transformation,

$$\alpha(\omega) = \frac{1}{k}\int_0^T dt p(t) e^{i\omega t}. \quad (2)$$

In practice, we calculate for a finite interval up to time T. To get a spectrum with energy resolution ΔE, we need to calculate up to $T \sim \hbar/\Delta E$.

As an example of the calculation, we show the optical response of C_{60} molecule [4]. The C_{60} molecule has a high symmetry of icosahedron, and its optical absorption shows a few, well separated, electronic excitations. Figure 1 shows the optical response in time domain. At $t = 0$, all electrons move coherently to the direction of applied field. Superposition of many electronic excitations makes complicated oscillation in time. The real and imaginary parts of the dynamic polarizability as

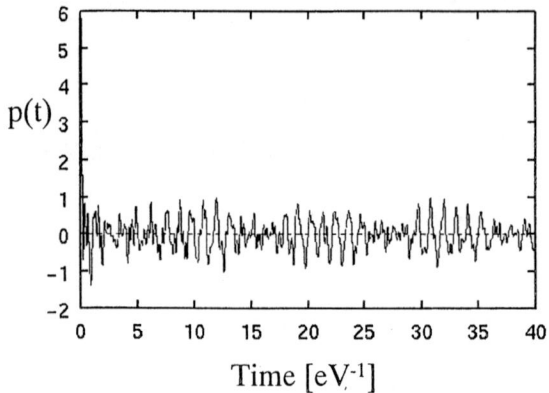

FIGURE 1. The optical response of C_{60} molecule in time domain.

functions of excitation energy are shown in Fig.2. The imaginary part is multiplied with ω and is proportional to the photoabsorption cross section. The ionization threshold of C_{60} molecule appears at 7.6 eV. Below the threshold, there appears four distinct transitions in the calculation. Three structures are known in the measurement. There are rather good correspondence between theory and measurement in the position and oscillator strength [4]. Above the threshold, the calculation shows many discrete structures. However, it is an artifact of the calculation achieved in a finite box area. Reflection of the electrons at the box boundary makes spurious peaks. The issue of treating continuum boundary condition will be discussed later.

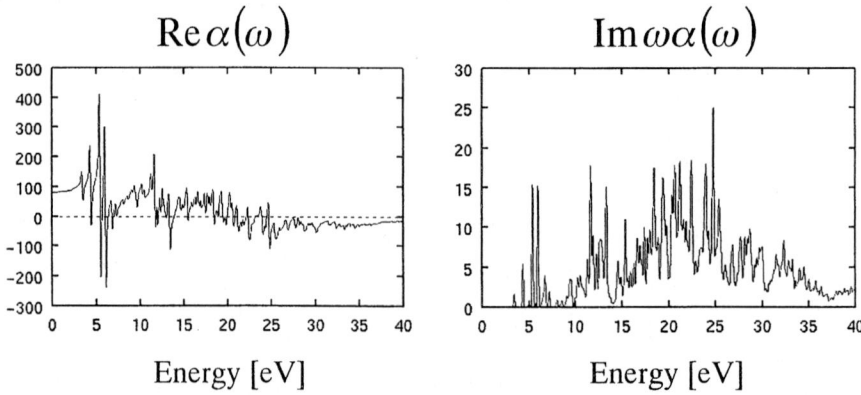

FIGURE 2. The real part (left) and the imaginary part multiplied with frequency (right) of the dynamic polarizability of C_{60} calculated from the polarizability in time shown in Fig.1

COLLISION OF HIGHLY CHARGED ION

As in nuclear physics, the application of the time-dependent mean-field theory is not restricted to the linear response properties but extended to the collision phenomena. Responses of many-electron systems to strong electromagnetic field, induced either by the short-pulse laser or the highly charged ion, have been studied. We here show some results of our analysis for collisions between highly charged ion and atom [13,14].

We consider a collision of Ar^{8+} and Ar as an example. We calculate dynamics of eight valence electrons of Ar atom, treating Ar^{8+} as an inert core. Collisions with incident ion velocity around one atomic unit will be considered; this means the relative velocity between ion and atom is comparable to the velocity of electronic motion in the atom. At this energy region, we may assume straight line trajectory for the ion. The time-dependent Kohn-Sham equation to be solved is given by

$$\left\{ -\frac{\hbar^2}{2m}\nabla^2 + V_{ion}(\mathbf{r} - \mathbf{R}_A(t)) + V_{ion}(\mathbf{r} - \mathbf{R}_B(t)) \right. $$
$$\left. + e^2 \int d\mathbf{r}' \frac{n(\mathbf{r}',t)}{|\mathbf{r}-\mathbf{r}'|} + \mu_{xc}(n(\mathbf{r},t)) \right\} \psi_i(\mathbf{r},t) = i\hbar \frac{\partial}{\partial t}\psi_i(\mathbf{r},t), \qquad (3)$$

where $\mathbf{R}_{A,B}(t)$ represents the coordinates of ions A and B at time t, respectively. V_{ion} is the interaction between electron and Ar^{8+} ion. The initial condition is given by a static Kohn-Sham solution in which all eight electrons occupy orbitals of the target atom.

We show in Fig.3 the contour plot of electron density during the collision for three different incident energies, 18.4 keV, 400 keV, and 3200 keV, respectively. The impact parameter is fixed at 4Å. We take a coordinate frame where two ions move with the same magnitude of velocity in opposite direction. The relative velocity between ions in atomic unit is given by 0.14, 0.64, and 1.80, respectively. We see from the figure that the electronic dynamics changes much at around the ion relative velocity of about 1 a.u. When the ion relative velocity is lower than 1 a.u., the internal electronic motion is faster than the ion relative motion. Therefore, the electronic dynamics is adiabatic. We see the collision at 18.4 keV gives density distribution of almost axially symmetric around the axis connecting two ions. The axial symmetry is to some extent violated at 400 keV collision. The dominant reaction process in these two cases is the multiple electron transfer from the target atom to the projectile ion. At 3200 keV, the electronic dynamics changes much. Since the ion relative motion is faster than the internal motion in the atom, the electrons removed from the target atom cannot follow the rapid change of ion field. It then induces the electron flow to the transverse direction. The reaction dynamics thus changes from the transfer to the emission into the continuum.

In 18.4 keV and 400 keV collisions, the electron distribution around the projectile ion is spatially extended after collision. The highly excited ion is thus formed. As a phenomenological model of multiple-electron transfer, the molecular classical

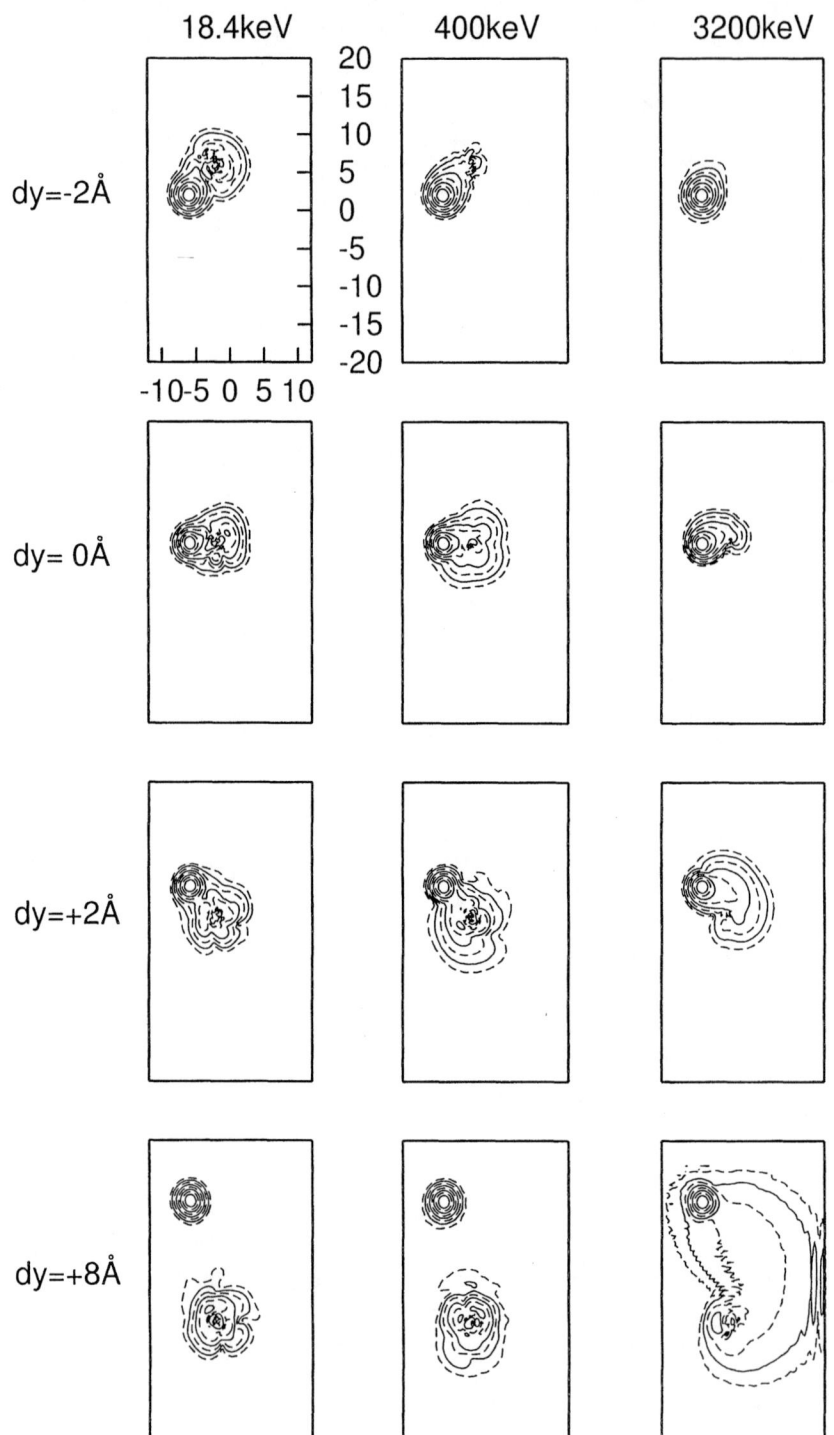

FIGURE 3. Electron density distribution in the collision of Ar-Ar^{8+} at three incident energies. The distance of two ions in longitudinal direction is specified by dy.

overbarrier model has been successful to describe basic aspects of the charge transfer processes [15]. In the model, the formation of the highly excited ion is explained as follows: the electrons in the atomic orbital transfer to the orbitals of highly charged ion at around the same orbital energy. The occupation of these orbitals in the highly charged ion results in the formation of the highly excited ion. A detailed comparison of our TDDFT results with measurements and with molecular classical overbarrier model is discussed in [13].

SOME RECENT PROGRESSES

A Treatment of continuum response

As discussed in the previous section, optical response above the first ionization threshold cannot be described adequately in the real-time method. The reflection of electrons at the box boundary brings spurious discrete structures above the ionization threshold. We have recently developed two methods in real-space to manage the continuum boundary condition [16]. The first method is to place the absorptive potential at the box boundary in the real-time calculation. By choosing appropriately the shape and the strength of the absorptive potential, it is possible to erase the reflected wave almost completely. The spectrum constructed by the Fourier transform includes a width due to the coupling to the continuum.

The second method is based on the Green's function method. A method employing a radial Green's function for spherical systems is well known and has been applied to the responses of nuclei, atoms, and metallic clusters. We have succeeded to extend the Green's function method for systems without any spatial symmetry. The detail of the method as well as the application to some clusters and molecules will be published elsewhere [16].

B Real-time, real-space method for the bulk periodic systems

Optical response of bulk periodic materials is characterized by the frequency-dependent dielectric function. The calculation of the dielectric function is usually achieved in the frequency and momentum representation. We have recently succeeded to extend the real-time, real-space method for the periodic systems [7]. In the method, the long-range polarization field induced by the surface charge is represented by a spatially-uniform, time-dependent vector potential. The dielectric response is described by the coupled equations among the time-dependent single-electron wave functions with periodic boundary condition (Bloch wave functions) and the time-dependent uniform vector potential. The detail of the method as well as applications to Li and diamond are reported in [7].

C Molecular hyperpolarizability

The nonlinear polarizability of a molecule is called the molecular hyperpolarizability and is a basic observable characterizing response of a molecule under a strong electric field. The microscopic calculation of the hyperpolarizability has long been studied with quantum chemistry methods. In the TDDFT framework, a nonlinear response function theory has been developed and applied to atoms [17]. We have recently extended the framework employing the real-space representation and applied the method for clusters and molecules as large as C_{60} molecules. Some preliminary results are shown in [18].

CONCLUSION

The time-dependent mean-field description for electronic dynamics is now developing rapidly as an extention of the static density-functional theory. To solve the time-dependent Kohn-Sham equation, we have applied a real-space and real-time method which was originally developed in nuclear theory. This real-space, real-time method offers a unified and efficient scheme to investigate the electronic dynamics with a first principle Hamiltonian.

We have applied the method for optical response properties of various finite systems such as clusters and molecules. The method has recently been extended to treat optical responses of infinite periodic systems as well. The applicability is not restricted to the linear response properties but is extended to the collision phenomena such as the charge transfer processes induced by an ion.

ACKNOWLEDGMENT

This report consists of my works under collaborations. I would like to thank my collaborators: G.F. Bertsch for optical response calculations and T. Tazawa, Y. Abe, and R. Nagano for charge transfer calculations.

REFERENCES

1. E. K. U. Gross, J. F. Dobson, and M. Petersilka, in *Density Functional Theory*, ed. R.F. Nalewajski, Springer Series Topics in Current Chemistry (Springer, Heidelberg, 1996).
2. A. Zangwill and P. Soven, *Phys. Rev.* **A21**, 1561 (1980)
3. M. E. Casida, C. Jamorski, K. C. Casida, and D. R. Salahub, *J. Chem. Phys.* **108**, 4439 (1998).
4. K. Yabana and G. F. Bertsch, *Int. J. Quantum Chem.* **75**, 55 (1999).
5. W. Ekardt, *Phys. Rev. Lett.* **52**, 1925 (1984).
6. Z. H. Levine, D. C. Allan, *Phys. Rev.* **B43**, 4187 (1991).
7. G. F. Bertsch, J.-I. Iwata, A. Rubio, and K. Yabana, *Phys. Rev.* **B62**, 7998 (2000).

8. K. Yabana, T. Tazawa, P. Bozek, and Y. Abe, *Phys. Rev.* **A57**, R3165 (1998).
9. H. Flocard, S. Koonin, and M. Weiss, *Phys. Rev.* **C17**, 1682 (1978).
10. K. Yabana and G. F. Bertsch, *Phys. Rev.* **B54**, 4484 (1996).
11. J. Chelikowsky, N. Troullier, K. Wu, and Y. Saad, *Phys. Rev.* **B50**, 11355 (1994).
12. K. Yabana and G.F. Bertsch, *Phys. Rev.* **A60**, 1271 (1999).
13. R. Nagano, K. Yabana, T. Tazawa, and Y. Abe, *J. Phys.* **B32**, L65 (1999).
14. R. Nagano, K. Yabana, T. Tazawa, and Y. Abe, *Phys. Rev.* **A**, in press.
15. A. Niehaus, *J. Phys.* **B19**, 2925 (1986).
16. T. Nakatsukasa and K. Yabana, submitted to *J. Chem. Phys.*
17. G. Senatore and K. Subbaswamy, *Phys. Rev.* **A35**, 2440 (1987).
18. J.-I. Iwata, K. Yabana, and G. F. Bertsch, *Nonlinear Optics* **26**, 9 (2000).

Formation of New Materials in Fullerenes by Using Nuclear Recoil

T. Ohtsuki*, K. Ohno**, K. Shiga[†], Y. Kawazoe[†], Y. Maruyama[‡],
K. Shikano[¶], K. Masumoto[§],

*Laboratory of Nuclear Science, Tohoku University, Mikamine, Taihaku, Sendai 982-0826, Japan.
**Department of Physics, Faculty of Engineering Yokohama National University, Tokiwadai, Hodogaya, Yokohama, 240-8501, Japan.
[†]Institute for Materials Research, Tohoku University, Katahira, Aoba-ku, Sendai 980-8577, Japan.
[‡]National Industrial Research Institute of Nagoya (NIRIN), 1-1 Hirate-cho, Kita-ku, Nagoya 462-8510, Japan
[¶]NTT Opto-Electronics Laboratories, Tokai, Ibaraki, 319-11, Japan.
[§]Radiation Science Center, KEK, Tanashi, Tokyo 188-8501, Japan.

Abstract. The formation of Sb or Te atom-incorporated fullerenes has been investigated by using radionuclides produced by nuclear reactions. From the trace of radioactivities of ^{120}Sb (^{122}Sb) or ^{121}Te after High Pressure Liquid Chromatography (HPLC), it was found that the formation of endohedral fullerenes or heterofullerenes in atoms of Sb or Te is possible by a recoil process following the nuclear reactions. To confirm the produced materials, *ab initio* molecular-dynamics simulations based on an all-electron mixed-basis approach was carried out. We present possibility of the formation of endohedral fullerenes or substitutional heterofullerenes incorporated with Sb or Te atoms.

Keyword: atom-incorporated fullerene, nuclear recoil, MD simulation

INTRODUCTION

Chemical interaction between C_{60} and a variety of atoms is becoming a very new field of cluster research. So far, numerous experimental studies for endohedrally doped [1–10] or exohedrally doped [11–13] fullerenes with foreign atoms have been undertaken by resorting to arc-desorption or laser-vaporization techniques. On the other hand, it has become possible to synthesize the heterofullerenes, where the foreign atom is incorporated into the carbon cage. Experimentally, heterofullerenes doped with foreign atoms, such as boron(B) [14,15], nitrogen(N) [16,17], silicon

(Si) [22,23] have been reported. In our previous studies, we have studied not only the endohedral doping of Kr and Xe [26] but also the substitutional doping of ^{11}C [18,19], ^{13}N [20], ^{69}Ge and ^{72}As [21] by a recoil-implantation process following nuclear reactions. In spite of the intense research, only partial facts for the formation process and the produced materials have been unveiled on the nature of the chemical interaction between a foreign atom and a fullerene cage. Therefore, it is very important and intriguing to synthesize new plastic materials, such as several atom-incorporated fullerenes, and their properties should be investigated for getting in quantities and opening up a new application in the future.

In this paper, we show evidence of Sb (or Te) atom-incorporated fullerenes on the collision between a C_{60} cage and an Sb (Te) atom, which was generated from a recoil process following nuclear reactions. We performed *ab initio* molecular-dynamics (MD) simulations: whether the Sb (Te) atom can be incorporated in the fullerene with the endohedral doping; Sb@C_{60} (Te@C_{60}), or the substitutional doping; SbC_{59} (TeC_{59}). Fuerthermore, the doping process of the Sb (Te) atom in a fullerene is compared with that of an As atom or a noble-gas atom.

EXPERIMENTAL PROCEDURE

According to the source nuclide used, high-energy bremsstrahlung or charged particle irradiation was used. In Table I, nuclide produced, characteristic γ-ray, half-life and reaction are listed for each material used here. About 10 mg of C_{60} fullerene powder was mixed homogeneously with 10 mg of Sb_2O_3, and used to the target material. (1) For production of ^{120}Sb(^{122}Sb)-doping fullerene, the samples were irradiated with bremsstrahlung of E_{max}=50 MeV which originated from the bombardment of a Pt plate of 2 mm in thickness with an electron beam which was provided by a 300 MeV electron linac, Laboratory of Nuclear Science, Tohoku University. Two radioisotopes of ^{120}Sb and ^{122}Sb can be produced by photonuclear reaction, (γ,n) reactions, by an irradiation on a natural Sb. Irradiation time was set to about 8 hour and the average beam current was typically 120 μA. The sample was cooled with water bath during the irradiation. (2) For production of ^{121}Te-doping fullerene, deuteron irradiation with beam energy of 16 MeV was performed at the Cyclotron Radio-Isotope Center (CYRIC), Tohoku University. Radioisotopes of ^{121}Te can be produced by (d,2n) reaction by an irradiation on a natural Sb. The beam current was typically 5 μA and the irradiation time was about 1 hour. The sample was cooled with He-gas during irradiation.

After the irradiation, the samples were left for one day to cool down the several kinds of short-lived radioactivities of byproducts. After the one-day cooling, radioactivities, such as ^{11}C or ^{13}N (*e.g.*, ^{11}C decays to ^{11}B with $T_{1/2}$=20 min), the radioactivities of ^{120}Sb (^{122}Sb) or ^{121}Te could be measured with its characteristic γ-rays(see Table I).

The fullerene samples were dissolved in *o*-dichlorobenzene after being filtrated to remove insoluble materials through a membrane filter (pore size=0.45 μm). The

TABLE 1. Nuclear data and experimental condition for the radioactive fullerenes

Nuclide produced	γ-ray*	Half-life	Reaction	Material** and abundance (%)
^{120}Sb	197 keV	5.76 d	^{121}Sb$(\gamma, n)^{120}$Sb	^{121}Sb, 57.4
^{122}Sb	564 keV	2.70 d	^{123}Sb$(\gamma, n)^{122}$Sb	^{123}Sb, 42.6
^{121}Te	573 keV	16.80 d	^{121}Sb$(d, 2n)^{121}$Te	^{121}Sb, 57.4

*: γ-ray used for the analysis [27].
**: Irradiated material as a target (both for the case of Sb and Te): Sb_2O_3.

soluble fraction was injected into a high-pressure liquid chromatograph (HPLC) equipped with a 5PBB (silica-bonded with the pentabromobenzyl group) column of 10 mm (inner diameter)× 250 mm (length), at a flow rate of 3 ml/min. The eluted solution was passed through a UV detector, the wavelength of which was adjusted to 290 nm in order to measure the amount of fullerenes and their derivatives.

The fraction was collected at 30 sec intervals, and the γ-ray activities of each fraction were measured with a Ge-detector coupled to the 4096-channel pulse-height analyzer whose conversion gain was set to 0.5 keV per channel. Therefore, the existence of ^{120}Sb (^{122}Sb) or ^{121}Te could be confirmed by their characteristic γ-rays [27].

RESULTS AND DISCUSSION

Figure 1 shows three elution curves of the C_{60} sample irradiated by bremsstrahlung of E_{max}=50 MeV, open circles for ^{120}Sb and solid circles for ^{122}Sb radioactivities, respectively, and by a UV detector (solid line). The horizontal axis indicates the retention time after injection into the HPLC and the vertical one the counting rate of the ^{120}Sb or ^{122}Sb radioactivities.

A strong absorption peak was observed at the retention time of 6.5-7 min in the elution curve (solid line) which was measured by the UV detector. This peak position corresponds to the retention time of C_{60} which was confirmed by the calibration run using the C_{60} sample before the irradiation. Following the first peak, two peaks at around 9-9.5 min and 13-16 min were consecutively observed in the UV chromatogram. For characterization of the components, the fraction corresponding to the second peak in the sample of C_{60} irradiated was collected and examined with MALDI TOF (matrix-assisted laser-desorption ionization time-of-flight) mass spectrometry in a separate run. The mass spectrum of the fraction exhibited a series of peaks at m/z 1440-24n (n=1-4) corresponding to the molecular ion peak of C_{120}-nC_2 in addition to the peak for C_{60} as a base peak [24]. This fact indicates that the second and smaller third peaks can be assigned to C_{60} dimers and C_{60} trimers, respectively. These materials can be produced by the interaction between C_{60}'s in coalescence reactions after ionization by incident γ-rays or produced charged particles [26,28].

Three peaks appeared in the curve of the radioactivities ^{120}Sb (^{122}Sb) in the

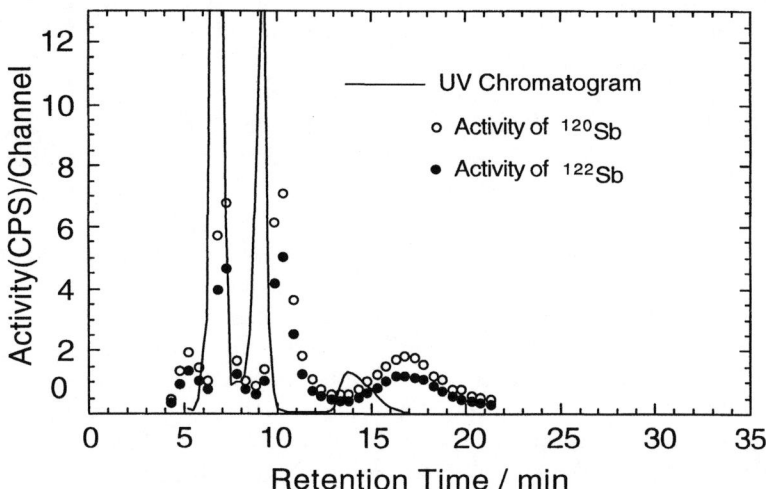

FIGURE 1. HPLC elution curves of the soluble portion of the crude extracted in the γ-ray-irradiated sample of C_{60} mixed with Sb_2O_3. The horizontal axis indicates retention time, and the vertical axis represents the counting rate of the radioactivities of ^{120}Sb or ^{122}Sb measured with a Ge-detector.

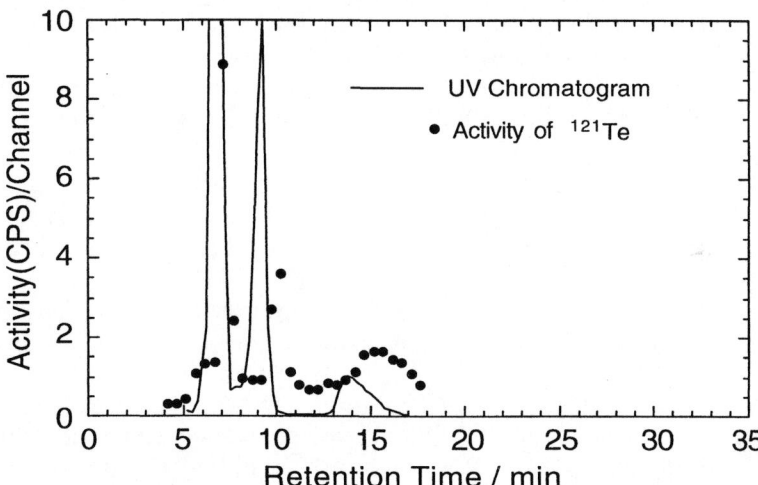

FIGURE 2. HPLC elution curves of the soluble portion of the crude extracted in the deuteron-irradiated sample of C_{60} mixed with Sb_2O_3. The horizontal axis indicates retention time, and the vertical axis represents the counting rate of the radioactivities of ^{121}Te measured with a Ge-detector.

radiochromatogram. Aside from a slight delay, the first peak (7 min) corresponds to the C_{60} UV absorption peak. The second as well as the relatively broad third peaks were observed at the retention time of 9-11 min, and of 14-20 min, respectively. Though there is a delay in the elution peaks of the radioactivities against that of the UV absorption peaks, it seems that the elution behavior is similar. This result indicates that the radioactive fullerene monomers and their polymers (dimers and tetramers) labeled with ^{120}Sb (^{122}Sb) possibly exist in the final fractions. In our previous study, a similar trend was also observed in the elution curve of Kr or Xe case [26]. The amount of the Sb-incorporated radioactive fullerenes produced here is estimated to be about $10^{10} \sim 10^{11}$ molecules.

The elution curves shown by solid line and solid circles in Fig. 2 indicate the absorbance monitored continuously by a UV detector and γ-counting rate of ^{121}Te measured by a Ge-detector, respectively. The horizontal and the vertical axes are also same as Fig. 1. From the mass measurements, three components in the UV chromatogram can be also attributed to C_{60} monomers, their dimers and trimers, respectively. Three populations of ^{121}Te are appeared at retention times of 7 min, 10 min and 14-18 min in Fig. 2. The amount of the radioactive fullerenes seems to be the same order of magnitude of that of the Sb case.

Here, it should be noted that no evidence of exohedral molecules with a covalent nature has been presented so far by an extraction in the soluble portion. Such molecules can be remove out during the solvation process if they are exohedral. Therefore, two possibilities should be considered in the present results; (1) endohedrally Sb (Te) atom-doped fullerenes, Sb@C_{60} (Te@C_{60}), (2) substitutionally Sb (Te) atom-doped heterofullerenes as a part of the cage, SbC_{59} (TeC_{59}).

In order to understand the present experimental results, *ab initio* molecular-dynamics simulations were carried out. The method, which is used here, is based on the all-electron mixed-basis approach [25,26,28–31] using both plane waves (PW's) and atomic orbitals (AO's) as a basis set within the framework of the local density approximation (LDA). In the present study, all the core atomic orbitals are determined numerically by a standard atomic calculation based on Herman-Skillman's framework with logarithmic radial meshes [25]. For the present system, we use 313 numerical AO's and 4,169 PW's corresponding to a 7 Ry cutoff energy. For dynamics, we assume the adiabatic approximation where the electronic structure is always in the ground state. We utilize a supercell composed of $64\times64\times64$ meshes, where one mesh corresponds to 0.196 Å. We set the basic time step as $\Delta t = 0.1$ fs and perform five steepest descent (SD) iterations after each updation of atomic positions. We do not impose any velocity control, so that the system is almost microcanonical with a little energy dissipation from the SD algorithm.

We performed the following three types of simulations for As or Te case; (A) insertion between one Sb atom and one C atom in C_{60} cage, and (B) structural stability of SbC_{59}. (C) insertion of Te atom through a six-membered ring of C_{60} (u-C_6).

Here, we describe the results of the present simulations. (A) First, we shift one of the C atoms of C_{60} outward by 1.3 Å and put additionally one Sb atom on the same radial axis by 1.3 Å inward from the original C position (see Fig.3). Then, starting the simulation with zero initial velocity, we found that there is a force acting on the Sb atom to move innerward to encapsulate (~135 fs). On the other hand, the C atom of C_{60} placed outward by 1.3 Å induced a force acting to move innerward as if it would create one of the membering u-C_6, and finally u-C_6 recovered its original configuration of C_{60}. This results may indicate the formation of $Sb@C_{60}$. (B) Second, one Sb atom was put at 1.3 Å outward from the cage sphere, instead of one C atom of u-C_6. Then, starting the simulation with zero initial velocity, we found that there is a slight moving force acting on the Sb atom against the cage, but still staying near by the initial position even after full relaxation(t=500 fs) with some rotational inertia in the system of SbC_{59}. Therefore, it seems that the Sb atom, when put outside the cage, can be stable to create a heterofullerene

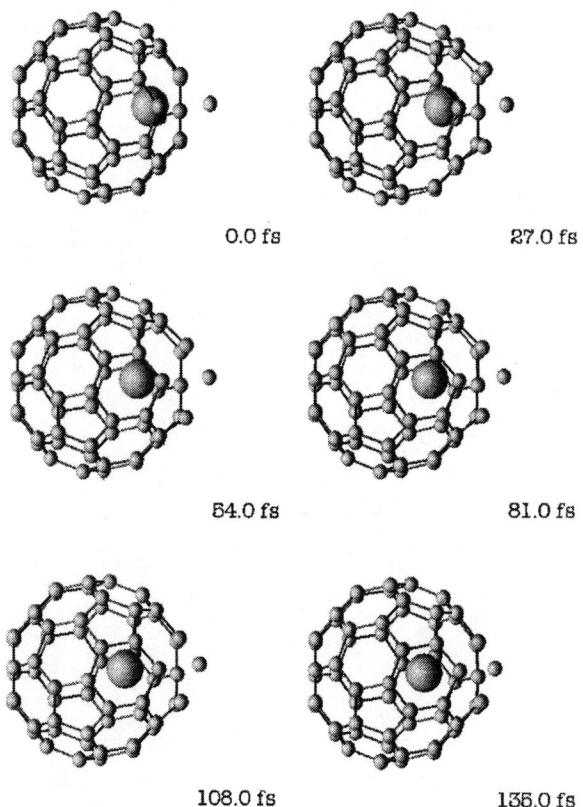

FIGURE 3. Simulation of the structural stability of $Sb@C_{60}$ or SbC_{59}: Change from an unstable innerside(Sb atom) and outer side(C atom) with an initial kinetic energy of 0 eV to a final stabilized configuration.

such as SbC_{59}. (C) Third, in the case of Te, Te atom with the initial kinetic energy (K.E.) of 40 eV can penetrate into the cage of C_{60} through the center of u-C_6 without difficulty, Figure 4 shows several snapshots (~150 fs) of the Te atom insertion with 40 eV K.E. In the figure, after the Te atom first touches u-C_6, carbon atoms are pushed to open u-C_6 and goes through. But the u-C_6 recovers soon its initial configuration. Finally the Te atom bounced at the other side and come back towards the center of the cage. The result of simulations changes of course according to the impact energy, impact point and angle. For relatively low initial K.E. C_{60} shows a tendency to recover its original shape within the simulation period. For higher initial kinetic energies six C_2 losses occur simultaneously from the upper side of C_{60}. If an atom is inserted toward off-center positions of a six- or five-membered ring, the damage suffered on C_{60} increases significantly.

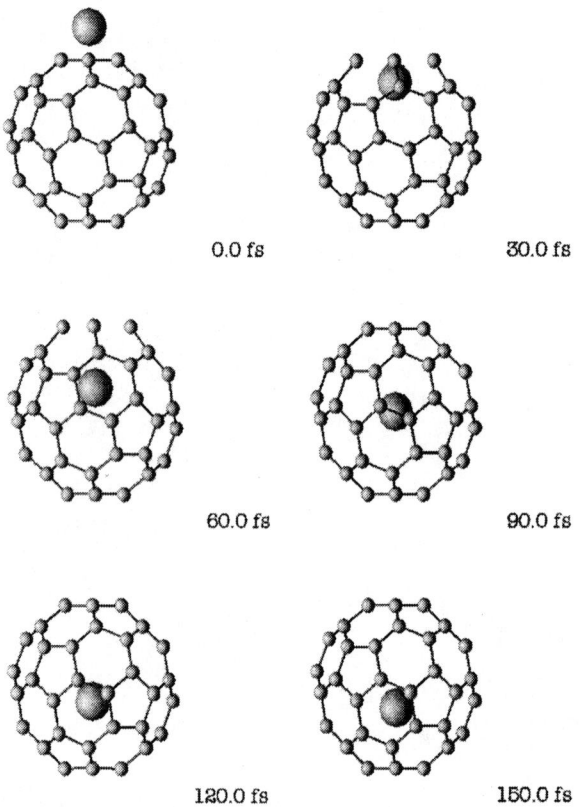

FIGURE 4. Simulation of Te hitting the center of a six-membered ring of C_{60} with a kinetic energy of 40 eV. Here, local skeleton disappears from the figure when the bond-length is elongated more than by 1.5Å.

It is interesting to note that the nature of the doping process of the Sb atom in the C_{60} cage can be compared with that of As atom, because the As and Sb atoms are the same 5B group element. In our previous study, we also performed similar simulations in the case of As atom [21]. From the simulations, the As atom put inside the cage is quite unstable and has a strong tendency to repel the closest C atom of C_{60} and, stabilized slightly outside the cage sphere, to create AsC_{59} (even if the As atom is put outside the cage). Therefore, we confirmed that a heterofullerene, such as AsC_{59}, may exist stably under realistic conditions. In the present results, however, it seems that the formation of $Sb@C_{60}$ is rather more likely than the formation of SbC_{59} for the interaction between the Sb atom and the C_{60} cage. But it may still remain the local stable point around the cage sphere to create a heterofullerene (SbC_{59}). In the case of Te, we found that the insertion of the Te atom through u-C_6 seems to be rather easier than that in the case of a noble-gas atom [26]: the Te atom can penetrate the u-C_6 in the K.E. of 40 eV, which is relatively lower K.E. than the case of Kr or Xe (Kr; 80–150 eV, or Xe; 130–200 eV). The difference can be due mainly to the magnitude of the covalent bonding between the C atom and the Sb (or As, Te) atom; it seems that the magnitude of the covalent bonding in the case Sb (Te) is weaker than that in the case of As.

In our previous study, several fourth~fifth-cycle elements in periodic table have been investigated by a recoil implantation following nuclear reactions. Schematic view of a periodic table is shown in Fig. 5 (section,§, indicates the elements investigated in the present/previous studies). In the figure, element, which is experimentally and/or theoretically confirmed as an atom-incorporated fullerene, can be limitted to (1) a small atom (like Li, Be), (2) a noble-gas atom (~Kr, Xe) and (3) a 4B~6B element. While other elements, like alkali, alkali-earth and transitional metals (Na, Ca, Sc, etc.), may destroy most of the fullerene cage in the same process, due to strong chemical reactivity between atoms and fullerenes. It is interesting to note that the group elements such as 4B~6B, even in fifth-cycle

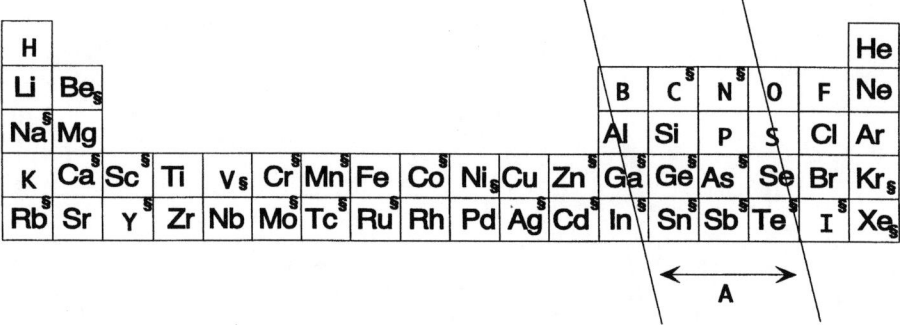

FIGURE 5. Schematic view of periodic table. In the figure, section(§) indicates the elements investigated in the present and/or previous experiments. "A" denotes the region of atom-incorporated fullerene investigated in 4B~6B elements.

elements(like Sb or Te), can be possible for a formation of endohedrally doping fullerene or a substitutionally doping heterofullerene.

The results of analyses of the present work have to be further supported by some other experimental data such as direct mass measurements by a Time-of-Flight Mass Spectrometry(TOFMAS). Finally, we briefly comment on the experimental works now in progress.

CONCLUSION

In this study, the formation of atom-incorporated fullerenes has been investigated by the traces of radioactivity of ^{120}Sb (or ^{122}Sb, ^{121}Te) produced by nuclear reactions. It was found that 5B~6B elements, like Sb or Te, remained in the final C_{60} portion after a HPLC process. This fact suggests that the formation of endohedral fullerenes, Sb@C_{60} (Te@C_{60}) and their polymers, (or substituted heterofullerenes, SbC_{59} and their polymers), can be possible by a recoil process following nuclear reactions. Carrying out *ab-initio* molecular-dynamics (MD) simulations on the basis of the all-electron mixed basis approach, we confirmed that endohedral fullerenes of Sb (Te) atoms inside the C_{60} cage can be possible. From the difference between the case of Sb (Te) and the case of As, the chemical nature of a doping atom seems to play an important role in the process of endohedrally doping or substitutionally doping in fullerenes.

Acknowledgments

The authors are grateful to the staff of the Laboratory of Nuclear Science(LNS) for handling the beam, and are grateful to the technical staffs, working at IMR, Tohoku University, for their continuous support, for the supercomputing facilities of HITAC S3800. This work was supported by the Grants-in-Aid for Co-operative Research No. 12640532 from the Ministry of Education, Science and Culture of Japan.

REFERENCES

1. Chai Y. *et al.*, *Phys. Chem.*, **95**, 7564 (1991).
2. Johnson R.D. *et al.*, *Nature (London)*, **355**, 239(1992) .
3. Weaver J. H. *et al.*, *Chem. Phys. Lett.*, **190**, 460(1992).
4. Shinohara H. *et al.*, *Nature (London)*, **357**, 52(1992).
5. Takata M. *et al.*, *Nature (London)*, **377**, 46(1995).
6. Sato W. *et al.*, *Phys. Rev. Lett.*, **80**, 133(1998).
7. Saunders M. *et al.*, *Science*, **271**, 1693(1996).
8. Braun T. and Rausch H., *Chem. Phys. Lett.*, **288**, 179(1998).
9. Braun T. and Rausch, H., *Chem. Phys. Lett.*, **237**, 443(1995).

10. Gadd, G.E. et al., *J. Am. Chem. Soc.*, **120**, 10322(1998).
11. Roth, L.M. et al., *J. Am. Chem. Soc.*, **113**, 6298(1991)
12. Huang, Y. and Freiser, B.S. , *J. Am. Chem. Soc.*, **113**, 9418(1991).
13. McElvany, S.W. et al., *J. Phys. Chem.*, **96**, 4935(1992).
14. Guo, T. et al., *J. Phys. Chem.*, **95**, 4948(1991).
15. Muhr, H.J. et al., *Chem. Phys. Lett.*, **249**, 399(1996).
16. Pradeep, T. et al., *J. Phys. Chem.*, **95**, 10564(1991).
17. Christian, J.F. et al., *J. Phys. Chem.*, **96**, 10597(1992).
18. Ohtsuki, T. et al., *J. Am. Chem. Soc.*, **117**, 12869(1995).
19. Ohtsuki, T. et al., *Material Sci. Eng.*, **217/218**: 38(1996).
20. Ohtsuki, T. et al., *J. Radioanal. Nucl. Chem.*, **239**, 365(1999).
21. Ohtsuki, T et al., *Phys. Rev.*, **B60**, 1531(1999).
22. Pellarin, M. et al., *Chem. Phys. Lett.*, **277**, 96(1997).
23. Ray, C. et al., *Phys. Rev. Lett.*, **80**, 5365(1998).
24. Ohtsuki, T et al., *Chem. Phys. Lett.*, **300**, 661(1999).
25. Ohno, K. et al., *Phys. Rev.*, **B56**, 1009(1997).
26. Ohtsuki, T. et al., *Phys. Rev. Lett.*, **81**, 967(1998).
27. Firestone, R.B. et al., Eds. *Table of Isotopes*, 8th Ed. Vol.I. John Wiley & Sons, Inc. 1996.
28. Ohtsuki, T. et al., *J. Chem. Phys.*, **112**, 2834(2000)
29. Ohtsuki, T. et al., *Phys. Rev. Lett.*, **77**, 3522(1996).
30. Ohno, K. et al., *Phys. Rev. Lett.*, **76**, 3590(1996).
31. Shiga, K. et al., *Modelling Simul. Mater. Sci. Eng.*, **7**, 621(1999).

Electronic Properties of Nanotube Based Materials

Susumu Saito

Department of Physics, Tokyo Institute of Technology
Oh-okayama, Meguro-ku, Tokyo 152-8551, Japan

Abstract. The electronic transport properties of nanotube-based materials obtained by the theoretical band-structure study are reviewed. Although fullerenes and nanotubes consist of only one element, carbon, they show rich electronic properties depending on their network topology and the details of the interunit interaction. In the case of metallic carbon nanotubes, the intertube interaction in its crystalline bundle phase can significantly alter the electron transport properties by opening the pseudogap at the Fermi level. Also their conducting behavior can be altered by doping. The K-doped metallic (10,10) nanotube lattice K_2C_{80} shows a rather simple upward shift of the Fermi level without modifying the band dispersion upon doping. Since the Fermi level goes above the second conduction-band minumum and the density of states at the Fermi level becomes much larger, K_2C_{80} should show much better conductivity than the pristine phase. Finally, the importance of the nearly free electron states being present in nanotube-based materials is discussed.

INTRODUCTION

During the last decade, the macroscopic production of a new form of carbon, i.e., fullerenes and nanotubes, have been achieved [1,2] and they have been studied intensively in many scientific and technological fields. Fullerenes are now considered as atom-like building blocks of solids and possess a variety of electronic properties depending on their network topology. C_{60} is known to have the closed-shell electronic structure with a finite gap between the highest-occupied state and the lowest-unoccupied state. Due to an interfullerene interaction, both electronic states of C_{60} form Bloch states with considerable dispersion (about 0.5 eV). Still the fundamental gap remains finite and solid C_{60} is therefore *semiconducting* [3]. Likewise solids of other extractable fullerenes, C_{76}, C_{84}, etc., are also semiconducting while the gap value itself depends on the size and the topology of the fullerene. Hence, fullerene solids are electronically completely different from the traditional form of sp^2 carbon, graphite, which is *metallic*.

In the case of nanotubes, it is now well-known that nanotubes can be *either metallic or semiconducting* depending on their network topology [4,5]. Although the topologically homogeneous nanotube materials have not been produced yet, their bulk properties have been studied theoretically and have proved to be highly interesting. In this article, some of our recent theoretical studies on nanotubes are reviewed.

EFFECT OF THE INTERTUBE INTERACTION

Effect on the Electronic Structure

Fullerene and nanotube solids can be classified as "materials of hierarchy": The constituent unit possesses electronic states which are well delocalized on the whole unit and they form Bloch states in the solid phase as in the case of solid C_{60}. In this sense, graphite can also be considered as a hierarchical solid consisting of two-dimensional units stacked along the perpendiclar direction. Although its interlayer distance is too long (about 3.35 Å) to form interlayer covalent bond, the interlayer interaction gives a very important effect on its electronic transport properties. A single-layer graphite sheet ("graphene") is known to be a zero-gap semiconductor while graphite can possess the semimetallic electronic band structure due to the interlayer interaction.

The importance of such an interunit interaction in hierarchical carbon materials has also been demonstrated in alkali-doped C_{60} superconductors [6,7]. Their superconducting transition temperatures depend strongly on the interfullerene distance which governs the conduction-band width and the Fermi-level density of states [8]. In the case of carbon nanotubes, therefore, the intertube interaction may also give considerable effects on their transport properties. Since the crystalline bundles of carbon nanotubes has been produced and the average-size control has also been achieved, both theoretical and experimental studies of the effect of the intertube interaction on the electronic transport properties are now of high importance.

Actually, an interesting and important role of the interlayer interaction on the electronic transport has been pointed out from the electronic band structure study on the crystalline phase of the (10,10) nanotubes [9,10]. In an isolated (10,10) nanotube system, there are two bands which have nearly linear dispersion and cross each other just at the Fermi level. Hence, they give a constant density of states around the Fermi level. In the crystalline (10,10)-nanotube bundle, however, the intertube interaction gives rise to an appearance of the pseudogap. The same pesudogap is also found to appear in the case of the (6,6) nanotube rope (Figure 1). The width of the pseudogap is in the same order as the intertube transfer energy. Therefore, the intertube interaction decreases the Fermi-level density of states. Although actual nanotube samples produced so far contain not only metallic but also semiconducting nanotubes, topologically homogeneous nanotube materials will become available in the future and the experimental study of the pseudogap will be performed as well.

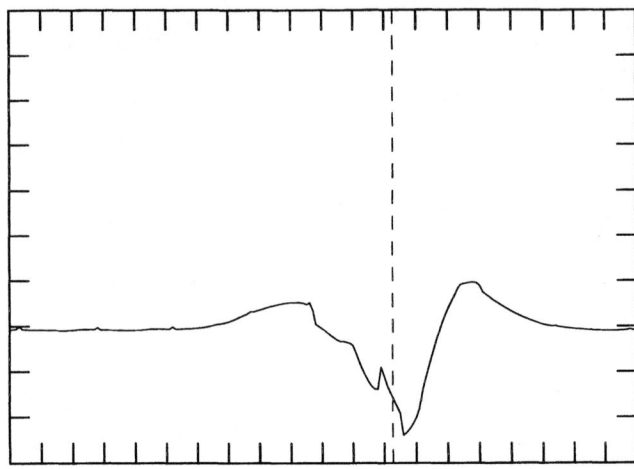

FIGURE 1. Pseudogap appearing in the electronic density of states of the crystalline (6,6) nanotube bundle (in arbitrary units). The Fermi level is denoted by the vertical line.

Effect on the Energetics

The effect of the intertube interaction on the energetics has also been studied theoretically, by using the density-functional theory with the local-density approximation (LDA). The enerngy gain for an isolated (10,10) nanotube to form the crystalline bundle phase is estimated to be about 9 meV per atom [10].

In the crystalline phase, the intertube interaction can give rise to an interesting orientation dependence of the total energy. In the case of solid C_{60}, melting of the orientational degree of freedom is known to take place around 260 K. In the crystalline bundle of homogeneous carbon-nanotube materials, also the orientational melting is expected to take place at even lower temperature [10].

DOPING INTO CARBON NANOTUBES

After a macroscopic production of fullerenes, doping K atoms into solid C_{60} resulted in the discovery of various interesting compounds including superconducting

K_3C_{60}, polymerized KC_{60}, semiconducting K_6C_{60}, etc. Therefore, the K-doped carbon nanotube should be one of the most interesting nanotube-based materials to be studied in detail.

Experimentally, K-doped nanotubes were produced and a large increase of the conductivity was reported [11,12]. Samples used, however, inevitably contain various different-topology nanotubes. Therefore, transport properties of homogeneous pristine and doped crystalline nanotubes are not yet studied experimentally. On the other hand, theoretical studies of the doped homogeneous crystalline nanotube have been done [13,14] as in the case of the pristine material discussed in the previous section.

The electronic band structure of the K-doped (10,10) nanotube lattice, K_2C_{80}, obtained in the LDA is shown in Figure 2. A comparison with its pristine-phase band structure indicates that dispersions of bands around the Fermi level are not modified upon the K doping but a constant upward shift of the Fermi level by about 1 eV has been observed. Therefore, the material is considered to be a rather simple charge-transfer system. Hence, the conduction electrons should be mostly on C sites.

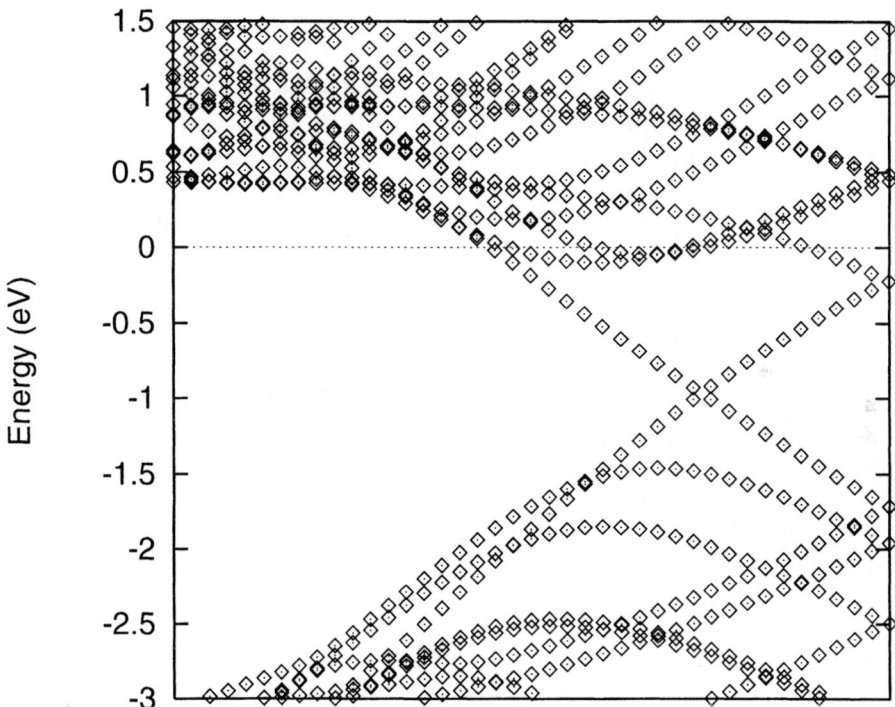

FIGURE 2. Electronic band structure of the K-doped (10,10) nanotube lattice K_2C_{80} along the Γ–A line of the hexagonal-lattice Brillouin zone [14]. Energy is measured from the Fermi level.

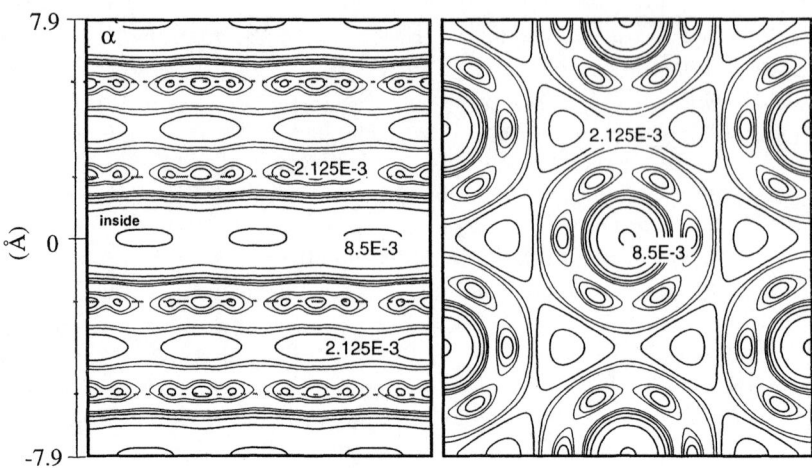

FIGURE 3. Spatial distribution of the nearly free electron state of the (6,0) nanotube lattice at the Γ point shown as a contour map of the squared amplitude of the wavefunction. The plane shown on the left pannel includes axes of three nanotubes at the top, the middle and the bottom of the pannel, and is perpendicular to the plane on the right pannel.

As can be seen from Figure 2, the Fermi level is now above the second-conduction-band bottom. Hence, the density of states at the Fermi level should be much larger than the pristine phase, and the material should show much better conductivity as well.

One of the reasons why the high superconducting transition temperature can be realized in alkali-doped fullerides is believed to be a strong electron-phonon coupling due to deep scattering potentials of C ions for the $2s$ and $2p$ electrons. Since the doped nanotube K_2C_{80} is actually having conduction bands mostly consisting of C states, the material is a good candidate for nanotube-based superconductors.

NEARLY FREE ELECTRON STATES

Interestingly, the lowest conduction band of graphite at the Γ point is known to be the interlayer state of which wave function has its maximum amplitude not around any atomic site but just at the mid space between layers. This interlayer state possesses a nearly free electron (NFE) character and plays an interesting role in electron-doped graphite intercalation compounds [15]. Since carbon nanotubes consist of graphitic layers as well, a similar NFE states can appear as a conduction-band state.

In the case of an isolated carbon nanotube, the NFE state is associated both on the inner and the outer surface of the nanotube with almost cylindorical spatial distribution. In crystalline bundle phase, on the other hand, the spatial distribution of the NFE state at the intertube region can possess triangular shape (Figure 3) [16]. They are located at about 3 to 4 eV above the Fermi level, which is still below the vacuum level. These NFE states are expected to be highly conductive and should again play a very interesting role also in electron-doped nanotube systems as well.

CONCLUDING REMARKS

Fullerenes and nanotubes are now classified as new forms of carbon being different from traditional form of sp^2 carbon, graphite, due to their different dimensionalities, i.e., zero and one-dimensional covalent-bond networks in fullerenes and nanotubes, respectively. The dimensionality is of essential importance for physical properties of materials and therefore many interesting properties have already been observed in these materials during the last decade. Still, it is surprising that just one element, carbon, can show such a variety of properties. In this sence, fullerenes and nanotubes are really the materials of topology. Since one can consider essentially an infinite number of topologically different carbon-network materials, the predictive theoretical study will continue to play a very important role in the field.

Acknowledgements: The work reported was supported by the Grant-in-Aid for Scientific Research on the Priority Area "Fullerenes and Nanotubes" by the Ministry of Education, Science, and Culture of Japan, JSPS under contract No. RFTF96P00203, and The Nissan Science Foundation. Numerical calculations were in part performed at Reserch Center for Computational Science, Okazaki National Institute, and at the Supercomputer Center of the Institute for Solid State Physics, University of Tokyo.

REFERENCES

1. Krätschmer, W., Lamb, L. D., Fostiropoulous, K., and Hoffman, D. R., *Nature* **347**, 354 (1990).
2. Thess, A., Lee, R., Nikolaev, P., Dai, H., Petit, P., Robert, J., Xu, C., Lee, Y. H., Kim, S. G., Rinzler, A. G., Colbert, D. T., Scuseria, G. E., Tománek, D., Fisher, J. E., and Smalley, R. E., *Science*, **273**, 483 (1996).
3. Saito, S. and Oshiyama, A., *Phys. Rev. Lett.* **66**, 2637 (1991).
4. Hamada, H., Sawada, S., and Oshiyama, A., *Phys. Rev. Lett.* **68**, 1579 (1992).

5. Saito, R., Fujita, M., Dresselhaus, M. S., and Dresselhaus, G., *Appl. Phys. Lett.* **60**, 2204 (1992).
6. Hebard, A. F., Rosseinsky, M. J., Haddon, R. C., Murphy, D. W., Glarum, S. H., Palstra, T. T. M., Ramirez, A. P., and Kortan, A. R., *Nature* **350**, 600 (1991).
7. Tanigaki, K., Ebbesen, T. W., Saito, S., Mizuki, J., Tsai, J. S., Kubo, Y., and Kuroshima, S., *Nature* **352**, 222 (1991).
8. Oshiyama, A. and Saito, S., *Solid State Commun.* **82**, 41 (1992).
9. Delaney, P., Choi, H. J., Ihm, J., Louie, S. G., and Cohen, M. L., *Nature* **391**, 466 (1998).
10. Kwon, Y.-K., Saito, S., and Tománek, D., *Phys. Rev. B* **58**, R13314 (1998).
11. Lee, R. S., Kim, H. J., Fischer, J. E., Thess, A., and Smalley, R. E., *Nature* **388**, 255 (1997).
12. Rao, A. M., Eklund, P. C., Bandow, S., Thess, A., and Smalley, R. E., *Nature* **388**, 257 (1997).
13. Saito, S., in *Recent Advances in the Chemistry and Physics of Fullerenes and Related Materials Vol.4* (Proc. 191st Elechtorochemical Soc. Meeting, Montréal, May 1997), edited by K. M. Kadish and R. S. Ruoff (Electrochemical Soc., Pennington, 1997) p.1055.
14. Saito, S., *Mat. Res. Soc. Symp. Proc.* Vol.593 (MRS, Warrendale, 2000) p.161.
15. Posternak, M., Baldereschi, A., Freeman, A. J., Wimmer, E., and Weinert, M., *Phys. Rev. Lett.* **50**, 761 (1983).
16. Okada, S., Oshiyama, A., and Saito, S., *Phys. Rev. B* **62**, 7634 (2000).

PHYSICS FOR NUCLEAR TRANSMUTATION (PNT)

Radiochemical Measurements Of Nuclear Data For Transmutation Of Minor Actinides

N. Shinohara

Japan Atomic Energy Research Institute
Department of Materials Science
Research Group for Innovative Nuclear Science
Tokai, Ibaraki 319-1195, Japan

Abstract. In order to understand quantitatively the transmutation of minor actinides in irradiation location, the nuclear data of uranium, neptunium, plutonium, americium and curium nuclides have been measured by radiochemical method: Several samples of the actinide nuclides irradiated in thermal and fast neutron reactors have been analyzed to determine the contents of actinides and fission products. Yields of the fission products in proton-induced fission of the minor actinides have been also measured by using a tandem accelerator. From the viewpoint of nuclear waste management, the transmutation process of the actinides in irradiation field is discussed quantitatively in this paper.

INTRODUCTION

One of the options for management of high-level radioactive waste is to transmute the actinide wastes into short-lived fission fragments by neutron-induced fission. A transmutation system with a very hard neutron energy spectrum and high neutron flux is efficient and effective for the actinide wastes. A concept of accelerator-driven transmutation system has been developed at Japan Atomic Energy Research Institute (JAERI). Figure 1 shows the schematic diagram of the proposed transmutation system concept [1]. In order to understand quantitatively the nuclear transmutation of minor actinides in neutron irradiation location, it is essential to obtain precise nuclear data on their neutron reactions.

In this study, neutron capture cross sections and fission yields of americium nuclides have been measured by a radiochemical method. Several enriched-samples of uranium, neptunium, plutonium, americium and curium nuclides irradiated in a fast neutron reactor have been also analyzed radiochemically to determine the contents of actinides and fission products formed in the samples. Yields of the fission products in proton-induced fission of neptunium, americium and curium nuclides have been, moreover, measured by using JAERI tandem accelerator. From the viewpoint of nuclear waste management, the transmutation processes of the actinides in irradiation fields are discussed quantitatively in this paper.

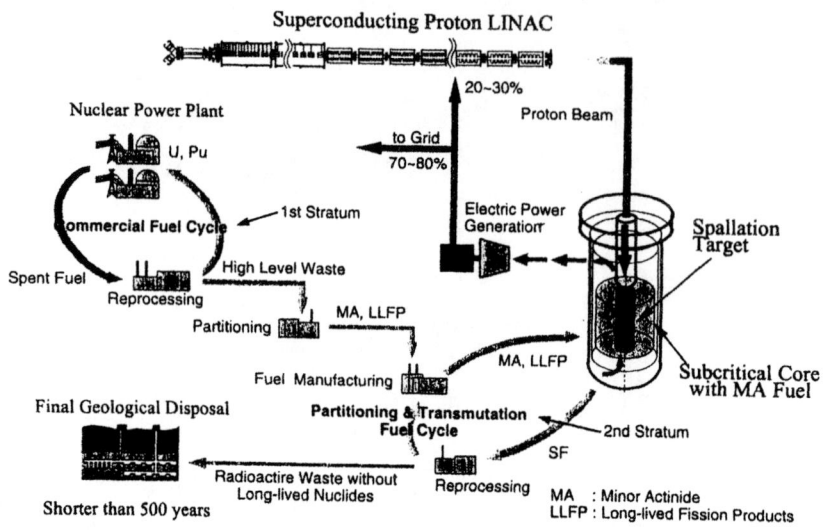

FIGURE 1. Concept of accelerator-driven transmutation system.

EXPERIMENTAL

Irradiation Of Americium Nuclides In Research Reactors

Highly purified targets of ^{241}Am were irradiated in Japan Material Testing Reactor at JAERI for 4 days. The thermal neutron fluxes and the epithermal neutron fractions were determined by measuring γ rays of ^{60}Co and ^{198}Au from the irradiated neutron flux monitors of Co/Al and Au/Al alloys. The neutron capture cross sections of ^{241}Am were decided by measuring growth and decay curves of the α-ray activity ratios of ^{242}Cm/^{241}Am, where the nuclide ^{242}Cm is the daughter of ^{242g}Am produced by the neutron capture of ^{241}Am. Analyzing the γ-ray spectra of both the irradiated americium targets and the chemically separated fractions measured with a HPGe detector, the yields of the fission products were obtained. These radiochemical procedures are given in detail elsewhere [2,3].

To determine the capture cross section of ^{243}Am, highly purified targets of ^{243}Am were irradiated in Japan Research Reactor-3M at JAERI during 10 hours. After the irradiation, γ rays of ^{244g}Am on the targets were measured with the HPGe detector. Measurements of the α-ray spectra for the irradiated samples were also carried out to determine the contents of ^{244}Cm in the samples. The thermal-neutron capture cross section of ^{243}Am was thus obtained from the contents of ^{244g}Am and ^{244}Cm in the target [4].

Irradiation Of Actinide Samples In A Fast Reactor

Enriched nuclides of uranium (^{233,234,236,238}U), neptunium (^{237}Np), plutonium (^{238,239,240,242,244}Pu), americium (^{241,243}Am) and curium (^{243,244,246,248}Cm) had been irradiated in the Dounreay prototype fast reactor (PFR) in Scotland [5]. Parts of the enriched samples were brought to JAERI for analyzing them. Isotopic concentration measurements of actinides and fission products formed by the irradiation of fast neutrons were performed by chemical separation, α- and γ-ray spectrometries and mass analysis. Flow sheet of the analysis is shown in Fig. 2. The detailed procedure related to the chemical analysis will be described in a future report. In this report, the results on the plutonium samples are presented.

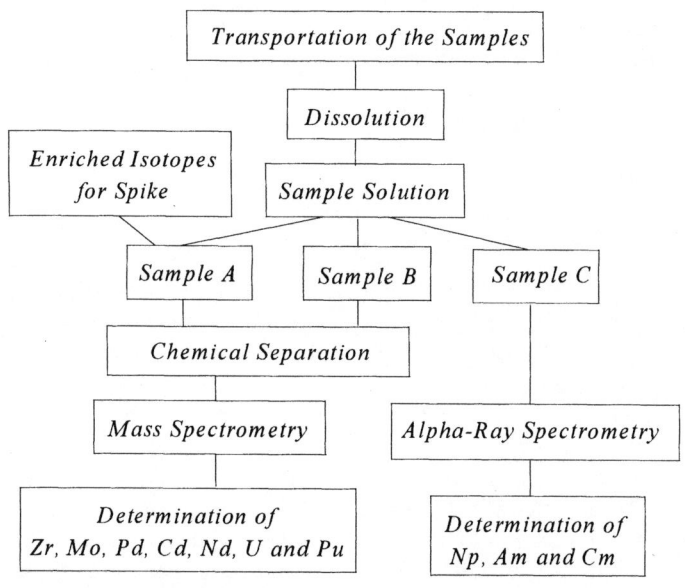

FIGURE 2. Flow sheet for the analysis of the irradiated actinide samples.

Proton Irradiation Of Minor Actinides In Tandem Accelerator

The minor actinide targets of ^{237}Np, ^{241,243}Am and ^{248}Cm were bombarded with proton beams in the energy range of 10-30 MeV at the JAERI tandem accelerator. After bombardment, the fission products caught in an aluminum foil (which was set behind the target) were identified and determined by γ-ray spectrometry. Their yields were derived from the data measured. The detailed procedure has been described in Ref. [6].

RESULTS AND DISCUSSION

Neutron Capture Cross Sections And Fission Yields of Americium Nuclides

The neutron capture cross sections of ^{241}Am and ^{243}Am obtained in this study are given in Table 1, where σ_0 and I_0 are the thermal cross sections and the resonance integrals, respectively. The differences between present results and the evaluated values [7,8] are 38-59% for ^{241}Am and 25% for ^{243}Am. This discrepancy affects greatly the estimation of the actinide waste generated from nuclear plants and the evaluation of nuclear transmutation system. The accurate data with accuracy of 10-15% on the capture cross sections of ^{241}Am and ^{243}Am have been required as registered in the World Request List for Nuclear Data [9]. It is very important to obtain the precise values with a view to treating the nuclear waste, because ^{241}Am and ^{243}Am are the starting nuclides in the buildup chain of transplutonium elements.

TABLE 1. Neutron Capture Cross Sections Of ^{241,243}Am.

Reaction	σ_0, b	I_0, b
^{241}Am$(n,\gamma)^{242m}$Am	85.7 ± 6.3	114 ± 7
^{241}Am$(n,\gamma)^{242g}$Am	768 ± 58	1,694 ± 146
^{243}Am$(n,\gamma)^{244m+g}$Am	56 ± 3.4	

Fission product yields of actinides are needed in calculating waste disposal inventory, and then it is substantial to measure the precise yield data. However, data on the fission yields of minor actinides are limited so far. Table 2 shows the fission yields of the products from the neutron-induced fission of ^{241}Am, where Y and Y' are the fission yields by thermal and epithermal neutrons, respectively. In this study, contribution of the fission of ^{242m}Am, ^{242g}Am and ^{242}Cm (that are produced by the neutron capture reactions of ^{241}Am) to the total fission cannot be ignored. The fission fractions of ^{241}Am, ^{242m}Am, ^{242g}Am and ^{242}Cm to the number of total fission were calculated [10] and given in Fig. 3. In a few days of thermal neutron irradiation, the contribution of the ^{241}Am fission decreases rapidly as seen from the figure, because the capture cross sections of ^{241}Am is large and the fission cross sections of ^{242m}Am and ^{242g}Am are also large. We may claim that much careful attention should be paid to the secondary reactions of the nucleus to be measured, such as neutron capture reaction and fission.

Isotopic Compositions Of The Plutonium Samples Irradiated In PFR

The plutonium nuclides change their atom numbers during the fast neutron irradiation, because of the nuclear reactions (such as neutron capture and fission) and the decays. One of the experimental results is shown in Table 3, where the values are the measured isotopic composition of the ^{238}Pu sample after 2,200-days irradiation in PFR. The nuclear transmutation of other plutonium samples (^{238,239,240,242,244}Pu) is depicted in Fig. 4 by comparing the isotopic compositions of the samples before and after the irradiation. The basic difference in the neutron binding energy between the even-even nuclei (^{238,240,242,244}Pu) and the odd-N ones (^{239,241}Pu) can be observed in Fig. 4. Buildup of the higher actinides such as ^{241}Am and ^{243}Am strongly depend on the inclination toward the capture cross sections, but the decays of the produced nuclides and the short-lived target (^{241}Pu) cannot be also neglected both in irradiation duration and in cooling of the irradiated samples. On the basis of the data measured in this study, we can verify the cross section libraries and the calculation codes for buildup and decay of actinides in fuel cycle of nuclear reactors.

TABLE 2. Fission Yields In The Neutron-Induced Fission Of ^{241}Am.

Nuclide	Fission Yield Y, b	Fission Yield Y', b
^{95}Zr	4.40 ± 0.40	4.68 ± 0.34
^{99}Mo	7.87 ± 0.75	8.19 ± 0.70
^{103}Ru	8.58 ± 0.78	8.11 ± 0.60
^{106}Ru	5.63 ± 0.74	7.2 ± 1.1
^{127}Sb	0.535 ± 0.078	<0.14
^{129m}Te	0.68 ± 0.26	<0.11
^{131}I	2.25 ± 0.20	4.30 ± 0.32
^{132}Te	6.20 ± 0.60	5.92 ± 0.67
^{133}Xe	6.50 ± 0.59	8.89 ± 0.61
^{140}Ba	7.28 ± 0.66	6.33 ± 0.47
^{141}Ce	8.98 ± 0.80	6.70 ± 0.48
^{147}Nd	3.84 ± 0.38	3.60 ± 0.28
^{156}Eu	0.050 ± 0.009	<0.33

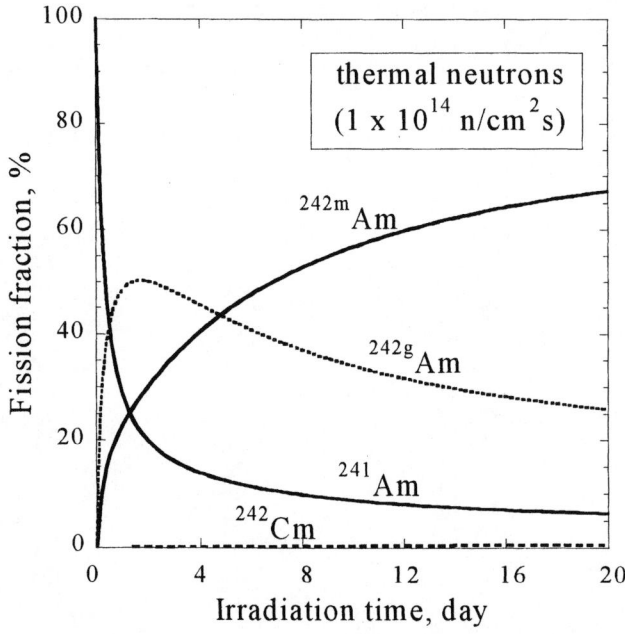

FIGURE 3. Fission fractions of ^{241}Am, ^{242m}Am, ^{242g}Am and ^{242}Cm in the neutron irradiation of ^{241}Am target.

TABLE 3. Nuclear Transmutation Of The ^{238}Pu Sample After 2,200-Days Irradiation In PFR.

Nuclide	Composition, % [Before Irradiation]	Composition, % [After Irradiation]
^{232}Th		0.6
^{233}U		0.2
^{234}U	12.5	13.4
^{235}U		1.1
^{236}U		0.1
^{238}U		0.04
^{238}Pu	86.3	59.3
^{239}Pu	0.3	8.5
^{240}Pu	0.3	0.9
^{241}Pu		0.03
^{242}Pu		0.02
Fission*		16.3

*FIMA%

Mass Yield Distributions Of The Minor Actinides In Proton-Induced Fission

Formation cross sections of the fission products in 20-MeV proton-induced fission of ^{248}Cm are shown in Table 4. Mass yield distribution is constructed from the cross sections observed in this study. The resulting mass yield distribution is typically asymetric. The mean weighted mass numbers of the asymetric mass yield peaks and the widths of the heavier asymetric mass yield peaks are independent of the proton energy range of 10-20 MeV in the p + ^{248}Cm reaction [6]. Other reaction systems of the proton-induced fission of ^{237}Np and ^{241,243}Am will be reported in a future paper.

TABLE 4. Cross Section Of The Fission Products In The p (20 MeV) + ^{248}Cm Reaction.

Nuclide	Cross Section, b	Nuclide	Cross Section, b
^{87}Kr	6.8 ± 0.9	^{132}Te	48 ± 13
^{88}Kr	7.2 ± 0.9	^{133m}Te	11 ± 2
^{91}Sr	14 ± 2	^{133g}I	41 ± 13
^{92}Sr	14 ± 2	^{135}I	20 ± 3
^{95}Zr	20 ± 3	^{136}Cs	10 ± 2
^{97}Nb	28 ± 4	^{138g}Cg	34 ± 4
^{103}Ru	44 ± 5	^{139}Ba	38 ± 5
^{104}Tc	41 ± 5	^{140}La	31 ± 4
^{105}Ru	46 ± 5	^{141}Ba	27 ± 3
^{107}Rh	50 ± 6	^{141}Ce	36 ± 4
^{113}Ag	49 ± 6	^{142}La	33 ± 4
^{115}Cd	33 ± 5	^{143}Ce	29 ± 4
^{117g}Cd	13 ± 2	^{147}Nd	18 ± 3
^{126}Sb	7.3 ± 0.9	^{149}Nd	18 ± 2
^{127g}Sn	12 ± 2	^{151}Pm	18 ± 2
^{127}Sb	26 ± 3	^{153}Sm	9.8 ± 1.3
^{128g}Sn	9.3 ± 1.8	^{155}Sm	8.3 ± 1.1
^{129}Sb	15 ± 2	^{156}Sm	7.3 ± 1.0
^{131}I	29 ± 3		

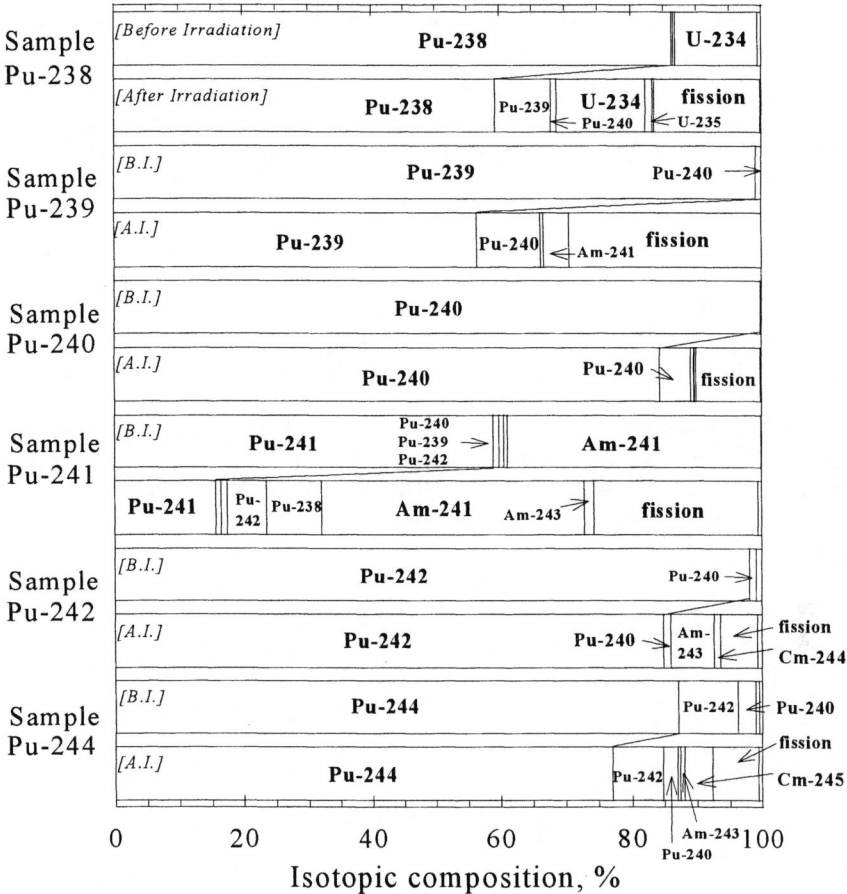

FIGURE 4. Isotopic composition of the enriched plutonium samples irradiated in the fast neutron reactor PFR.

Nuclear Transmutation Of Americium By Neutron Irradiation

In this section, nuclear transmutation process of americium is quantitatively considered by a mathematical procedure: We define that $N_A(t)$ is the atom number of a nuclide A at irradiation time t. The nuclide A is transmuted to different nuclides B, C, and so forth by neutron capture reaction. Time-derivatives of the atom numbers during irradiation are given by

$$\frac{dN_A(t)}{dt} = -N_A(t)\left(\lambda_A + \phi\sigma_A^{abs}\right), \tag{1}$$

$$\frac{dN_B(t)}{dt} = -N_A(t)\phi\sigma_A^{cap} - N_B(t)(\lambda_B + \phi\sigma_B^{abs}), \quad (2)$$

$$\frac{dN_C(t)}{dt} = -N_B(t)\phi\sigma_B^{cap} - N_C(t)(\lambda_C + \phi\sigma_C^{abs}), \quad (3)$$

where σ^{cap}, σ^{abs} and λ are the capture cross section, the absorption (capture + fission) cross section and the decay constant for each nuclide, respectively. By solving these differential equations [10], the formation and decay processes of the minor actinides can be calculated using the nuclear data measured in this study and previously reported ones.

Figure 5 shows the nuclear transmutation process of ^{241}Am by thermal-neutron irradiation. On the basis of the calculation mentioned above, the nuclear transmutation of ^{241}Am by the neutron-induced reaction can be understood quantitatively: 93.3% of the atom number of ^{241}Am (target) is transmuted to other nuclides after one year irradiation of thermal neutrons with flux of 10^{14} n/cm^{2}s in a reactor. The transmuted amount goes to ^{238}Pu (20.1%), ^{239}Pu (0.0001%), ^{240}Pu (0.001%), ^{242m}Am (0.08%), ^{242g}Am (0.04%), ^{243}Am (4.3%), ^{242}Cm (22.8%), ^{243}Cm (0.52%) and ^{244}Cm (0.13%), while the remaining percentage (45.5%) can be attributed mainly to the fission products. In order to consider quantitatively the transmutation of the minor actinides, it is essential to obtain the precise nuclear data, because the uncertainties of the calculated values of the transmuted amounts strongly depend on the nuclear data used.

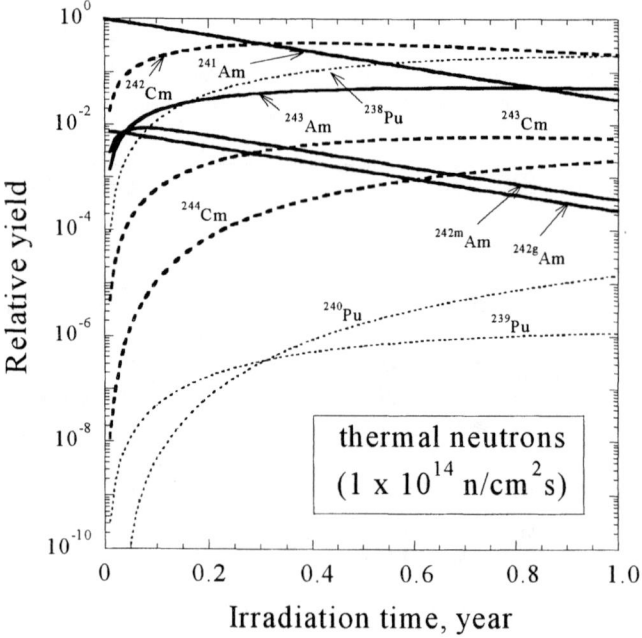

FIGURE 5. Calculated yields of the actinides produced by neutron irradiation of ^{241}Am.

CONCLUSION

With the precise nuclear data, the transmutation process of actinides in irradiation field can be discussed quantitatively and understood correctly by calculating the yields of the actinides for nuclear waste management. In this study, the nuclear data of the minor actinides on neutron capture cross sections and fission product yields were measured by a radiochemical method. The buildup of the minor actinides in the irradiation of the enriched actinide samples in PFR was also measured chemically to determine the isotopic compositions of the minor actinides and the fission products produced by neutron capture reactions and fission, respectively. In considering nuclear transmutation process of an actinide nuclide in irradiation field of high-flux particles, much careful attention should be paid to the secondary reactions of other actinides produced simultaneously by the particle capture reaction of the actinide target.

ACKNOWLEDGMENTS

The author wishes to express his deep gratitude to M. Andoh, W. S. Charlton, K. Gunji, H. Haba, K. Hata, Y. Hatsukawa, S. Ichikawa, J. Inagawa, M. Itoh, K. Katoh, N. Kohno, T. Mukaiyama, R. Nagaishi, Y. Nagame, Y. Nakahara, I. Nishinaka, M. Ohnuki, T. A. Parish, Z. Qin, S. Raman, T. Sakurai, H. H. Saleh K. Tsujimoto, K. Tsukada, K. Watanabe, and Y. L. Zhao for their help in carrying out this study.

REFERENCES

1. Matsuura, S. , *Nucl. Phys.* **A654**, 417c-435c (1999).
2. Shinohara, N., Hatsukawa, Y., Hata, and K., Kohno, N., *J. Nucl. Sci. Technol.* **34**, 613-621 (1997).
3. Shinohara, N., Hatsukawa, Y., Hata, and K., Kohno, N., *J. Nucl. Sci. Technol.* **36**, 232-241 (1999).
4. Hatsukawa, Y., Shinohara, N., and Hata, K., *JAERI-Conf 98-003*, 1998, pp. 221-224.
5. Walker, R.L., Botts, J.L., Hydzik, R.J., Keller, J.M., Dickens, J.K., and Raman, S., *ORNL-6837* (1994).
6. Qin, Z., Tsukada, K., Shinohara, N., Zhao, Y. L., Nishinaka, I., Hatsukawa, Y., Ichikawa, S., Hata, K., and Nagame, Y., *Radiochim. Acta* **84**, 115-120 (1999).
7. Mughabghab, S.F., "Neutron cross sections," in *BNL 325*, (4th ed.), vol. 1, Part B (1984).
8. Firestone, R.B. and Shirley, V.S., *Table of Isotopes*, (8th ed.), John Wiley and Sons, New York, 1996.
9. Kocherov, N. and McLaughlin, P.K. (ed.), *INDC (SEC)-104* (1993).
10. Shinohara, N., Hatsukawa, Y., Hata, K., Kohno, N., Andoh, M., Saleh, H.H., Charlton, W.S., Parish, T.A., and Raman, S., *J. Radioanal. Nucl. Chem.* **239**, 631-638 (1999).

Basic Nuclear Research Involvment in the French Research Programme on the Back-End of the Nuclear Fuel Cycle

Hubert C. Flocard

Theory Group, Institut de Physique Nucléaire
F-91406 Orsay CEDEX

Abstract. Approximately eighty percent of french electricity is from nuclear origin. Associated with this generation of electricity a substantial amount of radioactive waste is also produced, the destination of which is still unclear. In 1991, the french parliament has voted a law which organizes a systematic programme of research over a period of fifteen years along three different axes. The first axis investigates the partitioning of waste with a view to elaborate methods and materials appropriate for storage as well as dedicated targets for a transmutation programme. This axis is also concerned with transmutation methods whether in conventional or innovative reactors. The second axis is devoted to reversible and irreversible geological storage. The third axis considers chemical matrices for long term conditioning as well as the design of interim storage containers and facilities. In the first part of this text, we give a brief overview of the organization of the national programme of research supervised by the ministry of research and technology. The second half of this presentation focuses on the researches conducted, within the framework of this national programme, by physicists working in laboratories of public institutions whose core activity is basic nuclear research.

INTRODUCTION

In the public opinion as it reflects in polls and questions asked at hearings, the safety issues associated with the nuclear production of electricity are mostly associated with three different types of potential dangers. First, there is the release of radioactivity by plants whether in normal operation mode or in accidental situations. In the former case, the pollution is expected to be extremely limited and in any case to remain confined within the plant or its immediate neighbourhood. In the latter situation, the example of Tchernobyl has shown that the level of contamination depends on many geographical and meteorological factors and can take a world global dimension. The second danger which is mentioned is proliferation. Then, plutonium (Pu) is generally said to carry the major risk because of its current use in nuclear weapons. Finally, there are the dangers associated with the radioactive wastes of the electricity production, whether they remain in the spent fuel assemblies or appear in the waste streams from the partitioning industry. It is often stated that this waste generates a threat which extends to the entire region surrounding the site where it is temporarily or ultimately stored. This threat is described as being associated either to an accidental direct intrusion into the site or to the residual pollution consecutive to the slow diffusion of radioactive (or chemically dangerous) nuclides into the geosphere and in particular into the water table. In addition, it is recalled that the long periods associated with some radioactive decays add a time dimension to this third problem. As a consequence, the extension of the potential danger to generations beyond those which directly benefit from the electricity production is sometimes said to be a major ethical issue.

The present text presents research activities which deal with some aspects of the last of these three concerns (waste management).

THE FRENCH SITUATION

Although the USA are still today the largest nuclear energy producing country, the numbers illustrating this section will be taken from the french context. This context is definitely original if not on an absolute scale, certainly in relative terms when one considers the total national electricity production. In France, eighty percent of the produced electricity is nuclear. The installed power is 63GWe (in GWe, the letter 'e' stands for electric to distinguish from the thermal power). Taking into account the duty factor of the french plants, this corresponds to a total production of 388TWhe (1998). Now that Superphenix has been stopped (and neglecting the potential marginal energy production of the fast reactor Phenix), the french park consists entirely of pressurized water reactors (PWR). There are 59 operational units with a nominal electric power ranging from 0.9 to 1.45GWe. These reactors all operate with a neutron spectrum in the thermal range. Four out of five burn only uranium oxide (UOx) with an enrichment of 3.7% in ^{235}U. The remaining plants work with an heterogenous core with 70% UOx and 30% of a mixed uranium-plutonium core (MOx). The plutonium in the MOx results from the partitioning of a fraction of the spent UOx fuel.

Let us consider the yearly production of nuclear waste from a typical french power plant. As reference we take a 0.9GWe plant working with UOx fuel burnt at the rate of 33GWj per ton of heavy metal (U in this case). Taking a thermal to electric conversion efficiency of 33% and a duty factor of 70%, the yearly input of enriched uranium is 21.5T. The ouput consists of electricity (5.52TWhe) but also of spent fuel which contains 20.4T of uranium (U), o.21T of Pu, 16kg of Minor Actinides (MA), mostly ^{237}Np and ^{241}Am and 745kg of fission products. The long lived higly radioactive waste comprises all the actinides (U, Pu, MA) and 10% of the fission products (FP). These numbers scale in proportion with the delivered power. As a consequence, taking into account the present UOx-MOx policy of the french utility (EDF), the yearly french production of high activity nuclear waste (normalized to an electricity production of 390TWh) corresponds to 1500T of discharged metallic

Weak and Medium Activity, Short Lived	1 350 000 m³
Medium Activity Long Lived	56 000 m³
High Activity (MA+FP)	3 500 m³
Unprocessed Spent Fuel	15 000 T

TABLE 1. Inventory at end of present french nuclear park. (indicative: first evaluation from inventory national evaluation committee)

fuel. Most of this spent fuel is uranium. The rest contains Pu (11.6T), MA (1.55T) and 3.5T of long lived FP.

It is not clear when a new nuclear power plant is going to be built in any of the major developed countries. This holds also for France. Nevertheless, it seems plausible that most of the existing plants will be used for production as long as they are considered safe by the national competent authorities. Today, it is generally accepted that under present regulations, the lifetime of a plant can extend over 40 years. Based on this duration, the presently operational french nuclear park appears rather young. Indeed, the first stop of a plant would occur in 2017 and the last after 2040. From these figures, an estimate of the waste inventory has been made by a national evaluation committee. Working with the hypothesis of a continuation of the present UOx-MOx policy at today level, it has derived the numbers listed in table 1. Even when one sets aside the short lived waste of weak and medium activity which can be stored in the already opened surface repositary (or extensions of it), the volumes are large. Still, they do not appear out of reach of the presently available technology and of its foreseeable extensions. However, volume is only one part of the problem and presumably not that which weighs the most in the eyes of the public opinion. Moreover, the emphasis on the volume dimension more or less implies that the geological repository option is the one which should necessarily prevail in the end. In the eighties, such an early and (as it appeared so to the opponents) a priori choice turned out to be a dead end.

THE NATIONAL ORGANIZATION OF THE RESEARCH

The Law of 1991

The nuclear waste management issue in France was seen from a rather different perspective after the vote of the law Bataille (after the name of his promoter) by the parliament on 30 December 1991. Before, the official solution to the problem of waste was that which had been defined by the experts almost from the start of the ambitious french nuclear programme in 1973. It involved the partitioning of the spent fuel to separate the most dangerous radiochemical elements, their insertion into a specific glass containment matrix and their irreversible storage in a geological repository following the cooling down at an interim storage facility. The difficulty arose, when the moment came to select a site for the repository. In every of the areas which had been selected based on their geological qualities, the public adamantly, even sometimes violently, turned down the project. In public hearings, it was often stated that the storage solution had been selected in a technocratic manner without much attention being given to alternatives. Whether this criticism was justified or not, it appeared that the situation was utterly blocked. The parliament chose to acknowledge this clear public rejection and decided to change both the strategy and the calendar.

Abandoning the ambition to immediately create a repository, the lawmakers moved one step backward and decided instead to establish a 15 year research programme. In their formulation of the law, they stated three principles which they deemed essential for the definition of a proper solution. First, a technically and economically feasible solution should ensure an adequate protection of the persons and of the environment. Second, because of the long lifetimes, any solution must also take into account the right of future generations (note that the practical implications of this notion are left open by the law). Finally, it was said that the acceptance of any solution would be enhanced if the public could be convinced that this decision was based on the results of a comprehensive research programme on a broad spectrum of potential solutions and the consideration of both reversible and irreversible options. In practice, the law enforces this latter principle by means of an organization of the research along three axes :

- Axis 1 : Partitioning and transmutation
- Axis 2 : Geological storage
- Axis 3 : Waste conditionning and interim storage

In addition the law establishes a standing national evaluation committee (CNE) whose tasks involve a continuous review of the research and the production of a yearly evaluation report submitted both to the government and the parliament. At the end of the fifteen year period, it is expected that the parliament should have in its hands the information necessary to take appropriate measures and in particular decide on the construction of a repository.

Organization of the French Research

While the law implies that the research should involve all french competences, it specifically designates two agencies to take a leading role. The axes 1 and 3 are to be managed by the atomic energy commission (CEA) while axis 2 is to be supervised by the national agency for radioactive waste (ANDRA).

Few years later, it was thought appropriate that the entire programme of research should also be coordinated at the level of the the ministry of research and technology. This coordination takes place within a standing committee (COSRAC) in which all participants to the research are represented. In particular, the french utility EDF, the partitioning and fuel reprocessing firm COGEMA and the reactor company FRAMATOME seat with the CEA and ANDRA. From the start, the national center for scientific research (CNRS) and the universities have been associated with the activities of the COSRAC. This committee defines the national strategy and checks that the programmation of the research by the agencies stays in line. The COSRAC also prepares the interaction with the CNE, in particular the annual progress report. It updates both the calendar and the strategy taking into account the outcome of the research or the political orientations as they come (for instance: the government decision to shutdown Superphenix or its expression of the need for a stronger emphasis on reversibility).

The Involvement of CNRS and Universities

While CEA and ANDRA had a natural and longstanding implication in problems related with the back end of the nuclear fuel cycle, the same was certainly not true for basic research institutions such as CNRS and universities. Indeed it can be fairly said that the french organization of nuclear energy as it had been defined after the second world war, if not intentionally at least in the facts, created a clearcut separation between the worlds of basic research and of nuclear energy. Therefore, although it was evident that the contribution of CNRS/Universities laboratories on some specific subjects could be immediatly valuable, the overall knowledge of the status of nuclear energy science by the scientists working in basic research institutions quite naturally suffered from some significant gaps. Taking this situation into account, the CNRS decided that any work he would choose to get involved in, had to be organized in close relation with the well established major actors of the nuclear energy : CEA, ANDRA and the industry. On the other hand, it was felt essential to preserve the original characteristics of a basic science institution. These imply that the research be oriented by, but nevertheless remain upstream from technology concerns. In addition, it has to be performed in the usual academic open framework and in particular its results should be made widely accessible via publication in international journals.

An already existing structure called "research group" (GDR) turned out to be appropriate. In a GDR, on a specific subject, partners from the public and private sectors choose to organize a collaboration for a limited period (typically 4 years). Every year, a board evaluates the progress of research in relation with the subject, checks the budget of the year past and decides on a budget and on the support of a programme of research for the coming year. This programme has been prepared by the director of the GDR who has been initially appointed by the partners. Decisions are based on the recommendations of a scientific committee which includes some outside experts. This committee evaluates the quality of proposals resulting from an open call for proposals (laboratories outside partners institutions can submit proposals).

Four GDRs have been set up in relation with the back end of the nuclear fuel cycle :

- **PRACTIS** on the behavior of ions of actinide elements in a solution or at the interfaces with solids. The partners are CNRS, CEA and ANDRA. This GDR concerns both the axis 1 in relation with partitioning whether aqueous or pyrometallurgy with molten salts. It contributes also to axis 2 on questions related to the migration of nuclides in a storage or in the geological environnment.

- **GEDEON** on innovative options for the management of nuclear waste. The partners are CNRS, CEA, EDF and FRAMATOME. GEDEON contributes to the axis 1 in relation with the transmutation of MA and PF and on the analysis of low radioactive waste producing cycles such as the thorium cycle.

- **FORPRO** on deep geological formations. The partners are CNRS, CEA and ANDRA. It contributes to axis 2 on questions of geology in the short term (ground behavior near the repository during, and within few centuries, after the construction) as well as in the long term (evolution of the global geological environment).

- **NOMADE** on innovative materials for confinement. The partners are CNRS, CEA, ANDRA with a potential commitment of COGEMA. This GDR contributes to the three axes by working on the definition of improved transmutation targets and of confinement matrices with performances exceeding that of the presently retained solutions (glasses).

At the level of the CNRS a light structure (programme for the electronuclear back end : **PACE**) coordinates the overall commitment of the agency (budget and manpower) into these four GDRs and follows the actions of the CNRS in relation with le law of 1991.

It should also be noticed that, independently from the 1991 law, (and therefore today outside the competence of COSRAC), on a less formally organized level, some research is being performed on the following subjects:

- Biological and environmental radiotoxicology and toxicology of nuclear wastes. The actors are the life science department of CEA (DSV), the public medical institute INSERM, the national agronomical institute INRA, the CNRS and the Universities.

- Sociology studies involving, for instance, the CNRS, the Ecole des Mines and some independant groups.

- Economic studies on the nuclear energy which consider also back-end aspects. The actors are the Industry Ministery, concerned industries and several independant groups.

ACTIVITIES OF BASIC NUCLEAR RESEARCH GROUPS

From now on, this text will only report on the activities of the laboratories of IN2P3/CNRS participating to GEDEON. The scientific programme of the three other GDRs requires competences which exist in the chemistry and earth science departments of the CNRS. Within GEDEON, the CNRS nuclear physicists collaborate with the basic science division and the reactor division of CEA as well as with a team of the NOVATOME department of FRAMATOME. Most of the actions are also supported by the Euratom programme of the European Union.

The scientific programme of GEDEON aims at covering the spectrum of innovative options for management of nuclear waste. It is mostly concerned with the definition of transmutation schemes such that the radiotoxicity of the present and future waste can be considerably decreased in terms of both quantity and lifetime. Schematically speaking, one could say that the ultimate goal of transmutation is to ensure that after a "short" period (typically few centuries), the remaining radiotoxicity of the waste does not excede that of the uranium ore from which the nuclear fuel was made. Another line of studies concerns energy production scenarios whose waste stream would, from the start, be orders of magnitude smaller than the present UOx-MOx one. Such scenarios are not necessarily completely new. However, they have been partly overlooked in the past since back then the waste issue was generally not among the most important concerns of nuclear energy scientists and engineers.

The GEDEON research themes are :

1. Physics of a Spallation Target

2. Nuclear Data Acquisition

3. Neutronics of Subcritical media

4. Physico-chemistry of molten salts and metals

5. Irradiation Damages in materials

6. Generic Studies of Thorium cycle

7. High intensity acceleration

8. Nuclear scenarios and nuclear systems

Apart from the fourth theme which, within the CNRS, involves only chemists, CNRS and university nuclear physicists and engineers can contribute to all aspects of the GEDEON programme. The remainder of this section is devoted to a brief review of their involvment. Let me state again that the work which is reported below only concerns that which is conducted in french (CNRS and CEA) **basic** nuclear physics laboratories. Other qualitatively and quantitatively important activities take place within the specialized departments of CEA, ANDRA and the industry.

Physics of the Spallation Target

The experiments on thick targets were the last to be performed with the SATURNE accelerator before it closed at the end of 1997. The corresponding data has been published since. In the mean time, complementary experiments on the neutron production (rates and multiplicity) and on the H and He generation in materials have been performed at CERN.

The presently active part of this programme has moved to the laboratory GSI in Darmstadt. It focuses on a detailed measurement of spallation products in thin targets. Indeed, it is crucial to know the nature and the rate of production of foreign elements in the target since their increasing concentration during the operation may endanger the window (the interface between target and accelerator) through embrittlement and recoil damages, as well as the walls of the liquid target container by fostering sensitivity to corrosion. In addition, the presence of these spallation products is bound to increase the radioactive content of the material in the target and generate unwanted thermal effects.

At GSI, by colliding a heavy ion beam on a fixed liquid H or D target, one creates inverse kinematics conditions which allow the use of the powerful SHIP separator to extract the production rates of the complete set of spallation products (above a hundred) in a single experiment. The results will provide the very strict constraints required for the calibration of the simulation codes used to predict spallation cross sections at all energies.

The programme at GSI which started two years ago should reach its conclusion in 2001. By then, a complete coverage of spallation cross sections will have been achieved with a precision better or equal to 10% for a comprehensive set of projectiles which includes lead for the target material, uranium for the fissile core material and iron as representative of the structure materials. The selected energies, 0.5GeV, 0.8GeV and 1GeV cover the range considered as acceptable for a spallation target.

Data Acquisition

In the domain of reactor physics, the data accumulated over half a century are mostly limited to neutron cross sections in the low energy range (typically $E_n < 20$MeV. The knowledge at higher energies is very scarce. Since a determination of the neutron induced reactions rates on all the elements of interest appears out of reach, ultimately one will have to rely on theoretical simulation. Through interaction with experts from the OECD nuclear energy agency (NEA), a set of experiments has been selected which, once it is completed, should be adequate for testing the quality of simulation codes. The lack of appropriate facilities in France has led the french physicists to establish collaborations with belgian (Louvain), dutch (Groningen) and swedish (Uppsala) groups to perform experiments with neutron or proton beams for energies ranging from 20MeV to 200MeV.

A new world unique tool will become available at the end of year 2000 to study neutron induced capture and fission cross sections. The CERN time of flight neutron beam (nTOF) will provide neutrons with an energy resolution at least equal to 10^3 for a range of energy between 0 and 200MeV. In addition the time structure of the pulse (6ns every 1.4s) should allow a significant reduction of the background as compared to existing similar facilities. nTOF is an international collaboration in which Russia and USA are involved along with many european countries. It receives a funding from the Euratom programme and benefits from an important technical support from CERN. Within nTOF, the IN2P3 group at Orsay has taken the responsability for the construction of a set of parallel plate avalanche counters which will be the backbone of the equipment for the measurement of fission cross sections. The first experiments are scheduled in 2001.

Irradiation and Thermophysics of a Liquid Metal Target

On theme five, an ambitious project MEGAPIE plans the first study of the thermohydraulics of liquid metals with and without irradiation. It involves the building of a target to be installed in the beam of the cyclotron at the swiss laboratory PSI (Villingen). IN2P3 collaborates with PSI, and with the italian and german institutes ENEA and FZK. Thanks to PSI, european scientists have access to a proton beam whose performances (E_p=0.65GeV, I_p=1.5mA) are close to those required for the accelerator sustaining the chain reaction in a subcritical hybrid system. With MEGAPIE, one will be able to study the production of neutrons in realistic conditions and the thermal and fluid properties of several liquid metals. In addition one will obtain a direct observation of the modifications of mechanical properties of steels in contact with a metal, the damages induced by irradiation to all the components of a target and the corrosion effects. MEGAPIE will also be a demonstration of a target operating in realistic conditions and as such should provide invaluable information on both operation and safety aspects. MEGAPIE is expected to be operational in 2004 and should yield its first results within the two following years.

High Intensity Proton Accelerator

The subcritical hybrid systems envisaged for transmutation will require intense and energetic proton beams. Typically, one considers that a monoenergetic accelerator with an energy lying in the range 0.6GeV< E_p <1.2GeV and capable of delivering intensities between 2 and 50mA will cover all the needs of hybrid systems with a reactivity index k such that $0.75 \leq k \leq 0.98$. From the presently accumulated experience, these performances appear reachable in a near future. An additional challenge is the very high reliability which is requested by the reactor operation and safety (typically less than 100 trips per year). Moreover, the flexibility in the beam power delivery is an issue as the accelerator plays an important rôle in the regulation and control of hybrid systems. These requirements define a promising while challenging field for accelerator research.

In France, the competence is for the most part concentrated in basic research (nuclear and particle physics) laboratories. Three years ago, with the budget support of the ministery of research, of CNRS and of CEA, a programme on high intensity accelerator has been launched by IN2P3 and DAPNIA/CEA. The LINAC technology, which offers the most potential for an extrapolation towards industrial applications, has been selected. Today a 100mA source has been built and shown to work with the requested reliability. A project for the construction of the low energy section is underway (RFQ) or planned (DTL) which should lead to a 500kW beam by the year 2002. Simultaneously, a programme on low-β superconducting cavities for the high energy ($E_p > 100$MeV) sections has started. It has already achieved major breakthroughs with respect to the accelerating fields. A collaboration framework with similar projects in the USA (Los Alamos) and in Italy (INFN) has been established.

Neutronics of Subcritical Systems

The CEA center at Cadarache operates a research facility with unique features : MASURCA. Schematically, it can be described as a "zero-power, LEGO-like" reactor with neutron spectrum flexibility. At its inception, MASURCA was designed as a test ground for concepts for the future generations of critical fast reactors. Few years ago, members of the ISN/IN2P3 laboratory at Grenoble suggested that MASURCA could also be used to investigate the dynamics of subcritical systems. This was the start of the MUSE programme. The physicists of Grenoble had become acquainted with the field of reactor neutronics by participating to the seminal experiments FEAT and TARC initiated and led by C.Rubbia at CERN.

The already completed MUSE 3 experiment has tested the detection systems and the analysis tools. In the MUSE 4 phase due to start in 2001, MASURCA will be coupled with the accelerator GENEPI built in Grenoble. GENEPI delivers a pulsed deuton beam at 250keV which generates a neutron source by colliding on a tritium target placed at the center of MASURCA. The specific feature of GENEPI is the time structure of its pulse which is defined within a few ns. This precision allows to study even the fastest transients in the type of fast spectrum core anticipated for a class of hybrid systems. Several coolants (Na, Pb/Bi, gas) and fuels (U, Th) are going to be tested in the future phases of the MUSE programme.

Numerical Simulation of Systems and Scenarios

The CNRS contributes to a programme headed by CEA and EDF aimed at studying various scenarios for the production of electricity and transmutation. Some of these studies have been defined at the OECD level as part of an international evaluation. A similar analysis has also been requested by the national evaluation committee.

The IN2P3 groups at Grenoble and Orsay are developing tools for the analysis of the neutronics of specific systems (reactors) as well as that of the implications of the operation of critical reactors on the nature and the amount of waste. The simulations rely on Monte Carlo calculations allowing a detailed description of the reactor geometry for the neutron flow and on standard differential or global methods for the long term evolution of the reactions and decays in the fuel and in the waste. The specific contribution of the IN2P3 concerns mostly systems and scenarios associated with the thorium cycle. In addition, in the framework of collaborations with the industry (EDF and FRAMATOME), analysis of molten salt systems and of specific gas cooled design are also performed.

CONCLUSIONS

I hope that this presentation has given a brief but nevertheless fair idea of the extent of the involvment of the CNRS and Universities basic nuclear physics laboratories in the french research programme on nuclear waste management. It concerns important aspects where it introduces original approaches which complements the traditional techniques elaborated in the nuclear energy domain over half a century.

More specifically, as most of the effort deals with transmutation issues in innovative systems, this work should be understood as participating to the construction of the scientific platform with a view to build a demonstration facility for an hybrid subcritical system. In this perspective, the CNRS/University effort must be considered as a part of the national contribution to an enterprise conducted at the european level. Indeed, a technical working group impulsed and directed by C. Rubbia and involving Austria, Belgium, France, Germany, Italy, Spain, Sweden and Switzerland is presently working on the clarification of the original contributions of hybrid systems in energy production and waste management scenarios. From that point on, within a year, it will produce a document stating the goals of a demonstration facility, its main characteristics, the associated major technical features, a possible agenda for the construction, global budget indications and a description of the necessary project oriented research programme. This work aims at getting a specific support from national agencies and from the Euratom VIth framework programme.

To conclude, I would like now to turn to the questions which are sometimes asked by colleagues in our own laboratories : why should we get involved? is it not preferable and more efficient to leave the task entirely to people working in specialized labs? I think that a first answer could be that the subjects considered are new, challenging and that they will enrich our overall knowledge of the world of physics. The corresponding work is not very different in spirit from the upstream research aimed for instance at improving oil detection, electronic transmission, solar cells or medicine scanning. It fits into the general definition of research as can be checked by the interest they generate in some young (and not so young) nuclear physicists. Moreover, over the last fifty years, basic nuclear physics has steadily improved its techniques in the domains of the detection, the analysis and the particle acceleration. It thus matters that our community keeps an interest in the transfer of this knowledge into applications. Finally, there is another consideration which first concerned only nuclear energy but now extends to other domains such as, for instance, genetics. The growing sensitivity of the public to health or environment issues makes it essential that decisions be based on the widest possible information basis. Therefore, on such subjects it is important to open the world of expertise beyond the boundaries of specialized agencies. The academic system whose status, operation and knowledge distribution methods are not specifically tied to economic considerations should in that respect bring an invaluable contribution to the public debate.

Acknowledgement. The author thanks the Institute of Nuclear Theory at Seattle where part of this work has been completed.

PHYSICS WITH EXOTIC NUCLEI (PEN)

Nuclear Halo and Molecular States

N.A. Orr

Laboratoire de Physique Corpusculaire,
IN2P3–CNRS, ISMRa et Université de Caen,
14050 Caen Cedex, France

Abstract.
Significant advances have been made in recent years in the exploration of clustering in light nuclei. This progress has arisen not only from the investigation of new systems, but also through the development and application of novel probes. This paper will briefly review selected topics concerning halo and molecular states in light nuclei through examples provided by the neutron-rich Be isotopes.

INTRODUCTION

Clustering within nuclei is a widespread phenomenom which takes on many guises across the nuclear landscape. Until relatively recently cluster studies have been confined to on or near the line of beta stability where the rôle of α-clustering has long been established [1]. As clustering is expected to manifest itself most strongly near thresholds [2], exotic structures might be expected to form in very neutron (or proton) rich systems. Over the last decade the exploration of clustering in nuclei far from stability has become technically feasible as demonstrated most clearly by the discovery and subsequent probing of the nuclear halo [3]. Whilst an excess of neutrons (or protons) may naïvely be expected to dilute any underlying α-cluster structures, theoretical [4,5] and experimental work [6,7] indicate that molecular type structures such as α-chains "bound" by valence neutrons may exist.

In the present paper a number of selected topics concerning the study of halo and molecular states in light, neutron-rich nuclei will be reviewed. As they exhibit many of the facets of clustering and structural evolution far from stability the neutron-rich Be isotopes have been chosen as examples. In parallel, the techniques which have been developed to aid in probing the structure of such nuclei far from stability will also be discussed.

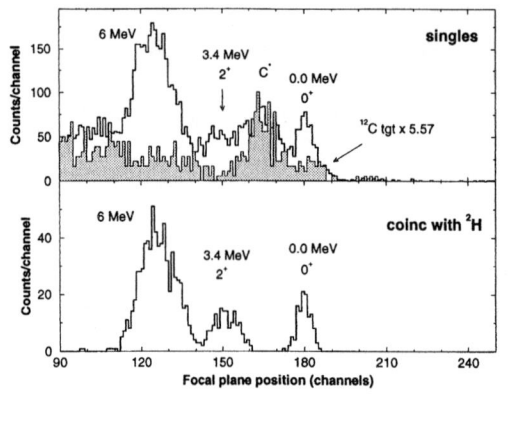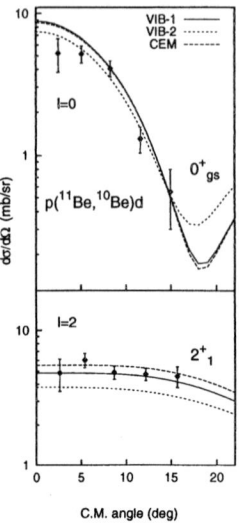

FIGURE 1. Left: Position spectra for ^{10}Be ions from the p(^{11}Be,^{10}Be)d reaction on a $(CH_2)_n$ target. Right: Angular distributions for transfer to the ^{10}Be ground and 2^+ states. Comparison is made with various core coupling descriptions for ^{11}Be [10,11].

REACTION SPECTROSCOPY

Single-Nucleon Transfer Reactions

One of the issues of prime importance in the study of nuclei far from stability is the extraction of reliable spectroscopic information. For nuclei on or near stability light-ion (p, d, ...) induced single-nucleon transfer reactions have long been one of the tools of choice. The application of such reactions to nuclei far from stability presents a number of experimental challenges [8,9]. Most significantly, whilst the cross sections are moderately high ($\sim$1-10 mb/sr) the short halflives of the nuclei of interest dictate that beams of these nuclei be used in inverse kinematics reactions. The consequent constraints on the final excitation energy and angular resolution limit the target thickness to the order of 1 mg/cm^2 and beams of at least some 10^4 pps are required in order to obtain angular distributions in a reasonable measurement time.

A recent study [10,11] of the structure of the single-neutron halo nucleus ^{11}Be via the p(^{11}Be,^{10}Be)d reaction at 35 MeV/nucleon provides a prototypical example of some of the features inherent in such reaction studies. Experimentally a magnetic spectrometer operated in a dispersion matched mode was employed to detect the heavy ejectile (^{10}Be). In order to remove the contamination arising from reactions on the C in the $(CH_2)_n$ target an array of large area silicon detectors was used to detect the coincident deuterons (figure 1). With a beam intensity of $\sim 3 \times 10^4$ pps angular distributions of reasonable quality were acquired in some 72 hours of running (figure 1).

In addition to the experimental features illustrated by this experiment it should be stressed that the analysis is not model independant. For example, not only did the weakly bound nature of ^{11}Be and the d have to be taken into account but realistic wavefunctions had to be employed – here coupling to the strongly deformed ^{10}Be core needed to be properly accounted for. Indeed, the use of the standard separation energy approach to derive the radial form factors lead to an artificially high core excited state admixture. As detailed in refs [10,11], careful analysis indicates that the ground state of ^{11}Be is dominated (85%) by the admixture corresponding to the valence neutron occupying the $2s_{1/2}$ orbital together with a modest contribution (15%) from the core excited ($2^+ \otimes \nu 1d_{5/2}$) configuration.

Single-Nucleon Removal Reactions

Measurements of one-nucleon removal (or "knockout") reactions on light targets have recently been proposed as a spectroscopic tool for high-energy radioactive beams [12,13]. This approach has arisen from the development of reaction calculations for halo nuclei in which the strong absorption limit [14] and core excited states are accounted for [13]. More specifically, the cross sections for the population of a given state of the core fragment (I_c^π) may be related to spectroscopic factors ($C^2S(I_c^\pi, nlj)$) using an extended version [13,15] of the spectator-core model [16] to calculate the cross section (σ_{sp}) for the removal of the nucleon (nlj).

$$\sigma(I_c^\pi) = \sum_{nlj} C^2S(I_c^\pi, nlj) \sigma_{sp}(nlj, S_n^{eff}) \qquad (1)$$

The corresponding momentum distributions are derived within the same eikonal formalism [17] or in a simpler fashion using the opaque limit of the Serber model [18,19]. As noted above, the integrated cross sections for the population of the core excited states are directly related to the associated spectroscopic factors. In analogy with transfer reactions, the shape of the core fragment momentum distributions plays the rôle of the angular distributions in specifying the l of the removed nucleon. Experimentally the core fragment states are identified by the de-excitation γ-rays emitted in-flight, whilst a high acceptance magnetic spectrograph is used to determine the momenta.

One of the principal virtues offered by high energy nucleon removal is the applicability very far from stability where beam rates are low. This ability to function with intensities as low as 1pps is a consequence of the large cross sections (~10-100 mb), coupled with the high beam energies ($\gtrsim$40 MeV/nucleon) which allow for the use of thick targets (~100 mg/cm^2).

A particularly clear example[1] of the application of the technique may be found in the measurement of single-neutron removal from ^{12}Be [22], whereby the only

---

[1] Studies of other near dripline and halo nuclei may be found in ref's [12,20,21,23].

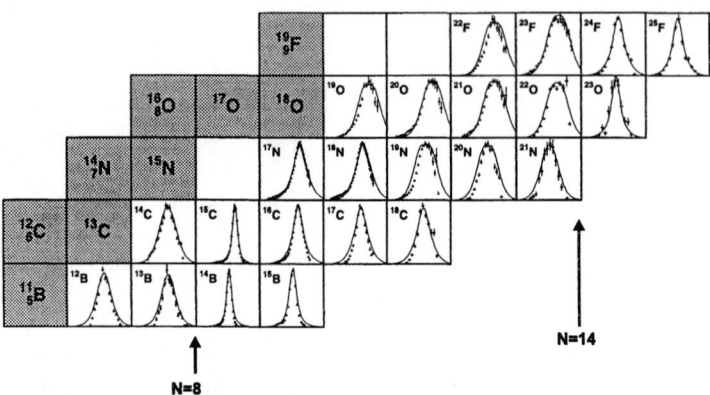

FIGURE 2. Core fragment longitudinal momentum distributions for one-neutron removal reactions on C. The solid lines correspond to Glauber model calculations [17].

bound core states are the ground ($J^\pi=1/2^+$) and 320 keV ($1/2^-$) levels in ^{11}Be. In simple terms, the population of the ^{11}Be ground state provides a measure of the $2\hbar\omega$ admixture present in ^{12}Be. The results of the experiment, which indicate only a ~30% admixture of the p^2-configuration in ^{12}Be, provide direct confirmation [24,25] of the breakdown in the N=8 shell closure.

The potential power of the technique is further illustrated by a recent systematic investigation of a broad range of light, neutron-rich psd-shell nuclei [17]. The inclusive longitudinal momentum distributions which were obtained in a single rigidity setting are displayed in figure 2 whereby a number of features are immediately apparent. Most notably, the crossing of the N=8 shell and N=14 sub-shell closures are associated with a marked reduction in the widths of the core momentum distributions (viz, ^{14,15}B, ^{15,16}C, ^{23}O and ^{24,25}F). The former effect arises from the large $\nu 2s_{1/2}$ admixtures expected in the ground states of the Z=4-6, N=9 isotones [23,26–28], which also persists for N=10, as suggested by recent studies of ^{14}Be [29,30] (see below). A narrowing of the momentum distributions may also be expected for N=15 and 16, as in a simple shell model picture the valence neutrons occupy the $\nu 2s_{1/2}$ orbital. Such results demonstrate that coupled with a high acceptance, broad range spectrograph, high energy single-nucleon removal reactions offer a powerful means to survey structural evolution over a wide range of isospin in a single experiment.

Finally, in the context of high energy reactions, it should also be noted, as described by Tostevin and Al-Khalili [15,31], that few-body Glauber model analyses of total reaction cross sections can also provide important constraints on the ground state wavefunctions of halo-like nuclei. As presented in the contribution to these proceedings by Suzuki, such analyses are now becoming more widespread in their application.

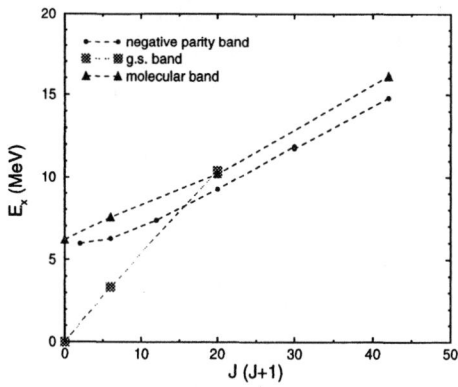

FIGURE 3. Spin-energy systematics for states observed in ^{10}Be (from [34]). The trajectories for the postulated positive and negative parity molecular bands are indicated.

NUCLEAR MOLECULAR CLUSTERS

As alluded to in the introduction, the α-particle plays an important rôle in the structure of light α-conjugate (A=4n) nuclei [1]. This is a direct consequence of the strongly bound character of the ^{4}He nucleus and the weakness of the α-α interaction, as evidenced by the unbound nature of ^{8}Be. The persistence of cluster structures for systems lying away from the line of beta-stability is well illustrated, as will be discussed here, by the beryllium isotopes, for which the α-α system may be regarded as the basis.

From a theoretical point of view, prescriptions such as the Molecular-Orbital Model (MO) [4] or the Two-Centre Shell Model (TCSM) [32], in which valence nucleons are added to the single-particle orbits arising from the two-centre potential, provide a successful means to describe the properties of these nuclei. Moreover these orbits may be viewed as the analogues of the σ and π-orbitals associated with the covalent binding of atomic molecules. The development of fully fledged Antisymmeterised Molecular Dynamics calculations (AMD) [5] is of particular interest as the N-nucleon system is modelled without any *a priori* imposition of an underlying cluster structure. Recent calculations, in particular, suggest the existence of two-centred structures in the Be, B and C isotopic chains with valence neutron density distributions exhibiting the features of molecular orbitals [5].

From an experimental perspective, von Oertzen [6] has compiled systematic evidence for the existence of dimers in $^{9-11}$Be and $^{9-11}$B. In the case of ^{9}Be, for example, the presence of a valence neutron results in a bound (Borromean) system, the ground and excited states of which may be understood in terms of a three-body α:n:α molecular structure. In particular, the rotational bands based on the ground and low-lying states exhibit large deformations consistent with the associated molecular configurations.

In the case of ^{10}Be, the experimental evidence for molecular configurations is

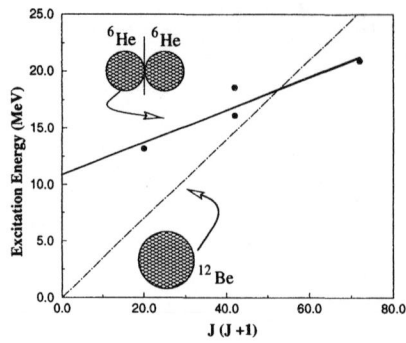

FIGURE 4. Spin-energy systematics for ^{12}Be → ^{6}He + ^{6}He (from [7]).

rather less well documented. Beyond the established 0_2^+, 2_2^+ and 1_1^- – 4_1^- states, the locations of the J=5 [6] and 6 members of the of the negative parity band, as well as the J=4 and 6 members of the positive parity band have been postulated following recent studies of the α-^{6}He breakup of ^{10}Be* [33,34]. As displayed in figure 3 [34], the spin-energy trajectories for the bands based on the 0_2^+ and 1_1^- states at ~6 MeV are consistent with large deformations as expected for molecular-like α:2n:α structures. Moreover, the location of the bandheads just below the threshold for α + ^{6}He decay is in accordance with the considerations of Ikeda describing the formation of clusters [2]. Further support for the postulated molecular states may be found in the recent AMD calculations of Kanada-En'yo and collaborators [5], whereby well developed α:2n:α configurations are predicted for the 0_2^+ and 1_1^- bands.

Given the existence of such molecular-type structures in ^{10}Be, the question naturally arises as to the existance of similar structures even further from stability. In this context the dripline nucleus ^{12}Be has been investigated. In a recent study[2] employing the inelastic excitation of an energetic (35 MeV/nucleon) secondary beam of ^{12}Be, evidence has been found in the breakup into ^{6}He+^{6}He of rotational states (J=4, 6, 8) in the excitation energy range 10-20 MeV [7]. As illustrated in figure 4, the inferred momenta of inertia ($\hbar^2/2\Im$=0.15±0.04 MeV) and bandhead energy (10.8±1.8 MeV) of the observed states are consistent with the cluster decay of a molecular structure which may be associated with α:4n:α configurations.

Further experimental support for the molecular nature of these states would be the observation of large spectroscopic factors for the associated clusters. The measurement of partial decay widths represents, however, formidable experimental challenges, though the presence of relatively few decay channels for the states in question may facilitate the measurements. Of additional interest is the search for in-band gamma transitions which should also furnish information on the degree of

---

[2] The existence of such structures was hinted at in an earlier study by Korsheninnikov *et al.* [35].

clustering [36]. Such measurements may be possible in the near future through α-pickup reactions – such as $^{12}\text{C}(^{6,8}\text{He},^{10,12}\text{Be}^*)^8\text{Be}$ – carried out in conjunction with high efficiency charged particle and gamma arrays.

HALO STATES

Perhaps the most extreme form of clustering is that exhibited by halo nuclei, whereby one or more nucleons reside on average well beyond the core potential [3]. In the present discussion the two-neutron halo nucleus ^{14}Be will be used as an illustrative example. In particular, the spectroscopy of $^{13,14}\text{Be}$, continuum excitations of ^{14}Be and the spatial configuration of the halo neutrons will be addressed.

The tool chosen to investigate ^{14}Be in the work described here was a kinematically complete measurement of the fragments (^{12}Be and two neutrons) from the dissociation of a 35 MeV/nucleon beam of ^{14}Be on C and Pb targets. Such a measurement is now a relatively standard technique and allows the two-neutron removal cross sections, neutron angular distributions and invariant mass spectra to be extracted, and the neutron-neutron correlations to be explored.

The details of the experiment will not be repeated here as they have already been described in refs [29,37,38]. It should be stressed, however, that one of the principle problems confronting such measurements is the detection with the highest possible efficiency of two beam velocity neutrons at very forward angles with small relative momenta and minimal cross-talk. As described in refs [37,38], the use of a highly granular array arranged in a staggered configuration coupled with off-line cross-talk rejection algorithms (based on kinematic conditions) permit such measurements to be undertaken.

Structure and Continuum Excitations

The results obtained for the two-neutron removal cross sections, σ_{-2n}, the single-neutron angular distributions[3], Γ_n, and the associated angle integrated cross sections, σ_n, are displayed in Table 1. The average neutron multiplicities ($\overline{m}_n = \sigma_n/\sigma_{-2n}$), which have also been deduced, are instructive in terms of the mechanisms leading to dissociation [39]. For a light target the reaction is expected to proceed via single-neutron removal (absorption or diffraction) followed by the in-flight decay of ^{13}Be. As approximately equal contributions are expected for absorption and diffraction [39] the average neutron multiplicity should be 1.5, in accordance with that measured. This scenario is also supported by the single-neutron angular distribution for the C target which is well reproduced assuming passage via a low-lying resonance in ^{13}Be [29,40]. In the case of a heavy target, nuclear and

---

[3] The angular distributions were well characterised by a Lorentzian lineshape.

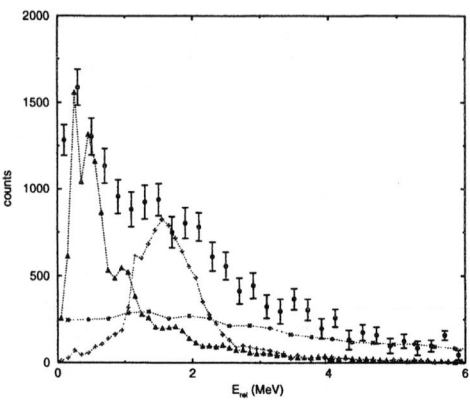

FIGURE 5. ^{12}Be + neutron relative energy spectrum from the dissociation of ^{14}Be on C at 35 MeV/nucleon [42]. The results of a simulation including states in ^{13}Be at 0.5 (triangles) and 2.0 MeV (crosses), with Γ=0.5 MeV are also displayed.

Coulomb dissociation occur. Given that Coulomb dissociation results in a multiplicity of 2, the average multiplicity for dissociation on Pb should be between 1.5 and 2, as observed.

TABLE 1. Measured cross sections, average neutron multiplicities and neutron distribution momentum widths for the dissociation of ^{14}Be at 35 MeV/nucleon [29].

	σ_{-2n} [b]	σ_n [b]	$\overline{m}_n$	Γ_n [MeV/c]
C	0.46±0.04	0.75±0.10	1.6±0.3	75±3
Pb	2.3±0.4	4.0±0.3	1.7±0.2	77±4
Pb(EMD)	1.45±0.40	2.7±0.4	1.9±0.6	87±6

The possible existence of a low-lying state in ^{13}Be is of considerable importance in modelling the structure of ^{14}Be [41]; in particular in fixing the ^{12}Be-n interaction. Beyond the single-neutron angular distributions, the relative energy for ^{12}Be + n events may be reconstructed from the measured momenta. The preliminary results of such an analysis [42] are displayed in figure 5 together with the results of Monte Carlo simulations (see below) in which levels at 0.5 and 2.0 MeV above threshold are assumed with widths of Γ=0.5 MeV. The level at 2.0 MeV is that observed in multi-nucleon transfer reaction studies [43,44], whilst the former is consistent with a preliminary result reported by the MSU group [45]. Given, as discussed above, the appearance of ground states dominated by a valence $\nu 2s_{1/2}$ configuration for the neighbouring N=9 isotones ^{14}B and ^{15}C [17,21,23], together with the predicted continuation of such an inversion for ^{13}Be [26], it appears that the ground state is most probably J^π=1/2$^+$. Analysis of the ^{12}Be - n angular correlations should shed further light on this conjecture.

The enhanced cross section for dissociation on the Pb target (Table 1) is indicative of a large EMD contribution. Assuming that the nuclear–Coulomb interference is small, the C target data may be scaled to estimate the nuclear contribution to breakup on Pb [29]. Given a root-mean-square radius of 3.2 fm for ^{14}Be [30], $\sigma_{-2n}^{nucl}(Pb) = 0.85\pm0.07$ b and, consequently, $\sigma_{-2n}^{EMD}(Pb) = 1.45\pm0.40$ b.

The invariant mass spectra, reconstructed from the measured momenta of the beam and fragments (^{12}Be and two neutrons) from breakup, are displayed in figure 6a and b for the C and Pb targets. The EMD spectrum (figure 6c) has been deduced following subtraction of the estimated nuclear contribution to reactions on Pb and exhibits enhanced strength around 1.5 MeV decay energy (E_{decay}). Given the complex nature of the response function of the present setup, a detailed Monte Carlo simulation, including the influence of all nonactive materials, was developed based on the GEANT package [29]. The results shown in figure 6 were obtained following the descriptions for dissociation on C and Pb outlined above and after filtering through the simulation. In the case of the nuclear induced reactions only the lowest low-lying state in ^{13}Be ($E_0 = 0.5$ MeV, $\Gamma_0 = 0.5$ MeV) was assumed to be populated following the diffraction of one of the halo neutrons. The EMD was simulated under the assumption that the energy sharing between the ^{12}Be and the two neutrons was governed by 3-body phase space. As shown in figure 6c, the observed EMD decay energy spectrum could be reproduced assuming a Breit-Wigner lineshape with $E_0 = 1.8 \pm 0.1$ MeV and $\Gamma_0 = 0.8 \pm 0.4$ MeV. Furthermore, the corresponding simulations of the single-neutron angular distributions were in good agreement with those observed [29].

Thompson and Zhukov have examined ^{14}Be within the framework of a 3-body model in which the ^{12}Be core was treated as inert[4] [41] and a number of trial wavefunctions developed. Based on the binding energy and matter radius of ^{14}Be, together with the known d-wave resonance at 2.0 MeV in ^{13}Be, two ^{14}Be wavefunctions were favoured (both of which required an s-wave state near threshold in ^{13}Be as suggested above): the so-called D4 wavefunction – 86% $\nu(2s_{1/2})^2$ and 10% $\nu(1d_{5/2})^2$; and C7 – 29% $\nu(2s_{1/2})^2$ and 67% $\nu(1d_{5/2})^2$. The EMD decay energy spectra calculated from the corresponding E1 strength functions [41] are compared in figure 6 with that of the empirical Breit-Wigner lineshape deduced from the measurements. The corresponding integrated two-neutron removal cross sections are 1.05 b (D4) and 0.395 b (C7) [41], compared to the measured value of 1.45±0.40 b. Although the strength is predicted to be concentrated at a somewhat lower energy than that observed, a large $\nu(2s_{1/2})^2$ admixture to the valence neutrons wavefunction is favoured. This conclusion is also supported by the total reaction cross section measurement of Suzuki et al. [30] reported in these proceedings.

A microscopic cluster model has also been used to explore ^{13,14}Be [46]. In the case of ^{13}Be an s-wave state is predicted very close to threshold, whilst the energy of the d-wave resonance is well reproduced. Significantly, a strong E1 transition [B(E1)$\approx$1.2e^{2}fm^2] centred at $E_{decay} = 1.5$ MeV is predicted in ^{14}Be, very close to the structure observed experimentally. Analysis of the corresponding energy surface

---

[4]) An inert core precludes, ab initio, the existence of any simple negative parity resonances.

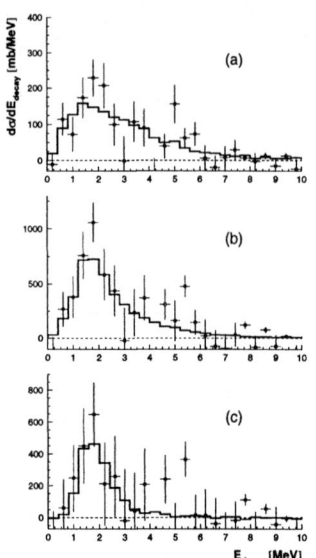
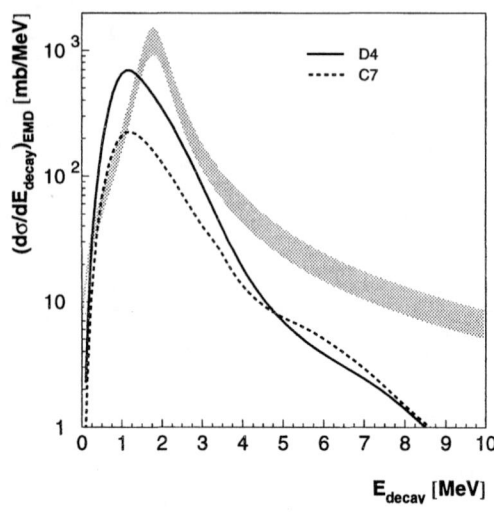

FIGURE 6. Left: Reconstructed decay energy spectra for the dissociation of ^{14}Be on (a) C, (b) Pb and (c) deduced for EMD on Pb [29]. Right: Comparison of the EMD decay energy spectrum deduced from the measurement (shaded region) with 3-body calculations (see text).

suggests, however, that this transition is not associated with a true resonance [46]. Consequently, the enhanced strength observed near threshold in the EMD invariant mass spectrum may be qualified as a nonresonant "soft" dipole excitation.

In light of the breakdown in the neutron p-shell closure in ^{12}Be discussed previously, the importance of p-sd cross shell excitations in the ^{14}Be ground state should be explored. In a first instance, the search for core excited states (^{12}Be*) in the dissociation of ^{14}Be may provide indications for the presence of such configurations, whilst theoretically the inclusion of a deformed core [25], in three-body calculations should be undertaken.

Neutron–Neutron Correlations

One of the most intriguing questions regarding the description of two-neutron halo systems is the degree of correlation between the neutrons. Considering only the intrinsic momentum distributions, a spatially compact dineutron will be characterised by $p_{core} = 2p_n$, whereas for a system with no correlations $p_{core} = \sqrt{2}p_n$. Unfortunately no unperturbed experimental measure of the momentum distributions is accessible.

Perhaps the method the most well adapted to probing spatial correlations is intensity interferometry [47]. In this approach the relative motion of the two outgoing neutrons is governed by the neutron–neutron final-state interaction and quantum statistics (Fermi-Dirac), both of which are related to the spatial characteristics of the source. Providing that any other effects on the neutron momenta may be neglected or eliminated in the construction of the correlation function, interferometry provides a means to probe the spatial correlations of the halo neutrons [37].

Importantly, owing to the low momentum content of the halo neutrons, the standard approach to constructing correlation functions (applied, for example, in the measurement of Ieki et al. [48]) is no longer valid [38]. In particular, the narrow momentum distributions result in strong residual correlations and consequently a significant overestimate of the source size. As a result a new iterative technique has recently been developed and applied to the dissociation on a Pb target of beams of ^{6}He, ^{11}Li and ^{14}Be [37]. Assuming that the timescale between the emission of the two neutrons was short ($\tau_{nn} \lesssim 100$ fm/c), as is most probably the case for EMD, neutron-neutron RMS separations of 5.9±1.2 (^{6}He), 6.6±1.5 (^{11}Li), and 5.4±1.0 fm (^{14}Be) have been deduced. These results appear to exclude the presence of any significant dineutron configurations. By way of comparison it is interesting to note that the RMS proton-neutron distance in the deuteron is some 3.8 fm. Future high statistics experiments emplying a well modelled system such as ^{6}He should allow the emission timescale and source sizes to be extracted simultaneously from the longitudinal and transverse neutron-neutron relative momenta as well as coherent analyses of the neutron-neutron and core-neutron correlations.

CONCLUSIONS

In this paper some of the facets of nuclear halo and molecular states in light nuclei have been reviewed. In particular, various illustrative examples provided by the neutron-rich beryllium isotopes have been discussed. In addition, a number of related experimental probes have also been presented. Clearly no single technique can furnish a complete description of these nuclei and it will require the application of a broad range of theoretical and experimental tools to obtain a better understanding of the phenomena outlined here.

It is a pleasure to thank my colleagues in the Groupe Exotiques at LPC and in the DEMON, CHARISSA and E264/281/295 collaborations who have contributed to the work described here.

REFERENCES

1. M. Freer, A. Merchant, J. Phys. **G23** (1997) 261
2. K. Ikeda, Prog. Theor. Phys. (Japan) Supplement (1968) 464
3. P.G. Hansen, A Jensen, B. Jonson, Ann. Rev. Nucl. Sci. **45** (1995) 591
4. M. Seya et al., Prog. Theor. Phys. (Japan) **65** (1968) 205
5. Y. Kanada-En'yo et al, Phys. Rev. **C60** (1999) 064304 and refs therein
6. W. von Oertzen, Z. Phys. **A354** (1996) 37; **A357** (1997) 355
7. M. Freer et al., Phys. Rev. Lett. **82** (1999) 1383

8. J.S. Winfield, W.N. Catford, N.A. Orr, Nucl. Instr. Meth. **A396** (1997) 147
9. W.N. Catford, Proc. of RNBV, Nucl. Phys. A (in press)
10. S. Fortier et al, Phys. Lett. **B461** (1999) 22
11. J.S. Winfield et al., nucl-ex/0009015, Nucl. Phys. A (in press)
12. A. Navin et al., Phys. Rev. Lett. **81** (1998) 5089
13. J.A. Tostevin, J. Phys. G: Nucl. Part. Phys. **25** (1999) 735
14. J. Hüfner, M.C. Nemes, Phys. Rev. **C23** (1981) 2538
15. J.A. Tostevin, Proc. of the 2nd Int. Conf. on Fission and Properties of Neutron-rich Nuclei (World Scientific, Singapore, 2000) p429
16. M.S. Hussein, K.W. McVoy, Nucl. Phys. **A445** (1985) 123
17. E. Sauvan et al., nucl-ex/0007013, Phys. Lett. B (in press); E. Sauvan, Thèse, Université de Caen (2000), LPC Report LPCC T-00-01
18. P.G. Hansen, Phys. Rev. Lett. **77** (1996) 1016
19. H. Esbensen, Phys. Rev. **C53** (1996) 2007
20. T. Aumann et al., Phys. Rev. Lett. **84** (2000) 35
21. V. Guimarães et al., Phys. Rev. **C61** (2000) 064609
22. A. Navin et al., Phys. Rev. Lett. **85** (2000) 266
23. V. Maddalena et al., NSCL Report MSUCL-1171, August 2000
24. R. Sherr, H.T. Fortune, Phys. Rev. **C60** (1999) 064323 and refs therein
25. H. Iwasaki et al., Phys. Lett. **B481** (2000) 7
26. Z. Ren et al., Z. Phys. **A357** (1997) 137
27. D.E. Alburger, D.R. Goosman, Phys. Rev. **C10** (1974) 912
28. F. Ajzenberg-Selove, Nucl. Phys. **A449** (1986) 1
29. M. Labiche et al., nucl-ex/0006003; M. Labiche, Thèse, Université de Caen (1999), LPC Report LPCC T-99-03
30. T. Suzuki et al., Nucl. Phys. **A658** (1999) 313
31. J. Al-Khalili, J. Tostevin, Phys. Rev. Lett. **76** (1996) 3903
32. J.M. Eisenberg, W. Greiner, *Nuclear Theory* (North Holland, Amsterdam, 1975)
33. N. Soić et al., Europhys. Lett. **34** (1996) 7
34. M. Freer et al., submitted to Phys. Rev. C
35. A.A. Korsheninikov et al., Phys. Lett. **B343** (1995) 53; RIKEN Report AF-NP-175
36. Y. Alhassid et al., Phys. Rev. Lett. **49** (1982) 1482
37. F.M. Marqués et al., Phys. Lett. **B476** (2000) 219
38. F.M. Marqués et al., Nucl. Instrum. Meth. **A450** (2000) 109
39. F. Barranco et al., Phys. Lett. **B319** (1993) 387
40. F. Barranco et al., Z. Phys. **A356** (1996) 45
41. I.J. Thompson, M.V. Zhukov, Phys. Rev. **C53**, 708 (1996); *priv. comm*
42. K.L. Jones, Thesis, University of Surrey (2000)
43. A.N. Ostrowski et al., Z. Phys. **A343** (1992) 489
44. M. Belozyorov et al., Nucl. Phys. **A636** 419 (1998) 419
45. M. Thoennessen, Proc. Int. School of Heavy-Ion Physics, 4th Course: Exotic Nuclei, Ed. R.A. Broglia, P.G. Hansen (World Scientific, 1998) p269
46. P. Descouvemont, Phys. Rev. **C52**, 704 (1995); *priv. comm*
47. D.H. Boal et al., Rev. Mod. Phys. **62** (1990) 553
48. K. Ieki et al., Phys. Rev. Lett. 70 (1993) 730

Measurements of interaction cross sections for nuclei far from stability at relativistic energies

T. Suzuki for the GSI–RIKEN–Kurchatov–Comenius collaboration [1]

Department of Physics, Niigata University, Niigata 950-2181, Japan
GSI, Planckstrasse 1, D-64291 Darmstadt, Germany
RIKEN, Wako, Saitama 351-0198, Japan
Kurchatov Institute, Kurchatov sq. 1, 123182 Moscow, Russia
Department of Physics, Tohoku University, Miyagi 980-8578, Japan
Faculty of Mathematics and Physics, Comenius University, 84255 Bratislava, Slovak Republic

Abstract. We measured interaction cross–sections (σ_I) of $^{12-14,16-20}$C, $^{14-16,18,19,21-23}$N, $^{16-19,21-24}$O, and $^{18-21,23-26}$F on a carbon target at energies of around 950 A MeV. We derived effective matter radii of the nuclei by a Glauber–model analysis. Based on the assumption of a core–plus –neutron structure, we have applied Glauber–model calculations for weakly–bound systems to some C, N, O, and F isotopes. A one–neutron halo structure is suggested for ^{23}O as well as for ^{19}C. We have surveyed the neutron separation energies (S_n) and σ_I for the neutron–rich p-sd and the sd shell region. A neutron–number dependence of S_n shows clear breaks at $N=16$ near the neutron drip line ($T_Z \geq 3$), which is indicative of a new magic number. The neutron–number dependence of σ_I shows a large increase of σ_I at $N=15$ for $T_Z \geq 3$, which supports the new magic number.

INTRODUCTION

Projectile fragmentation of high–energy heavy–ions was pioneered at the Bevalac in Berkeley [1,2]. The studies showed that a wide variety of nuclei including extremely neutron–rich ones are produced through that process. This opened new possibilities of using beams of radioactive nuclei for studies of nuclear properties. After the first radioactive beam experiment, which measured interaction cross sections for He isotopes [3], measurements have been extended to almost all known p–shell nuclei [4-8] and to some F, Ne, and Na isotopes [9,10]. Measurements of interaction cross sections(σ_I) at relativistic energies allow us to derive nuclear matter radii ($\tilde{r}_m \equiv < r_m^2 >^{1/2}$) using Glauber model.

Since the isotope dependence of $\tilde{r}_m$ is sensitive to the halo–structure [11], the

---

[1] E-mail address:suzuki@nuexne.sc.niigata-u.ac.jp

measurements of σ_I have been used to search for new halo nuclei. Recently, Niigata [12] and Surrey groups [13] have developed few–body Glauber–model calculations for weakly–bound systems (FB). They took into account the correlation between the valence nucleon (nucleons) and the localized core. Such a FB description leads to smaller calculated reaction cross sections than those with the optical limit (OL) approximation of the Glauber–model, with significant implications for the deduced size and ground state structure of the halo. Therefore, we should take into account the above differences in Glauber–model calculations to derive $\tilde{r}_m$.

There are some potential candidates for neutron–halo nuclei among the light neutron–rich nuclei ($A \geq 14$), Recent experimental studies show a one–neutron halo structure for ^{19}C [14], but no halo structure for ^{17}C [15]. A narrow width of the $p_\parallel$ spectrum of ^{12}Be after breakup of ^{14}Be indicates a two–neutron halo structure in ^{14}Be [16]. However, for other nuclei, signatures for halo–structures were not observed experimentally up to now. We investigated halo nuclei in the light neutron–rich region by measuring σ_I of $^{12-14,16-20}$C, $^{14-16,18,19,21-23}$N, $^{16-19,21-24}$O, and $^{18-21,23-26}$F on a carbon target. Such measurements of σ_I can be performed at or near drip–line in many cases because of recent improvements of secondary beam–technique. We derived the effective $\tilde{r}_m$ value, the effective density distributions with assumptions of a core–plus–neutron structure, and some spectroscopic informations. Such information is very valuable for halo nuclei since no spectroscopic information is available, sometimes even not the ground–state spin and parity.

In addition, we surveyed the neutron separation energies (S_n) and σ_I for the neutron–rich p–sd and the sd shell region to investigate a possible change in the shell structure near the neutron drip line.

In this contribution, our experimental setup is shown in Section 2. Analysis procedure and experimental results are described in Section 3. In Section 4, we derive the effective $\tilde{r}_m$ values by Glauber–model calculations. We also derive the effective nucleon density distributions and some spectroscopic informations. In Section 5, a new magic number for neutron–rich $N=16$ nuclei is discussed based on the neutron–number dependence of S_n and σ_I. Finally, we conclude in Section 6.

MEASUREMENTS OF σ_I

Measurements of σ_I were performed by the transmission method. The typical experimental setup is FRS at GSI, which is described in [18]. Secondary beams were produced through projectile fragmentation of a primary beam accelerated to around 1 A GeV by the heavy–ion synchrotron SIS. The first half of the FRS, down to the intermediate focal plane (F2), was used to separate and identify the incident beams. Reaction targets of carbon were placed at F2. The second half of the FRS was used as a spectrometer to transport the non–interacting secondary beams, which were identified and counted at the final focal plane of the FRS (F4). Particles were identified by the Bρ–ΔE–TOF method in front of and behind the reaction targets, thus providing a measurement of σ_I by the transmission method.

The measured σ_I are shown as a function of the neutron number (N) of the boron, carbon, nitrogen, oxygen and fluorine isotopes in Fig. 1. The observed σ_I increase monotonically with neutron number. The solid lines in Fig. 1 show σ_I calculated by the equation

$$\sigma_I = \pi[R_I(^{12}C) + r_o A^{1/3}]^2 \qquad (1),$$

where $R_I(^{12}C)$ is the interaction radius of ^{12}C (2.61 fm) and r_o is selected to reproduce σ_I for ^{10}B in boron, ^{12}C in carbon, ^{14}N in nitrogen, ^{16}O in oxygen and ^{19}F in fluorine, respectively.

EFFECTIVE ROOT–MEAN–SQUARE RADII

The effective $\tilde{r}_m$ values of the nucleon distributions were derived from σ_I using Glauber–model calculations in the OL approximation, the details of which are described in Ref. [7]. The $\tilde{r}_m$ values were derived by assuming a Harmonic–Oscillator (HO)–type density distribution with $1s$, $1p$, $1d$ and $2s$ orbitals. These $\tilde{r}_m$ values are shown in Fig. 2.

Recently, groups at Niigata and Surrey have developed few–body Glauber–model

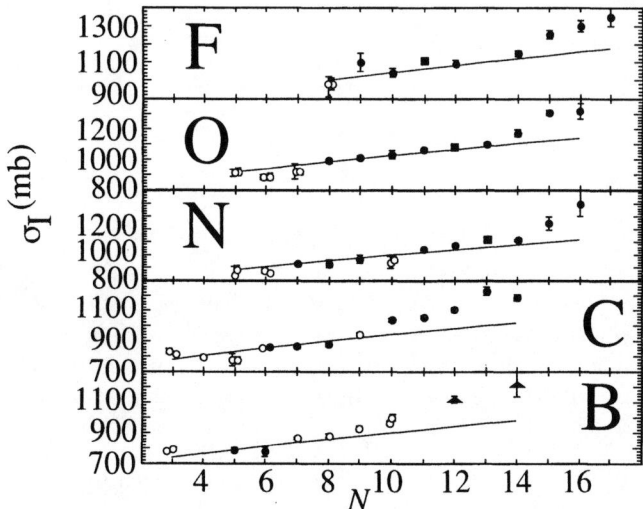

FIGURE 1. Interaction cross–sections (σ_I) for boron, carbon, nitrogen, oxygen and fluorine isotopes on a carbon target in mb. The closed circles are data points obtained in this work. The open circles are data from previous studies at LBL [7]. The closed triangles and closed squares are data from previous studies at GSI [17,10] The solid line shows the σ_I calculated by Eq. (1).

calculations for weakly–bound systems (FB) to deduce $\tilde{r}_m$ values. The Glauber–model with FB approach gives larger $\tilde{r}_m$ values than by the Glauber-model with OL approximation, specially for neutron–halo cases such as ^{11}Be and ^{11}Li [12,13]. We performed calculations similar to those in Ref. [8] to deduce $\tilde{r}_m$ values for a Glauber–model with FB approach. Note that we used the static–density approach (OL) for those cases where we do not expect clusterization and that we performed FB approach for nuclei where we expect cluster (halo) structure..

Here, we assume a core–plus–neutron structure for the nuclei. The nucleon density distribution of the core (^{A-1}Z) has been parametrized to reproduce the observed σ_I of ^{A-1}Z on a carbon target. The density distribution for the valence neutron of ^{A}Z was calculated by the code WAVEFUNC [8], where single–nucleon density distributions for the s, p and d orbitals in a core–plus –nucleon potential were calculated for given values of S_n. We also assumed a Woods–Saxon shape for

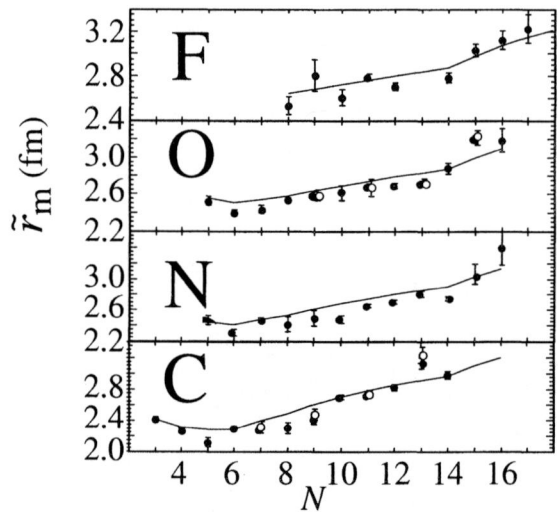

FIGURE 2. Effective RMS matter radii ($\tilde{r}_m$) for carbon, nitrogen, oxygen and fluorine isotopes in fm. The closed and open circles are radii derived by Glauber–model calculations with the OL approximation and the FB approach, respectively. The solid line shows radii calculated by Relativistic–Mean–Field (RMF) calculations [18].

the potential, fixed the radius parameter to be parametrized by $1.22A^{1/3}$ (fm) and the diffuseness parameter to be 0.70 fm, and adjusted the depth of the potential so as to reproduce the S_n of ^{A}Z. The resulting single–neutron distribution was added to the core density distribution to obtain the nucleon density distribution of ^{A}Z, taking into account the difference in the center of mass for the composite system. We considered the configuration allowed by the J^π for ^{A}Z, if it is known, in the above steps. We then mixed configurations so as to reproduce σ_I for ^{A}Z. Thus, we derived the corresponding $\tilde{r}_m$ values. With this procedure we can also derive some spectroscopic information as well as the effective nucleon densities for ^AZ. It is also noted that the results for ^{11}Be and ^{19}C in our Glauber–model with FB approach are the same as those of the Surrey group [13] within experimental uncertainties. The thus–determined $\tilde{r}_m$ values from FB analysis are also displayed for some of the nuclei in Fig. 2.

NUCLEON DENSITIES AND SPECTROSCOPIC FACTORS

We applied the method described in the previous section to ^{11}Be to check the consistency of the nucleon density distributions. The density distribution of ^{11}Be has been deduced by the energy dependence of the reaction cross sections (σ_R) [21]. For ^{11}Be, $J^\pi=1/2^+$ of the ground state and $S_n=504\pm6$ keV are experimentally well

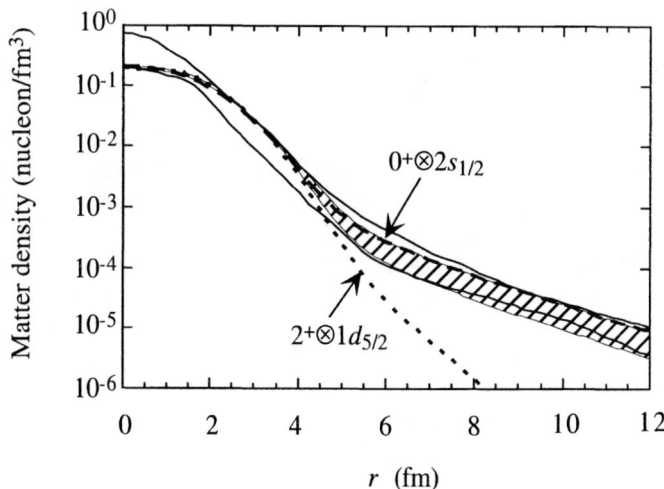

FIGURE 3. Nucleon–density distributions for ^{11}Be. The thick–dashed and thick–dotted lines show the density distributions deduced from pure $0^+ \otimes 2s_{1/2}$ and $2^+ \otimes 1d_{5/2}$ configurations, respectively. The hatched area shows the density distribution required to reproduce the observed σ_I for ^{10,11}Be and S_n of ^{11}Be. The two thick solid lines show the maximum and minimum density distributions deduced from the energy dependence of σ_R [20].

known. Due to $J^\pi=1/2^+$, we considered $0^+\otimes 2s_{1/2}$ and $2^+\otimes 1d_{5/2}$ configurations, where 0^+ (2^+) means the ground state (the first excited state; $E_x=3.368$ MeV) of ^{10}Be. The nucleon density distributions deduced for the above configurations are shown by the hatched area in Fig. 3. We needed to mix $0^+\otimes 2s_{1/2}$ and $2^+\otimes 1d_{5/2}$

configurations to reproduce the observed σ_I of ^{11}Be. The resulting $2s_{1/2}$ spectroscopic factor is 0.80±0.11. Nucleon density distributions deduced from the energy dependence of σ_R are shown by the solid lines (maximum and minimum). Both densities agree perfectly each other as seen in Fig. 3. The spectroscopic factor deduced from the ^{10}Be$(d,p)^{11}$Be reaction is 0.73±0.06 and 0.77 [22], which is consistent with that deduced by our FB approach. Very recently, the $p(^{11}$Be, ^{10}Be$)d$ reaction has been studied using a radioactive ^{11}Be beam at 35 A MeV [23]. Their spectroscopic factor deduced from a DWBA analysis is 0.67–0.79, which is also consistent with our value.

Next we applied the same method to ^{15,17,19}C and ^{23}O. The derived $2s_{1/2}$ spectroscopic factor for ^{15}C is 0.49±0.22. The deduced nucleon density distribution for ^{15}C has a long neutron tail, as shown in Fig. 4 (a). However, the tail density is about one order magnitude lower than that in ^{11}Be. The small tail component supports the small halo structure for ^{15}C. As seen in Fig. 4 (b), the deduced nucleon density distribution for ^{17}C is close to that for a pure $0^+\otimes 1d_{5/2}$ configuration. However, the tail component is slightly smaller than that of the pure configuration, which implies the $1d_{5/2}$ spectroscopic factor is less than 1.0. Although J^π for ^{17}C cannot be determined in this analysis, a predominance of d–wave, not s–wave, for the valence neutron in ^{17}C is clearly shown. This conclusion agrees with that in Ref. [15].

The derived $2s_{1/2}$ spectroscopic factor for ^{19}C is 0.74±0.28. The deduced nucleon density for ^{19}C, shown in Fig. 4 (c), has a long neutron tail, that is almost comparable to that of ^{11}Be. This clearly shows the one–neutron halo structure for ^{19}C Although J^π for ^{19}C cannot be determined in this analysis, a predominance of s–wave, not d–wave, for the valence neutron in ^{19}C is clearly shown.

As can be seen in Fig. 4 (d), the deduced nucleon density for ^{23}O is consistent with that deduced for a pure $0^+\otimes 2s_{1/2}$ configuration. The density deduced for a pure $0^+\otimes 1d_{5/2}$ configuration does not reproduce the observed σ_I. Again, a dominance of s–wave configuration, not a d–wave, for the valence neutron in ^{23}O is clearly shown. The presently deduced density distribution for ^{23}O has a long neutron tail, which is almost comparable to that of ^{19}C. It suggests an one–neutron halo structure of ^{23}O.

NEW MAGIC NUMBER $N = 16$, NEAR THE NEUTRON DRIP LINE

Nuclear shell structure is one very important characteristic of the nuclear structure. The persistence of the magic numbers in the neutron–rich region have been studied extensively. The breaking of the shell closure at N=20 has been evidenced in terms of a low–lying 2^+ level [24] and a large $B(E2;0_1^+ \to 2_1^+)$ value in ^{32}Mg [25], for example. Mixing of $2s_{1/2}$ and $1p_{1/2}$ orbitals indicates a weakening of the N=8 magic number in neutron–rich Li and Be isotopes [26,27]. Though the disappear-

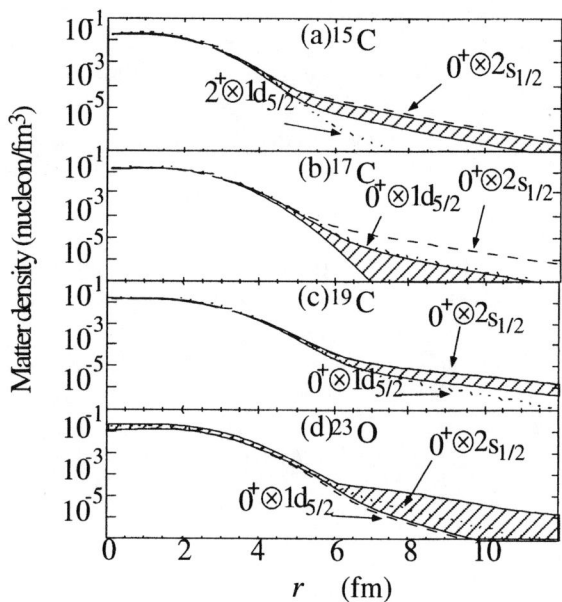

FIGURE 4. Nucleon–density distributions for ^{15,17,19}C and ^{23}O. The dashed lines show the density distributions deduced from a pure $0^+ \otimes 2s_{1/2}$ or $2^+ \otimes 1d_{5/2}$ configurations. The hatched areas show the density distributions required to reproduce the observed σ_I for the core nuclei and halo nuclei under the constraint that S_n is reproduced.

ance of magic numbers has been discussed no appearance of a new magic number has so far been shown experimentally.

A neutron number dependence of experimentally observed S_n [28] for nuclei with odd N and even Z (odd N and odd Z) is displayed in Figs. 5(a) and 5(b), respectively. As seen in Fig. 5, a magic number appears as a decrease of S_n with increasing N [29]. The traditional magic numbers (N=8,20) are clearly seen for nuclei close to β–stabiliy as breaks in the lines for small T_z. However, the break at $N = 8$ ($N = 20$) disappears at T_z=3/2 (T_z=4), which is also known in connection with other experimental quantities, as mentioned above. On the other hand, a break in the S_n line appears at N=16 for $T_z \geq 3$, which indicates the formation of a new magic number at N=16 near the neutron drip line. Note that this was predicted already 25 years ago by Beiner et al. [30].

As seen in Fig. 1, we observe a steep increase of σ_I from N=14 to N=15 for N to F isotopes. However, no steep increase in σ_I is observed for Ne to Mg isotopes as seen in Fig.2 of Ref. [31]. It is noted that a clear difference occurs at T_z=3, which suggests some correlation with the new magic number N=16.

For further quantitative discussions concerning the increase of σ_I, we performed calculations similar to those described in Section 3. We assumed a core–($N=14$ nuclei)–plus–neutron structure for the $N=15$ nuclei. It is found that the observed σ_I can be reproduced only by using a pure $2s_{1/2}$ orbital and that the observed σ_I cannot be reproduced for the case of a pure $1d_{5/2}$ orbital even if we use $S_n=0$ MeV [31]. Therefore, we conclude a dominance of the $2s_{1/2}$ orbital for the valence neutron in ^{22}N, ^{23}O, and ^{25}F. On the other hand, the increase from ^{25}Na to ^{26}Na can be explained by a mixing of $2s_{1/2}$ and $1d_{5/2}$ orbitals.

For nuclei with $N=15$ near the stability line, ^{27}Mg, ^{28}Al, and ^{29}Si, spectroscopic factors were investigated by (d,p) reactions [30]. The angular distributions show an s-orbital structure for these nuclei. However as shown in the inset of Fig. 5, the deduced spectroscopic factors of $2s_{1/2}$ are 0.3–0.8, which indicates some mixing of other orbitals for the valence neutron in these nuclei. Thus, a valence neutron in the $N=15$ chain shows the following tendency: The purity of the $2s_{1/2}$ orbital is larger when T_z is larger and drastically increases at $T_z=3$. This conclusion supports the creation of a new magic number at $N=16$ near the neutron drip line.

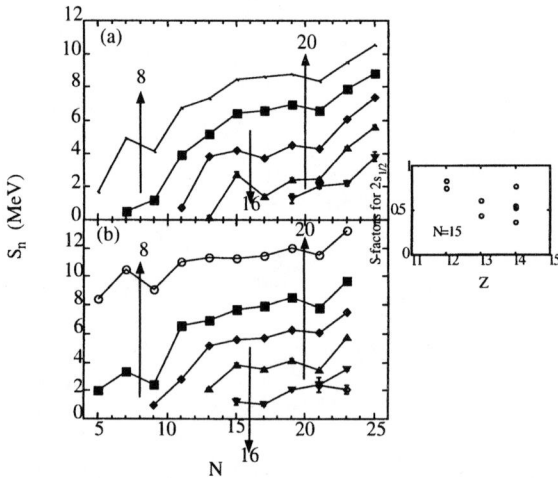

FIGURE 5. (a) Neutron number dependence for experimentally observed neutron separation energies S_n [27] for nuclei with odd N and even Z. The closed circles, squares, diamonds, triangles, and inverse triangles show different isospin numbers (T_z) from 1/2 to 9/2. (b) Same as (a) but for nuclei with odd N and odd Z, from $T_z=0$ to 5. Experimental uncertainties below ±10 keV are not shown. The solid lines are guides to the eye. The breaks corresponding to magic numbers are shown by arrows. The inset shows deduced spectroscopic factors for nuclei with $N=15$, which are data deduced from (d,p) reactions [29].

CONCLUSION

In summary, we have measured interaction cross–sections for light–nuclei at relativistic energies. We derived effective Root–Mean–Square matter radii of light–nuclei by a Glauber–model analysis. Based on the assumption of a core + neutron structure, we applied the Glauber–model with the few–body approach to the one–neutron halo nucleus ^{11}Be and to halo–candidates ^{15,17,19}C, and ^{23}O. The deduced nucleon density for ^{11}Be shows a long tail and is consistent with that deduced from the energy dependence of the reaction cross-sections. The deduced density for ^{15}C shows a long tail, but it is not so pronounced as that for ^{11}Be. The deduced densities for ^{19}C and ^{23}O also show a long tail, almost comparable to that of ^{11}Be, and clearly show a one–neutron halo structure.

We have surveyed the neutron separation energies(S_n) and the interaction cross sections (σ_I) for the neutron–rich p–sd and the sd shell region. A neutron number dependence of S_n shows clear breaks at $N=16$ near the neutron drip line ($T_z \geq 3$). A neutron number dependence of σ_I shows a large increase in $N=15$ nuclei ($T_z \geq 3$), which shows that the purity of the $2s_{1/2}$ orbital is larger when T_z is larger in our analysis. These two facts indicate the appearance of a new magic number at $N=16$ near the drip line.

REFERENCES

1. D.E. Greiner et al., Phys. Rev. Lett. 35 (1975) 152.
2. A.S. Goldhaber and H.H. Heckmann, Ann. Rev. Nucl. Part. Sci. 28 (1978) 161.
3. I. Tanihata et al., Phys. Lett. B 160 (1985) 380.
4. I. Tanihata et al., Phys. Lett. B 206 (1988) 592.
5. I. Tanihata et al., Phys. Lett. B 287 (1992) 307.
6. A. Ozawa et al., Phys. Lett. B 334 (1994) 18.
7. A. Ozawa et al., Nucl. Phys. A 608 (1996) 63.
8. M. Obuti et al., Nucl. Phys. A 609 (1996) 74.
9. T. Suzuki et al., Phys. Rev. Lett. 75 (1995) 3241.
10. L. Chulkov et al., Nucl. Phys. A 603 (1996) 219.
11. I. Tanihata et al., Phys. Rev. Lett. 55 (1985) 2676.
12. Y. Ogawa et al., Nucl. Phys. A 543 (1992) 722.
13. J.S. Al-Khalili and J.A. Tostevin, Phys. Rev. Lett. (1996) 3903.
14. D. Bazin et al., Phys. Rev. Lett. 74 (1995) 3569.
15. D. Bazin et al., Phys. Rev. C 57 (1998) 2156.
16. M. Zahar et al., Phys. Rev. C 48 (1993) R1484.
17. T. Suzuki et al., Nucl. Phys. A 658 (1999) 313.
18. Y. Sugahara et al., Prog. Theor. Phys. 96 (1996) 1165.
19. J.S. Al-Khalili, J.A. Tostevin, and I.J. Thompson, Phys. Rev. C 54 (1996) 1843.
20. M. Fukuda et al., Phys. Lett. B 268 (1991) 339.
21. F. Ajzenberg-Selove, Nucl. Phys. A 506 (1990) 1.
22. S. Fortier et al., Phys. Lett. B 461 (1999) 22.

23. D. Guillemaud-Mueller *et al.*, Nucl. Phys. A 426 (1984) 37.
24. T. Motobayashi *et al.*, Phys. Lett. B 346 (1995) 9.
25. H. Simon *et al.*, Phys. Rev. Lett. 83 (1999) 496.
26. H. Keller *et al.*, Z. Phys. A 348 (1994) 61.
27. G. Audi and A.H. Wapstra, Nucl. Phys. A 595 (1995) 409.
28. A. Bohr and B.R. Mottelson, *NuclearStructure* (Benjamin, New York, 1969), Vol. 1, p.192.
29. P.M. Endt, Nucl. Phys. A 521 (1990) 1.
30. M. Beiner, R. J. Lombard and D. Mas, Nucl. Phys. A 249 (1975) 1.
31. A. Ozawa *et al.*, Phys. Rev. Lett. 84 (2000) 5493.

Shell structure of neutron-rich light nuclei: New Vista

Faisal Azaiez

Institut de Physique Nucléaire, IN2P3-CNRS, 91406 Orsay Cedex France

Abstract. The structure of neutron-rich light nuclei around N=20,28 has been investigated at GANIL by means of in-beam gamma spectroscopy using fragmentation reactions of ^{36}S and ^{48}Ca beams on a Be target. Gamma-decay of relatively high-lying excited states have been measured for the first time in nuclei around ^{32}Mg and ^{44}S. Level schemes are proposed and discussed for a large number of these neutron-rich nuclei around N=20 and N=28.

INTRODUCTION

From the study of the structure of light neutron-rich nuclei, it has been recently suggested that some major shell-gaps are weakened when large isospin values are encountered. The typical cases of ^{32}Mg and ^{44}S, where a large quadrupole collectivity has been found [1-3] have brought some evidence for such a shell-gap weakening at large neutron excess. Though, information on the excitation energies of the first 2^+ states and on the B(E2) values of the 0^+ to 2^+ transitions is not sufficient to fully understand the structure of these nuclei. For instance the measurement of higher lying excited states and for example the $E(4^+)/E(2^+)$ ratio should shed some light on the origin of the large quadrupole collectivity observed. In order to bring more spectroscopic information on nuclei around ^{32}Mg and ^{44}S, a novel experimental method has been used. This method is based on the production of very neutron-rich nuclei in relatively higher excited states, through the projectile fragmentation process and on the detection of their in-beam γ-decay. Such experiments have been recently performed at GANIL. In a first experiment aiming for the study of neutron-rich nuclei around ^{32}Mg, a ^{36}S beam, at 77 MeV/u was used on a 2.77 mg/cm^2 Be target. The second experiment used a ^{48}Ca beam at 60 MeV/u on the same Be target in order to study nuclei around ^{44}S and ^{46}Ar. The target was located at the entrance of the SPEG spectrometer which was used to analyze the different fragments produced in the reaction. Many neutron rich nuclei have been produced and identified in a time of flight versus energy-loss plot. Gamma-spectroscopy for all the produced fragments is obtained by performing coincidences between the analyzed fragments and γ-rays emitted in flight during their decay to the ground state. For that purpose a highly efficient (25% at 1.33MeV) γ-array consisting of 74 BaF$_2$ crystals was used around the target covering symmetrically the upper and lower hemispheres (roughly 80% of the solid angle around the target is covered). This array is supposed to provide γ-fragment as well as γ-γ-fragment coincidences. The latter is needed to build-up a level scheme for each fragment. It is worth having in mind that all γ-rays of interest are emitted from projectile fragments with their full velocities (about 40% of the speed of the light). Typical Doppler corrected spectra from the BaF$_2$ detectors exhibit gamma lines with FWHM varying from 200 keV at 1 MeV up to 500 keV at 3 MeV. In addition to the BaF$_2$ array, two different Ge detectors setup have been used in the two different experiments in order to help identifying more complex gamma spectra. In the first experiment performed with the ^{36}S beam, four 70% high resolution Ge detectors were used at the most backward angles with an efficiency of 1.2 10^{-3} at 1.3 MeV. Typical Doppler broadening of γ-lines detected in these Ge detectors is 30 keV at 1.3 MeV. In the second experiment with the ^{48}Ca beam, three segmented clover detectors have been used at about 20cm from the target and at three different angles 85 degrees, 122 degrees and 136 degrees with respect to the beam axis. Utilizing the possibilities provided by this type of Ge detectors, we could enhance the efficiency by using the add-back method (the overall photopeak efficiency for this system is more than one order of magnitude higher than the efficiency of the 4 coaxial Ge detectors used for the first experiment) and reduce the Doppler broadening by using the segments information (a FWHM of 35 keV has been obtained for a typical 1.5 MeV γ-ray emitted at 35% of the speed of the light).

With the used Be target and typical 15 nAe beam intensity, production rates going from few 100/s (for the most populated fragments) to 0.1/s (for fragments very far from stability such as ^{28}Ne and ^{44}S). In these conditions, individual Ge crystal counting rates of the order of 10 to 15 kHz have been obtained. After gating on the proper fragment and on the true γ-fragment coincidences (subtracting the random coincidences contribution), Doppler corrected γ-spectra are produced together with γ-γ matrices for coincidence analysis. Typical γ-spectra from Ge and BaF$_2$ detectors have been presented in previous publications [4-6]. In the following, I will limit my talk to the presentation of the general properties and features of this new method based on the in-beam γ-spectroscopy with fragmentation reactions and to the discussion of some of the results related to the structure of neutron-rich nuclei around N=20 and N=28.

GENERAL FEATURES OF IN-BEAM γ-SPECTROSCOPY WITH FRAGMENTATION REACTIONS

This is the first time that in-beam γ-spectroscopy combined with projectile fragmentation reactions at intermediate energies, has been used [4]. Some general features could be extracted from the analysis of the feeding intensities of different excited states in different fragments, from the use of the 74 BaF$_2$ detectors as a multiplicity filter and from the analysis of the angular distributions of different γ transitions in both experiments. These features that will be summarized as follows has been used, in addition to coincidence relationships and intensity arguments to propose level schemes for many of the populated neutron-rich fragments: i) In all cases the obtained gamma spectra exhibit the same features. Namely, an exponential background with few discrete γ-lines. States with spin values up to 4 have been populated in most of the cases.

ii) In all known cases, the Yrast states were found to be favored among all other states in the population mechanism of the different fragments. This supports the idea that after the fragmentation-like reactions occur, different primary fragments undergo statistical decay. For example, in the case of ^{20}O where low-lying excited states are known from previous studies, the first 2^+ and 4^+ excited states are equally and more populated than the second excited 2^+ and 0^+ states.

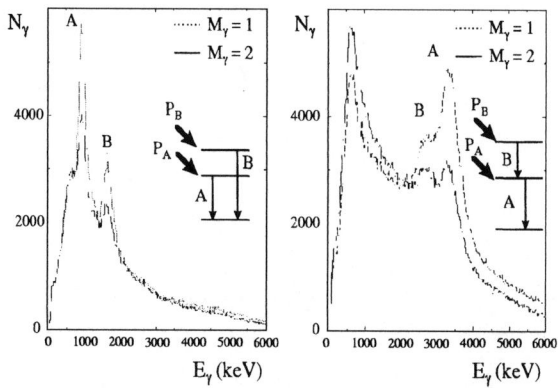

FIGURE 1. BaF$_2$ spectra with the condition (M$_\gamma$=1) that only one BaF$_2$ detector fired and the condition (M$_\gamma$=2) that two BaF$_2$ detectors fired, obtained for ^{12}B (left) and ^{10}Be (right). Notice the different behavior when two γ transitions are in coincidence (belong to the same cascade) and when they are not in coincidence (belong to two different cascades)

iii) A clear correlation has been found between the number of fired BaF$_2$ detectors (related to γ multiplicity) and the fragment position in the focal plane of the SPEG spectrometer. This is indicating for instance that low velocity fragments are correlated to more dissipative collisions and consequently to higher excitation energy and angular momentum. The comparison of spectra with high BaF$_2$ multiplicity (low velocity fragments) and low BaF$_2$ multiplicity (higher velocity fragments) conditions was used in many cases in order to help establishing the position of a given γ-ray in the level schemes(see figure 1).

iv) The analysis of the gamma-ray angular distributions using the BaF$_2$ detectors was found, within the error bars, to be independent of the multipolarity of known gamma-rays from the data. This is suggesting that the orientation of the fragment angular momenta is not larger then 15%. Thus from the first experiment (that used ^{36}S beam) no information on the multipolarity of the γ transitions could be extracted. The second experiment (that used ^{48}Ca beam) has the advantage of using three segmented clover Ge detectors placed at three different angles suitable for angular correlation measurements. As shown in figure 2, the observed γ-ray intensity ratios are fairly different for typical $\Delta L=1$ and $\Delta L=2$ γ-transitions allowing multipolarity assignments. Despite a very low degree of orientation of the fragment angular momenta, this has been made possible because of the much higher resolving power of segmented clover Ge detectors compared to the BaF$_2$ detectors.

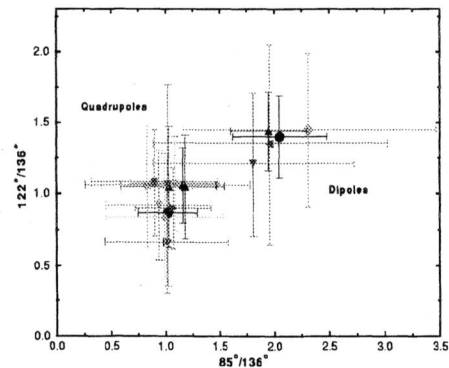

FIGURE 2. Systematics of the measured γ-ray intensity ratios between the two angles pairs of segmented clover Ge detectors. Note the significantly different clustering of the measured values for dipole and quadrupole transitions.

STRUCTURE OF NEUTRON-RICH NUCLEI AROUND N=20

The fragmentation ^{36}S beam into a Be target produced a big number of neutron-rich nuclei up to A=32 for Mg isotopes, A=31 for Na isotopes, A=28 for Ne isotopes, A=25 for F isotopes and A=22 for O isotopes. Many of the obtained gamma spectra have been shown and discussed elsewhere [4–6]. In the following I am going to focus on the discussion of some of the obtained level schemes and their comparison to shell model calculations. In agreement with the β-decay study of ^{32}Na [8], the two lines: the 885 keV (the well known 2^+ to 0^+ transition in ^{32}Mg [7]) and the 1430 keV, were found to be in coincidence. The second experiment that used the fragmentation of ^{48}Ca beam had also produced ^{32}Mg. The analysis of these two gamma-ray angular distributions from the collected data in the second experiment

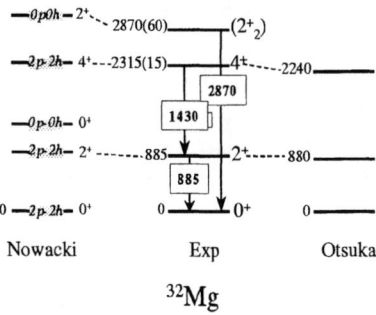

FIGURE 3. Level schemes of ^{32}Mg from shell model calculations by Utsuno et al. [9] (right) and Nowacki et al. [10] (left). The proposed level scheme from the experiment is also shown (middle).

indicated that both are of quadrupole character. An other weaker γ-line have also been observed at 2870(40) keV. The analysis of the intensity dependence with BaF$_2$ multiplicity (point iii of the previous section) suggests that the 2870 keV γ-transition corresponds to the direct decay of an excited state to the ground state. Based on these observations a level scheme is proposed (see figure 3). As it is shown in this figure, the agreement with shell model calculations including the fp shell in the valence space [9], [10] is very good. Both calculations suggest that the Yrast states (up to spin 4) 0f ^{32}Mg are deformed. The calculations by Nowacki et al. [10] also show that some excited states, such as the second 0^+ and 2^+ states, are spherical. From the proposed level scheme one can see a candidate for an excited spherical 2^+ state. this can be taken as an indication that deformed and spherical states co-exist in ^{32}Mg.

Very low statistics have been obtained for ^{31}Na. Only information from the BaF$_2$ spectrum could be obtained. This spectrum has mainly three gamma lines at 770 keV, 2470 keV and 2850 keV. Because of its intensity, the lowest energy γ-ray transition (770 keV) could correspond to the decay of the lowest excited state in ^{31}Na. Though, recently a Coulomb excitation experiment performed at MSU [11] has reported that the lowest excited state is at 350 keV. Because of the high energy threshold (approximately 300 keV) of the BaF$_2$ detectors, such a low energy γ-ray transition could not be observed. Therefore the 770 keV transition would be very likely followed by the 350 keV in the level scheme of ^{31}Na.

From the collected data on ^{26}Ne, a level scheme was built. It is shown in figure 4 together with shell model calculations [12] including only the sd shell in the neutron valence space. The agreement between the experimental and the theoretical level schemes is indicating that ^{26}Ne is spherical and is not part of the so-called "isle of inversion".

Gamma-ray spectra have been obtained for both ^{25}Ne and ^{27}Ne. In contrast with ^{25}Ne the partial level scheme proposed for ^{27}Ne was found to be in total disagreement with shell model calculations including only the sd shell [12]. This is also suggesting that in Neon isotopes deformation may already sets in at N=17 and that the neutron fp shell has to be included in the valence space in order to reproduce the data. Two gamma lines were observed in the BaF$_2$ spectrum of

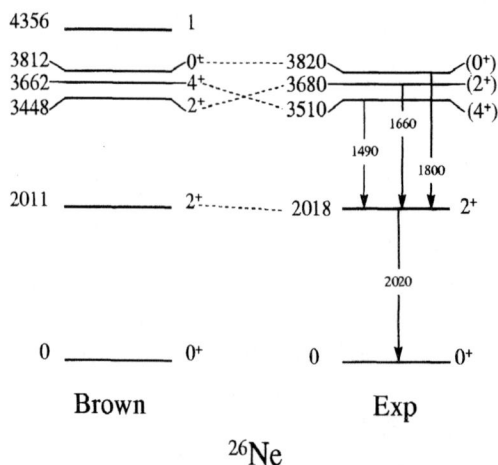

FIGURE 4. theoretical and experimental level schemes of ^{26}Ne

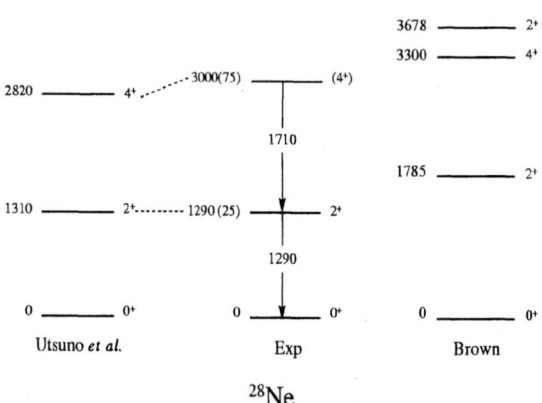

FIGURE 5. theoretical and experimental level schemes of ^{28}Ne

^{28}Ne. From intensity arguments, the γ-line at 1320 keV is assigned to the 2$^+$ to 0$^+$ transition which shows for the first time that, approaching N=20, the 2$^+$ energies in the Ne isotopes decrease dramatically. This result has been interpreted [9] as due to an increase of deformation in neutron rich Ne isotopes similar to the one already observed in the Mg isotopes. This has been recently, confirmed by Pritychenko et al. [13] in a Coulomb excitation experiment at MSU. The obtained level scheme for ^{28}Ne is shown in figure 5, to be in striking agreement with Monte-Carlo shell model calculations including both the sd and fp neutron shells [9].

The γ-spectra obtained for the first time in ^{22}O from both the BaF$_2$ and Ge detectors indicate that the γ-line at 3200 keV represents the 2^+ to 0^+ transition which extends the systematic of the 2^+ transition energies of Oxygen isotopes up to N= 14. One can see from this systematics that Oxygen isotopes exhibit a rapid increase of the 2^+ energy at N=14. This appearance of a spherical shell effect at N=14 has been recently confirmed by a Coulomb excitation experiment at MSU [14]. The increase in the oxygen isotopes of the 2^+ energy at N=14 is very similar to the one observed in the Neon and Magnesium isotopes. This

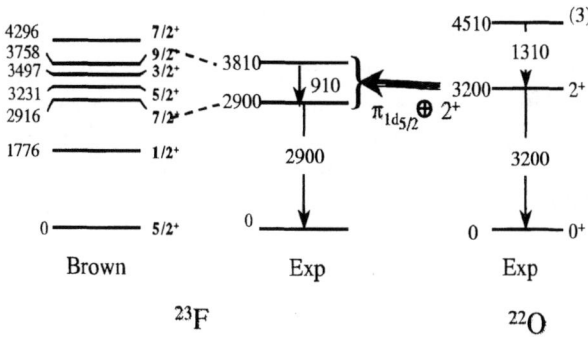

FIGURE 6. Level schemes of ^{23}F from experiment (middle) and from shell model calculations including only the sd neutron valence space [12] (left). For comparison the experimental level scheme of ^{22}O is also shown (right)

together with the fact that ^{26}O and ^{28}O isotopes seem to be unbound [15–17], is suggesting that a similar shell effect weakening is maybe developing at N=20 for oxygen isotopes. The measurement of the 2^+ excitation energy in ^{24}O (provided it is particle-bound)is therefore very crucial. The obtained level schemes from the data for ^{23}F and ^{25}F seem to suggest that ^{22}O and ^{24}O would have their first 2^+ state at about the same excitation energy, just like ^{24}Ne and ^{26}Ne do. The experimental level schemes for both ^{23}F and ^{25}F (see figure 6 and 7) are in very good agreement with the sd-shell model calculations [12] (Again one has to keep in mind that only Yrast states are populated in the reaction we used). This is indicating that both odd Fluorine isotopes are spherical. Furthermore the excited states in ^{23}F can be interpreted as due to the weak coupling of the 1d$_{5/2}$ single proton to the 2^+ phonon state in ^{22}O. If we assume that the same weak-coupling picture holds for

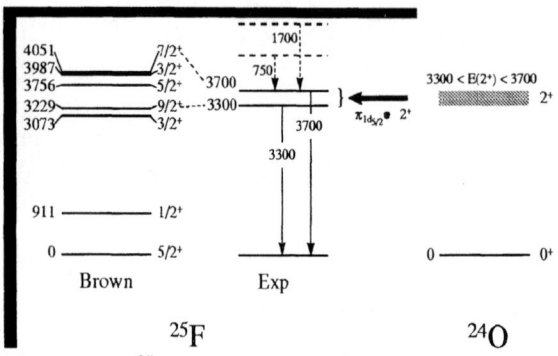

FIGURE 7. Level schemes of ^{25}F from experiment (middle) and from shell model calculations including only the *sd* neutron valence space [12] (left). A suggestion of what would be the level scheme of ^{24}O is indicated (right).

^{25}F, then the first 2^+ state in ^{24}O should be expected to have about 3.5 MeV excitation energy. From the tendency that have been seen when going from Mg isotopes to Ne Isotopes, namely that Neon isotopes start to be deformed earlier (at N=18) than the Magnesium isotopes (at N=20), one also could expect Oxygen isotopes to start to be deformed even earlier. If we assume what just have been suggested for ^{24}O, from the measured level scheme in ^{25}F, this tendency does not seem to be followed by Oxygen nuclei. In that case the Z=8 strong spherical shell gap would be responsible of not allowing deformation in oxygen isotopes. This could also explain why oxygen isotopes are particle-unbound so early (at N=18). In other words, in the case of Ne and Mg isotopes, deformation could push further the neutron drip-line whereas in the case of O isotopes it stays where it should be for spherical nuclei. Nevertheless, the question whether there is a weakening of the N=20 spherical shell gap in oxygen isotopes is still an open question. It may be answered in a near future by the study of the spectroscopy of ^{23,24}O isotopes. All together these results show that ^{32}Mg is not an isolated case with regard to the 2^+ excitation energy when approaching N=20 and that shape coexistence as it has been recently suggested [18] is responsible for the structure of these nuclei in the so-called island of inversion. Why for these neutron-rich nuclei the N=20 deformed shell effect become more effective than the N=20 spherical shell effect? This is still an open question.

STRUCTURE OF NEUTRON-RICH NUCLEI AROUND N=28

The second experiment that used the fragmentation of ^{48}Ca allowed the population of many of the neutron-rich nuclei around N=28 and also nuclei around N=20 such as ^{32}Mg. Revisiting the region of nuclei around N=20 benefit from the multipolarity determination made possible by the use of segmented clover Ge detectors in this experiment. The analysis of this experiment is still in progress [19]. Though, very interesting results on S and Ar isotopes have been already obtained. Using the relative intensities of the γ-rays and energy balance, level schemes for ^{40,42,44}S and ^{46}Ar have been proposed. The assignment of spins and parities have been based on the measured multipolarities of the transitions. The energies of the previously known first 2^+ excited states in ^{40,42,44}S and ^{46}Ar [2,3] as well as the energy of the second excited state in ^{40}S are confirmed. Several new states such as 4^+ states and excited 2^+ and 0^+ states have been identified in many of these nuclei. The proposed level schemes are shown in figure 8 in comparison with recent shell model calculations [10]. It can be seen from the $E(4^+)/E(2^+)$ ratios that although the ground state and the first excited 2^+ in these nuclei are suggested by shell model calculations [10] to be deformed, only ^{42}S has a value that corresponds to a rotor-like behavior. The same shell model calculations suggest that the second excited 2^+ and 0^+ have spherical shapes. The overall nice agreement between the

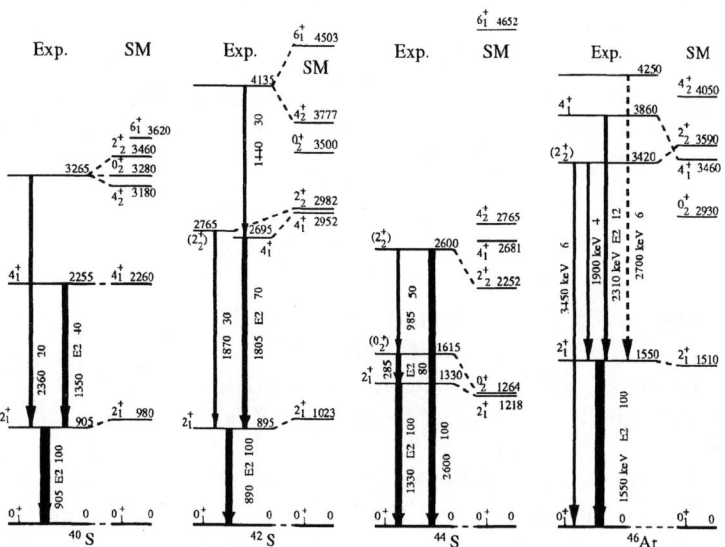

FIGURE 8. Proposed level scheme for ^{40,42,44}S and ^{46}Ar extracted from the in-beam γ-spectroscopy method using the fragmentation of ^{48}Ca beam on a Be target. For comparison level schemes from shell model calculations [10] are also shown.

experimental and theoretical level schemes should be taken as a strong indication of shape co-existence in these nuclei. while their ground and first excited states are deformed, they exhibit spherical higher lying excited states.

CONCLUSION

Beside the importance of the obtained results on neutron-rich nuclei around N=20 and 28, both experiments showed that the in-beam-γ-spectroscopy from fragmentation reactions is very promising for exploring nuclear structure far from stability. It also highlights the needs for a dedicated gamma detection system from the point of view of efficiency, resolution and Doppler broadening reduction. These three features are the basic requirements for which EXOGAM [20] was built and thus make it the ideal gamma detection system for such experiment.

ACKNOWLEDGMENT

The results discussed in this talk are from experiments performed at GANIL by a large team of physicists from many European laboratories. Both experiments using in-beam gamma spectroscopy with fragmentation reactions benefit from the availability of ^{36}S and ^{48}Ca isotopes kindly provided by our colleagues from DUBNA and from the smooth running of the accelerator from the GANIL crew. The use of the segmented clover detectors was made possible thanks to the EXOGAM Collaboration. I have a special acknowledgment to F. Amorini, M. Belleguic, D. Sohler, J. Timar, C. Donzaud, S. Leenhardt, M. J. Lopez and M. Stanoiu who worked hard on analyzing the data. I also would like to thank O. Sorlin, Zs. Dombradi, M.G. Saint-Laurent F. Nowacki and T. Otsuka for valuable discussions on the interpretation of the data.

REFERENCES

1. T. Motobayashi et al., Phys. Lett. **B346**, 9 (1995)
2. H. Scheit et al., Phys. Rev. Lett. **77**, 3967 (1996)
3. T. Glasmacher et al., Phys. Lett. **B395**, 163 (1997)
4. F. Azaiez et al. in proceedings of the Int. Conf. "Nuclear Structure 98", Gatlinburg (1998).
5. F. Azaiez in proceedings of the Int. Conf. "ENPE99", Sevilla, Spain (1999).
6. M. Belleguic et al., Phys. Scripta. T88, 122 (2000)
7. C. Detraz et al., Phys.Rev. **C19**, 164 (1979)
8. G. Klotz et al., Phys. Rev. **C47**, 2502 (1993)
9. Y. Utsuno et al., Phys. Rev. **C60**, 054315 (1999)
10. F. Nowacki *private communication*
11. B. Pritychenko et al. in proceedings of the Int. Conf. "ENPE99", Sevilla, Spain (1999).
12. B. A. Brown *private communication*
13. B. V. Pritychenko et al. Phys. Lett. **467B**, 309 (1999)
14. B. P.G. Thirolf et al. Phys. Lett. **485B**, 16 (2000)
15. D. Guillemaud-Mueller et al., Nucl. Phys. **A426**, 37 (1984)

16. M. Fauerbach et al., *Phys. Rev.* **C53**, 647 (1996)
17. O. Tarasov et al, *Phys. Lett.* **B409**, 64 (1997)
18. P.-G.Reinhard et al, *Phys. Rev.* **C60**, 014316 (1999)
19. D. Sohler et al, in proceedings of the Int. Conf. " ENS200", Debrecen, Hungary. (2000)
20. F. Azaiez and W. Korten., *Nucl. Phys. News*, **Vol.7**, No.4 (1997)

Proton drip-line nuclei studied at intermediate energies

B. Blank

CEN Bordeaux-Gradignan, Le Haut-Vigneau, F-33175 Gradignan Cedex, France

Abstract. In experiments at the SISSI/LISE3 facility of GANIL, the doubly-magic nucleus ^{48}Ni has been observed for the first time and decay information was gained for very proton-rich nuclei like ^{42}Cr, ^{45}Fe, and ^{49}Ni, as ^{48}Ni possible candidates for two-proton radioactivity. In the lighter-mass region, detailed studies of ^{21}Mg and ^{25}Si were performed and complete decay schemes for allowed decays were established.

I INTRODUCTION

Doubly magic nuclei enabled nuclear physicists to interpret many properties of nuclei. They are spherical and exhibit a remarkable stability compared to neighboring isotopes. However, one of the questions of current interest is whether or not the role of the magic numbers, well established along the valley of stability, remains important when an extreme excess of protons or neutrons is present in a nucleus. The doubly-magic nucleus ^{48}Ni is of specific interest for mirror symmetry studies since it is the only case of a doubly-magic nucleus for which the mirror nucleus, ^{48}Ca, is bound. With an isospin projection of $T_z = -4$, ^{48}Ni is the most proton-rich nucleus. Together with ^{45}Fe and ^{54}Zn, ^{48}Ni is also predicted by theory being a good candidate for the yet unobserved two-proton radioactivity [1-4]. In addition, with ^{48}Ni, ^{56}Ni and ^{78}Ni, nickel is probably the only element with 3 doubly-magic isotopes, that makes this element interesting for a study of the evolution of shell structure with the isospin quantum number.

Concerning two-proton (2p) decay, several experimental attempts did not succeed in producing any evidence for a simultaneous correlated 2p emission of the candidate nuclei. For this purpose, β-delayed decays were studied [5,6] in the case of e.g. ^{22}Al or ^{31}Ar. In these decays, the mechanism was clearly identified to be a sequential emission via an intermediate state. A similar picture was revealed in the decay of an excited state in ^{17}Ne [7,8].

2p emission from the ground state has been studied in ^{6}Be and ^{12}O. However, the decay of these nuclei could be explained with a direct three-body decay [9,10]. ^{45}Fe and ^{48}Ni are believed to be the most promising candidates for a correlated 2p emission, as the one-proton emission is energetically forbidden.

In our spectroscopic studies, we investigated ^{42}Cr, ^{45}Fe, and ^{49}Ni to search for direct 2p emission from the ground state of these nuclei. In a similar experiment, we studied ^{21}Mg and ^{25}Si as test cases for the investigation of the β2p emitters ^{22}Al, ^{26}P, and ^{27}S.

II EXPERIMENTAL PROCEDURE

The medium-mass exotic nuclei of interest were produced by the fragmentation of a 74.5 MeV/nucleon ^{58}Ni primary beam in a nickel target followed by a carbon stripper foil. This target was located between the two superconducting solenoids of the SISSI device in order to increase the acceptance of the spectrometer. The secondary beams were transported through the ALPHA spectrometer to the LISE3 separator.

The light-mass nuclei were produced by the fragmentation of a 95 MeV/nucleon ^{36}Ar in a carbon target. In both cases, a shaped beryllium degrader, located at the intermediate focal plane of LISE3, provided an achromatic tuning for a refined selection of projectile fragments. In addition, a final selection was performed with a Wien filter after the second dipole of LISE3.

The fragments were implanted in a stack of silicon detectors. Together with time of flight (TOF) measurements, these detectors are used to identify the selected ions yielding their energy loss and their residual energy.

The silicon detectors were equipped with two electronic chains for implantation events and decay events. Thus, besides the identification of implanted ions, their decay via charged-particle emission could be studied. An efficient γ-detection setup close to the implantation detectors completed the device.

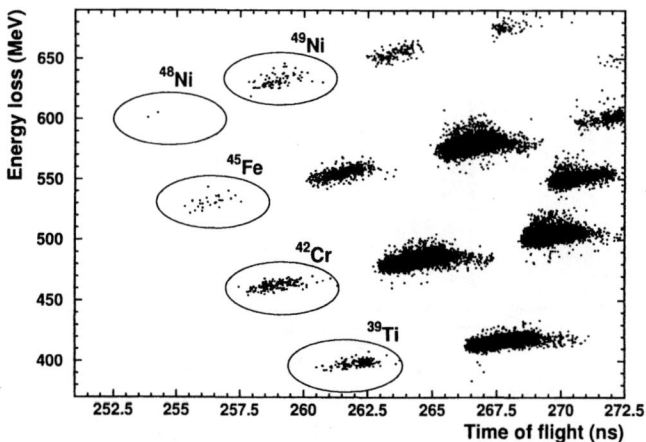

FIGURE 1. Identification plot representing the TOF versus the energy loss in a silicon detector. Two events of ^{48}Ni are clearly visible.

III IDENTIFICATION OF ^{48}NI

The identification plot of energy loss in the first silicon detector versus time of flight is shown in Fig. 1. This plot shows 2 events of ^{48}Ni, 77 of ^{49}Ni, 29 ^{45}Fe and 164 of ^{42}Cr, which fullfil the crossed conditions on all the parameters. There is essentially no background on this plot, since no count appears where unbound isotopes would be located. The spectrum clearly demonstrates the unambiguous observation of ^{48}Ni. Details of the analysis may be found in Ref. [11].

From the time of flight of ^{48}Ni between the target and the detection setup of 1.32 μs, we estimate a lower limit for its half-life of about 0.5 μs by comparing the observed ^{48}Ni rate to the expected one as extrapolated from neighboring nuclei. ^{48}Ni is predicted to be a good candidate for two-proton emission from the ground state. As an even-even nucleus, it is bound with respect to one-proton emission. The barrier penetration half-life of this nucleus for two-proton emission is shown in Fig. 2 as a function of the two-proton separation energy. For the calculations, we used a simple picture of a ^{2}He particle tunneling through the Coulomb barrier assuming a spectroscopic factor of unity. The experimental lower limit of the half-life indicates that Q_{2p} may be lower than 1.5 MeV. This result is compared to half-lives deduced from Q_{2p} values from different mass predictions [1–4,12–17].

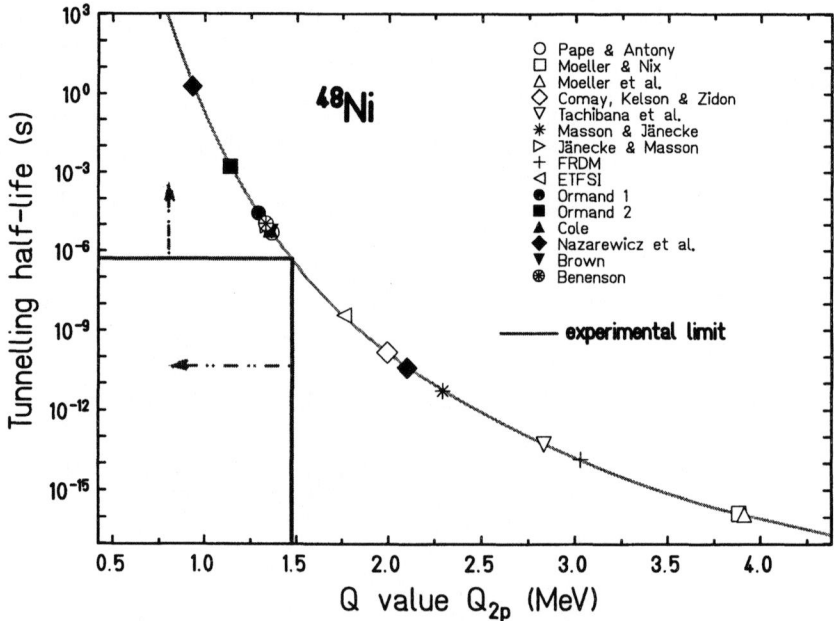

FIGURE 2. ^{48}Ni half-life calculated from the tunnelling of an ^{2}He particle through the Coulomb barrier as a function of the two-proton separation energy. The experimental lower limit is compared to several mass-model predictions.

IV THE DECAY OF ^{49}NI AND ^{45}FE

As mentioned above, ^{42}Cr, ^{45}Fe, and ^{49,48}Ni are possible candidates for 2p radioactivity. The theorectical predictions tend to give less chance for a 2p emission in the case of ^{42}Cr and ^{49}Ni. This is due to the available energy for this emission, the Q_{2p} value. However, as in the case of ^{48}Ni shown above, different mass models predict Q_{2p} values ranging from 1.5 MeV to 0 MeV.

An elegant way to test the decay mode of these nuclei is their implantation in a silicon detector and the observation of their decays. This procedure was applied to ^{42}Cr, ^{45}Fe, and ^{49}Ni, where we had reasonable statistics to search for 2p decay.

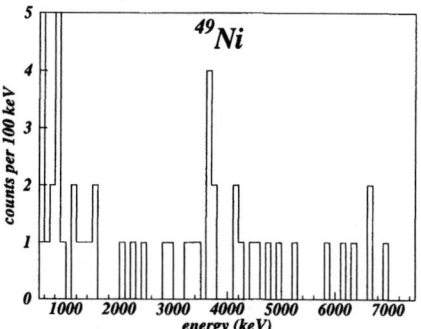 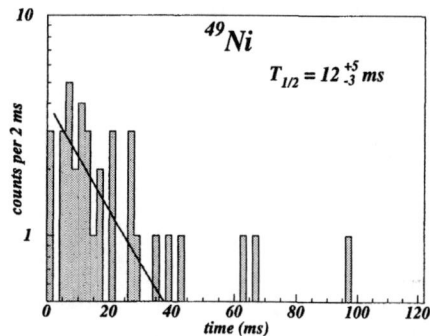

FIGURE 3. Left-hand side: Charged-particle spectrum from the decay of ^{49}Ni. Right-hand side: Decay-time distribution for ^{49}Ni gated on events with an energy above 1 MeV.

As a result, the experimental decay spectrum observed for ^{49}Ni is shown on the left-hand side of Fig. 3. A charged-particle peak is clearly observed at about 3.7 MeV. The decay-time histogram determined when gating on events above 1 MeV is shown on the left-hand side. The experimental value obtained is $T_{1/2} = 12^{+5}_{-3}$ ms. For two reasons, these experimental observations allow to conclude that ^{49}Ni decays by β emission: i) If the peak at 3.7 MeV were a 2p emission peak, the barrier penetration half-life would be of the order of 10^{-15}s, orders of magnitude lower than the flight time of the fragments from the target to the detection setup. ii) The observed half-life is typical for a β-decay half-life in this region of the chart of the nuclei. We therefore conclude that ^{49}Ni decays with a branching ratio of close to 100% by a β-delayed mode. Due to low statistics, the detailed origin of the 3.7 MeV activity could not be determined. No coincident γ rays were observed.

A similar conclusion can be drawn for the decay of ^{42}Cr where a charged-particle peak was observed at 1.9 MeV and a half-life of $13.4^{+3.6}_{-2.4}$ ms was measured. Therefore, as ^{49}Ni, this nucleus decays predominantly by β decay.

In the case of ^{45}Fe, the picture is less clear. First of all, due to the lower production cross section, the statistics is lower. No pronounced peak appears in the charged-particle spectrum. The half-life analysis of events with a deposited energy above 1 MeV yields a value of 6^{+17}_{-3} ms which has to be compared to β-decay half-life predictions ranging from 3-12 ms. For example, shell model approaches predict a half-life of 7 ms [2]. This result could be interpreted as ^{45}Fe decaying mainly by β decay with a small component of a faster 2p emission. However, higher statistics measurements are needed to get a clear picture of the decay of this nucleus. A 2p branch, even though small, should be clearly visible in the decay-energy spectrum.

V THE DECAY OF ^{25}SI

The decays of ^{21}Mg and ^{25}Si have been studied as test cases for the investigation of the more exotic nuclei ^{22}Al, ^{26}P, and ^{27}S. The decay of the former two nuclei has been studied in the past, mainly with helium-jet techniques [18–20]. The absolute branching ratio was determined from a comparison of the observed activity from the isobaric analogue state with the calculated decay rate for a superallowed Fermi transition. No γ-ray measurements were performed in the past. The feeding of the particle-bound levels has been determined by means of mirror symmetry using the decays of ^{21}F and ^{25}Na.

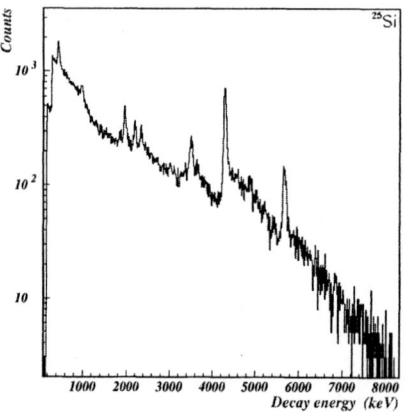 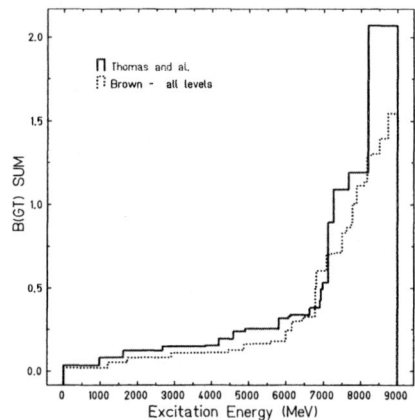

FIGURE 4. Right-hadn side: Charged-particle spectrum from the decay of ^{25}Si. All proton lines are identified and allow to establish the decay of this nucleus. The inset is the γ-ray spectrum in coincidence with protons. Left-hand side: Comparison of the summed Gamow-Teller distribution from shell-model theory and experiment.

In our experiment, we performed for the first time β-proton-γ coincidence measurements which enabled us to establish a complete decay scheme using mirror symmetry only for the decay to the ground state. The decay-energy spectrum for ^{25}Si and the γ-ray spectrum are shown in figure 4. The identified transitions result in the decay scheme shown in figure 5. The experimental decay scheme may be compared to shell-model calculations using the OXBASH code and the USD effective interaction. Reasonable agreement is achieved demonstrating that the sd shell is nicely understood in term of the nuclear shell model.

The summed Gamow-Teller strength is plotted as a function of the excitation energy in figure 4. A remarkable agreement is found between the experimental and the shell model result at low excitation energy. At higher excitation energy, the agreement is less good, as part of the strength is missing in the experimental data due to a lack of efficiency at these high energies.

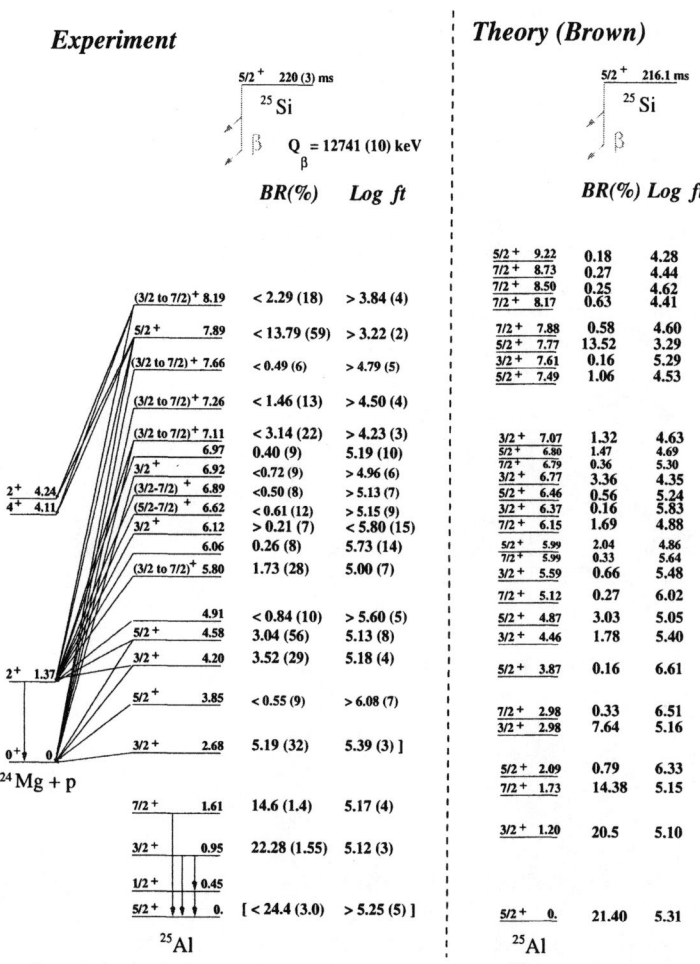

FIGURE 5. Comparison of the experimental decay scheme of ^{25}Si and a prediction from the nuclear shell model with the OXBASH code and the USD force.

Similar agreement could also be achieved for ^{21}Mg, the second test case. The analysis of the more exotic nuclei mentionned above is in progress. High-quality data are available for these nuclei, however, due to limited statistics for ^{27}S, the experimental information on this nucleus will be somewhat scarce.

VI CONCLUDING REMARKS

An unambiguous observation of doubly-magic ^{48}Ni was achieved at the GANIL/LISE3 facility from the projectile fragmentation of an intense ^{58}Ni beam. From this observation, we deduce a lower limit of its half-life of about 0.5 μs.

In the same experiment, first spectroscopic data have been obtained for nuclei in the vicinity of ^{48}Ni. The decay spectra of ^{42}Cr and ^{49}Ni show no evidence for 2p emission and can rather be interpreted as a pure β-delayed charged-particle decay. In the case of ^{45}Fe, higher-statistics data are needed to yield a clearer picture of its decay.

The preliminary results on sd-shell nuclei again demonstrate the predictive power of the nuclear shell model in the sd shell. Nice agreement was obtained for the $T_z = -3/2$ nuclei ^{21}Mg and ^{25}Si.

ACKNOWLEDGEMENT

The work and the contribution from collegues from CEN Bordeaux-Gradignan, GANIL, Warsaw, Bucharest, Jyväskylä, Lyon, LPC Caen, Strasbourg, and Edinburg is greatfully acknowledged.

REFERENCES

1. B. A. Brown, Phys. Rev. C **43**, R1513 (1991).
2. W. E. Ormand, Phys. Rev. C **53**, 214 (1996).
3. W. E. Ormand, Phys. Rev. C **55**, 2407 (1997).
4. B. J. Cole, Phys. Rev. C **54**, 1240 (1996).
5. M. D. Cable et al., Phys. Rev. C **30**, 1276 (1984).
6. H. Fynbo et al., Phys. Rev. C **59**, 2275 (1999).
7. C. Bain et al., Phys. Lett. **373B**, 35 (1996).
8. M. Chromik et al., Phys. Rev. C **55**, 1676 (1997).
9. O. V. Bochkarev et al., Sov. J. Nucl. Phys. **55**, 955 (1992).
10. R. A. Kryger et al., Phys. Rev. Lett. **74**, 860 (1995).
11. B. Blank et al., Phys. Rev. Lett. **84**, 1116 (2000).
12. W. Nazarewicz et al., Phys. Rev. C **53**, 740 (1996).
13. W. Benenson, Nukleonika **20**, 775 (1975).
14. P. Haustein, At. Data Nucl. Data Tab. **39**, 185 (1988).
15. P. Möller et al., At. Data Nucl. Data Tab. **59**, 185 (1995).
16. J. Duflo and A. Zuker, Phys. Rev. C **52**, R23 (1995).
17. Y. Aboussir et al., At. Data Nucl. Data Tab. **61**, 127 (1995).
18. R. Sextro, R. Gough, and J. Cerny, Phys. Rev. C **8**, 258 (1973).
19. J. Robertson et al., Phys. Rev. C **47**, 1455 (1993).
20. S. Hatori et al., Nucl. Phys. **A549**, 327 (1992).

Nuclear structure in the vicinity of shell closures far from stability

H. Grawe[1] in collaboration with
M. Górska[1,2] J. Döring[1], C. Fahlander[3], M. Palacz[4], F. Nowacki[5],
E. Caurier[5], J.M. Daugas[6], M. Lewitowicz[6], M. Sawicka[7],
M. Pfützner[7], R. Grzywacz[8], K. Rykaczewski[9], O. Sorlin[10],
S. Leenhardt[10], F. Azaiez[10], M. Rejmund[11], K. Hauschild[11]
J. Uusitalo[12]

[1] *GSI, Darmstadt, Germany;* [2] *IKS University of Leuven, Belgium;* [3] *Dept. of Physics, Lund University, Sweden;*
[4] *Heavy Ion Laboratory, Warsaw University, Poland;* [5] *IreS, Strasbourg, France;* [6] *GANIL, Caens, France;*
[7] *IEP, Warsaw University, Poland;* [8] *Department of Physiscs, University of Tennesse, Knoxville, USA;*
[9] *ORNL, Oak Ridge, USA;* [10] *IPN Orsay, France;* [11] *DAPNIA/SPhN CEA Saclay, France;*
[12] *JYFL, University of Jyväskylä, Finland.*

Abstract. The status of experimental approach to ^{100}Sn and ^{78}Ni is reviewed. Revised single particle energies for neutrons are deduced for the N=Z=50 shell closure and evidence for low lying $I^\pi = 2^+$ and 3^- states is presented. Moderate E2 polarization charges of 0.1 e and 0.6 e are found to reproduce the experimental data when core excitation of ^{100}Sn is properly accounted for in the shell model. For the neutron rich Ni region no conclusive evidence for an N=40 subshell is found, whereas firm evidence for the persistence of the N=50 shell at ^{78}Ni is inferred from the existence of seniority isomers. The disappearance of this isomerism in the mid $\nu g_{9/2}$ shell is discussed. The spectroscopy of ^{216}Th disproves the existence of a Z=92 shell gap as predicted by some recent mean field calculations. Inversion of the $\pi h_{9/2}$ and $f_{7/2}$ orbitals at Z=90 is ascribed to the coupling of the $f_{7/2}$ (and $i_{13/2}$) protons to the low-lying 3^- state ($\hbar \omega_3$=1.69 MeV).

I INTRODUCTION

Doubly magic nuclei in exotic regions of the nuclidic chart are key points to determine shell structure as documented in single particle energies (SPE) and residual interaction between valence nucleons. The parameters of schematic interactions, entering e.g. the numerous versions of Skyrme-Hartree-Fock calculations, are not

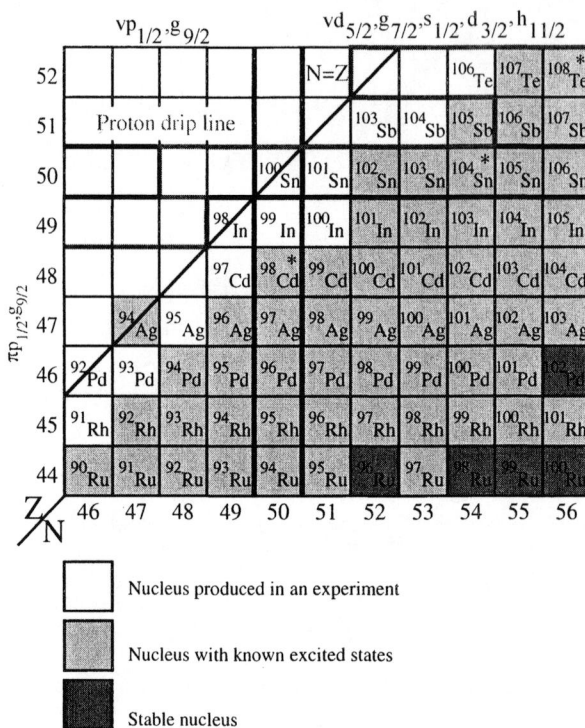

FIGURE 1. Experimental approach to ^{100}Sn. The most neutron deficient compound nuclei that can be reached in stable beam-target combinations are marked by asterisks.

well determined by the spectroscopy of nuclei close to β-stability [1]. Relativistic mean field predictions are hampered by the delicate balance of single particle and spin-orbit potentials, which are determined by the small difference, respective large sum of the contributions from vector and scalar mesons [2]. Experimental SPE on the other hand, are often distorted by quadrupole and octupole correlations, which are not accounted for in mean field calculations. This prevents a direct comparison between experiment and theory. Modern realistic interactions as deduced from NN phase shifts by G-matrix many-body theory [3] account very well for the spectroscopy of nuclei [4–6] but need to be modified [7] to reproduce saturation. It has been pointed out recently that there is an intimate relation between correct reproduction of the spectroscopy of nuclei and the physics of neutron stars, mediated by the realistic interaction [8].

As a mandatory prerequisite to study the relation between nuclear spectroscopy and realistic interactions large scale shell model codes were developed, which enable a detailed comparison not only of level schemes but also of electromagnetic transition rates, Gamow-Teller strengths and even shape evolution up to ^{100}Sn [9,10].

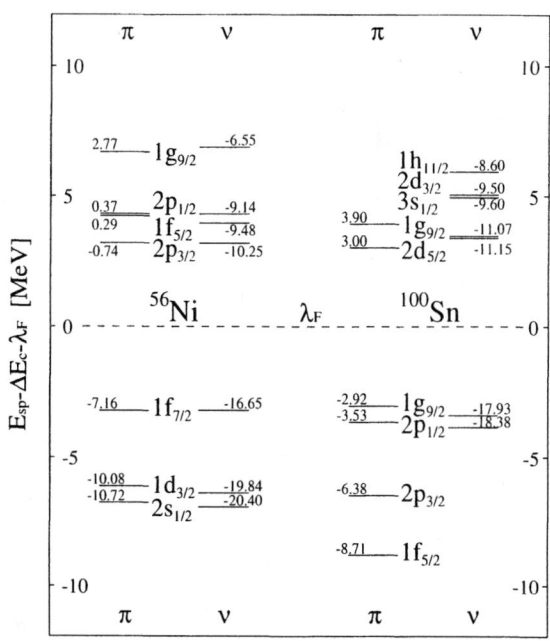

FIGURE 2. Experimental resp. extrapolated single particle (hole) energies for ^{56}Ni and ^{100}Sn.

Monte-Carlo shell model calculations have also become feasible for the region between ^{56}Ni, ^{78}Ni and ^{100}Sn [11,12]. For the semimagic N=126 isotones unrestricted shell model calculations in the full proton space beyond ^{208}Pb have been performed recently up to ^{218}U [10].

Spectroscopic studies around the doubly magic nuclei ^{100}Sn, at N=Z and the proton dripline, and ^{78}Ni at N/Z=1.78 and beyond ^{208}Pb provide therefore an excellent probe for shell structure, residual interaction, spin-isospin and shape correlations over the full range of nuclei accessible with state of the art experimental techniques. Recent progress in the development of efficient γ-arrays with highly selective ancillary detectors as operative in EUROBALL and GAMMASPHERE and the effective combination of highly efficient γ-detection with in-flight and on-line separators have enabled spectroscopy of these exotic regions of the Segré chart. ^{100}Sn and ^{78}Ni, situated at key points of the rp– and r–paths, respectively, cover the full range in the ratio $1.0 \leq N/Z \leq 1.8$ that is available to probe the isovector part of the nuclear mean field, which is not determined by the large body of data available for less exotic nuclei. The shell structure beyond ^{208}Pb is a key testing ground for mean field predictions of super-heavy elements (SHE) shell predictions. The single particle structure in this region is highly distorted by coupling to L=2 and 3 phonons [13] and deformation [14].

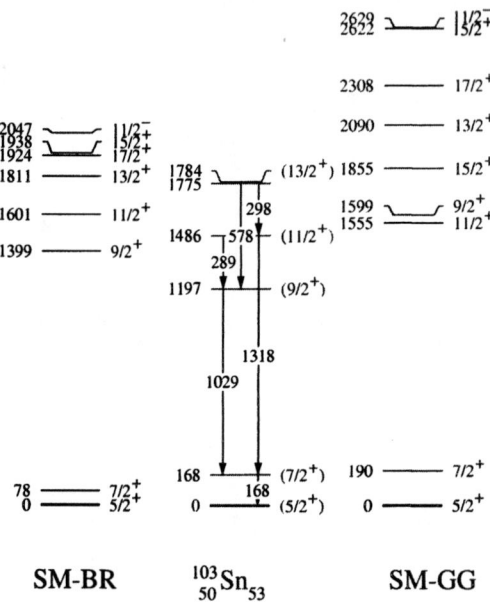

FIGURE 3. Preliminary level scheme for ^{103}Sn [21] in comparison to different shell model approaches (BR) [22] and (GG) [23–26].

II THE ^{100}SN REGION

At the doubly magic shell closure N=Z=50, the single particle structure in comparison to mean field predictions, the residual interaction as deduced from NN potentials by G-matrix many-body theory and the response to core excitations of type E2, E3 and spin-flip are the main study objects. Due to the similar shell structure with an *ls*-open core, separated by one major shell a close resemblance to the N=Z=28 closure at ^{56}Ni is expected. The experimental approach to ^{100}Sn is landmarked by the in-flight identification in a few events in fragmentation [15,16], and by spectroscopy of excited states in their neighbours ^{98,99}Cd [17–19], ^{102,103}Sn [20,21] and ^{105}Sb [19] (fig.1).

A Single particle energies

As the single particle (hole) neighbours are not yet accessible to experimental studies, the shell structure of ^{100}Sn is inferred from shell model studies of less exotic neighbouring nuclei as described in ref. [23]. The results are published in previous papers [23,24]. Recently an improved fit was made to the N=51 quasi-particle states [25,26]. Using identical interaction and model space $\pi(p_{1/2},g_{9/2})$ $\nu(d_{5/2},g_{7/2},s_{1/2},d_{3/2},h_{11/2})$ the neutron single particle energies (SPE) of the hypothetical ^{88}Sr core were modified to reproduce the trend of measured states. The results are shown in fig.2 in comparison to ^{56}Ni. The mutual correspondence of orbitals, separated by a major oscillator shell, is apparent. This implies very similar structure for the two doubly-magic nuclei including their susceptibility to quadrupole and spin-flip core excitations. Additionally ^{100}Sn should exhibit octupole correlations due to the $p_{1/2}^{-1}$ $d_{5/2}$ stretched ($\Delta l = l_1 + l_2$) and $p_{1/2}^{-1}$ $g_{7/2}$ anti-stretched ($\Delta l = l_1 - l_2$) excitations.

Recently ^{103}Sn was identified and studied for the first time in an in-beam experiment at EUROBALL [21]. In fig.3 the preliminary level scheme is compared to two different shell model predictions in the full pure neutron space $\nu(d_{5/2}, g_{7/2}, s_{1/2}, d_{3/2}, h_{11/2})$ employing different interactions and SPE (BR) [22] and (GG) [23-26]. Note the good agreement obtained for the $d_{5/2}$ - $g_{7/2}$ splitting, obtained with the SPE of fig.1.

B Effective E2 charge, E3 transitions, core excitation

Ever since the first observation of ^{98}Cd [17] and ^{102}Sn [18] in-beam and the measurement of the respective B(E2;$8^+\to6^+$) and B(E2;$6^+\to4^+$) the small proton and the large neutron polarization charges needed in pure proton hole and neutron particle model spaces have been a puzzle. For ^{98}Cd meanwhile two deviating values are known, namely B(E2) = 14^{+7}_{-4} e^2fm^4 [17] and 30 ± 4 e^2fm^4 [27], which points to an experimental complication, but even the larger value demands a proton polarization charge of only $\delta e_\pi \simeq 0.3$ e. On the other hand $\delta e_\nu \geq 1.6$ e depending on the shell model approach is required for ^{102}Sn, as deduced from the experimental value B(E2) = 115^{+70}_{-30} e^2fm^4 [18]. Recent large scale shell model calculations in the complete $\pi\nu$(s,d,g) orbitals using a G-matrix based realistic interaction tailored for this model space [28] (SDG) shed light on these controversy results. Using equal polarization charge δe= 0.5 e for protons and neutrons B(E2) values of 57 e^2fm^4 and 73 e^2fm^4 are calculated for ^{98}Cd and ^{102}Sn allowing up to 4p-4h and 3p-3h excitations of the ^{100}Sn core, respectively, which is the present computational limit. For ^{102}Sn this is an increase by a factor of 10 compared to the 0p-0h approach whereas for ^{98}Cd only a moderate increase by 50 % is obtained. To reproduce the experiment moderate polarization charges of $\delta e_\pi \simeq 0.1$ e (for the larger experimental value) and $\delta e_\nu \geq$ 0.6 e are derived. They are in remarkable agreement with recent core polarization calculations for ^{100}Sn [4], which supports the existence of a large isovector effect. The SDG calculations imply for ^{100}Sn a low lying I^π=2^+ state, which shows a similar convergence behaviour with increasing np-nh excitations as for ^{56}Ni. This has been extrapolated from the experimental evolution of I^π=2^+ states in Sn isotopes and N=50 isotones as compared to the ^{56}Ni region, implying $E_x(2^+) \simeq 3$ MeV [24]. Similarly the strong E3 enhancement found in ^{104}Sn [29] points to a low lying $I^\pi = 3^-$ state at about the same energy [24].

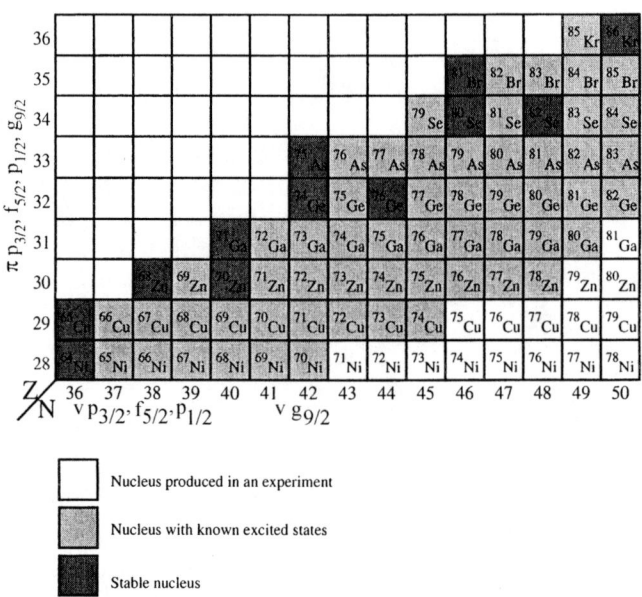

FIGURE 4. Experimental approach to ^{78}Ni.

III FROM $^{68}_{28}$NI$_{40}$ TO $^{78}_{28}$NI$_{50}$

At $N \gg Z$ the disappearance of the familiar Woods-Saxon shell closures and their reappearance as harmonic oscillator magic numbers is a burning question, which is strongly related to the dilute neutron density and the decoupling of proton and neutron motion in weakly bound systems. Recent experimental data provide evidence on the existence of the N=40 and N=50 shell gaps at ^{68}Ni and ^{78}Ni, respectively. The experimental approach to ^{78}Ni is landmarked by the in-flight identification in a few events in relativistic fission [30] and by isomer and β-decay spectroscopy of ^{70}Ni [31], ^{73}Cu [32] and ^{78}Zn [33] (fig.4).

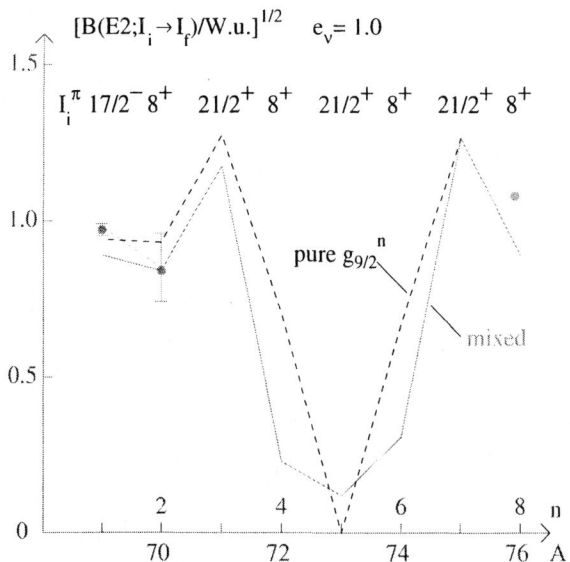

FIGURE 5. B(E2) values for the $I_{max} \to I_{max}-2$ transitions between ^{68}Ni and ^{78}Ni for pure $\nu g_{9/2}^n$ configuration (dashed line) and the full shell model calculation (full line). The experimental point shown for ^{76}Ni is taken from the isotone ^{78}Zn.

A The N=50 shell at ^{78}Ni

Beyond $^{68}_{28}$Ni$_{40}$ the filling of the $\nu g_{9/2}$ shell should give rise to $I^\pi=8^+$ isomers, as known from the valence mirror N=50 isotones (see ref. [17] for a summary). Such isomer was identified in ^{70}Ni [31]. Shell model calculations using a realistic interaction [34] in the full $\pi\nu$(p$_{3/2}$, f$_{5/2}$, p$_{1/2}$, g$_{9/2}$) model space were performed for Ni, Cu and Zn isotopes and account well for the decay scheme and the B(E2;$8^+\to 6^+$) using polarization charges of $e_\nu = 1.0\ e$ and $\delta e_\pi = 0.5\ e$ (fig.5) [25,33,35]. For details of the calculations see refs. [25,33,35]. The related isomers $I^\pi=17/2^-$ in ^{69}Ni (fig. 5) and $I^\pi=19/2^-$ in ^{71}Cu [25,35], obtained by coupling a $\nu p_{1/2}$ hole and a $\pi p_{3/2}$ particle to the 8^+ isomer are equally well reproduced. The systematics of shell model predictions of the B(E2;$I_{max} \to I_{max}-2$) for the $\nu g_{9/2}^n$ Ni isotopes are shown in fig.5.

Recently an extensive search for the 8^+ isomer in ^{72}Ni was performed with negative result [36]. Precisely, an isomer can be excluded in the range 20 ns $\leq t_{1/2} \leq$ 2.5 ms. This is at variance with shell model predictions (fig.5) as well as evidence from seniority schemes j^n for $\pi g_{9/2}$ (N=50), $\pi h_{11/2}$ (N=82), $\pi h_{9/2}$ (N=126), $\nu h_{11/2}$ (Z=50) and $\nu i_{13/2}$ (Z=82). The general trend shown in fig. 5 is true only in lowest

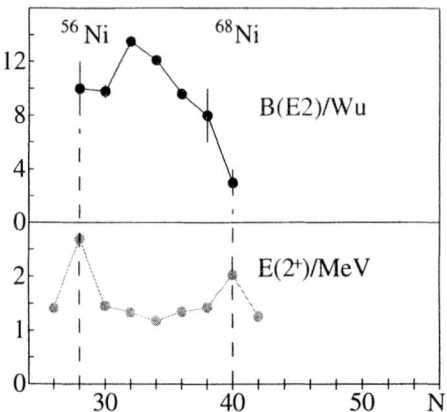

FIGURE 6. Excitation energy $E_x(2^+)$ and transition strength $B(E2; 2^+ \rightarrow 0^+)$ for Ni isotopes. The values for ^{66,68}Ni are preliminary [42].

seniority v=2,3. Extracting an empirical $\nu g_{9/2}^2$ interaction from ^{70}Ni in the spirit of ref. [37] (ESM), we have calculated a ^{72}Ni level scheme [35]. In contrast to the full shell model calculation using a realistic interaction (S3V) [34] the $I^\pi=6^+$, v=4 state in the ESM is calculated below the $I^\pi=8^+$, which would enable a fast B(E2) resulting in a predicted half life of about 20 ns, whereas 20 μs are expected in the S3V shell model (see ref. [35] fig. 7). This feature of the empirical interaction, which is unique in the Segré chart, except for the $1f_{7/2}$ shell, might be induced by $\pi f_{7/2}$ core excitations from outside the present model space or, alternatively, by a reduction of the $\nu g_{9/2}^2$ pairing matrix element.

Towards the end of the $\nu g_{9/2}$ shell for ^{76}Ni the isomerism should reappear. ^{76}Ni is not accessible yet for spectroscopic studies, but in a recent experiment an $I^\pi=8^+$ isomer was established in its isotone ^{78}Zn [33]. In fig.5 the B(E2) for ^{78}Zn is shown nd the good agreement with the ^{76}Ni prediction is a clear indication that "a fortiori" the isomer in ^{76}Ni exists. This is to date the most direct proof that the N=50 shell closure persists at ^{78}Ni. These features are well reproduced in the full shell model calculation shown in fig.5.

B The N=40 subshell at ^{68}Ni

It has been frequently invoked that the N=40 shell closure is well developed in ^{68}Ni [37–39]. More recently experimental evidence from β-decay was discussed [40,41] and a subtle balance of a possible shell gap and the pairing gap was made responsible for the controversial experimental evidence. The experimental facts are summarized as follows. ^{68}Ni has a large $I^\pi=2^+$ excitation energy and a small $B(E2;2^+ \rightarrow 0^+)$ [42] (fig.6). In fact this value is one of the smallest found in

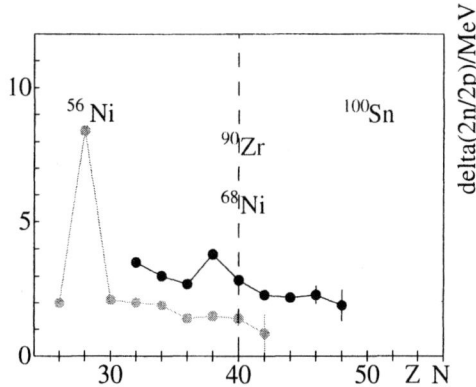

FIGURE 7. Derivative of the two-neutron and two-proton separation energies δ_{2n} for Ni isotopes and δ_{2p} for N=50 isotones.

semimagic nuclei and a factor of three smaller than the value for ^{56}Ni [43]. On the other hand the derivative δ_{2n} of the two-neutron separation energy S_{2n}, which is free from pairing effects, does not show any sign of a peak near N=40 (fig.7). This is different from the behaviour of the N=50 valence mirror nuclei around ^{88}Sr (Z=38) and ^{90}Zr (Z=40) (fig.7). Moreover, mean field calculations [44] show a distinct subshell closure for N=40 and Z=28. The most simple explanation for the contradictory conclusions from the experimental facts is the existence of the high spin intruder orbital $\nu g_{9/2}$ above the odd-parity (f,p) orbitals. Below N$\simeq$40 recoupling of a pair of nucleons is sufficient to reach the 2^+ state from the g.s. At N=40 due to the parity change besides breaking a pair two particles have to be excited to $\nu g_{9/2}$ to have even parity. This is more costly in energy, and moreover the back-decay to the g.s. is particle forbidden. This is exactly what the shell model with a ^{56}Ni core reproduces without any distinct shell gap between $\nu p_{1/2}$ and $\nu g_{9/2}$ [41,42]. There is also clear evidence that the apparent shell gap as calculated from ^{67,68,69}Ni binding energies equals 2Δ, with the pairing gap Δ calculated as $\Delta_n^{(4)}$ [45] from ^{63,64,65}Ni. In contrast, for the N=50 valence mirror region the "gap" exceeds 2Δ by 0.5 MeV.

To check this interpretation a schematic shell model calculation was performed, employing a surface delta interaction (SDI) with strength parameter A_1=0.53 MeV and monopole term B_1=0.50 MeV [46] and assuming degenerate $\nu p_{3/2}$, $f_{5/2}$ and $p_{1/2}$ orbitals in ^{57}Ni and varying the distance ϵ of the $\nu g_{9/2}$ orbital in multiples of A_1. In fig.8 the results for $E_x(2^+)$ and $B(E2; 2^+ \to 0^+)$ are shown for ϵ/A_1=5,6 and 7 are shown. The pattern of fig.6 is clearly reproduced qualitatively. Note, that ϵ/A_1=5.0 (long dashed) corresponds to complete degeneracy, i.e. no ordering, of all orbitals at N=40, while ϵ/A_1=7.0 (short dashed) reproduces roughly the experimental $p_{1/2}$ - $g_{9/2}$ distance in ^{67}Ni. The results of fig.8 are qualitative in the sense that the SDI parameters used [46] overestimate pairing, which is found to be essential for the reduced shell gap at N=40 [40,35].

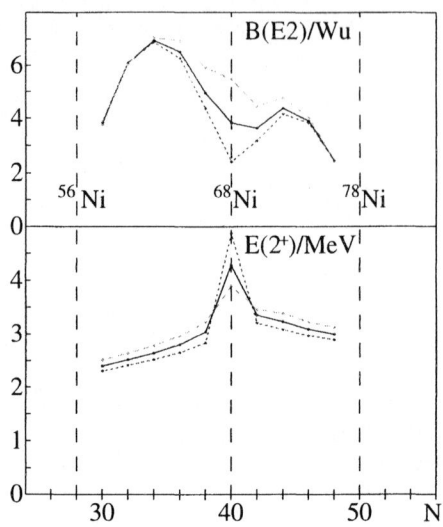

FIGURE 8. As fig. 6, but for a schematic shell model approach using SDI (see text).

IV BEYOND ^{208}PB – THE N=126 ISOTONES

The doubly-magic nucleus $^{208}_{82}$Pb$_{126}$ marks the last and best studied shell closure in the Segré-chart. Beyond this region and on the way to the island of super-heavy elements (SHE) shell structure estimates rely entirely on theoretical predictions. Due to the subtle balance of liquid drop and shell structure and the coupling of single particle motion to surface vibrations, reliable predictions become increasingly difficult. Experimentally a detailed spectroscopy is hampered by the increasing fissility of compound nuclei formed in fusion reactions. In recent years the recoil-decay tagging method (RDT) has been developed into a powerful tool to identify and study evaporation residues in the presence of an excessive amount of fission products. Exploiting the selectivity of α-decay (see fig.9) spectroscopy up to Z=102 (Nb) isotopes and down to the nb cross section level has become feasible [14]. Systematic mean field calculations beyond ^{208}Pb aiming at shell structure of SHE predict for some of the interactions used a substantial shell gap for Z=92, N=126 [47]. This is at variance with extrapolations from experiments performed up to $^{215}_{89}$Ac$_{126}$ [48].

In an RDT experiment at the gasfilled separator RITU of the JYFL cyclotron, ^{216}Th, the two-hole nucleus in the predicted Z=92 subshell, was studied [49]. The resulting level scheme along with the systematics of N=126 isotones and the results of large scale shell model calculations [10] is shown in fig. 10. The two I^π=8$^+$ states of predominant and hardly mixed configuration $\pi h^n_{9/2,v=2}$ and $\pi h^{n-1}_{9/2,v=1}$ f$_{7/2}$

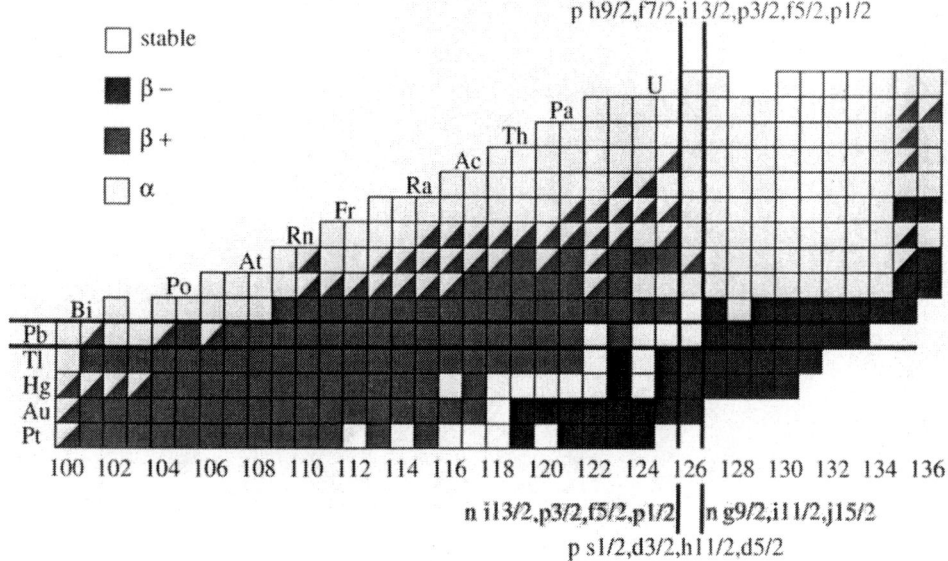

FIGURE 9. Nuclidic chart beyond ^{208}Pb. Note the numerous α-decaying isotopes accessible to the RDT method.

are expected to be degenerate in ^{216}Th. From the B(E3) of the feeding of the observed $I^\pi=(8^+)$ isomer ($t_{1/2} = 128(8)\mu s$) by the $I^\pi=(11^-)$ isomer ($t_{1/2} = 615(55)$ ns) and the slow B(E2,$(8^+) \to (6^+)$) the configuration of the (8^+) isomer is inferred to be $\pi h_{9/2,v=1}^{n-1} f_{7/2}$. It decays by α's and γ's, which show identical halflife within experimental errors. The second expected 8^+ state cannot be yrast, as it would be an yrast trap decaying by α's only, which is not observed. Hence $\pi h_{9/2}$ and $f_{7/2}$ have swapped positions in ^{216}Th, which disproves existence of a major shell gap at Z=92. As a consequence the $\pi h_{9/2}^n$ seniority scheme is largely deviating from the ideal behaviour shown in fig.5 for $\nu g_{9/2}^n$. The reason for this distortion is the lowering of the $\pi f_{7/2}$ (and $i_{13/2}$) levels by their favorable coupling to the low-lying 3^- state ($\hbar\omega = 1.687$ MeV, see fig. 10), as known from the ^{208}Pb neighbours [13].

V SUMMARY AND CONCLUSION

The recent developments in detector technology and in-beam and online identification techniques have enabled detailed spectroscopy of exotic nuclei at N≃Z up to ^{100}Sn and at N≫Z close to ^{78}Ni. For ^{100}Sn the single particle structure can be inferred, the adequacy of realistic interactions is established and first evidence on the rôle of core excitations is obtained implying low lying 2^+ and 3^- states in this doubly magic nucleus. The importance and need for advanced shell model codes

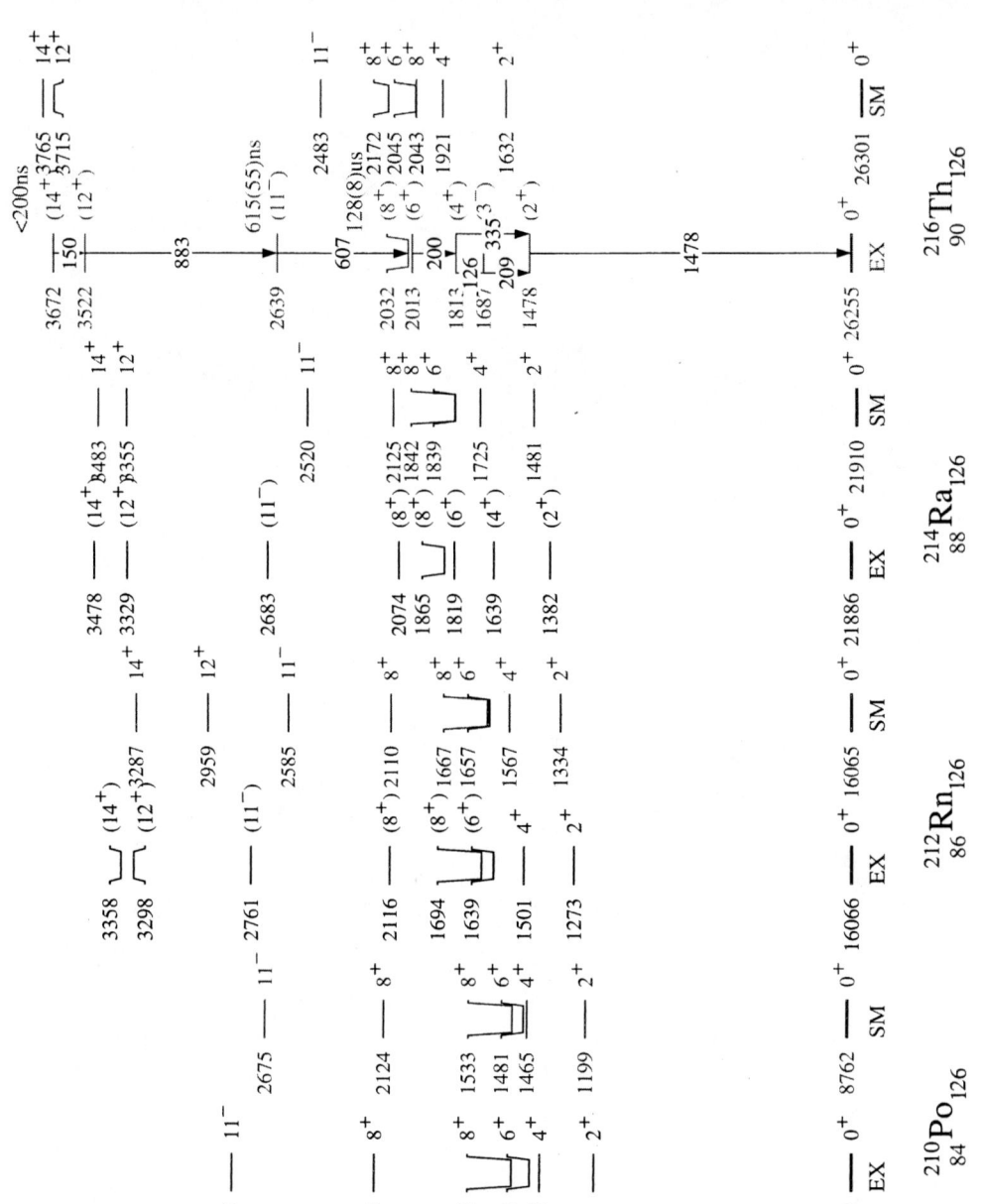

FIGURE 10. Near-yrast levels in the N=126 isotones of Po - Th in comparison to results of a large scale shell model calculation in the full proton space Z=82-126 [10].

and techniques is demonstrated. For neutron rich Ni isotopes there is clear evidence for the persistence of the N=50 shell, for an only very weak N=40 subshell as compared to the N=50, Z=40 valence mirror region and very likely for a distortion of the seniority scheme in the $\nu g_{9/2}$ midshell. For N=126 there is no evidence for a Z=92 shell gap. The coupling of the $\pi f_{7/2}$ (and $i_{13/2}$) orbital(s) to the low-lying octupole state ($\hbar\omega$=1.69 MeV) reduce the $f_{7/2} - h_{9/2}$ gap to near-degeneracy and causes a distortion of the $h_{9/2}^n$ seniority scheme. It should be noted, however, that the modern experimental techniques in many respects are still in there infancy and essential improvements are to be expected, which includes the exploit of radioactive ion beams.

Acknowledgment This work was supported in part by the Polish Committee of Scientific Research under Grant No. KBN 2 P03B 036 15, by the Program for Scientific Technical Collaboration (WTZ) under Project No. POL 99/009 and by the European Community under Contract No. ERBFMGECT950083.

REFERENCES

1. B.A. Brown, Phys. Rev. **C58**, 220 (1998)
2. P. Ring, Progr. Part. Nucl. Phys. **37**, 193 (1996)
3. M. Hjorth-Jensen et al., Phys. Rep. **261**, 125 (1995)
4. T. Engeland et al., Phys. Rev. **C61**, 021302(R) (2000)
5. A. Covello et al., Proc. 6th Int. Spring Seminar on Nuclear Physics - " Highlights of modern nuclear structure", ed. A. Covello, World Scientific, Singapore, 1999, p.129
6. E. Caurier et al., Phys. Rev. Lett. **75**, 2466 (1995)
7. A. Poves and A.P. Zuker, Phys. Rep. **70**, 235 (1981)
8. H. Heiselberg, M. Hjorth-Jensen, Phys. Rep. **328**, 237 (2000)
9. F. Nowacki, Proc. SM2K, RIKEN 2000, Nucl. Phys. **A**, in print
10. E. Caurier, Proc. SM2K, RIKEN 2000, Nucl. Phys. **A**, in print
11. S.E. Koonin et al., Phys. Rep. **278**, 1 (1997)
12. T. Otsuka et al., Phys. Rev. Lett. **81**, 1588 (1998)
13. M. Rejmund et al., Eur. Phys. J. **A8**, 161 (2000)
14. M. Leino et al., Eur. Phys. J. **A8**, 63 (1999)
15. R. Schneider et al., Z. Phys. **A348**, 241 (1994)
16. M. Lewitowicz et al., Phys. Lett. **B332**, 20 (1994)
17. M. Górska et al., Phys. Rev. Lett. **79**, 2415 (1997)
18. M. Lipoglavšek et al., Phys. Lett. **B440**, 246 (1998)
19. C. Baktash, Proc. SM2K, RIKEN 2000, Nucl. Phys. **A**, in print
20. M. Lipoglavšek et al., Phys. Rev. Lett. **76**, 888 (1996)
21. M. Palacz et al., Proc. SM2K, RIKEN 2000, Nucl. Phys. A, in print, and C. Fahlander et al., submitted to Phys. Rev. Lett.
22. B.A. Brown, K. Rykaczewski, Phys. Rev. **C50**, R2270 (1994)
23. H. Grawe et al., Physica Scripta **T56**, 71 (1995)

24. H. Grawe et al., Proc. 6th Int. Spring Seminar on Nuclear Physics - "Highlights of modern nuclear structure", ed. A. Covello, World Scientific, Singapore, 1999, p. 137
25. H. Grawe et al., Proc. Workshop on "The beta-decay, from weak interaction to nuclear structure", March 17 - 19, 1999, IReS Strasbourg, France, edts. Ph. Dessagne, A. Michalon, C. Miehe, p. 211
26. M. Górska et al., Proc. of the Int. Conf. on Nucl. Phys. in Europe (ENPE 99) Facing the Next Millennium, Sevilla, Spain (eds. B. Rubio, M. Lozano, W. Gelletly), AIP Conf. Proc. **495**, 217 (1999)
27. R. Grzywacz, ENAM98, AIP Conf. Proc. **481**, 257 (1999)
28. F. Nowacki et al., submitted to Phys. Rev. Lett.
29. M. Górska et al., Phys. Rev. **C58**, 108 (1998)
30. Ch. Engelmann et al., Z. Phys. **A352**, 351 (1995)
31. R. Grzywacz et al., Phys. Rev. Lett. **81**, 766 (1998)
32. S. Franchoo et al., Phys. Rev. Lett. **81**, 3100 (1998)
33. J.M. Daugas et al., Phys. Lett. **B476**, 213 (2000)
34. J. Sinatkas et al., J. Phys. **G18**, 1377 and 1401 (1992)
35. H. Grawe et al., Proc. SM2K, RIKEN 2000, Nucl. Phys. **A**, in print
36. M. Sawicka, R. Grzywacz et al., to be published
37. T. Ishii et al., Phys. Rev. Lett. **81**, 4100 (1998)
38. R. Broda et al., Phys. Rev. Lett. **74**, 868 (1995)
39. J.I. Prisciandaro et al., Phys. Rev. **C60**, 031301 and 054307 (1999)
40. W.F. Mueller et al., Phys. Rev. Lett. **83**, 3613 (1999)
41. W.F. Mueller et al., Phys. Rev. **C61**, 054308 (2000)
42. O. Sorlin et al., to be published
43. G. Kraus et al., Phys. Rev. Lett. **73**, 1773 (1994)
44. M. Bender et al., *Nuclear Astrophysics - Proc. Int. Workshop XXVI on Gross properties of Nuclei and Nuclear Excitations*, Hirschegg, Austria, January 11-17, 1998, p. 59
45. Aa. Bohr and B.R. Mottelson, *Nuclear structure* (World Scientific, Singapore 1998), vol. I, p. 169
46. J.E. Koops and P.W.M. Glaudemans, Z. Phys. **A280**, 181 (1977)
47. K. Rutz etal., Nucl. Phys. **A634**, 67 (1998)
48. D.J. Decman et al., Z. Phys. **A310**, 55 (1983)
49. K. Hauschild et al., to be published

SUPERHEAVY ELEMENTS (SHE)

Heaviest Elements (Synthesis and Decay Properties)

Yuri Ts. Oganessian[1]

Flerov Laboratory of Nuclear Reactions, Joint Institute for Nuclear Research, 141980 Dubna, Moscow region, Russia

Abstract. For the 60 years that have passed after the discovery of the first artificial elements Np and Pu, the investigations of the properties of new elements have become one of the fundamental and quickly developing fields of nuclear physics and nuclear chemistry.

The transition from the traditional method of producing transuranium elements, where continuous and pulsed neutron fluxes have been used, to nuclear reactions induced by heavy ions has made it possible to synthesize 12 new elements heavier than fermium ($Z = 100$).

The theoretical description of the masses and fission barriers of the new nuclei led in the mid-1960s to the prediction of "islands of stability" for the very heavy and superheavy nuclides in the vicinity of the closed proton and neutron shells. The experimental data that demonstrate the enhanced stability of nuclei, close to the deformed shells $Z = 108$ and $N = 162$, relative to different decay modes and also the reactions of their synthesis are discussed from the point of view of advancing into an unexplored region of heavier (superheavy) and significantly longer-lived nuclides close to the spherical shells $Z = 114$ and $N = 184$.

Experiments are described and first results are presented on the synthesis of superheavy nuclei in ^{48}Ca-induced reactions. Presented are also the observed decay chains of individual atoms consisting of sequential α-decays and terminated by spontaneous fission. The energies and half-lives are in agreement with the predictions of theoretical models describing the structure of heavy nuclei. They are considered as a first evidence of the existence of the hypothetical region of stability of superheavy elements.

The experiments were carried out at the FLNR heavy ion accelerator in the framework of a large collaboration with LLNL (Livermore), GSI (Darmstadt), RIKEN (Saitama), the Institute of Physics and Department of Physics of the Comenius University (Bratislava) and the Department of Physics of the University in Messina.

"ISLANDS OF STABILITY" IN THE REGION OF SUPERHEAVY NUCLIDES

One of the fundamental consequences of modern nuclear theory is the prediction of the islands of stability in the region of superheavy elements.

According to the classical approach, the transition into the region of extremely heavy nuclear masses and high charges is connected with a substantial loss of their stability against

---

[1] The present paper is a revised version, containing new data, of the talk "The synthesis and decay properties of the heaviest elements", presented at the VII International Conference on Nucleus-Nucleus Collisions, Strasbourg, July 3-7, 2000.

spontaneous fission. The existence of regions of stability of extremely heavy nuclei is caused entirely by the quantum effect of nuclear shells. Calculations of nuclear masses and deformations lately made using macro-microscopic approaches [1-3] and the determination of the nuclear shell amplitude using Strutinsky's method have confirmed the predictions made earlier on the existence of a quite stable doubly magic spherical nucleus $^{298}_{184}114$ which follows right after $^{208}_{82}$Pb [4]. The results of calculations presented in Figure 1 demonstrate the influence of the shell effects on the stability of heavy nuclei.

We shall consider the spontaneous fission probability. The probability of spontaneous fission strongly depends on the amplitude of the shell correction. The significant rise in $T_{SF}(N)$ when moving away from the $N = 152$ shell, which manifests itself noticeably in the radioactive properties of the actinide nuclei, is due to the influence of another neutron shell at $N = 162$. These two shells are related to deformed nuclei. The maximum stability with respect to spontaneous fission is expected for the nucleus $^{270}108$ ($N = 162$) for which the predicted T_{SF} value can amount to several hours. When the neutron number increases, the ground-state deformation of the nucleus decreases due to the moving away from the deformed shell $N = 162$. At $N > 170$ a significant rise of T_{SF} is expected for nuclei up to $^{292}108$ ($N = 184$), whose partial half-life with respect to spontaneous fission is as long as $T_{SF} \approx 3 \cdot 10^4$ years.

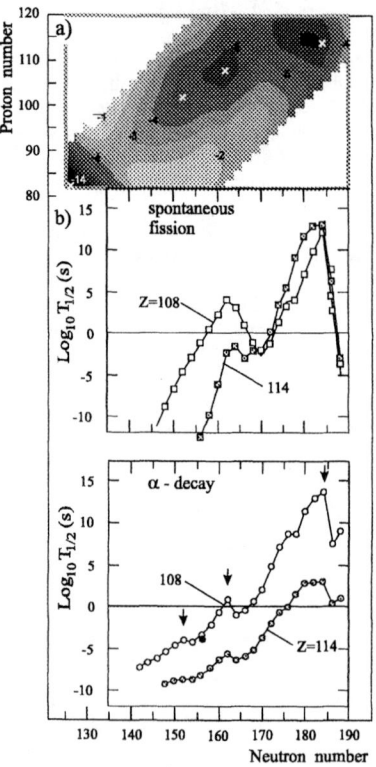

FIGURE 1. a) Map of the shell corrections (in MeV) to the nuclear liquid drop potential energy.
b) The open squares and circles denote calculated half-lives with respect to α-decay and spontaneous fission, the black points – experimental data.

Here we come across a very interesting situation.

If superheavy nuclei possess high stability with respect to spontaneous fission, they will undergo other modes of decay – α-decay and, perhaps, β-decay. The probability for these modes of decay will be determined by the nuclear masses in the ground state. The latter can be calculated by different models, which are based on different assumptions about the fundamental properties of nuclear matter. In these circumstances any experimental result becomes extremely informative for verifying the theoretical models. Following the macro-microscopic calculations [1] the deformed nucleus $^{268}106$ (N = 162) should undergo α-decay with a half-life of $T_\alpha \approx 2$ h. (According to the calculations in Ref. [3] – a few days). For the heavier spherical nucleus $^{294}110$ (N = 184), T_α increases to several hundred years. It is worthwhile noting that in the absence of nuclear structure (in the liquid drop model) this nucleus should fission spontaneously with $T_{SF} < 10^{-19}$ s. The difference is, as we can see, about 30 orders of magnitude!

Calculations of the energy of the nucleus as a many-body system, carried out in the Hartree-Fock approach, as well as calculations in the relativistic mean field model also indicate a significant increase in the binding energy of the nucleus when approaching the closed neutron shell N = 184.

There is yet no agreement among theoreticians regarding the magic proton number for which the maximum binding energy of the spherical nucleus is realized, while the neutron number considered is N = 184. In macro-microscopic models the shell correction amplitude has a maximum value for the nucleus $^{298}114$, irrespective of the variation of the parameters used in the calculations. On the contrary, the calculations performed using the method of Hartree-Fock employing Skyrme effective interaction of the Sly4 (HF+Sly4) type [5], or using a self-consistent relativistic mean-field model with the parameterization of the NL-2Z type (RMF+NL-72) [6] the proton shells are predicted at Z = 120 and 126. This, however, does not change the main conclusion that in the region of very heavy nuclei there may exist "islands of stability", which in turn substantially can extend the limits of existence of superheavy elements.

PRODUCTION OF SUPERHEAVY NUCLEI

It is well known that the first artificial elements heavier than uranium were synthesized in reactions of sequential capture of neutrons during long exposures at high-flux reactors. The long lifetime of the new nuclides made their separation and identification possible using radiochemical methods followed by the measurement of their radioactive decay properties. This pioneering work, which was performed in the Lawrence Berkeley National Laboratory (USA) by Prof. G.Seaborg and colleagues in the period of 1940–1953, led to the discovery of 8 artificial elements with Z = 93 ÷ 100. The heaviest nucleus was ^{257}Fm ($T_{1/2} \sim 100$ d). The further advance into the region of heavier nuclei was blocked by the extremely short lifetime of ^{258}Fm ($T_{SF} \sim 0.3$ ms). The attempts to overcome this barrier by means of pulsed neutron fluxes (underground nuclear explosions) also did not go beyond observing ^{257}Fm.

The transfermium elements with mass A > 257 were produced in heavy-ion induced reactions. The interaction of massive nuclei turned out to be a complicated process, characterized by a large number of reaction channels. Only one of them - complete fusion - can lead to the formation of a superheavy nucleus.

The formation cross section of the final nuclei in this channel is defined by the probability of formation and survival of the compound nucleus in the whole available energy region. Having this in mind, fusion reactions can be conditionally divided into two kinds:
- "cold fusion reactions", based on the use of ^{208}Pb or ^{209}Bi targets and characterized by a low excitation energy ($E_x \sim$ 15-20 MeV);
- "hot fusion reactions" in which heavier nuclei, such as isotopes of uranium or transuranium elements, are used as targets ($E_x \sim$ 30-50 MeV).

We shall start our discussion with the final results. For this purpose in Figure 2 the results are presented on the production cross section of evaporation residues (EVR), obtained in the two reaction types. We shall try to estimate the potentialities inherent in both methods for the synthesis of nuclei with $Z \geq 112$.

As a whole – this is a difficult task. The theoretical models based on different assumptions on the fusion dynamics of massive nuclei leading to the formation of new elements are tested only by the final results – the rare events of radioactive decay of evaporation residues. The separation of the two stages - the compound nucleus formation and its decay by neutron emission - would be very helpful in providing more information for the understanding of the reaction process. The compound nucleus formation cross section can be in principle determined by the characteristics of its main decay mode – the fission into two fragments. Such attempts were made recently in the experiments carried out in FLNR (Dubna) in measurements of the angular correlations, mass and energy distributions of the fission fragments as well as, in some cases, of the fission neutrons [7,8].

As a reference reaction, we shall consider the reaction ^{48}Ca + ^{206,208}Pb, leading to the formation of isotopes of No (Z = 102), whose fission barrier height is determined mainly by the amplitude of the shell correction to the nuclear deformation energy.

The two-dimensional mass and kinetic energy spectra presented in Figure 3a for the correlated fragments indicate that there are two different channels for their formation: asymmetric mass distribution due to the massive transfer of nucleons from the target nucleus to the projectile and symmetric distribution arising as a result of the fission of the compound nucleus with Z = 102. Comparing the dependence $\sigma_{CN}(E_x)$ with the cross sections σ_{EVR} in the

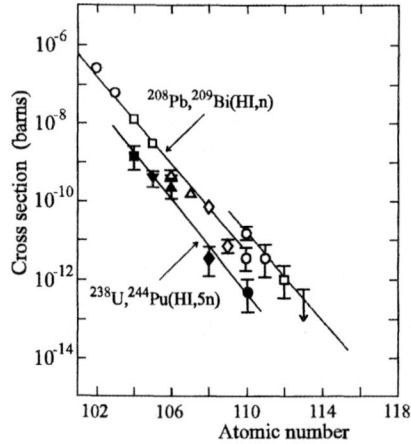

FIGURE 2. EVR cross sections for cold and hot fusion reactions.

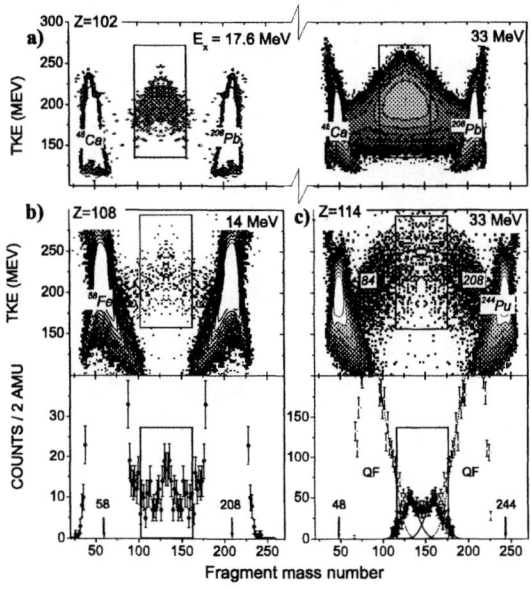

FIGURE 3. TKE vs A_f for correlated fragments, obtained in the reactions:
a) ^{48}Ca + ^{208}Pb, b) ^{58}Fe + ^{208}Pb, c) ^{48}Ca + ^{244}Pu.
The events marked with a square are attributed to the fission of a compound nucleus.

xn-channels (x = 1– 4), it is possible to determine the survival probability of the compound nuclei ^{254,256}No [9] in a wide excitation energy range (Figure 4). On the other hand, the survival probability could also be calculated within the statistical model. Further, the dependence $B_f(E_x)$ can be determined from the best fit between calculation and experiment. The use of cold fusion reactions for the advance into the region of heavier elements brings forth a considerable drop in σ_{CN}. When the projectile mass increases by 10 a.m.u. (i.e., ^{58}Fe instead of ^{48}Ca), only a small part of the fragments in the region of symmetric masses can be attributed to the fission of the compound nucleus ^{266}Hs (Z = 108) (Figure 3b). At the same time, the experimentally observed drop by 3 orders of magnitude in the cross section for ^{265}Hs in comparison to ^{255}No (both nuclei are formed in the 1n-evaporation channel at E_x ~15 MeV) is due to the small survival probability of the heavy nucleus (Figure 5). The survival probability, as it is well known, is defined by the neutron binding energy in the nucleus ^{266}Hs (B_n = 8.03 MeV and the height of its fission barrier B_f ~ 5 MeV).

In the synthesis of heavier nuclei this trend is kept. In the reaction ^{86}Kr + ^{208}Pb only the upper limit of σ_{CN} was obtained for ^{294}118 at the level of $\sigma_{CN} \leq 0.5$ µb [10]. We should note that this limit is approximately 200 times less than the cross section for the formation of the compound nucleus ^{266}Hs [12] at the same excitation energy E_x ~ 14 MeV.

The other means - the transition to more asymmetric reactions - is connected to the increase of the target mass. The transition from ^{208}Pb to the isotopes of actinide elements such as ^{238}U and ^{244}Pu (the target mass is increased by 30-36 a.m.u.) qualitatively changes the picture of the fusion process.

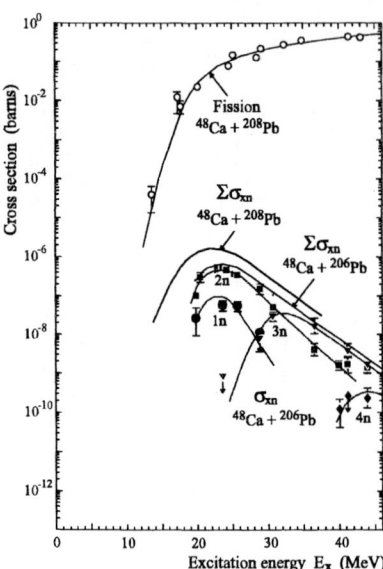

FIGURE 4. Cross sections for compound nucleus (fission) and evaporation residues, obtained in the ^{48}Ca + ^{206,208}Pb reaction. The lines are drawn through the experimental points to guide the eye.

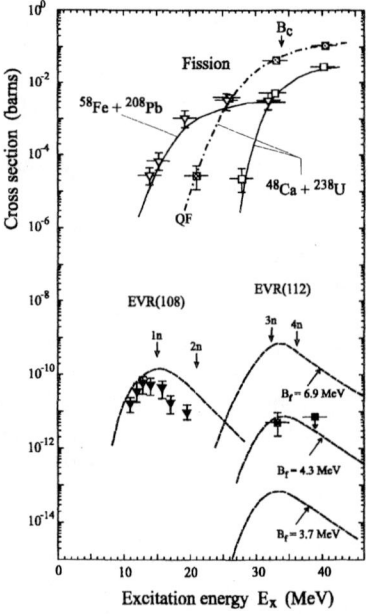

FIGURE 5. Experimental cross sections, obtained in fusion-fission reactions with ^{208}Pb-targets and ^{48}Ca, ^{58}Fe-beams at different excitation energies. Open symbols denote symmetric fission; black symbols - EVR; dot-dashed line – quasifission; dashed lines - calculated cross sections for EVR.

The main part of the fission fragments, as it is shown in Figure 3c, is connected to the asymmetric quasifission of the nucleus $^{292}114$ into two fragments with masses close to $A_1 = 208$ and $A_2 = 84$. Only a small part of the fragments in the region of the medium mass nuclei can be attributed to the fission of the compound nucleus $^{292}114$. In this region the fission fragment mass distribution is also asymmetric with maxima close to $A_{f1} = 132$ and $A_{f2} = 160$. The asymmetry in the fission of the superheavy nucleus is most probably due to the shells $Z = 50$ and $N = 82$ in the ^{132}Sn nucleus, formed as a light fragment. A similar situation, as it is well known, is observed also in the fission of actinide elements ($Z = 90 - 98$) with the only difference in that the shell effect manifests itself in the formation of the heavy fragment.

In asymmetric hot fusion reactions, where the compound nucleus excitation energy even at the Coulomb barrier amounts to 45-50 MeV, the survival probability strongly depends on E_x. For this reason, the choice of the nuclei participating in hot fusion reactions leading to compound nuclei with minimum excitation energy is of great importance. Most promising from this point of view seem to be the fusion reactions induced by ^{48}Ca-ions. Due to the considerable mass excess of the doubly magic nucleus ^{48}Ca, the excitation energy of the compound nuclei with $Z = 112$, 114 at the Coulomb barrier amounts to about 30 MeV. However, even in these favourable conditions σ_{EVR} is very sensitive to the fission barrier of the compound nucleus. Following the calculations of Meyers and Swiatecki [11] predicting for the $^{286}112$ nucleus a fission barrier of about 6.9 MeV, we can expect that the maximum cross sections σ_{EVR} in the reaction ^{238}U(^{48}Ca,3-4n)$^{283,282}112$ may reach hundreds of *pb*. However, σ_{EVR} decreases by almost 3 orders of magnitude for $B_f \sim 3.7$ MeV, obtained in the macro-microscopic calculations of R.Smolanczuk [2]. This may be the reason why previous attempts to produce superheavy elements in ^{48}Ca-induced reactions, for which the production cross sections amounted to hundreds of picobarn, have not be successful [12-14]. Therefore, for the synthesis of superheavy elements the sensitivity of the new experiments in comparison to earlier ones must be improved by hundreds of times.

EXPERIMENTAL APPROACH

The planning of experiments on the synthesis of heaviest elements is determined to a great extent by the radioactive properties of the atoms to be synthesized. The lifetime can vary in a wide range depending on how justified are the predictions concerning the influence of nuclear shells on the stability of heavy nuclides with different Z and A. Hence the experimental set-up should be sufficiently fast. On the other hand, the evaporation residues, whose yield is extremely small, should be quickly separated from the enormous background of incidental reaction products, which are formed with 8-10 orders of magnitude higher probability.

These conditions can be achieved if the separation of the products is performed in-flight (during $10^{-6} \div 10^{-5}$ s), taking into account the kinematical characteristics of the reaction channels.

The efficiency of the kinematic separators depends on the ratio of the masses of the interacting nuclei. For fusion reactions induced by relatively light projectiles ($A_p \leq 20$) it amounts to only a few percent, but increases to $30 \div 50\%$ when going to ions with mass $A_p \geq 40$. The experimental set-ups have also high selectivity. In the focal plane practically all the background from the primary beam is eliminated and the products of incomplete fusion reactions are suppressed by a factor of $10^4 \div 10^7$, depending on the kinematical characteristics

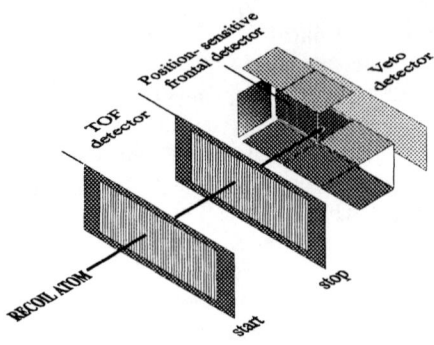

FIGURE 6. The detector array used as a focal plane detector.

of the different channels leading to their formation. This is, however, not enough for the identification of the extremely rare events corresponding to the production of atoms of a new element. Therefore, the selection of the sought nuclei is further accomplished with a sophisticated registration device (Figure 6).

The recoil atoms, which have reached the focal plane, are implanted into a multistrip silicon semiconductor detector with an active area of $\sim 40 \div 50$ cm^2. Each strip has longitudinal position sensitivity. The position resolution depends on the particle type (recoil nucleus, α-particle or spontaneous fission fragments). However, more than 95% of all particles, accompanying the decay of the implanted atom, are confined in an interval $\Delta_x \sim 1.0 - 1.5$ mm. The front detector is surrounded by side detectors so that the entire array has the shape of a box with an open front wall. The detection efficiency for particles resulting from the decay of the implanted nucleus (α-particles or two fission fragments) is increased to $85 \div 87\%$. For distinguishing between the signals of the recoil nucleus and those belonging to the particles from its decay, a time-of-flight (TOF) detector is situated before the front detector. The signals from the TOF detector are used also for determining the velocity of the implants.

The selection of the events according to their generic decay significantly improves the background conditions. The parent nucleus, implanted into the detector, can be reliably identified if the decay chain of its sequential α- and β-decays leads to nuclei with known properties. This method has been successfully used in the experiments on the synthesis of new elements with $Z = 107 \div 112$ whose isotopes have insignificant neutron excess (N-Z ≤ 53). Advancing into the region of spherical, more neutron-rich nuclei this advantage is lost. Here the decay of the parent nucleus results in the formation of hitherto unknown nuclei, whose properties can be only predicted with precision allowed by the theoretical calculations.

At the same time, if the basic theoretical prediction on the existence of the "island of stability" is justified, in any decay chain of sequential α- and β-decays the daughter nuclei move more and more away from the closed spherical shells. Finally the decay chains will be terminated by spontaneously-fissioning nuclei. In principle, such a decay scheme appears to be a reliable sign of the formation of a heaviest nucleus. From a technical point of view the appearance of such an event should strongly differ from other possible correlated decays.

ENHANCED STABILITY OF NUCLEI CLOSE TO THE DEFORMED SHELLS Z = 108, N = 162

The synthesis and the decay properties of nuclides in this region as a matter of fact provide the first test of the theoretical concepts on the role of shell effects in nuclei, where the liquid drop fission barrier is practically absent.

Most interesting for this purpose seem to be the even-even isotopes of Sg (Z = 106), which can be synthesized with relatively high cross section ($\geq$ 100 pb) in cold and hot fusion reactions. For nuclei with Z = 106 situated between the two shells N = 152 and N = 162, the increase of the neutron number should, according to theory [1], considerably enhance their stability with respect to both α-decay and spontaneous fission (Figure 7).

For the synthesis of the heavy isotopes of element 106 we chose the reaction ^{22}Ne + ^{248}Cm at a beam energy close to the Coulomb barrier, where the maximum cross section expected is for the 4n- and 5n-evaporation channels. In the irradiation of the ^{248}Cm-target, at the gas-filled separator two isotopes of element 106 with masses 265 and 266 were synthesized [15].

The α-decay energy of the even-even nuclide ^{266}Sg (Q_α = 8.76 MeV) determines its half-life $T_{1/2}$ = 20 ± 10 s. On the basis of 6 registered correlated events (α – SF), observed in the decay of the nucleus ^{266}Sg, the half-life of the daughter nucleus ^{262}Rf (Z = 104, N = 158) was also determined. It undergoes spontaneous fission with T_{SF} ~ 1.2 s. The radioactive properties of the new nuclides indicate a considerable rise in the stability of heavy nuclides with respect to spontaneous fission when approaching the closed shells Z = 108 and N = 162 (Figure 7) and quantitatively agree with the macro-microscopic calculations of Patik et al. [3].

The data obtained for the Z = 106 nuclei force the conclusion that for even heavier nuclei with Z > 106 the fission modes will not bring forth a revision of the optimistic theoretical predictions on the existence of a wide region of stability of heavy nuclei. They will predominantly undergo α-decay as long as the shell corrections to the deformation energy will restrain spontaneous fission. These conclusions are confirmed by the experimental investigations, carried out at GSI, where many new α-radioactive isotopes with atomic number

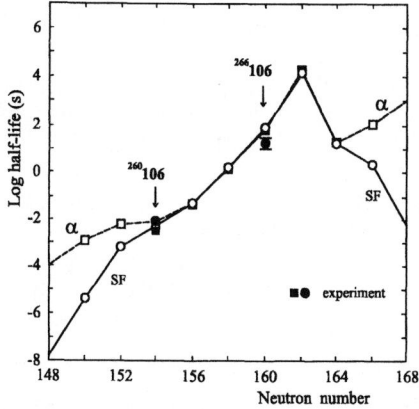

FIGURE 7. Half-lives for α-decay and spontaneous fission for the Sg-isotopes (Z = 106). Black symbols – experiment; open symbols – calculated values [1].

up to Z = 112 were synthesized and their properties determined [16-18]. To an equal extent these conclusions are true also for the heaviest isotopes with Z = 106 - 110, obtained in FLNR in hot fusion reactions [15,19,20].

With the increase of the neutron number (going away from N = 162), the spontaneous fission probability will be rising up to the moment when the stabilizing effect of the next shells will start to play a role.

TRANSITION TO THE NEUTRON-RICH NUCLEI IN THE ^{48}Ca + ^{244}Pu REACTION

According to the macro-microscopic theory, the maximum contribution of nuclear shells appears at Z = 114 and N = 184 (Figure 1). No combination of stable or even long-lived isotopes is able to bring us to the very neutron-rich isotopes of element 114. One may hope to approach at least the boundaries of this unknown region so as to come under the influence of the N = 184 spherical shell. For this purpose the fusion reaction ^{48}Ca + ^{244}Pu was chosen, as it leads to the formation of the compound nucleus ^{292}114 (N = 178). The experiments were performed at the Dubna Gas-Filled Recoil Separator.

The target consisted of the enriched isotope ^{244}Pu (98.6 %), supplied by our collaborators from LLNL (Livermore). The energy of the ^{48}Ca-beam in the middle of the target was chosen equal to 236 MeV; the excitation energy of the compound nuclei was in the range 31.5 to 39 MeV. The average intensity of the ^{48}Ca-beam on the target was about $4 \cdot 10^{12}$ pps at a rate of material consumption equal to about 0.3 mg h^{-1}.

At a beam dose of $1 \cdot 10^{19}$ ions two identical α-decay chains terminated by spontaneous fission were registered [21]. The genetic relation in the chain was determined according to their position. All 4 signals (EVR, α_1, α_2, SF) appeared within a position interval of 0.5 mm, which indicates that there is a strict correlation between the observed decays (Figure 8a). In the limits of the detector energy resolution and the statistical uncertainty in the decay times, the two events coincide in all measured parameters. The probability of random coincidences imitating recoil nuclei and their correlated decays is estimated to be less than $5 \cdot 10^{-13}$.

We should note that in this long-term experiment only two spontaneous fission events were observed: they are characterized by a large value of the fission fragment kinetic energy (TKE ~ 235 MeV) and both are preceded by the same α-decay sequences. The projectile energy, measured at the moment of registration of these events, corresponds to an excitation energy of the compound ^{292}114 nucleus equal to E_x = 37 ± 2 MeV. At this energy the most probable 4n-evaporation channel leads to the formation of the isotope ^{288}114 (N=174). Evaporation channels accompanied by the emission of charged particles (protons, α-particles) are strongly suppressed due to the high Coulomb barrier of the heavy nucleus. Channels leading to the formation of heavy spontaneously fissioning isotopes with Z ≥ 98 are also hindered.

The parent nucleus ^{288}114 undergoes α-decay with a decay energy Q_α= 9.98 ± 0.05 MeV and a half-life $T_\alpha = 1.8^{+2.1}_{-0.6}$ s. For the even-even nuclei the value of Q_α is directly connected with the mass difference between the parent and daughter nuclei. According to the Geiger-Nuttall rule, the energy and probability of α-decay, Q_α and T_α, determine the atomic number of the parent nucleus. Following the relation $T_\alpha(Q_\alpha)$ as a version of the Viola and Seaborg formula with constant coefficients, describing the α-decay of all known 58 even-even isotopes

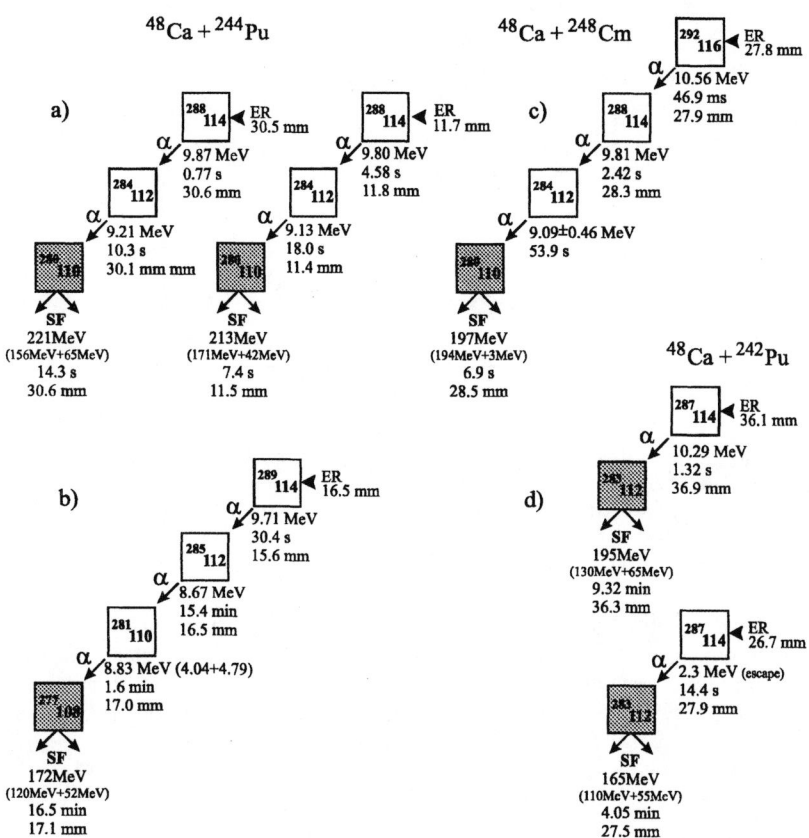

FIGURE 8. Decay sequences, observed in the ^{48}Ca + ^{242,244}Pu, ^{248}Cm reactions. For the SF fragments the energy deposit in the front and side detectors is shown. For all registered signals positions in the strip detector are presented.

with Z > 82 and N > 126 [2], it follows that the first-step α-decay in the observed decay sequences and the third event – from the decay of the ^{292}116 nucleus (see below) is due to the nucleus with Z = $114.4^{+1.2}_{-0.8}$ (Figure 9). The daughter nucleus, the even-even isotope ^{284}112 also undergoes α-decay with Q_α = 9.30 ± 0.05 MeV and a half-life T_α = $19.0^{+22.7}_{-6.7}$ s. Finally, the "grandchild", the nucleus ^{280}110 undergoes spontaneous fission with a half-life T_{SF} = $7.5^{+13.7}_{-2.9}$ s. For the two observed binary events the fission fragment energy release in the detectors is about 40 MeV higher than the value obtained in the reaction ^{208}Pb(^{48}Ca,2n) for the known spontaneously fissioning nucleus ^{252}No. Regardless of the relatively wide fragment kinetic energy distributions (TKE) manifested in spontaneous fission, this value also gives evidence that the "grandchildren" in the decay chain are the result of the fission of a sufficiently heavy nucleus.

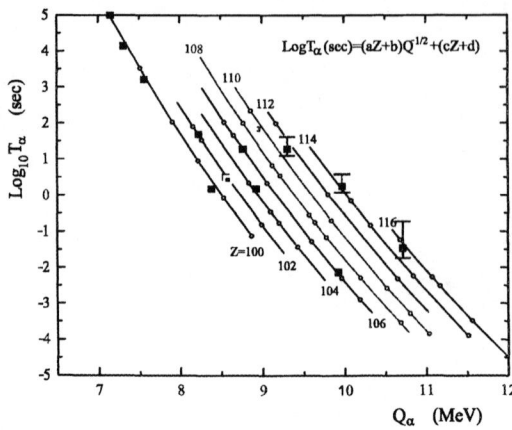

FIGURE 9. The experimental values of Q_α and T_α (black squares) for the even-even isotopes with $Z \geq 100$. The squares with error bars refer to the data, obtained in the experiments in which the ^{244}Pu, ^{248}Cm + ^{48}Ca reactions were used. The lines and dots denote the calculated dependence $T_\alpha(Q_\alpha)$, obtained using the formula of Viola and Seaborg with coefficients from ref. [2].

In a second experiment (chronologically, it was the first one), with a beam dose of $5.2 \cdot 10^{18}$ ions, another longer α-decay sequence was observed terminating by spontaneous fission [22]. Applying the same criteria as in the previous case, we came to the conclusion that this event was due to the decay of the neighbouring isotope of element 114. In this decay chain, all 5 signals (recoil nucleus, α_1, α_2, α_3, SF) appeared within a position interval of 1.5 mm, which also is a strong indication of the correlation between the observed decays (Figure 8b). The total time between the implantation of the recoil nucleus and the spontaneous fission amounts to 34 min. The probability of random coincidences imitating such a decay in any point of the working area of the detector was less than 0.6%. For that place, where the decay occurred (the given position window in the considered strip), this probability was even less ($\sim 10^{-4}$).

In the experiment ^{48}Ca + ^{248}Cm a long chain of sequential α-decays, terminated by spontaneous fission, was also observed. Unfortunately, the first α-transition, expected for the even-odd nucleus ^{293}116, did not give a signal in the detector, although the consequent 3 α-decays and the spontaneous fission were registered. The energies and the decay times in the chain are in good agreement with the ones given above for the ^{289}114 nucleus, which was produced in the ^{48}Ca + ^{244}Pu reaction. We present this information as a preliminary result of the analysis of the experimental data from the ^{48}Ca + ^{248}Cm experiment, which is still going on.

The decay properties of the neighbouring odd isotope ^{289}114 agree well with the above-mentioned properties of the even-even nucleus ^{288}114. For this nucleus, as it was expected, an increase is observed in the halflives T_α and T_{SF} that is due to the additional odd neutron.

The production cross section of the new nuclides amounts to about 1 pb. In spite of the small value of the cross section, obtained in both experiments, background from spontaneous fission is practically negligible. At a total beam dose of $1.5 \cdot 10^{19}$ ions, in addition to the observed decays only two decay events were registered, which could be attributed to the short-lived spontaneously fissioning nucleus ^{244m}Am ($T_{SF} = 0.9$ ms) with energy release in the detectors amounting to 149 and 153 MeV. From this point of view the sensitivity of the experiment can in the future be raised by increasing the intensity of the ^{48}Ca-beam.

EXPERIMENTS WITH THE ^{238}U AND ^{242}Pu TARGETS

If the identification performed in the preceding two experiments with the ^{244}Pu target is correct, it is not difficult to predict the properties of another isotope, viz. ^{287}114 (N = 173). It should predominantly undergo α-decay to the daughter nucleus ^{283}112, which was formed in a previous experiment using the ^{48}Ca + ^{238}U reaction [23]. In the given case we can expect a short decay sequence (α-SF), implying a few-seconds α-decay half-life, followed by spontaneous fission with considerably longer (of the order of minutes) half-life. This isotope of element 114 can be synthesized via the 3n-evaporation channel in the ^{48}Ca + ^{242}Pu reaction.

The experiments were performed at the VASSILISSA separator [24]. The rotating ^{242}Pu target was bombarded by a 235 MeV ^{48}Ca-beam with a total beam dose of $7.5 \cdot 10^{18}$ ions. The most probable deexcitation channel of the compound nucleus ^{290}114 ($E_x \approx 33.5$ MeV), corresponding to the emission of 3 neutrons, leads to the formation of the even-odd isotope ^{287}114 (N = 173).

In this experiment 4 spontaneous fission events were observed.

Two spontaneous fission events were registered 59 and 20 μs after the implantation of corresponding position-correlated recoil atoms, respectively. We could assign these events to the SF-isomer ^{241mf}Pu(24-μs), produced in the one-neutron transfer reaction on the ^{242}Pu-target nucleus. In the case of the two other events, the search for α-decays preceding the spontaneous fission revealed the two decay chains presented in Figure 8d.

In one of the sequences only one α-particle (E_α = 10.29 MeV) was detected by the front detector 1.32 s after the implantation of a heavy recoil. Spontaneous fission was observed 9.32 min later. All three signals (recoil nucleus, α, SF) appeared within a position interval of 0.82 mm, which indicated that there is a correlation between the observed decays. The second decay chain is similar to the previous one with the exception that the detector registered only a part of the α-particle energy emitted in the back hemisphere (the open window). The probability that both events are due to random coincidences imitating the decay chains (recoil nucleus, α, SF) in the given position intervals is less than 10^{-4}.

In both events the parent nucleus undergoes α-decay. The half-life of the parent nucleus, derived on the basis of the two events, amounts to $T_\alpha = 5.5^{+10}_{-2}$ s. The detected daughter nuclei undergo spontaneous fission. Their decay properties are comparable with the ones of the spontaneously-fissioning nuclide produced earlier in the ^{48}Ca + ^{238}U reaction [23]. All four spontaneous fission events, observed in these two reactions, within the experimental error, can be described by the same half-life $T_{SF} = 3.0^{+2.8}_{-1.0}$ min and can be attributed to the decay of one and the same nucleus. In the ^{48}Ca + ^{238}U reaction this nuclide was produced directly as an EVR in the 3n-evaporation channel, while in the ^{48}Ca + ^{242}Pu reaction it is the daughter of the α-decay of the parent ^{287}114 nucleus (E_α = 10.29 MeV).

The production cross section of the new isotope of element 114 amounts to about 2 pb.

Its half-life and the decay sequence are shorter than the ones of the previously observed heavier isotope ^{289}114, formed in the reaction ^{48}Ca + ^{244}Pu. Such a trend is expected, according to theory, with the decrease of the neutron number of the superheavy nucleus, or in other words with moving away from the closed N = 184 shell.

OBSERVATION OF THE DECAY OF ^{292}116

On June 14, 2000, an experiment aimed at the synthesis of superheavy nuclei with Z=116 in the complete fusion reaction ^{248}Cm + ^{48}Ca was started at FLNR.

The transition from the Pu to the ^{248}Cm target is not simple. Unlike the ^{48}Ca + ^{244}Pu reaction, in which close to the Coulomb barrier the formation of spontaneously fissioning nuclei - the isotopes ^{252,254}Cf - is strongly suppressed (this corresponds to the transfer of 4 protons and 6-8 neutrons from the projectile to the target nucleus), in the case of the ^{48}Ca + ^{248}Cm reaction, these isotopes can be produced with significantly higher probability via the channels where the transfer of ^{4}He and ^{6}He takes place.

Since in the earlier experiments, performed in LBL (Berkeley) and at GSI (Darmstadt), considerable background due to spontaneous fission had been observed, here we paid a lot of attention to this problem.

In Figure 10, the fission fragment spectrum, obtained during a 45-day irradiation of ^{248}Cm with ^{48}Ca-ions (beam dose ~9·10^{18}), is presented. In this experiment 11 spontaneous fission events were observed. After the experiment, the measurements were continued. During 78 days 11 more events were detected. This indicated that spontaneously fissioning nuclei with the half-life expected for ^{254}Cf ($T_{s.f.}$ = 60 d) had been produced. Thus the background due to fragments coming from spontaneous fission amounted to no more than 1s.f./4days, that should not disturb the observation of the decay chains of superheavy nuclei lasting up to 1 hour and even longer.

The decay products of the isotopes ^{292,293}116 are the isotopes ^{288,289}114, which had been observed in the experiments ^{48}Ca + ^{244}Pu as described earlier. Thus, if the nuclei with Z = 116 were formed, we should observe the emission of α-particles with an energy higher than 10 MeV and after that all the links of the chains connected with the decay of one of the two known isotopes with Z = 114.

For this reason, in the experiment aimed at the synthesis of element 116 a new algorithm was introduced in the measurements. After the detection of the recoil nucleus in the front detector and the consequent emission in a given position window of an α-particle with energy $E_\alpha \geq 10.2$ MeV the beam was switched off. The following decay was then observed in the

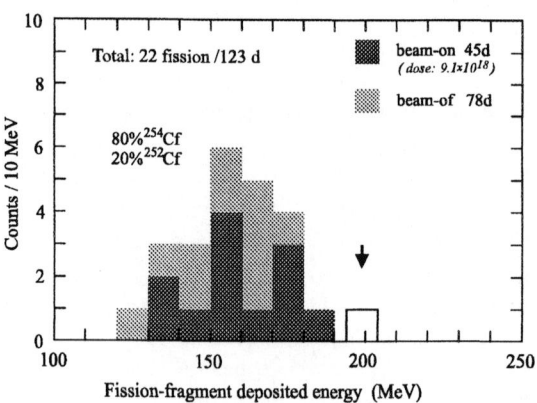

FIGURE 10. Total energy spectrum of the spontaneous-fission fragments, registered by the detector array in the ^{48}Ca + ^{248}Cm reaction.

absence of background signals, originating from the presence of the beam on the target. The counting rate of α-particles lying in the energy interval $8.0 \leq E_\alpha \leq 10.0$ MeV was less than 10^{-3} h^{-1}.

On the 35-th day of irradiation, with the accumulated beam dose of 6.6×10^{18} ions, the first event sequence was observed, that could be assigned to the implantation and decay of the isotope of element 116 with mass number 292, see Figure 8c.

Implantation of a heavy recoil in the focal-plane detector was followed, in 46.9 ms, by an α-particle with $E_\alpha=10.56$ MeV. This sequence switched the ion beam off, for one hour and further decays were detected under lower-background conditions. A second α-particle with $E_\alpha=9.81$ MeV was observed 2.42 s later. Then, in 53.87 s the third α-decay was registered by the side detector with the energy of 8.63 MeV. The energy deposited by this α-particle in the focal-plane detector was lower than the detection threshold of 0.92 MeV. Thus its total energy was determined with larger uncertainty, $E_\alpha=9.09\pm0.46$ MeV; the probability that the third α-particle appeared in the chain (Δt~1 min) due to random count could be estimated as ~1%.

Finally, 6.93 s after the last α-decay, two coincident fission fragments with sum energy of 197 MeV were registered by both the focal-plane and the side detectors. The low energy of one fission fragment measured by the side detector for this event meant large energy loss by this fragment in the dead layers.

The positions of the four events (EVR, $α_1$, $α_2$, and SF) were measured to be within a window of about 0.5 mm, all events appeared within time interval of 63.26 s, which pointed to a strong correlation between them. The probability of the random origin of the observed event chain is negligible ($<<10^{-6}$).

All the decays following the first 10.56-MeV α-particle agree well with the decay chains of ^{288}114, previously observed in the ^{244}Pu+^{48}Ca reaction, see Figure 8a.

Thus, it is reasonable to assign the observed decay to the nuclide ^{292}116, produced via evaporation of 4 neutrons in the complete-fusion reaction ^{248}Cm+^{48}Ca. The decay energy of the newly observed nuclide is Q = 10.71 MeV, its half-life estimated from one event is $T_\alpha = 33^{+155}_{-15}$ ms.

Experiments are in progress.

THEORY AND EXPERIMENT

The data on the radioactive decay of nuclides, produced in the given series of investigations, can be compared with the predictions of different theoretical models.

Unfortunately, in many cases the calculations go no further than look for the region of greatest stability of superheavy nuclides without giving the definite properties of nuclei comprising this region. Therefore we shall consider only those few cases, which can be directly applied for comparison with the experiment.

Figure 11 shows the half-lives of even-even isotopes with Z = 110, 112, 114 and 116, calculated in the macro-microscopic model [1,2] and the experimental values, obtained in cold fusion reactions and in reactions induced by ^{48}Ca ions.

First we would like to point out that the heaviest isotopes with Z = 110, 112 and 114, produced in the ^{48}Ca-induced reactions, undergo α-decay. Spontaneous fission in this region of nuclei is observed only for elements when (N-Z) ≤ 61.

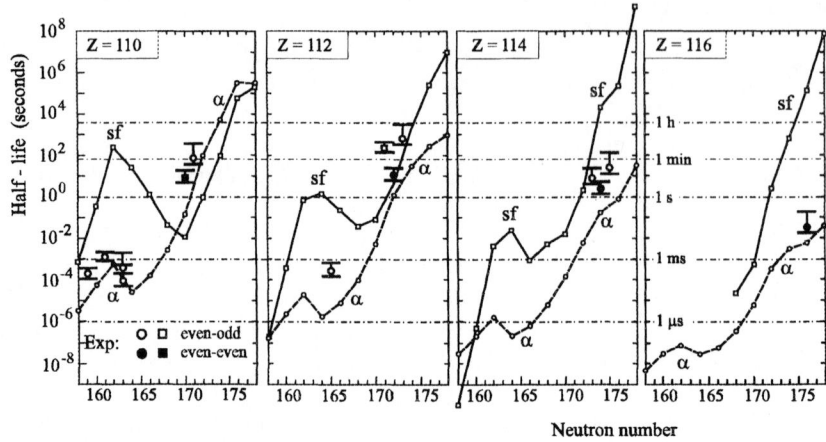

FIGURE 11. Calculated and experimental α-decay (dashed lines) and spontaneous fission (solid lines) half-lives for even-even isotopes of elements 110-116. Open and black squares denote spontaneous fission, the dots - α-decay.

For the even-odd nuclei ^{277}108 (N = 169) and ^{283}112 (N = 171) the spontaneous fission half-lives turn out to be respectively 5 and 3 orders of magnitude higher than the predicted values for the neighbouring even-even nuclei. Such a difference can be explained by the presence of the extra neutron, which significantly diminishes the probability for spontaneous fission of the heavy nucleus. For the even-even nucleus ^{280}110 (N = 170) the experimental value ($T_{SF} \sim 10$ s) is also about 3 orders of magnitude longer than the calculated one, obtained in ref. [1]. Keeping in mind the uncertainty in calculating the probability for spontaneous fission, connected with tunnelling through the potential barrier, this difference could be taken as an evidence of the larger contribution of the shells to the deformation energy of the nucleus.

Some conclusions can be drawn on the basis of the ground state properties of the observed nuclei (Table 1).

The experimental values of the α-decay energies of all known nuclides with $Z \geq 100$ and $N \geq 148$ are shown in Figure 12, together with the calculated values Q_α, obtained in the macro-microscopic theory, for all even-even isotopes of the same elements [1,2].

Certainly, the experimental results well reflect the expected by theory changes $Q_\alpha(N)$ for different N and Z, including the region of superheavy elements, where a transition from deformed to spherical shells is predicted. Quantitatively, for nuclei located in the transition region between the deformed neutron shells N = 152 and N = 162 a negligible difference $\Delta Q_\alpha \leq 0.2$ MeV is observed between the calculated and experimental values. When going to the spherical shell in the region N = 170 ÷ 175 the difference increases to $\Delta Q_\alpha \leq 0.5$ MeV.

Other calculations, carried out by S.Cwiok, W.Nazarewicz and P.H.Heenen [5] using the Hartree-Fock-Bogoliubov method with selected Skyrme forces, concerned the heaviest nuclide ^{289}114, produced in the ^{48}Ca + ^{244}Pu reaction. Within this approach quite good agreement was obtained between the calculation and experiment ($\Delta Q_\alpha \leq 0.25$ MeV) for the chain ^{289}114 – ^{285}112 – ^{281}110. The calculations of the α-decay energies of heavy nuclei, performed recently by M.Bender [6] in the relativistic mean field theory, where the spin-orbit

TABLE 1. Experimental and Calculated Q_α Values for the α-Decay Chains of the Isotopes Z=110-116.

Z	A	$Q_{exp.}$ (MeV)	$Q_{th.}$			
			YPE+WS	FRDM+FY	SHFB	RMF
110	273	9.87 11.24±0.02 11.36	-	10.09	-	10.82
110	280	≤9.4[a] (SF)	9.8	9.05	9.8	8.98
110	281	8.96±0.18	-	8.55	9.44	8.68
112	277	11.82 11.62±0.02 11.33	-	12.08	-	11.32
112	283	10.39±0.02	-	9.01	-	-
112	284	9.3±0.05	9.8	8.69	9.42	9.3
112	285	8.80±0.05	-	8.59	8.76	9.02
114	288	9.98±0.05	10.3	9.16	9.35	9.83
114	289	9.85±0.05	-	8.87	10.16	9.38
116	292	10.71±0.05	11.07	10.82	10.61	11.65

YPE+WS   – macroscopic-microscopic Yukawa-plus-exponential model with Woods-Saxon single-particle potentials [1,2].
FRDM+FY   – macroscopic-microscopic finite-range droplet model with folded Yukawa single-particle potentials [3].
SHFB   – self-consistent Skyrme-Hartree-Fock-Bogoliubov model with pairing [5].
RMF   – self-consistent relativistic mean-field model [6].
[a] The Q_α limit was calculated from experimental $T_{1/2}$ value using the formula by Viola and Seaborg with parameters [2].

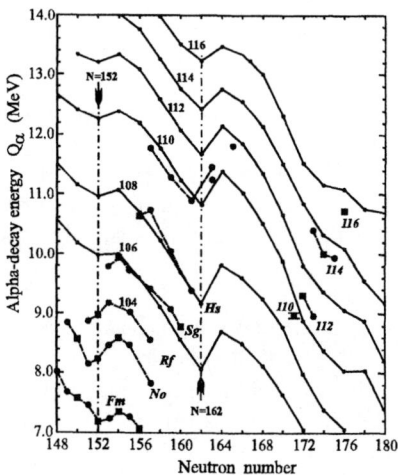

FIGURE 12. Solid lines and open symbols denote calculated values Q_α for even-even isotopes, obtained in the framework of the macro-microscopic theory [1,2]. The black symbols denote the experimental Q_α values; the points – even-odd nuclei, squares – even-even nuclei. The dashed line is drawn through the experimental points to guide the eye.

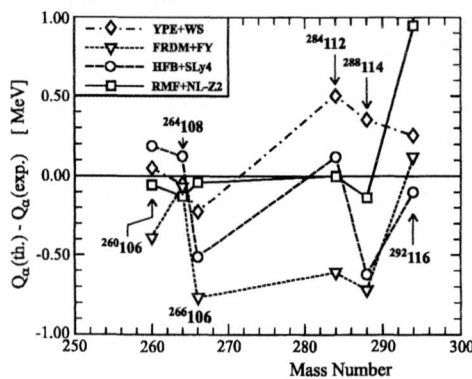

FIGURE 13. The difference $\Delta Q_\alpha = Q_\alpha(\text{th}) - Q_\alpha(\text{exp})$ for the heaviest even-even nuclei with Z = 106 – 114. The line through the points correspond to calculations performed in different models (indicated in the Figure).

interactions of the particles are calculated more precisely, very well agree with the experimental values of Q_α for the sequence of the even-even nuclei $^{288}114 - ^{284}112$, but greatly differs from $Q_\alpha(\text{exp})$ for $^{292}116$. For the decay sequences of the even-odd nucleus $^{289}114$ the calculated Q_α values differ from the experimental ones by $\Delta Q_\alpha \leq 0.3$ MeV.

More definitely a comparison between theoretical calculations within different approaches and experiment, from our point of view, can be made for the even-even isotopes, where ΔQ_α, as it is well known, corresponds to the mass difference in the ground state. As it can be seen in Figure 13, the variation in ΔQ_α for 5 known even-even nuclei with Z = 106 – 114 amounts to about ±0.7 MeV (this is approximately 7% of the calculated energy Q_α). It is clear that theoreticians have the possibility to improve the agreement with experiment, while the experimentalists can add new even-even nuclides to the already known ones. For this purpose, cold fusion reactions can be used as before using ^{207}Pb targets and hot fusion reactions using Th-Cm targets with ^{36}S and ^{48}Ca ion beams.

Finally, it is necessary to point out that the decay properties of the heaviest nuclei that have been synthesized up till now confirm the basic conclusions of the shell model on the decisive role of the nuclear structure on the stability of superheavy elements.

ACKNOWLEDGMENTS

This paper presents the results obtained by a large group of physicists, many of them are co-authors of the original publications. The experiments were performed at the heavy ion accelerator complexes of GSI (Darmstadt) and FLNR (Dubna) in collaboration with LLNL (Livermore), RIKEN (Saitama), the Institute of Physics and Department of Physics of the Comenius University (Bratislava) and the Department of Physics of the University of Messina.

I am taking the opportunity to express my warm gratitude to Profs. W.Greiner, M.Itkis, G.Münzenberg, S.Hofmann, E.K.Hulet, V.I.Zagrebaev, A.Sobiczewski and A.A.Goverdovsky for the interesting and fruitful discussions. I am also grateful to Dr. R.Kalpakchieva for her help in preparing this manuscript.

REFERENCES

1. Patyk, Z., and Sobiczewski, A., *Nucl. Phys.* **A533**, 132 (1991).
2. Smolanczuk, R., *Phys. Rev.* **C56**, 812 (1997).
3. Möller, P., Nix, J.R., *J.Phys.* **G20**, 1681 (1994).
4. Sobiczewski, A., Gareev, F.A., Kalinkin, B.N., *Phys. Lett.* **22**, 500 (1966). Meldner, H., *Ark. Fys.* **36** (1967).
5. Cwiok, S., Nazarewicz, W., Heenen, P.H., *Phys. Rev. Lett.* **83**, 1108 (1999).
6. Bender, M., *Phys. Rev.* **C61**, 031302(R) (2000).
7. Itkis, M.G., Kondratiev, N.A., Kozulin, E.M., Oganessian, Yu.Ts., Pokrovsky, I.V., Prokhorova, E.V., Rusanov, A.Ya., *Phys. Rev.* **C59**, 3172 (1999).
8. Itkis, M.G., Oganessian, Yu.Ts., Kozulin, E.M., Kondratiev, N.A., Pokrovsky, I.V., Prokhorova, E.V., Rusanov, A.Ya., *Nuovo Simento* **111** N 6-7, 783 (1998).
9. Oganessian, Yu.Ts., Utyonkov, V.K., Lobanov, Yu.V., Abdullin, F.Sh., Polyakov, A.N., Shirokovsky, I.V., Tsyganov, Yu.S., Iliev, S., Subbotin, V.G., Sukhov, A.M., Buklanov, G.V., Mezentsev, A.N., Subotic, K., Ivanov, O.V., Moody, K.J., Wild, J.F., Stoyer, N.J., Stoyer, M.A., and Lougheed, R.W., Proceedings of the 4th International Conference on Dynamical Aspects of Nuclear Fission, Casta-Papiernicka, Slovak Republic, 19-23 October 1998, World Scientific, 2000, p.334-348.
10. Itkis, M.G., Oganessian, Yu.Ts., Kozulin, E.M., Kondratiev, N.A., Kliman, J., Itkis, Yu.M., Juadel, M., Krupa, L., Pokrovsky, I.V., Prokhorova, E.V., Rusanov, A.Ya., Hanappe, F., Benoit, B., Rowley, N., Stuttge, L., Giardina, G., Moody, K.J., in *Fission and Properties of Neutron-Rich Nuclei*, edited by S.H.Hamilton, Proceedings of the Second Int. Conf., World Scientific, Singapore, 1999, p.268.
11. Myers, W.D., Swiatecki, W.J., Table of Nuclear Masses According to the 1994 Thomas-Fermi Model, Report No. LBL-36803, Berkeley, 1994.
12. Hulet, E.K., Lougheed, R.W., Wild, J.F., Landrum, J.H., Stevenson, P.C., Ghiorso, A., Nitschke, J.M., Otto, R.J., Morrissey, D.J., Baisden, P.A., Gavin, B.F., Lee, D., Silva, R.J., Fowler, M.M., Seaborg, G.T., *Phys. Rev. Lett.* **39**, 385 (1977).
13. Oganessian, Yu.Ts., Bruchertseifer, H., Buklanov, G.V., Chepigin, V.I., Choi, V.S., Eichler, B., Gavrilov, K.A., Gaeggeler, H., Korotkin, Yu.S., Orlova, O.A., Reetz, T., Seidel, W., Ter-Akopian, G.M., Tretyakova, S.P., Zvara, I., *Nucl. Phys.* **A294**, 213 (1978).
14. Armbruster, P., Agarwal, Y.K., Brüchle, M., Brügger, M., Dufour, J.P., Gaeggeler, H., Hessberger, F.P., Hofmann, S., Lemmertz, P., Münzenberg, G., Poppensieker, K., Reisdorf, W., Schädel, M., Schmidt, K.-H., Schneider, J.H.R., Schneider, W.F.W., Sümmerer, K., Vermeulen, D., Wirth, G., Ghiorso, A., Gregorich, K.E., Lee, D., Leino, M., Moody, K.J., Seaborg, G.T., Welch, R.B., Wilmarth, P., Yashita, S., Frink, C., Greulich, N., Herrmann, G., Hickmann, U., Hildenbrand, N., Kratz, J.V., Trautmann, N., Fowler, M.M., Hoffman, D.C., Daniels, W.R., Von Gunten, H.R., Dornhöfer, H., *Phys. Rev. Lett.* **54**, 406 (1985).
15. Lazarev, Yu.A., Lobanov, Yu.V., Oganessian, Yu.Ts., Utyonkov, V.K., Abdulin, F.Sh., Buklanov, G.V., Gikal, B.N., Iliev, S., Mezentsev, A.N., Polyakov, A.N., Sedykh, I.M., Shirokovsky, I.V., Subbotin, V.G., Sukhov, A.M., Tsyganov, Yu.S., Zhuchko, V.E., Lougheed, R.W., Moody, K.J., Wild, J.F., Hulet, E.K., McQuaid, J.H., *Phys. Rev. Lett.* **73**, 624 (1994).
16. Armbruster P., *Ann. Rev. Nucl. Part. Sci.* **35**, 135 (1985).
17. Münzenberg, G., *Rep. Progr. Phys.* **51**, 57 (1988).
18. Hofmann, S., *Rep. Progr. Phys.* **61**, 639 (1998).
19. Lazarev, Yu.A., Lobanov, Yu.V., Oganessian, Yu.Ts., Tsyganov, Yu.S., Utyonkov, V.K., Abdullin, F.Sh., Iliev, S., Polyakov, A.N., Rigol, J., Shirokovsky, I.V., Subbotin, V.G., Sukhov, A.M., Buklanov, G.V., Gikal, B.N., Kutner, V.B., Mezentsev, A.N., Sedykh, I.M., Vakatov, B.V., Lougheed, R.W., Wild, J.F., Moody, K.J., Hulet, E.K., *Phys. Rev. Lett.* **75**, 1903 (1995).
20. Lazarev, Yu.A., Lobanov, Yu.V., Oganessian, Yu.Ts., Utyonkov, V.K., Abdullin, F.Sh., Polyakov, A.N., Rigol, J., Shirokovsky, I.V., Tsyganov, Yu.S., Iliev, S., Subbotin, V.G., Sukhov, A.M., Buklanov, G.V., Gikal, B.N., Kutner, V.B., Mezentsev, A.N., Subotic, K., Wild, J.F., Lougheed, R.W., Moody, K.J., *Phys. Rev.* **C54**, 620 (1996).
21. Oganessian, Yu.Ts., Utyonkov, V.K., Lobanov, Yu.V., Abdullin, F.Sh., Polyakov, A.N., Shirokovsky, I.V., Tsyganov, Yu.S., Gulbekian, G.G., Bogomolov, S.L., Gikal, B.N., Mezentsev, A.N., Iliev, S., Subbotin, V.G., Sukhov, A.M., Ivanov, O.V., Buklanov, G.V., Subotic, K., Itkis, M.G., Moody, K.J., Wild, J.F., Stoyer, N.J., Stoyer, M.A., and Lougheed, R.W., *Phys. Rev.* **C62**, 041604(R) (2000).
22. Oganessian, Yu.Ts., Utyonkov, V.K., Lobanov, Yu.V., Abdullin, F.Sh., Polyakov, A.N., Shirokovsky, I.V., Tsyganov, Yu.S., Gulbekian, G.G., Bogomolov, S.L., Gikal, B.N., Mezentsev, A.N., Iliev, S., Subbotin, V.G.,

Sukhov, A.M., Buklanov, G.V., Subotic, K., Itkis, M.G., Moody, K.J., Wild, J.F., Stoyer, N.J., Stoyer, M.A., Lougheed, R.W., *Phys. Rev. Lett.* **83**, 3154 (1999).
23. Oganessian, Yu.Ts., Yeremin, A.V., Gulbekian, G.G., Bogomolov, S.L., Chepigin, V.I., Gikal, B.N., Gorshkov, V.A., Itkis, M.G., Kabachenko, A.P., Kutner, V.B., Lavrentev, A.Yu., Malyshev, O.N., Popeko, A.G., Rohac, J., Sagaidak, R.N., Hofmann, S., Münzenberg, G., Veselsky, M., Saaro, S., Iwasa, N., Morita, K., *Eur. Phys. J.* **A5**, 63 (1999).
24. Oganessian, Yu.Ts., Yeremin, A.V., Popeko, A.G., Bogomolov, S.L., Buklanov, G.V., Chelnokov, M.L., Chepigin, V.I., Gikal, B.N., Gorshkov, V.A., Gulbekian, G.G., Itkis, M.G., Kabachenko, A.P., Lavrentev, A.Yu., Malyshev, O.N., Rohac, J., Sagaidak, R.N., Hofmann, S., Saro, S., Giardina, G., Morita, K., *Nature* **400**, 242 (1999).

Recent Developments in the Synthesis of Super Heavy Elements

D. Ackermann

Gesellschaft für Schwerionenforschung mbH, Planckstraße 1, D-64291 Darmstadt, Germany[1]
Johannes Gutenberg-University of Mainz, Dept. of Physics, AG EXAKT, D-55099

Abstract. Throughout the passed two decades isotopes of the elements with atomic numbers 107-112 have been synthesized and unambiguously identified at the velocity filter SHIP at GSI. In a recent experiment at SHIP the results for element 112 have been confirmed and a third decay chain of the isotope $^{277}112$ has been observed. Cold fusion reactions using Pb- and Bi- targets and evaporation residue(ER)-α-α correlations together with an efficient separation and detection system are the major ingredients for the success of these experiments. The sensitivity limit of the set-up at GSI has reached the 1pb level. For a systematic investigation in this region of the chart of nuclei and to synthesize heavier nuclei this limit has to be pushed to even lower values. An extensive development program is pursued at SHIP in order to reach at least an order of magnitude lower cross sections. Systematic investigations, the construction of decay chain networks and mass measurements are some of the possible approaches to study the decay chains attributed to isotopes of the elements 114, 116 and 118 at Dubna and Berkeley, which are, in contrast to those observed at GSI, not connected to decays of known isotopes. For the Berkeley results, in particular, several trials of confirmation have been undertaken at various laboratories including GSI.

INTRODUCTION

The search for superheavy elements, predicted close to the double magic nucleus $^{298}114$ [1] - more recent theoretical results are found in [2,3] - was a substantial motivation for the construction of the UNILAC and the velocity filter SHIP [4] at GSI in Darmstadt. To reach the "island of superheavy elements" in the beginning of the experimental work at SHIP in 1976 only one method seemed possible: to jump across the "sea of instability". Although this method was tempting, it contained severe uncertainties. Decay properties of nuclei in the intended region, such as decay modes, decay energies and half-lives, were not known and could only be estimated on the basis of predicted mass excesses, shell effects, fission barriers etc., and were therefore extremely uncertain. The same held for the prediction of production cross sections using fusion-evaporation codes optimized to reproduce data in the region of known elements. Experiments, performed at SHIP, to produce superheavy elements in bombardments of ^{170}Er with ^{136}Xe or ^{238}U with ^{65}Cu [5], as well as by the reaction ^{48}Ca + ^{248}Cm [6] did not show positive results. It turned out to be more successful to approach the heavier elements step by step. Following the concept of "cold" fusion of lead or bismuth targets with medium heavy projectiles like ^{40}Ar or ^{50}Ti, first applied successfully by Oganessian et al. [7], the SHIP group succeeded to produce and identify about 25 new isotopes with atomic numbers from Z=98 up to Z=112. Mutual interaction of experimental results and theoretical calculations led to a better understanding of their stability, while measured excitation functions allowed for a reliable empirical extrapolation of optimum bombarding energies and cross sections for 1n deexcitation channels. Continuous technical development pushed the sensitivity of the set-up down to a cross section value of about 1 pb. To proceed towards higher Z an extensive development program is being followed at present. Recently the synthesis of isotopes of the elements 114, 116 and 118 has been reported at Dubna and Berkeley. The unambiguous assignment of those events, however, is not yet possible. An attempt to confirm the Berkeley results for element 118 at SHIP did not yield a positive result. A recent review on the discovery of the heaviest elements [8] gives a complete overview over the recent achievements in the field. There also a detailed description of the experimental set-up at GSI can be found.

EXCITATION FUNCTIONS

Complete fusion reactions appear as most successful method for the production of transactinide nuclei. The formation cross section of a specific nuclide in a given reaction, however, is strongly dependent on the excitation energy E* of the compound nucleus, according to the relation $E^* = E_{cm} + Q$ (where E_{cm} denotes the energy in the center-of-mass system and Q the Q-value of the reaction), and thus on the bombarding energy $E_{lab} = (mp + mt)/mt \times E_{cm}$. Since maximum production cross sections are decreasing rapidly with increasing atomic numbers, the choice of the optimum E_{lab} is crucial for the production of the heaviest nuclei. Measured excitation functions for reactions of ^{208}Pb, ^{209}Bi targets with various projectiles producing heavy nuclei in the range Z=104 to 112 are presented in fig. 1. Excitation energies were calculated using experimental mass excesses published by Audi and Wapstra [9] and values predicted by Myers and Swiatecki [10]. They were calculated for the center of the target using energy losses of the projectiles according to [11]. In all shown cases the cross section maxima are approximately centered between zero and the interaction barrier according to the Bass model [12].

RECENT RESULTS ON THE SYNTHESIS OF HEAVY ELEMENTS WITH Z=110-112 AT GSI

The elements with Z=107-112 have been synthesized and unambiguously identified at SHIP. The elements 107-109 have already been named and have been entered

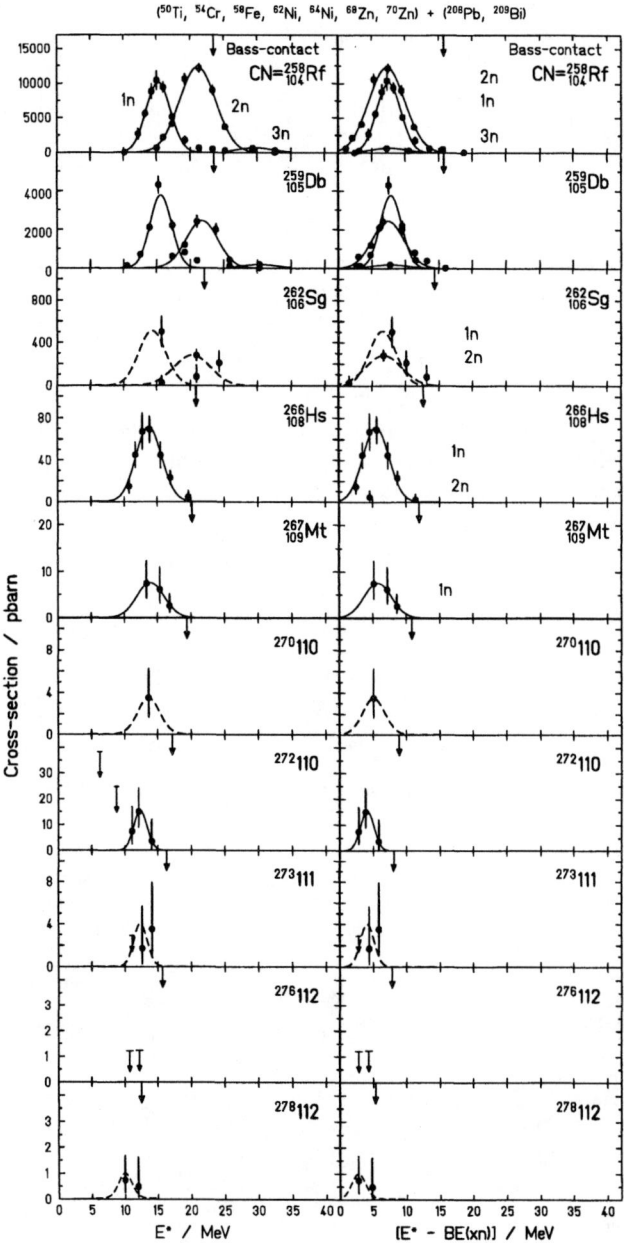

FIGURE 1. Measured excitation functions for Z=104 to 112. Cross sections are plotted as a function of the excitation energy (left panel) and the excitation energy lowered by the neutron binding energy according Myers and Swiatecki [10] for the various evaporation channels (right panel). The continuous curves are Gaussian fits through the data points, the dashed curves are interpolations. The arrows in both mark the interaction barriers of the reaction according to the Bass model [12].

as Bohrium (Bh, Z=107), Hassium (Hs, Z=108) and Meitnerium (Mt, Z=109) in the periodic table of elements. The properties found for the elements 110, 111 and 112 are presented in this section.

A linear extrapolation of the optimum excitation energies for the production of ^{257}Rf and ^{265}Hs (see fig. 1) resulted in an 'optimum' value of $E^* = 12.3$ MeV for the production of ^{269}110 via the reaction ^{62}Ni + ^{208}Pb. In an experiment in November 1994, where a total projectile dose of 2.2×10^{18} was collected, four α-decay chains were observed, which were attributed to the isotope with the mass number 269 of the new element 110 [13]. The assignment was based on the observation that the α-decays directly preceded the well established α-decay chain of ^{265}Hs and, therefore, have to origin from the α-mother ^{269}110. From the measured decay data an average decay energy of $E = (11.112 \pm 0.020)$ MeV and a half-life of $T_{1/2} = (170^{+160}_{-60})$ μs was obtained. The production cross section was $\sigma = (3.5^{+2.7}_{-1.8})$ pb.

Since it is well established in the region of transfermium nuclei that more neutron rich projectiles lead to higher formation cross sections, one could expect for the combination ^{64}Ni + ^{208}Pb a still higher ER cross section than for ^{62}Ni + ^{208}Pb. In a directly following experiment in November/December 1994 the ER production by the reaction ^{64}Ni + ^{208}Pb was investigated at $E^* = (8\text{-}13)$ MeV. Nine α-decay chains observed in this experiment could be attributed to ^{271}110. A maximum cross section of $\sigma = 15 \, \binom{+9}{6}$pb was measured at $E^* = 12.1$ MeV. Details of the decay chains can be found in ref. [16].

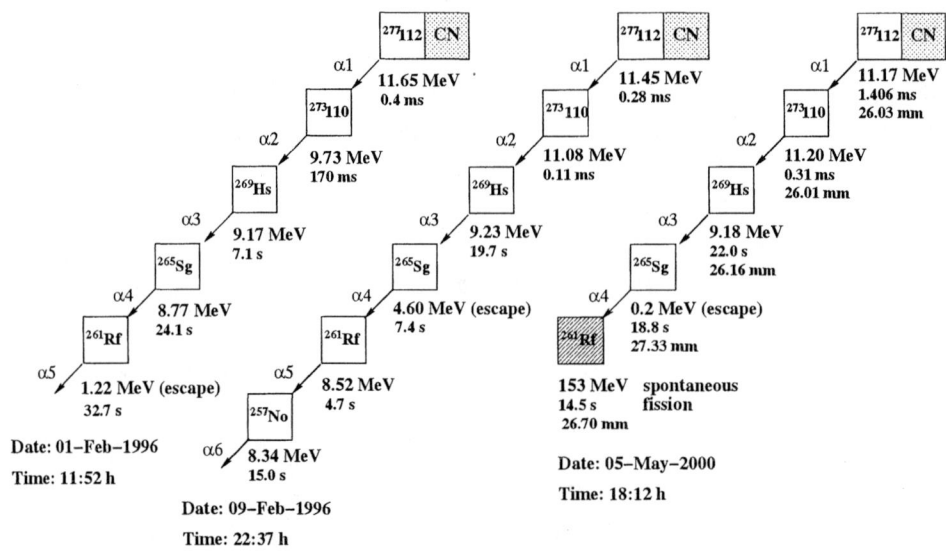

FIGURE 2. The three decay-chains observed for the isotope ^{277}112, including the chain observed in the confirmation run in May 2000. For this chain also the position in vertical direction on the 5 mm wide detector strip, where this event was observed, is given in mm.

On the basis of these encouraging results for the synthesis of element 110 in the reactions ^{62,64}Ni + ^{208}Pb the production of an isotope of element 111 by the reaction ^{64}Ni + ^{209}Bi was undertaken in an experimental run in December 1994. Three bombarding energies at 10.0 MeV, 11.6 MeV, and 13.0 MeV were chosen using the predicted mass excess of [10] for the compound nucleus ^{273}111 excitation energies. Projectile doses of 1.0×10^{18} at E* = 10.0 MeV, 1.1×10^{18} at E* = 11.6 MeV and 1.1×10^{18} at E* = 13.0 MeV were collected. While no decay chain that could be attributed to ^{272}111 was registered at E* = 10.0 MeV, one event was observed at E* = 11.6 MeV, and two events at E* = 13.0 MeV [14], referring to a cross section of $\sigma = 3.5^{+4.6}_{-2.3}$ pb.

In early 1996 the search for element 112 was undertaken using the projectile target combination ^{70}Zn + ^{208}Pb. A total projectile dose of 3.4×10^{18} was collected. Following the systematics on optimum excitation energies a bombarding energy according to E* = 10.1 MeV was chosen. Two decay chains which could be attributed to ^{277}112 were observed, the resulting production cross section was $\sigma = 1.0$ pb [15]. The most striking result, however, was the significant difference in the decay energies and lifetimes of the daughter isotope ^{273}110 of E = 9.73 MeV, $\Delta t = 170$ ms (chain 1) and $E_\alpha = 11.08$ MeV, $\Delta t = 110\mu$s (chain 2). Due to the large differences in lifetime the two transitions must be assigned to different levels in ^{273}110. In a recent experiment in May 2000 a third decay chain of ^{277}112 has been recorded. It is shown together with the first two chains in fig. 2. This latter chain has been observed at an excitation energy of about 2 MeV higher at E* = 12 MeV. During an irradiation time of 19 days a total of 3.5×10^{18} projectiles were sent onto the target. The resulting cross section at this energy is $(0.5^{+1.1}_{-0.4})$ pb. This value fits well into the cross section systematics shown in fig. 1. The first two α decays have energies of 11.17 and 11.20 MeV, respectively. They are succeeded by an α of only 9.18 MeV, an energy step of 2 MeV. Correspondingly, the lifetime increases by about five orders of magnitude between the second and third α decay. This decay pattern is in agreement with the one observed for the second chain in the first experiment and supports the explanation of a local minimum of the shell correction energy at neutron number N = 162, which is crossed by the α-decay of ^{273}110. The α energy of 9.18 MeV for ^{269}Hs is within the detector resolution identical to the one observed in the first chain. A new result is the occurrence of fission ending the new chain at ^{261}Rf, for which fission was not observed so far, but is likely to occur taking into account the high fission probabilities of the neighboring isotopes. For more details see ref. [8].

HINTS FOR THE SYNTHESIS OF ELEMENTS WITH Z=114, 116, AND 118

In spring 1999 the group running the newly built Berkeley gasfilled separator (BGS) [17] at the LBNL in Berkley reported the observation of three long α decay chains which were tentatively attributed to the decay of the isotope ^{293}118 [18].

The observed ER-α-α correlations indicated the possible synthesis of $^{293}118$ and $^{289}116$ in the reaction ^{86}Kr+^{208}Pb. The energy of the ^{86}Kr beam was 459 MeV. Before impinging on the 300-450 µg/cm^2 ^{208}Pb-target the projectiles had to pass through an entrance window into the gas-filled chamber of the separator of 0.1 mg/cm^2 carbon and the target backing of 40 µg/cm^2 carbon. The beam energy in the middle of the target thickness corresponded to a calculated excitation energy of about 13 MeV. The detector system was similar to the one used at SHIP. In a first five day experiment from April 8 to April 12, 1999 with a total beam dose of 0.7×10^{18} projectiles two chains were observed. The experiment was repeated in Berkeley from April 30 to May 5, 1999. In this experiment one additional chain was observed at a total beam dose of 1.6×10^{18} projectiles. The cross section resulting

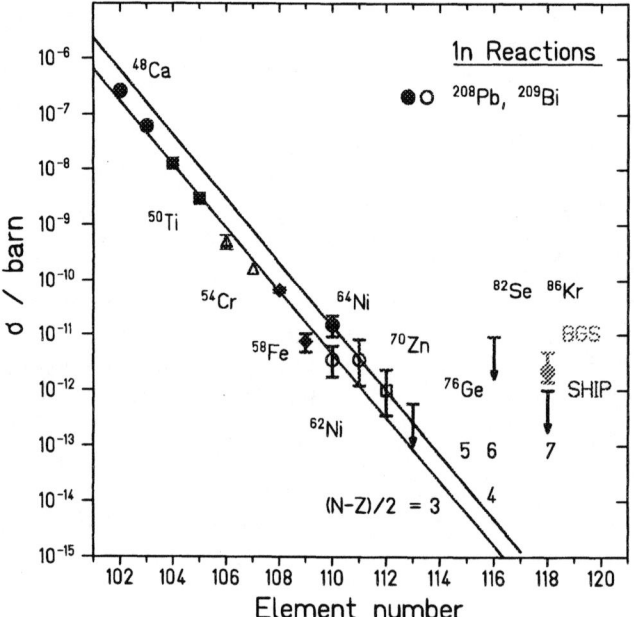

FIGURE 3. Maximum cross sections and cross section limits for heavy elements in fusion reactions with Pb and Bi targets for various projectiles at SHIP and the recent result from the BGS at the LBNL (see text).

TABLE 1. Main parameters for the experiments investigating the reaction ^{86}Kr+^{208}Pb at the LBNL and GSI.

	BGS	SHIP
Time	11.0 d	24 d
Current	300 pnA	224 pnA
Dose	2.3×10^{18}	2.9×10^{18}
Efficiency	75% d	50%
E*	13 MeV	13 MeV
Events	3	0
σ	$2.2^{+2.6}_{-0.8}$ pb	< 1.0 $(0.5^{+1.1}_{-0.4})$

from these two irradiations was given by Ninov et al. [18] as $(2.2^{+2.6}_{-0.8})$pb. In order to examine this result the same reaction was investigated at SHIP in summer 1999 [8]. No decay chain was observed which could have confirmed the LBNL result. With a slightly higher beam dose a cross section limit of 1pb was established on the basis of $0.5^{+1.1}_{-0.4}$ pb for one event at the given beam dose of 2.9×10^{18} collected projectiles. The parameters of both experiments are compared in table 1. In fig. 3 the values for the cross section measured at Berkeley and the limit obtained at SHIP are shown together with maximum cross sections for reactions with Pb and Bi targets leading to ER with Z$\geq$102.

A major problem, however, of the decay chains observed here is that they are not connected to α decays of known isotopes, as it is the case for the chains observed at SHIP for the elements 107-112. The same problem exists for the decay chains found at the FLNR in Dubna for reactions with ^{48}Ca projectiles on ^{242,244}Pu and ^{248}Cm. Here chains were seen which tentatively have been assigned to isotopes of the elements 114 and 116 [20]. In fig. 4 the chart of nuclides is shown for proton numbers Z$\geq$104 and neutron numbers N$\geq$149, including the isotopes which are contained by chains reported from Berkeley and Dubna. Here additional information is needed to identify mass and atomic charge of the detected ER. For direct mass measurements using time of flight techniques detector development is needed. Another possibility to attack the problem of the non-connected decay chains is the systematic investigation of decay chain networks. Redundant information for the same isotopes can be provided using different reactions and decay paths producing the same isotopes. A given chain can be entered at different entry points using different reactions, as e.g. in the case of the Berkley chain for 118 one could exchange the projectile of ^{86}Kr by ^{82}Se and produce via the 1n channel the daughter of ^{293}118: ^{289}116. Similarly, as what has been reported at this conference by Oganessian [20], the exchange of the ^{244}Pu with ^{248}Cm leads via the 4n channel to the mother ^{292}116 of ^{288}114 which is the 4n ER produced in the reaction ^{48}Ca+^{244}Pu. Varying excitation energy and isotope one can in principle synthesize the same nucleus via different

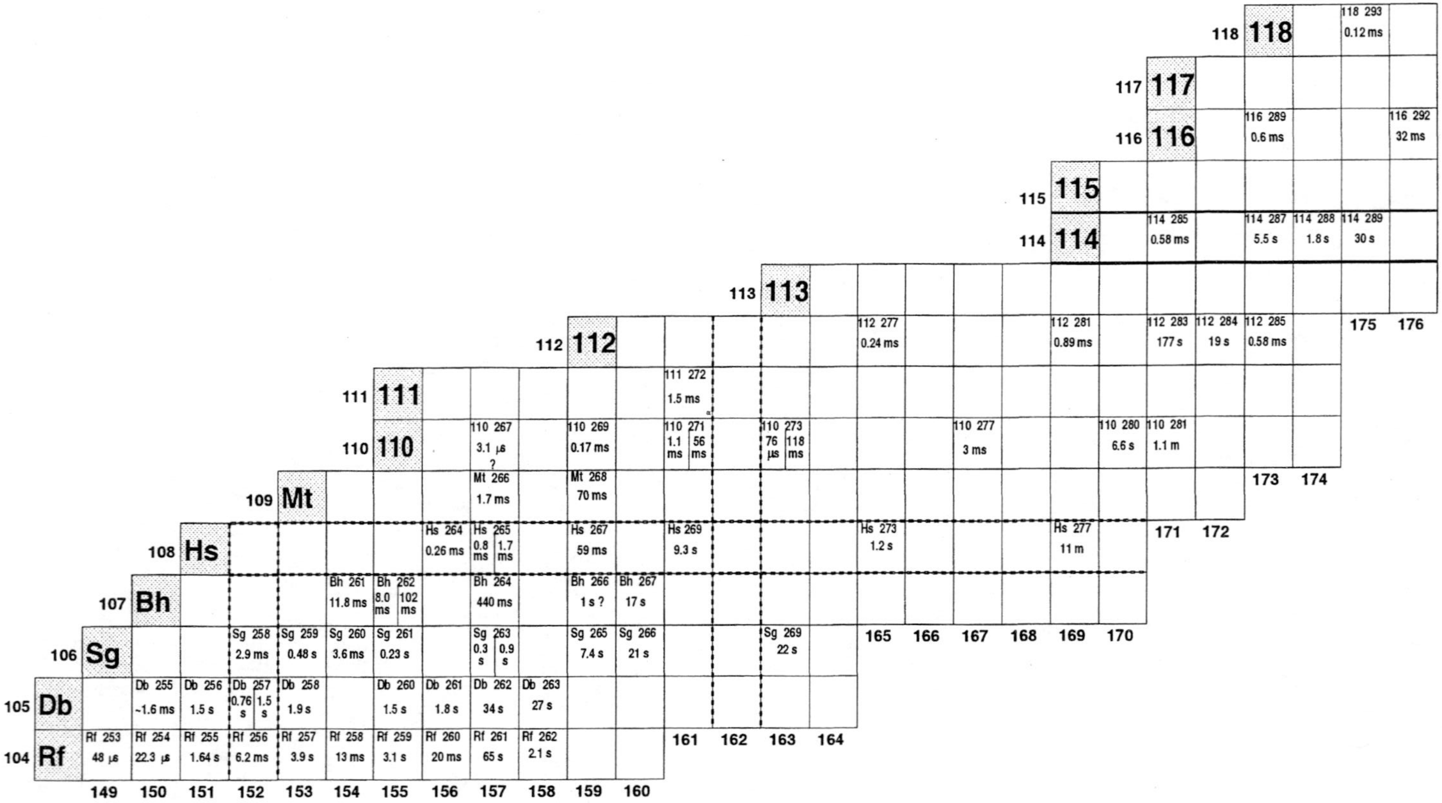

FIGURE 4. Excerpt of the chart of nuclides for $Z \geq 104$ and $N \geq 149$ including the nuclides contained in the α decay chains, tentatively assigned to isotopes of the elements 114, 116 and 118 in Dubna and Berkeley (see text).

evaporation channels. A limitation for this kind of investigations is certainly the low cross sections which are to expect. Therefore, the state-of-the-art sensitivity of the experimental equipment, which presently limits the accessible cross section regime to 1 pb, has to be lowered by orders of magnitude. This sensitivity improvement is required for a successful continuation of the Pb and Bi based reactions as well as for those using actinide targets.

TECHNICAL DEVELOPMENT AT SHIP

The three areas presently under technical investigation at SHIP are:

- beam development
- target development
- background reduction.

To access a region of lower cross section the number of interactions and, therefore, the number of projectiles has to be increased. The UNILAC at GSI delivers the beam with a duty cicle of about 28%. Apart from raising the beam current, the use of an accelerator with 100% duty cycle (DC) whould already provide a factor of 3.5 in higher beam intensity. The increased beam current, together with a higher Z of the projetiles in some cases, asks for measures to protect the Pb and Bi targets, both having a low melting point. A first step is to spread the beam as homogeneous as possible over a maximum area. With the target wheel presently in use we have already reached the limit for the presently availible beam intensities. The planned introduction of ion optical elements like octupole magnets in the UNILAC beamline will help to approach the desired optimum of a rectangular beam profile illuminating the target as uniformly as possible. Besides those "passive" measures also an "active" target cooling is now under development. A set-up providing a gas jet blown onto the spot where the beam hits the target is currently being developed. Chemical compounds of Pb or Bi with higher melting temperatures are also under investigation. The higher projectile rate required for a successful investigation of reactions with lower cross section will have as a consequence an increase of background per time unit. To improve the background suppression we test the use of foils to stop scattered beam particles which pass SHIP with low kinetic energy. With all those measures and an increase of the beam intensity from presently 3×10^{12} particles s^{-1} to 3×10^{13} particles s^{-1} a cross section regime of one order of magnitude lower than the present limit could be reached.

ACKNOWLEDGEMENTS

The recent experiments were performed together with P. Armbruster, H.-G. Burkhard, H. Folger, F.P. Heßberger, S. Hofmann, B. Kindler, B. Lommel, V. Ninov, S. Reshitko, H.-J. Schött, C. Stodel (GSI Darmstadt), A.N. Andreyev,

A.Yu. Lavrentev, A.G. Popeko, A.V. Yeremin (FLNR-JINR Dubna), S. Antalic, P. Cagarda, R. Janik, Š. Šaro (Uiversity of Bratislava), and M. Leino (University of Jyväskylä).

REFERENCES

1. Meldner, H., *Arkiv f. fysik* **36**, 593-598 (1967).
2. Möller, P., Nix, J.R., *J. Phys. G. Part. Phys* **20**, 1681-1747 (1994).
3. Cwiok, S., Sobiczewski, A., *Z. Phys. A* **342**, 203-213 (1992).
4. Münzenberg, G., FAust,W., Hofmann, S., Armbruster, P., Güttner, K., Ewald, H., *Nucl. Instrum. Meth.* **161** 65-82 (1979).
5. Münzenberg, G., Armbruster, P., Faust,W., Hofmann, S., Reisdorf, W., Schmidt, K.-H., Valli, K., Ewald, H., Güttner, K., Clerc, H.G., Lang, W., *GSI Jahresbreicht 1977* **GSI-J-1-78**, 75 (1978).
6. Agarwal, Y.K., Armbruster, P., Hofmann, S., Heßberger, F.P., Münzenberg, G., Poppensieker, K. , Reisdorf, W., Schmidt, K.-H., Schneider, J.R.H., Schneider, W.F.W., Vermeulen, G., Ghiorso, A., Leino, M., Moody, K.J., *GSI Scientific Report 1983* **GSI-84-1**, 79 (1984).
7. Oganessian, Yu.Ts.; editors Harney, H.L., Braun-Munzinger, P., Gelbke, C.K., *Lecture Notes in Physics Vol. 33*, Berlin, Heidelberg, New York: Springer, 1974, pp. 221-252.
8. Hofmann, S., and Münzenberg, G., *Rev. Mod. Phys.* **72**, 733 (2000).
9. Audi, G., and Wapstra, A.H., *Nucl. Phys. A* **565**, 409-480 (1993).
10. Myers, W.D., and Swiatecki, W.J., *Nucl. Phys. A* **601**, 141-167 (1996).
11. Hubert, F., Bimbot, R., Gauvin, H., *Atomic Data and Nuclear Data Tables* **46**, 1 (1990).
12. Bass, R., *Nucl. Phys. A* **231**, 45 (1974).
13. Hofmann, S., Ninov, V., Heßberger, F.P., Armbruster, P., Folger, H.,Münzenberg, G., Schött, H.-J., Popeko, A.G., Yeremin, A.V., Andreyev, A.N., Saro, S., Janik, R., Leino, M., *Z. Phys. A* **350**, 277-280 (1995).
14. Hofmann, S., Ninov, V., Heßberger, F.P., Armbruster, P., Folger, H.,Münzenberg, G., Schött,H.-J., Popeko, A.G., Yeremin, A.V., Andreyev, A.N., Saro, S., Janik, R., Leino, M., *Z. Phys. A* **350**, 281-282 (1995)
15. Hofmann, S., Ninov, V., Heßberger, F.P., Armbruster, P., Folger, H.,Münzenberg, G., Schött,H.-J., Popeko, A.G., Yeremin, A.N., Saro, S., Janik, R., Leino, M., *Z. Phys. A* **354**, 229-230 (1996)
16. Heßberger,F.P. ,Hofmann, S.,Ninov, V.,Armbruster, P.,Folger, H., Lavrentev, A.,Leino, M.E.,Münzenberg, G., Popeko, A.G.,Saro, S., Stodel, Ch., Yeremin, A.N., *Proceedings of the Tours Symposium on Nuclear Physics III, Tours, France, 2.-5. September 1997*, AIP Conference Proceedings 425, Woodbury, New York, 1998, pp. 3-17.
17. Ninov V., Gregorich, K.E., MacGrath, C.A., *Proceedings of the 2nd International Conference on Exotic Nuclei and Atomic Masses, ENAM-98*, AIP Conference Pro-

ceedings No.455, Bellaire, Michigan, June 23-27 1998, edited by B.M. Sherril, D.J. Morissey, and C.N. Davids, Woodbury, New York, 1998, p. 704.
18. Ninov, V., *et. al.*, *Phys. Rev. Lett.* **83**, 1104 (1999).
19. Oganessian, Yu.Ts., *et. al.*, *Nature* **400**, 242-245 (1999).
20. Oganessian, Yu.Ts., *et. al.*, *contribution to this proceedings*.

Fusion of Deformed Nuclei in the Vicinity of the Coulomb Barrier

H. Ikezoe[*], S. Mitsuoka[*], K. Nishio[*], K. Satou[*], and S.C. Jeong[†]

[*]Advanced Science Research Center, JAERI, Tokai, Ibaraki, 319-1195, Japan
[†]Institute of Particle and Nuclear Studies, KEK, Tsukuba, Ibaraki 305-0801, Japan

Abstract. The dependence of the fusion probability on the orientation of deformed nuclei was investigated for the ^{60}Ni + ^{154}Sm and ^{76}Ge + ^{150}Nd reactions. The fusion of the ^{82}Se + $^{nat.}$Ce (spherical) reaction was also measured to compare its fusion probability with that of the ^{76}Ge + ^{150}Nd reaction. Evaporation residues were measured for these reaction systems in the vicinity of the Coulomb barrier and the fusion probability was extracted as a function of bombarding energy. It was found that the fusion probability depends strongly on the orientation of the nuclear deformation. The fusion probability is considerably reduced when the projectiles collide at the tip of the deformed nuclei. On the other hand, when the projectiles collide at the side of the deformed nuclei, the fusion occurs without any hindrance. This phenomenon is understood qualitatively by comparing the distance between the mass centers of two colliding nuclei at touching with the position of the saddle point of the compound nucleus.

INTRODUCTION

Heavy-ion fusion reaction between massive nuclei has been extensively investigated so far. This is partly because heavy-ion fusion reaction is a unique method to synthesize the superheavy elements. The cold fusion and the hot fusion [1] have been successfully used for this purpose but the production cross sections of the superheavy elements are the order of pico barn or less as their atomic numbers increase. It is well known that the fusion probability between massive nuclei depends on the charge product Z_pZ_t of projectile and target. When the charge product is small ($Z_pZ_t < 1800$), the fusion process is well simulated by a one-dimensional barrier penetration model including a coupling effect of inelastic channels (for instance, surface vibrations and nucleon transfers). In this case, the fusion barrier distributes around a Coulomb barrier, and thus the enhancement of the fusion cross section is observed in the vicinity of the Coulomb barrier. When the charge product is larger than ~1800, the fusion probability decreases rapidly as Z_pZ_t increases. This means that even if the kinetic energy of projectile is enough large to surmount the fusion barrier, the compound nucleus is not always formed. An additional kinetic energy is needed for projectile to fuse together with target nucleus and to make the compound nucleus.

It is considered that the compound nucleus is formed only after the saddle point of the compound nucleus is surmounted in the course of the fusion process. This means that the compact touching shape, where the contact point is close to the saddle point, evolves more easily to the compound nucleus formation than an elongated touching shape. Thus the relative distance between the mass centers of two colliding nuclei at touching with respect to the saddle point plays essential role for fusion of massive reaction system. The contact point of massive reaction system with a large charge product (> ~1800) usually locates outside the saddle point. This situation is changed in the case of the fusion reaction between deformed nucleus and spherical projectile. In this case, the distance between the mass centers at touching depends on the orientation of deformed nucleus, that is, the distance becomes short when projectile collides at the side of deformed nucleus and long when projectile collides at the tip of deformed nucleus. This suggests that the colliding angle with respect to the symmetric axis of deformed nucleus influences an effect on the fusion process.

In order to investigate this effect, we measured evaporation residues (ERs) for the reactions ^{60}Ni + ^{154}Sm (Z_pZ_t = 1736) and ^{76}Ge + ^{150}Nd (Z_pZ_t = 1920), where the nuclei ^{154}Sm and ^{150}Nd have large nuclear deformations (β_2, β_4) as listed in Table 1. The fusion reaction between the projectile ^{82}Se and the spherical target nucleus $^{nat.}$Ce (^{140}Ce:85.48%, ^{142}Ce:11.08%, Z_pZ_t=1972) was also investigated to compare its fusion probability with that of the ^{76}Ge + ^{150}Nd reaction.

The fusion probability can be extracted from the cross section of the ERs. The ER cross section depends on the fusion probability and the survival probability in the de-excitation process of the compound nucleus. The de-excitation process can be simulated by the calculation based on the statistical model. There are several parameters to be fixed in the statistical model to simulate correctly the de-excitation process of excited compound nucleus. For this purpose, we also measured the ERs produced in the ^{32}S + ^{182}W and ^{28}Si + ^{198}Pt reactions. The former reaction system makes the same compound nucleus ^{214}Th as for the ^{60}Ni + ^{154}Sm reaction and the later reaction system makes the same compound nucleus ^{226}U as for the ^{76}Ge + ^{150}Nd reaction. Since the reaction systems ^{32}S + ^{182}W and ^{28}Si + ^{198}Pt have the small charge products Z_pZ_t of 1184 and 1092, respectively, we neglected the fusion hindrance at the entrance channel and assumed the complete fusion process for these reaction systems. We adjusted the statistical model parameters so as to fit measured cross sections of the ERs of these lighter reaction systems and then applied the statistical model with the fixed parameters to the heavier systems with the deformed nuclei. Finally we extracted the fusion probability for the heavier reaction systems and discuss the reaction mechanism between spherical projectile and deformed nucleus.

EXPERIMENTAL PROCEDURES AND RESULTS

The experiments were carried out at the tandem-booster facility of Japan Atomic Energy research Institute (JAERI). The experimental details are described elsewhere [2, 3] and only the essential points are written here.

The fusion cross sections for the lighter reaction systems of ^{32}S + ^{182}W and ^{28}Si + ^{198}Pt were obtained from the measurements of the their fission cross sections. Here we assumed the complete fusion-fission process for these reaction systems in the excitation energy range of 40 – 75 MeV and that the fission was the dominant decay mode in the excited compound nuclei ^{214}Th and ^{226}U. The angular distributions of the fission fragments were measured by a ΔE-E ionization chamber and the fission cross sections were obtained by integrating the angular distributions. The results are shown in Fig. 1 as a function of the center-of-mass (c.m.) bombarding energy E_{cm}.

The ERs produced in the ^{32}S + ^{182}W, ^{60}Ni + ^{154}Sm, ^{28}Si + ^{198}Pt, ^{76}Ge + ^{150}Nd, and ^{82}Se + $^{nat.}$Ce reactions were measured by the JAERI recoil mass separator (JAERI-RMS). The ERs emitted to the beam direction were separated in flight from the primary beams and various products of background reactions by the JAERI-RMS. The separated recoils were implanted into a double-sided position-sensitive strip detector (DPSD). Two large area timing detectors were used to obtain the time-of-flight (TOF) signal of incoming particles. The presence of a timing signal in the TOF detectors was used to separate ER implant events from the subsequent α-decays, which generate no TOF signals. The energy, the detection time, the TOF signal and two-dimensional positions of both the implanted ER and the subsequent α-decay particles were recorded.

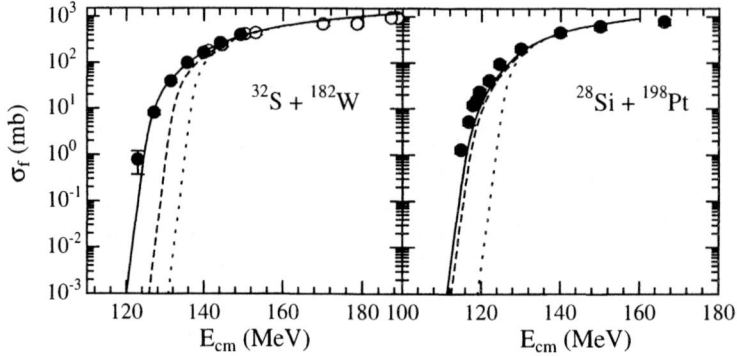

FIGURE 1. Measured fission cross sections (solid circles) for the ^{32}S + ^{182}W and ^{28}Si + ^{198}Pt reactions together with the data in [12] (open circles). The dotted curves are the calculated fusion cross sections based on the one-dimensional barrier penetration model. The dashed curves are the calculated results when the deformation effects of the projectile and target are taking into account. One obtains the solid curves when the couplings to the inelastic channels (2$^+$ and 3$^-$) are additionally considered.

The ERs were identified by measuring their α–decay energies with the typical energy resolution of 75 keV. The time difference between the ER implantation and the α_1 decay, namely, the lifetime of α_1-decaying nucleus, was used to improve the accuracy of the ER identification. The ER identification was also confirmed by the correlated α_1 - α_2 chains between the parent and the daughter α decays.

In order to obtain the absolute cross sections of the ERs, the detection efficiencies of various ERs were estimated according to the procedure described in [4]. The detection efficiency depends on the reaction system, the evaporation channel and also the bombarding energy. The typical efficiency for xn and αxn channels was 0.05 (E_{cm} = 120 MeV) and 0.02 (E_{cm} = 135 MeV), respectively, for ^{28}Si + ^{198}Pt, 0.25 for xn channel and

0.18 for αxn channel of ^{60}Ni + ^{154}Sm, and 0.2 (E_{cm} = 120 MeV) for αxn channel of ^{76}Ge + ^{150}Nd.

Evaporation residue cross sections are shown in Fig. 2 - 5 as a function of E_{cm}. The errors in the figures include both statistical contributions and the systematic error of 50 % coming from the transport efficiency of the ERs through the JAERI-RMS.

DISCUSSIONS

The experimental data were compared with the statistical model calculations using the code HIVAP [5]. Here, the fusion cross sections of the present reaction systems were estimated by the code CCDEF [6], which took into account the nuclear deformation and the coupling of inelastic excitations to the fusion process.

The nuclear deformation parameters used in this calculation are listed in Table 1, where the static nuclear deformations were assumed for ^{182}W, ^{198}Pt and ^{28}Si in addition to ^{150}Nd and ^{154}Sm. The couplings of the surface vibrations (2$^+$ and 3$^-$) were also taken into account. As shown in Fig. 1, the calculated fusion cross sections well reproduce the measured fission cross sections. The present result indicates that the nuclear deformation and also the inelastic couplings of 2$^+$ and 3$^-$ excitations to fusion cause the fusion enhancement in the vicinity of the Coulomb barrier. In the following discussion, the fusion cross sections calculated by the CCDEF code were used as the input of the statistical model calculation.

The details of the statistical model calculation are written in [2]. According to the prescription of [7, 8], the rotational and vibrational enhancements of the level density were taken into account. The fission barrier height was given as $B_f = B_{LD} - E_s$, where the liquid drop fission barrier B_{LD} was calculated by [9] and E_s is a shell correction energy. The shell damping factor of 18 MeV was assumed.

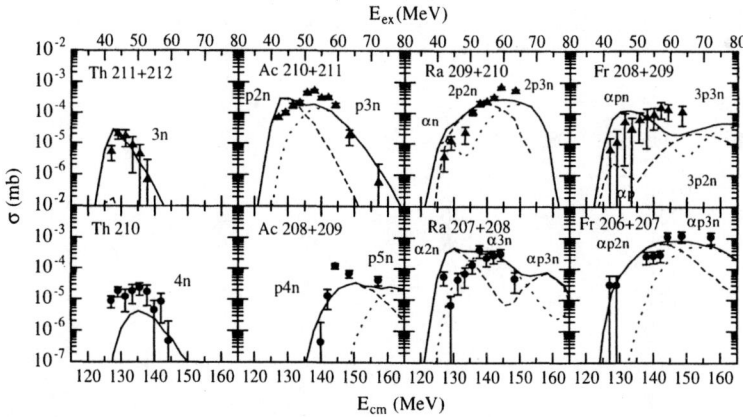

FIGURE 2. Measured ER cross sections for the ^{32}S + ^{182}W reactions as a function of the c.m. energy E_{cm}. The upper abscissa indicates the excitation energy E_{ex} of the compound.

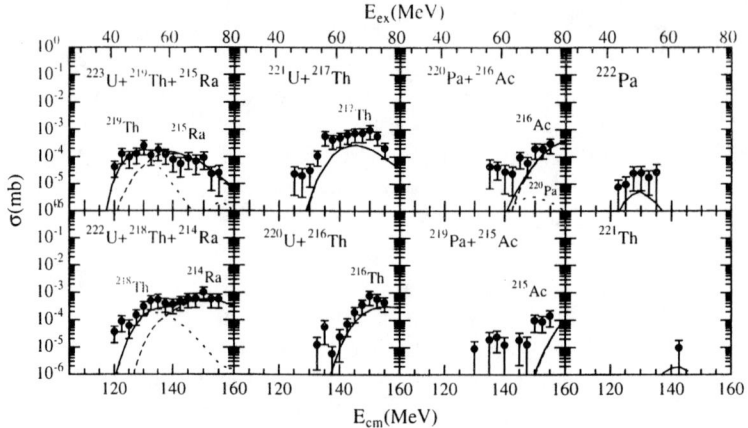

FIGURE 3. Same as Fig.2 except for the ^{28}Si + ^{198}Pt.

The good agreement within the factor 2 – 5 between the measured ER cross sections and the calculations is demonstrated in Fig. 2 for the ^{32}S + ^{182}W reaction and Fig. 3 for the ^{28}Si + ^{198}Pt reaction. The fusion hindrance was not observed for these reactions. This is because of the small charge products as mentioned above. In Fig. 4 and Fig. 5, the calculated ER cross sections (dashed curves) are compared with the measured cross sections for the ^{60}Ni + ^{154}Sm and ^{76}Ge + ^{150}Nd reactions. The calculated Coulomb barrier heights are shown in Fig. 6 as a function of the colliding angle θ_{coll} with respect to the orientation of deformed nucleus for the present two reaction systems. In the case of ^{60}Ni + ^{154}Sm, the calculated Coulomb barrier height is 172 MeV for the tip collision and 198 MeV for the side collision. Although the barrier distribution becomes wider due to the additional inelastic couplings which effects are minor but important, it is considered that the ER yields at E_{cm} ~180 MeV mainly come from the near tip collisions. As shown in

TABLE 1. Deformation parameters.

Nuclei	β_2	β_3	β_4
^{28}Si	0.408	0.283	
^{32}S	0.312	0.410	
^{60}Ni	0.207	0.208	
^{76}Ge	0.268	0.140	
^{82}Se	0.194	0.161	
^{140}Ce	0.101	0.127	
^{150}Nd	0.358	0.070	0.107
^{154}Sm	0.321	0.084	0.08
^{182}W	0.276	0.050	-0.089
^{198}Pt	-0.113	0.050	

Fig. 4, no ER yield was observed at E_{cm} = 175 and 182 MeV and the measured cross sections of $2n, p2n, \alpha n, \alpha 2n$ and αp channels below E_{cm} = 200 MeV are considerably smaller than the calculations. On the other hand, the measured cross sections above E_{cm} = 200 MeV are consistent with the present calculations. The fact that no ER was observed at the lowest two energies suggests that the fusion probability is significantly small at the near tip collision. Although the collisions for all nuclear orientation occur at the energy higher than E_{cm} ~ 200 MeV, it is considered that the near side collisions mainly contribute to the compound nucleus formation because of their larger solid angle than that of the tip collision.

In the case of ^{76}Ge + ^{150}Nd, the barrier height also varies significantly with the colliding angle as show in Fig. 6. It results in a large enhancement of the fusion cross section below the spherical Coulomb barrier of 209 MeV. However, no event was observed in E_{cm} < 205 MeV, whereas the measured cross sections above 210 MeV are consistent with the calculations. Especially a large discrepancy between the calculation and the experimental data is found in the ^{225}U channel at 185 – 190 MeV. The calculation predicts the cross section of ~1 μb, but the experimental upper limit of this cross section was about 3 – 4 nb. Here we see again the large fusion hindrance at the near tip collision (E_{cm} < 185 - 190 MeV) and no fusion hindrance at the energy higher than ~ 210 MeV where the side collision is dominant.

The fusion probabilities shown in Fig. 7 were obtained from the measured cross sections and calculated survival probabilities for the ^{76}Ge + ^{150}Nd and ^{82}Se + $^{nat.}$Ce reactions. Here the charge products of these reactions are similar (1920 for ^{76}Ge + ^{150}Nd and 1972 for ^{82}Se + $^{nat.}$Ce), while the former reaction has a deformed nucleus ^{150}Nd and the later has a spherical nucleus ^{140}Ce. In Fig. 7, the fusion probability for the ^{28}Si + ^{198}Pt

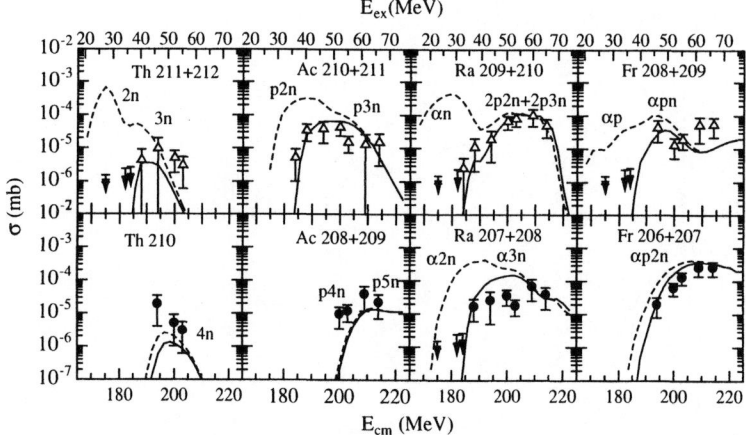

FIGURE 4. Same as Fig. 2 except for the ^{60}Ni + ^{154}Sm reaction. The dashed curves are the calculated results of HIVAP. The solid curves show the calculated results taking into account the extra-extra push energy (see [2]). The upper limits are shown as arrows.

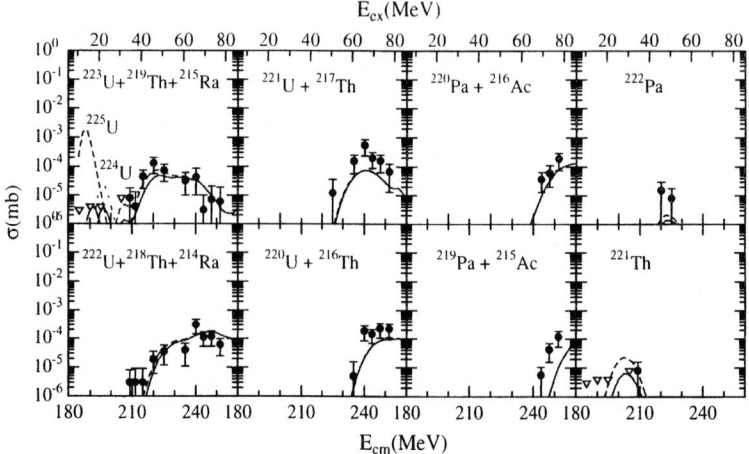

FIGURE 5. Same as Fig. 4 except for the ^{76}Ge + ^{150}Nd. The upper limits are shown as open triangles at the lowest four energies (see [3]).

reaction is also shown. The abscissa is the *c.m.* energy normalized by the Bass barrier for each reaction system. The fusion probability of ^{76}Ge + ^{150}Nd slowly decreases as decreasing E_{cm} down to the Bass barrier energy. This trend is similar to that of the ^{28}Si + ^{198}Pt reaction, where no fusion hindrance was observed. The upper limits of the fusion probability for ^{76}Ge + ^{150}Nd are shown as downward arrows below the Bass barrier energy. It is apparent that the ^{76}Ge + ^{150}Nd reaction has no fusion hindrance above the Bass barrier, showing marked contrast to that of ^{82}Se + $^{nat.}$Ce, where the fusion probability decreases with decreasing the *c.m.* energy above the Bass barrier energy.

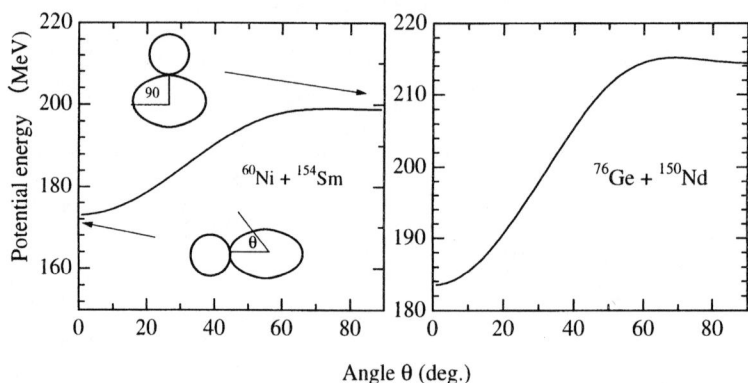

FIGURE 6. Calculated potential energy for ^{60}Ni + ^{154}Sm and ^{76}Ge + ^{150}Nd.

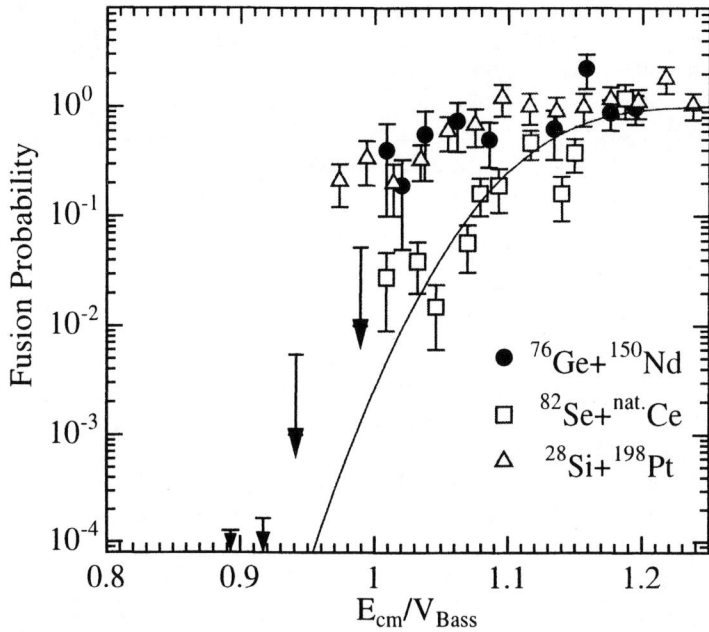

FIGURE 7. Fusion probabilities as a function of the c.m. energy normalized by Bass barrier for each reaction system. The arrows indicate the upper limit for the ^{76}Ge + ^{150}Nd reaction. The solid curve shows the prediction including the E_{xx} of 28 MeV and its variance s_B of 10 MeV for the ^{82}Se + ^{140}Ce.

The fusion hindrance is characterized by the extra-extra push energy E_{xx} [10, 11] and its variance σ_B [11] as a function of the charge product $Z_p Z_t$. The solid curve shown in Fig. 7 is the calculation with $(E_{xx}, \sigma_B) = (28$ MeV, 10 MeV$)$ and reasonably agrees with the measured fusion probability for ^{82}Se + $^{nat.}$Ce [13]. These values are slightly larger than the values $(E_{xx}, \sigma_B) = (21$ MeV, 8 MeV$)$ predicted by the systematics of [11] for ^{82}Se + ^{140}Ce. The systematics predicts $(E_{xx}, \sigma_B) = (25$ MeV, 9.5 MeV$)$ for ^{76}Ge + ^{150}Nd, which is close to the solid curve. Although the experimental result for ^{82}Se + $^{nat.}$Ce follows the systematics trend, the measured fusion probability of ^{76}Ge + ^{150}Nd deviates considerably from the systematics of [11]. This discrepancy can be ascribed to the large nuclear deformation of ^{150}Nd.

In the present reaction systems, it is concluded that the collision of projectile at the side ($\theta_{coll} > 50° - 60°$) of deformed nucleus results in a fusion without any hindrance, whereas the collision near the tip of deformed nucleus does not result in the formation of the compound nucleus even if the potential barrier at that point is surmounted.

This fact can be qualitatively understood by considering the position of the contact point relative to the saddle point of the compound nucleus. In the case of ^{60}Ni + ^{154}Sm, the minimum distance $R_{min}/R_0 = 1.40$ between the mass centers corresponds to the side collision and is shorter than the position of the saddle point $R_{saddle}/R_0 \sim 1.5$ of the compound nucleus ^{214}Th. Here the position of the saddle point was estimated from [14]. On the other hand, the maximum distance $R_{max}/R_0 = 1.71$ corresponding to the tip collision is well outside the saddle point. Here R_0 is the radius of the compound nucleus. In the case of ^{76}Ge + ^{150}Nd, the maximum distance of 1.76 at the tip collision is outside the saddle point $R_{saddle}/R_0 \sim 1.4$ of ^{226}U. The minimum distance of 1.43 at the side collision is close to the saddle point. If the compact configuration at the contact point tends to evolve into the formation of the compound nucleus than more elongated configuration, the compound nucleus may be more easily formed for the side collisions than the tip collision for the present reaction systems.

The present experimental result indicates that the nuclear deformation plays a significant role for the formation of the compound nucleus in the massive heavy-ion fusion reaction. This result is consistent with the theoretical predictions of the gentle fusion [15] and the hugging fusion [16] between two deformed nuclei. It is predicted in these fusion models that when two well-deformed nuclei approach under the condition of their symmetry axes orthogonal to each other, these two nuclei form the most compact configuration at the contact point and then tend to evolve into the compound nucleus formation with a high probability. In order to confirm the present conclusion and the theoretical prediction, it is necessary to continue further experimental studies of the fusion between heavier projectiles and deformed target nuclei at the vicinity of the Coulomb barrier.

ACKNOWLEDGMENTS

The authors thank the crew of the JAERI tandem-booster facility for the beam operation.

REFERENCES

1. S. Hofmann, *Rep. Prog. Phys.* **61**, 639-689 (1998) and see the references therein.
2. Mitsuoka, S., Ikezoe, H., Nishio, K., and Lu, J., *Phys. Rev. C (2000) in press.*
3. Nishio, K., Ikezoe, H., Mitsuoka, S., and Lu, J., *Phys. Rev. C* **62**, 014602-1-12 (2000).
4. Kuzumaki, T., Ikezoe, H., Mitsuoka, S., Ikuta, T., Hamada, S., Nagame, Y., Nishinaka, I., and Hashimoto, O., *Nucl. Instrum. Methods Phys. Res.* **A 437**, 107-113 (1999).
5. Reisdorf, W., and Schädel, M., *Z. Phys.* A **343**, 47-57 (1992).
6. Fernández, J. O., Dasso, C. H., and Landowne, S., *Comput. Phys. Commun.* **54**, 409-412 (1989).
7. Junghans, A. R., de Jong, M., Clerc, H. –G., Ignatyuk, A. V., Kudyaev, G. A., and Schmidt, K. –H., *Nucl. Phys.* **A629**, 635-655 (1998).
8. Iljinov, A. S., Mebel, M. V., Bianchi, N., De Sanctis, E., Guaraldo, C., Lucherini, V., Muccifora, V., Polli, E., Reolon A. R., and Rossi, P., *Nucl. Phys.* **A543**, 517-557 (1992).
9. Cohen, S., Plasil, F., and Swiatecki, W. J., *Ann. Phys. (N. Y.)* **82**, 557-596 (1974).
10. Bjørnholm, S., and Swiatecki, W. J., *Nucl. Phys.* **A391**, 471-504 (1982).

11. Quint, A. B., Reisdorf, W., Schmidt, K. -H., Armbruster, P., Hessberger, F. P., Hofmann, S., Keller, J., Münzenberg, G., Stelzer, H., Clerc, H. -G., Morawek, W., Sahn, C. -C., *Z. Phys.* A **346**, 119-131 (1993).
12. Back, B. B., Fernandez, P. B., Glagola, B. G., Henderson, D., Kaufman, S., Keller, K. G., Sanders, S. J., Viedebæk, F., Wang, T. F., and Wilkins, D., *Phys. Rev. C* **53**, 1734-1744 (1996).
13. Nishio, K., Ikezoe, H., Mitsuoka, S., Satou, K., and Jeong, S. C., submitted to *Phys. Rev C*.
14. Möller, P, and Nix, J. R., *Nucl. Phys.* **A272**, 502-532 (1976).
15. Nörenberg, W, "Fusion of Heavy Nuclei and a Novel fusion Path to Superheavies," in *the International Workshop on Heavy-Ion Fusion*, edited by Stefanini et al., World Scientific, Padova, Italy, 1994, pp.248-253.
16. Iwamoto, A., Möller, P., Nix, J. R., and Sagawa, H., *Nucl. Phys.* **A596**, 329-354 (1996).

Production of Superheavy elements at GANIL

C. Stodel[1], N. Alamanos[3], N. Amar[2], J.C. Angélique[2], R. Anne[1],
G. Auger[1], J.M. Casandjian[1], R. Dayras[3], A. Drouart[3], J.M. Fontbonne[2], A.
Gillibert[3], S. Grévy[2], D. Guerreau[1], F. Hanappe[2], R. Hue[1],
A.S. Lalleman[1], N. Lecesne[1], T. Legou[2], M. Lewitowicz[1],
R. Lichtenthäler[4], E. Liénard[2], L. Maunoury[5], W. Mittig[1], N. Orr[2],
J. Péter[2], E. Plagnol[6], G. Politi[7], M.G. Saint-Laurent[1], J.C. Steckmeyer[2],
J. Tillier[2], R. de Tourreil[1], A.C.C. Villari[1], J.P. Wieleczko[1], A. Wieloch[8].

[1] *GANIL, B.P. 5027, F-14076 Caen Cedex 5, France*
[2] *LPC-ISMRa, Bld Maréchal Juin, F-14050 Caen Cedex, France*
[3] *CEA-Saclay, DAPNIA-SPhN, F-91191 Gif sur Yvette Cedex, France*
[4] *Instituto de Física da Universidade de São Paulo, C.P. 66318, 05315-970 São Paulo, Brazil*
[5] *C.E.A.-DAM BIII, B.P. 12, F-91680 Bruyères-Le-Chatel, FRANCE*
[6] *I.P.N., IN2P3-CNRS, F-91406 OrsayCedex, FRANCE*
[7] *L.N.S., v.S. Sofia 44, I-95100 Catania, Italy*
[8] *Instytut Fizyki, Uniwersytet Jagellonski, ul. Reymonta 4, 30-059 KRAKOW 16*

Abstract. A long-term new experimental program has begun at GANIL, i.e. search for new super heavy nuclei and their structure. The first part consists in studying the structure of the $^{273}110$ isotope which involves the development of high intensity Se beam. In parallel, reactions involving Inverse Kinematics will be studied allowing to have a versatileness set-up. By adding germanium and electron detectors, spectroscopic studies could be made on trans-fermium elements. Preliminary results showed that the Wien Filter has a suppression of the incident beam with a 10^{10} factor, which is comparable with results elsewhere. We show recent results with the present set-up at GANIL in producing Fr isotopes in the Kr + Sb reaction. We present also the result of our Kr + Pb experiment, which tried to reproduce the Berkeley result of the element 118.

1 INTRODUCTION

During the last decades, the search for super heavy nuclei went on mainly at the Berkeley, GSI and Dubna laboratories. Searching for new elements is an attempt to answer questions of fundamental character: the investigation of nuclei at the limits of stability. GANIL and LPC (Caen), supported by CNRS and CEA, took in 1997 the opportunity to use the velocity filter LISE3 in order to investigate this field of research. For 3 years, technical developments and test experiments have demonstrated the capabilities of the GANIL set-up to pursue this quest of super-heavy elements. An attempt to produce the element 118 with the reaction ^{86}Kr + ^{208}Pb, in a repetition of the Berkeley experiment, has been performed in November-December 1999. In this experiment, as in those performed at GSI and RIKEN, no α decay channel corresponding to the Berkeley results has been observed. A new experimental program, using a beam of ^{82}Se is now suggested to pursue this research by producing "neutron rich" isotopes of the elements Z=109, 110, 111, 112, etc. Experiments on the study of the structure of the super-heavy elements and on the analysis of the possibilities offered by inverse kinematics with the set-up are also proposed and described in this article.

2 EXPERIMENTAL SET-UP

The super heavy elements are produced by complete fusion between an incident and a target ion. The beam is produced by the high intensity ECR Ion Sources of GANIL, it is then accelerated to low energy (4-5.5 MeV/u) in the CSS1 cyclotron and driven through the LISE spectrometer. The beam irradiates a target located in front of the Wien filter. After de-excitation at the target stage, the evaporation residues (ER) are separated from the incident beam using the LISE3 Wien Filter. After implantation in a double-stripped Si-detector, the ER's are identified by their α-decay chains.

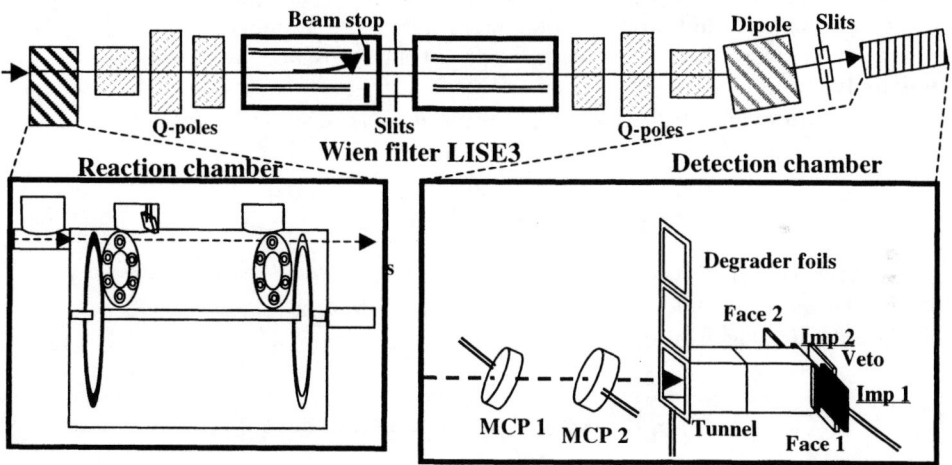

Figure 1. Experimental set-up.

Targets

For targets with a low melting point (Pb, Bi, ...), two wheels, with a diameter of 670mm, bearing 36 targets are mounted on a same axis to rotate in coincidence at

2000RPM. Targets (300 µg/cm^2) are mounted on the first wheel and carbon foils on the second one. The targets are "sandwiched" between two carbon foils of 40 and 10 µg/cm2. A Si detector continuously monitors the status of each target. The Carbon foils (≈50 µg/cm^2) are needed to equilibrate the charge state of the reaction products.

For materials with higher melting point, a single rotating target will be used.

Wien Filter LISE III

The incident beam is deflected out after the first half of the velocity filter. During the first tests, the distance between the beam axis and the upper electrode of the first section of the Wien filter was 5 cm and an opening of 10cm long has been built at its exit, allowing a better suppression of scattered incident particles. This suppression will be improved by increasing the distance between the beam axis and the upper electrode to 7 cm. A dipole magnet at the exit of the filter improves also the suppression of unwanted products.

Detectors

The tagging of implanted particles as well as their velocities are obtained with two micro channel plate detectors [1]. Their kinetic energy and localization are given by a X-Y silicon implantation detector. The energy of alphas and of the fission fragments escaping from the implantation detector is measured with a "tunnel" of 8 silicon detectors. A silicon veto detector is installed behind the implantation detector. In order to measure long half-life products without background, the implantation detector is moved out and replaced by a second one when a possible interesting event is registered.

Specific electronics with a double trigger data acquisition system have been developed. The dead time between two successive events is 10 µs.

Test experiments

Test experiments have begun in 1996, to get a response of the complete set-up in the fusion reaction conditions. A ^{58}Ni beam irradiated a natural tin target, with an intensity of few µAe. A rejection rate of the incident beam of 10^{10} was obtained.

3 SEARCH FOR ELEMENT 118

After tests in 1997 and 1998, the full experimental set up was used with the system ^{86}Kr + ^{121,123}Sb producing Fr isotopes. The excitation functions of these systems were measured at GSI [2]. With a beam intensity of 10 pnA, at 4.3 MeV/u, a rejection rate of the incident ^{86}Kr beam of 2x 10^9 was obtained. The α decay lines of Fr isotopes and their daughters were observed by the silicon detector (Figure 2).

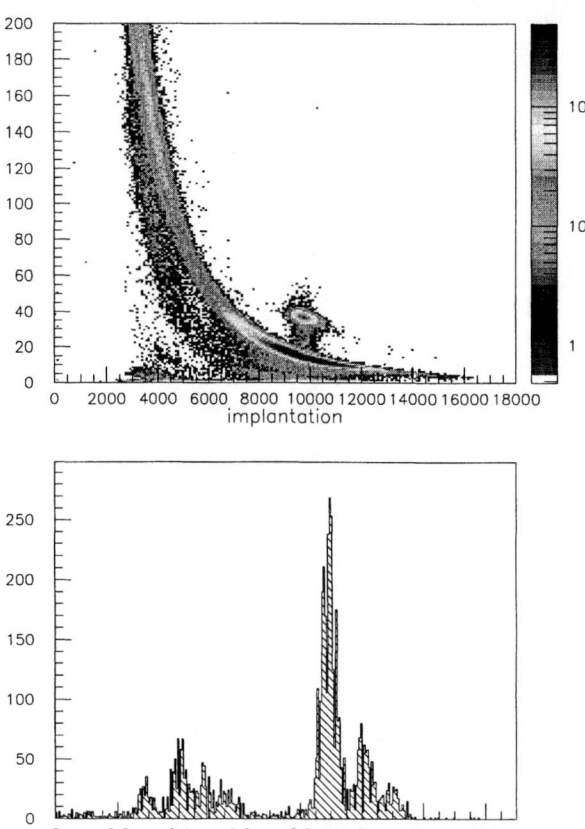

FIGURE 2. upper part: Energy versus Time-Of-flight spectrum for the system ^{86}Kr (4.3 MeV/u) + ^{121,123}Sb; lower part: α spectrum in anti-coincidence with the Time-Of-Flight.

The beam was then tuned to 5.27 MeV/u in order to make measurements on the ^{86}Kr + ^{208}Pb system for which long alpha chains attributed to element 118 were observed at Berkeley [3]. The upper part of figure 3 represents the raw energy spectrum obtained with the implantation silicon detector. Events in anti-coincidence with the Time-Of–Flight detector are plotted on the spectrum of the lower part. Figure 4 represents events detected in the two micro channel plate detectors and in the implantation Si detector. With a total dose of 1.1×10^{18} ions on a 300μg/cm^2 lead targets, no α chain within the expected energies was observed at GANIL, in agreement with the results of SHIP [4].

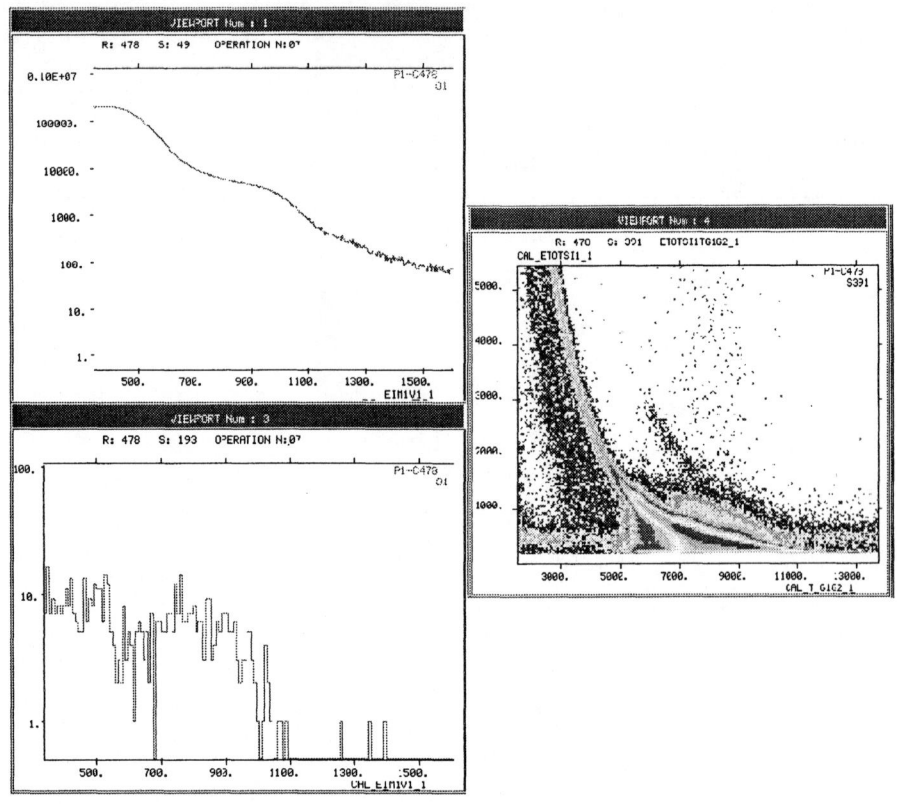

FIGURE 3 (LEFT). α spectra with and without Tof anti-coincidence for the system ^{86}Kr + ^{208}Pb at 5.27 MeV/u.

FIGURE 4 (RIGHT). Time of flight-Energy spectrum for the system ^{86}Kr + ^{208}Pb at 5.27 MeV/u

4 FUTURE

Study of super-heavy elements using a ^{82}Se beam.

Since 1987, the super-heavy element (107 to 112) were first synthesised by cold fusion reactions at GSI, Darmstadt, with the Wien filter SHIP: complete fusion of ^{208}Pb or ^{209}Bi targets with neutron rich natural projectiles up to ^{70}Zn at excitation energies lower than 20 MeV. Up to element 110, the identification was unambiguous via known α radioactive decay chains of the daughter nuclei. For the elements Z=111 [5] and 112 [6], the identification is not so clear because of unknown daughter nuclei:

The element 111 [5] produced in the reaction ^{64}Ni(^{209}Bi, 1n)^{272}111, decays by α emission on the unknown ^{268}109 and ^{264}107 isotopes and the chain finally ends on the known ^{260}105 and ^{256}Lr isotopes.

The element 112 [6] was produced in the reaction ^{70}Zn(^{208}Pb, 1n) ^{277}112. Two α decay chains were then observed, ending on the known ^{265}106, ^{261}104 and ^{257}No. Unfortunately, data on the daughter nuclei ^{273}110 and ^{269}108 are very scarce. Energy of 11.35 MeV was previously reported for the α decay of ^{273}110 by the Dubna-Livermore group [7]. However, in the two decay chains of element 112, the second α corresponding to the decay of the same ^{273}110 isotope was observed with energies of 9.73 and 11.08 MeV and with decay-times of 170 ms and 110 μs, respectively. These significant differences in energy and lifetime could only be imputed to different levels in the ^{273}110 isotope. A new experiment made at GSI recently [8] with the same reaction ^{70}Zn(^{208}Pb, 1n)^{277}112 lead to the observation of only one α-chain. The α-energies are 11.17, 11.20 and 9.18 MeV for the decay of the isotopes ^{277}112, ^{273}110 and ^{269}108, respectively, and therefore, there is no definite conclusion concerning the α-decay energy of the ^{273}110 isotope.

In order to check precisely this point, it seems obvious that a direct production of the ^{273}110 isotope is necessary. Moreover, this isotope of the element 110 was never produced directly by the cold fusion method; the most "neutron-rich" element thus produced being the isotope ^{271}110.

We propose to synthesize directly this isotope (^{273}110) in order to determine its decay properties and to, indirectly, confirm the existence of element 112, by the reaction ^{82}Se(^{192}Os, 1n)^{273}110. This system has a low fusion barrier, which enables the reaction to occur at low excitation energy for the compound nucleus (see Table 1). Compared to typical cold fusion reactions with lead or bismuth targets (^{64}Ni(^{208}Pb, 1n)^{271}110) (σ_{max}=15pb), the cross-section may be, on one hand lowered due to the decreased asymmetry of the system and the absence of closed shell in the reaction partners, but on the other hand, it should increase due to the more "neutron rich" compound nucleus. From systematic and extrapolations, a cross section in the range of picobarns can be expected for the production of element ^{273}110.

Before attempting to produce element Z=110, the entire set-up will be tested and calibrated using a reaction with a higher cross-section. For this purpose, the detection system will be checked with auxiliary targets from ^{138}Ba to ^{154}Sm (See Table 1) producing by emission of few neutrons, isotopes of Th to Cm which decay by α emission. The cross-sections are expected to be at the microbarn level. The system ^{82}Se +^{138}Ba is of particular interest because it gives a compound nucleus (^{220}Th*) already produced via other entrance channels [9]. The measured cross sections could then be used to study the influence of the entrance channel on sub-barrier fusion.

Table 1 shows other targets` which lead to isotopes of elements 104, 106 and 108:

a) Using ^{176}Yb and ^{180}Hf targets, the selenium beam enables to form the same compound nuclei (^{258}Rf* and ^{262}Sg*) as the reactions ^{50}Ti+^{208}Pb (σ_{max} (E*=15MeV)=10nb) [10] and ^{54}Cr+^{208}Pb (σ_{max} (E*=16MeV)=400pb) [11]. The results

TABLE 1. Compound nuclei produced with a selenium beam, E^*_{Bass} is the excitation energy at the Bass barrier [24].

Target	Compound nucleus	E^*_{Bass} (MeV)
^{197}Au	^{279}113	7.2
^{198}Pt	^{280}112	11.6
^{193}Ir	^{275}111	12.7
^{192}Os	^{274}110	14.6
^{187}Re	^{269}Mt (Z=109)	15.6
^{186}W	^{268}Hs (Z=108)	17.2
^{180}Hf	^{262}Sg (Z=106)	17.7
^{176}Yb	^{258}Rf (Z=104)	20.1

of the measured cross-sections of the one neutron evaporation channel will be useful to compare the entrance channels using lead targets or selenium beam leading to the same compound nuclei.

b) A comparative study can be made using a ^{186}W target giving the compound nucleus ^{268}Hs* (Z=108). The isotope ^{265}Hs was produced from the reaction ^{58}Fe +^{208}Pb [11], with a maximum cross section of 50 pb at 13 MeV of excitation energy. The measured cross-section will indicate the effects, on one hand of the absence of the closed shell in the entrance channel and the decreased asymmetry, and on the other, of the increased mass (N=N+2) of the compound nucleus.

Furthermore, the same selenium beam could be used in the future to search for new isotopes of the elements 111 (^{82}Se(^{193}Ir, 1n)^{274}111), 112 (^{82}Se(^{198}Pt, 1n)^{279}112). The advantage of a ^{82}Se beam is to produce isotopes richer by 2 neutrons than those produced with ^{208}Pb or ^{209}Bi targets for elements Z=108-112.

Inverse Kinematics

Experiments are also proposed to compare direct and inverse kinematics Fusion. In the fusion-evaporation reactions, the difficulty is to detect with a high efficiency the few ER's emitted close to 0 degree in a X-Y implantation detector and to reject the beam particles. At GANIL, the tool used for this rejection is Lise3. However there are some multi-scattered projectiles and other reaction products which have velocities and magnetic rigidities close to the ER's values. The importance of this "background" cannot be calculated and must be measured experimentally and reduced by adjustments of several parameter settings.

The proposal consists in using the method of complete fusion reactions, cold or hot, in inverse kinematics: beams of ^{208}Pb or other heavy nucleus. Compared to usual kinematics, this method has one drawback but several assets.

Drawback :

Due to the smaller velocity difference between the beam and the evaporation residues, larger magnetic and electric fields are needed in the Wien filter Lise3 to deflect the beam: the Wien filter is powerful enough for this.

Assets :
- The ER's are strongly focussed at forward angles and have a much larger velocity, which ensures a good transmission even after α-emission in the target or α-decay in flight
- In usual kinematics, the target thickness is limited by multiple scattering. A thickness of 300-400 µg/cm² corresponds to an excitation energy range $\Delta E^* $ = 2-3 MeV. In inverse kinematics, the velocity of the ER's is larger, the target mass is smaller, therefore the thickness of several mg/cm² could be used allowing one to cover a larger excitation range.
- The ionic charge state distributions of super-heavy nuclei are better estimated at higher energies [12, 13, 14, 15].
- The energy of an ER is large allowing to better control the depth inside the Si detector.

The purpose of this experiment is to find out how useful and convenient is the inverse kinematics method and then to apply it to the production of super-heavy nuclei.

Structure of super heavy elements

A new experimental research program at GANIL on the study of trans-fermium isotopes is proposed. The main goal is to achieve spectroscopic information of these nuclei in order to have a better understanding of their shell structure. It is proposed to initiate this program with the α-, e⁻-, and γ-spectroscopy of ^{251}Md and ^{251}Fm populated by the α-decay of ^{255}Lr.

The [521] 1/2- and [514] 7/2- orbitals are expected to be close to the Fermi surface in the Md isotopes. Experimentally a spin of 7/2- was assigned to the ground state of ^{255,257,259}Md [16, 17] and the 1/2- is expected as the first excited state but the prediction is reversed as compared to the other Md isotopes : 1/2- as ground state and 7/2- as first excited state at 300 keV [18] which should decay via an M3 transition. However, experiment performed on ^{251}Md indicate the presence of an excited state at 160 keV with a branching ratio of ~50% [19].

^{251}Fm was studied in 1971 by α-decay of ^{255}No [20, 21]. Two rotational bands were found and two gamma transitions were deduced. The ground state spin and parity was tentatively assigned to 9/2- by favored alpha transition to the 9/2- state in ^{247}Cf.

The ground state of the N=151 isotones is based on the [734] 9/2- Nilsson orbital. The 9/2- ground state character was firmly established in ^{249}Cf. The [622] 5/2+, [624] 7/2+ and [620] 1/2+ bandheads have also been firmly established in ^{249}Cf and tentatively in ^{251}Fm. The Z=100 is a gap in the proton single particle energy level and N=151 is one neutron below the stabilizing shell effect at N=152 so that ^{251}Fm is well suited to study neutron single-particle states.

The goal of the experiment is to measure precisely via e⁻- and γ-spectroscopy the single particle structure of ^{251}Md and ^{251}Fm. The compound system ^{255}Lr produced in the 2n channel of the ^{48}Ca + ^{209}Bi fusion evaporation reaction [22] is expected to decay by 85% to the 4 min half-life ^{251}Md. This nucleus should decay with less than 10% by α-decay to ^{247}Es and with more than 90% by EC to ^{251}Fm. Since the ground state parity and spin of the first level of ^{251}Fm are known, we should be able to determine the ground state of ^{251}Md and possibly the one of ^{255}Lr.

The experimental set-up should be the same as the one described above, additional detectors will be added. A tunnel of Si detectors will detect conversion-electrons and α particles escaping the Si strip detector. In this way also α-conversion-electrons angular correlations which are sensitive to the spin of the level involved can be measured. A set of Ge clover detectors from the EXOGAM collaboration positioned in a close geometry (at 15cm around the silicon strip detector and at 5cm at 0°) will detect γ- and X-rays (in particular those associated with the electron capture of ^{251}Md).

5 SUMMARY

A long-term new experimental program has begun at GANIL, i.e. search for new super heavy nuclei and their structure. The first part consists in studying the structure of the ^{273}110 isotope which involves the development of high intensity Se beam. In parallel, reactions involving Inverse Kinematics will be studied allowing to have a versatileness set-up. By adding germanium and electron detectors, spectroscopic studies could be made on trans-fermium elements. Preliminary results showed that the Wien Filter has a suppression of the incident beam with a 10^{10} factor, which is comparable with results elsewhere. The perspective of having very high intensity beam on the target [23] and the versatility of the Wien filter of LISE (unique in the world) allows us to contribute in this field in a worldwide frame.

Particularly to the development involving the Se beam, this program opens to GANIL the opportunity to confirm the new element 112 in a first step, and to go on with the synthesis of new elements by a channel which is different from the one used up to now at GSI, Berkeley or Dubna, i.e. with Lead, Bismuth or Actinide targets.

ACKNOWLEDGMENTS

The authors thank the accelerator staff for the excellent performance of the ^{76}Kr beam. We are indebted to the target laboratories of IFUSP, Brazil, LNS, Catania and IPN-Orsay, France for the skilful preparation of the large-area lead-targets. We also acknowledge the groups STCM (Service des Techniques de Cryogénie et de Magnétisme) and SIG (Service d'Insrtumentation Générale) from Saclay, France, for the conception and completion of the fast rotating target wheel and scattering chamber. We are grateful to the GIP (Groupe Informatique pour la Physique, GANIL) who developed the dedicated fast acquisition system.

REFERENCES

1. O. H. Odland et al, *Nuc. Instr. And Meth.* **A378**, 149-157 (1996).
2. C. Sahm, *Nucl. Phys.* **A441**, 316 (1985).
3. V. Ninov etal., *Phys. Rev.Letters* **83**, 1108 (1999).
4. S. Hofmann, *Proc. Conf. Exp. Nuclear Phys. In Europe*, 1999.
5. S. Hofmann et al., *Zeit. Phys.* **A 350**, 281-282 (1995).
6. S. Hofmann et al., *Zeit. Phys.* **A 354**, 229 (1996).
7. Yu.A. Lazarev et al., *Dubna Report* **No E7-95-552**.
8. S. Hofmann and G. Münzenberg, *GSI Preprint* 2000-02, Januar 2000, to be published in *Rev. Mod. Phys.*.
9. C. Stodel, *PhD University of Caen* (1998).
10. F.P. Hessberger et al., *Zeit. Phys.* **A 359**, 415-425 (1997).
11. S. Hofmann et al., *Rep. Prog. Phys.* **61**, 639-689 (1998).
12. E. Baron et al, *Nucl. Inst. Meth.* **A328**, 177 (1993).
13. V.S. Nikolaev and I.S. Dmitriev, *Phys. Letters* **28A**, 277 (1968).
14. K. Shima et al, *At. Data and Nucl. Data Tables* **51**, (1992), *phys. Rev.***A40, 3557** (1989).
15. R.N. Sagaidak and A.V. Yeremin., N*ucl. Ins. Meth.* **B93**, 103 (1994).
16. K. Eskola et al., *Phys. Rev.* **C4** (1971) 632.
17. I. Ahmad *et al.*, *Phys. Rev.* **C61** (2000) 044301.
18. C.E. Bemis et al., *ORNL-5137* (1976) 73, *ORNL-5111* (1976) 58.
19. S. Cwiok et al., *Nucl. Phys.* **A573** (1994) 356.
20. P. Eskola et al., *Phys. Rev.* **C2** (1970) 1058.
21. C.E. Bernis et al., *ORNL-4706* (1971) 62.
22. Gäggeler et al., *Nucl. Phys.* **A502** (1989) 561c.
23. Baron E., "Upgrading the GANIL facilities for High Intensity H. I. Beams (THI project)", *14th Int. Conf. Cycl. And Their Applications*, Cape Town, October 1995.
24. R. Bass, *Nucl. Phys.* **A 231**, 45 (1974).

Search for a Z = 118 Superheavy Nucleus in the reaction of Kr beam with Pb target at RIKEN

Kouji Morimoto, Kosuke Morita, Isao Tanihata, Naohito Iwasa,
Rituparna Kanungo, Toshiyuki Kato, Kenji Katori, Hisaaki Kudo,
Toshimi Suda, Isao Sugai, Satoshi Takeuchi, Fuyuki Tokanai,
Koji Uchiyama, Yoshiaki Wakasaya, Takayuki Yamaguchi,
Alexander Yeremin, Akira Yoneda and Atsushi Yoshida
RIKEN SuperHeavy Experimental Group

The Institute of Physical and Chemical Research (RIKEN), 2-1, Hirosawa, Wako-shi, Saitama 351-0198, Japan

Abstract. We performed two search experiments for a Z = 118 Superheavy nucleus in the reactions of ^{86}Kr + ^{208}Pb and ^{84}Kr + ^{208}Pb. We accumulated a dose of 2.1 x 10^{18} ions and 2 x 10^{18} ions, respectively. But we did not find any candidate of the evaporation residues with both experiments. A future plan proceeded by RIKEN Superheavy elements experimental group is also described.

INTRODUCTION

Recently, three decay chains were reported by a Lawrence Berkeley Laboratory group [1] in a reaction products of ^{86}Kr with ^{208}Pb. The observed chains are consistent with the formation of ^{293}118 and its sequential α decays. However, the decay chains are not connected to any known nuclei. The observed production cross section is 2.2 (+2.6, -0.8) pb. This experiment was motivated by a following recent prediction of Smolanczuk [2], which indicates that the cross section of a cold-fusion reaction of ^{86}Kr with ^{208}Pb is 670 pb. Although the experimental cross section reported is by 3 hundred times smaller than that of the theoretical prediction, we made a followup experiment of LBL group, because it is still much bigger than the cross section systematics with extrapolation of experimental values.

If the cross section of a cold-fusion reaction of ^{84}Kr with ^{208}Pb have almost the same as that of ^{86}Kr with ^{208}Pb, as suggested by T. Wada et al., we have a chance to observe ^{291}118 with decay chains connected to the known nuclei. From this view, we conducted an another experiment to observe the Superheavy nucleus ^{291}118 produced in the reaction of ^{84}Kr with ^{208}Pb. Figure 1 shows the decay chains of the Z = 118 Nucleus.

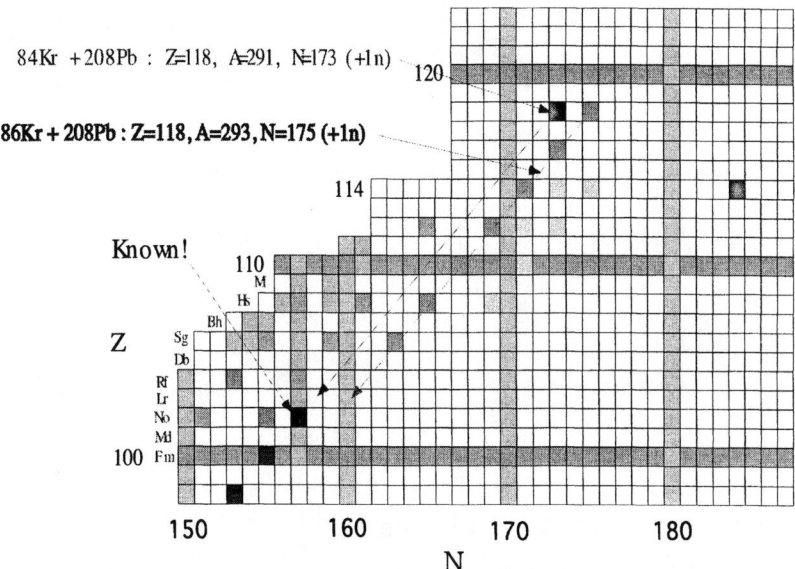

FIGURE 1. The reaction of ^{86}Kr with ^{208}Pb is the same reaction as the LBL's experiment. The reaction residues from the ^{84}Kr with ^{208}Pb have a possibility connecting to a known nuclei after 8th α decays.

EXPERIMENTAL SETUP

The experimental setup was in the following. The target thickness was 320-440 μg/cm^2; it was evaporated onto the downstream side of a 100 μg/cm^2 carbon foil. Furthermore, the target was covered by a 20 - 30 μg/cm^2 carbon with evaporation. Six segments of it were mounted on a wheel that rotated at 600 rpm. The beam energy was 457.6 MeV, and the beam current was 500 particle nanoamperes in average. The incident beam energy at the center of the target was 449 MeV. To avoid the primary beam from impinging on the silicon detector in the event of a broken target, a carbon foil of 30 μg/cm^2 thickness was mounted as a charge stripper of primary beams. The beam intensity was monitored through elastic scattering by a silicon detector installed at 45 degrees with respect to the incident beam. The reaction products recoiling out of the target were separated from the primary beam by using a gas-filled recoil separator, GARIS. [3] Figure 2 shows the schematic view of the GARIS system. The GARIS consists of three magnets which are arranged in a D-Q-Q configuration. The separator was filled with He gas at one torr. The optimal magnetic field setting was estimated by the formula proposed by Ghiorso et al. [4] The system was checked by the reaction of ^{86}Kr with ^{110}Cd. We have estimated the magnetic rigidity (Bρ) to be 2.1 Tm, which is the same value of the LBL's measurement. After passing through the separator the recoiling particles were implanted in a position-sensitive silicon detector (PSD), which had an active area of 60 x 60 mm with a thickness of 300 μm. Light charged particles passing through the PSD were detected with veto SSD mounted after the PSD for anti-coincidence. Total counting rate of the PSD was several counts/sec under the condition with 1000 particle nanoamperes of the beam intensity. The time-of-flight measurement was additionally performed using a micro-channel plate assembly (MCP) and the PSD. A dead time of electronics was reduced by using a double trigger system. It was 20 μs for less than 2 events and 800 μs for more than 3 events.

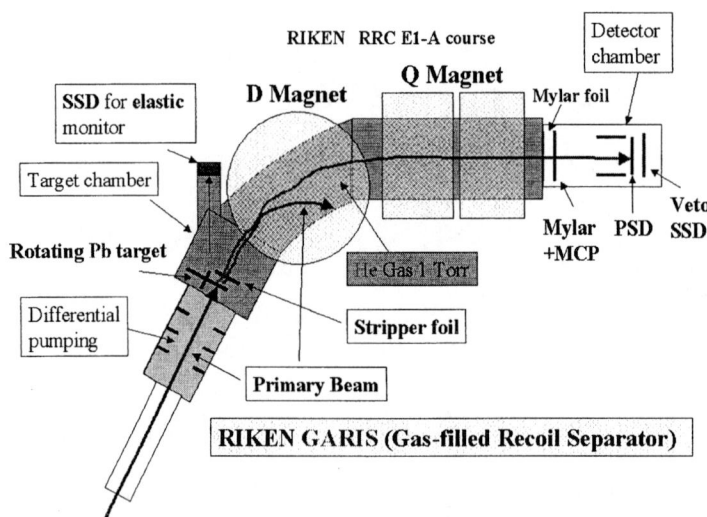

FIGURE 2. Experimental set up

RESULTS AND DISCUSSION

We had machine times of two weeks in Feb. 2000. The first 4 days was performed for a test of targets, an estimation of the magnetic rigidity (Bρ), calibrations of detectors and a check of the data-acquisition system. The last 10 days were performed to accumulate a dose of 2 x 10^{18} ions, but no candidate of the evaporation residue was found. In this beam dose, when we assume that the efficiency of GARIS was 0.8 and that one event was observed, the cross section corresponds to 0.6 pb and this also corresponds to the upper limit to be 1.2 pb.

Since we used a gas-filled separator with a helium gas as LBL group did, a problem of charge changing decay of isometric states in flight will be less significant than the production experiment with vacuum type kinematic separator as SHIP at GSI or as LISE at GANIL. We chose a value of the magnetic rigidity of the separator to be 2.1 Tm this time, in order to make the experimental condition as close as that of the LBL group. But we still have an ambiguity on this value. Because the mean values of the charge states of such atoms, as with its atomic number larger than 110, in helium gas, have not been well studied experimentally, we should perform a systematic study of them.

The other experiment was performed in the reaction of ^{84}Kr with ^{208}Pb in Sep. 1999. The experimental set up was the same as ^{86}Kr with ^{208}Pb experiment. The incident beam energy at the center of the target was 442.2 MeV. We accumulated a dose of 2×10^{18} ions, but no candidate of the evaporation residue was found.

FUTURE PLANS

The RIKEN Linear accelerator RILAC is being upgraded after these experiments. The GARIS system is moved just behind the RILAC from the Ring-Cyclotron Accelerator. In order to reduce a background another Dipole magnet is added just after the GARIS Quadrapole magnets. After the upgrade of the RILAC completed, the maximum energy will be up to about 6 A MeV from 4 A MeV. It is enough energy for inducing cold fusion reaction. Furthermore, the beam intensity is expected about more than 10 times larger than the previous one. The start operation of new system is expected in early 2001. Our coming experimental plans are as follows. We start from a follow-up experiment that has a large cross section reaction for checking the new system. In the next we measure an average charge state of Superheavy element in the Helium gas. Finally we continue trials to search the Superheavy element not yet observed.

REFERENCES

1. V. Ninov et al.: Phys. Rev. Lett. **83**, 1104 (1999).
2. R. Smolanczuk: Rhys. Rev. C **59**, 2634 (1999).
3. K. Morita et al.: Nucl. Instrum. Methods Phys. Res. B **70**, 220 (1992).
4. A. Ghiorso et al.: Nucl. Instrum. Methods Phys. Res. A **269**, 192 (1988).

Spectroscopic Studies of Mass-separated Heavy Nuclei

M. Asai[1], M. Sakama[1,2], K. Tsukada[1], S. Ichikawa[1], H. Haba[1],
I. Nishinaka[1], Y. Nagame[1], S. Goto[1,3], Y. Kojima[4], Y. Oura[2],
H. Nakahara[2], M. Shibata[5], and K. Kawade[5]

[1] *Advanced Science Research Center, Japan Atomic Energy Research Institute, Tokai, Ibaraki 319-1195, Japan*
[2] *Department of Chemistry, Tokyo Metropolitan University, Hachioji, Tokyo 192-0397, Japan*
[3] *Department of Chemistry, Niigata University, Niigata 950-2181, Japan*
[4] *Applied Nuclear Physics, Faculty of Engineering, Hiroshima University, Higashi-Hiroshima 739-8527, Japan*
[5] *Department of Energy Engineering and Science, Nagoya University, Nagoya 464-8603, Japan*

Abstract. EC and α decays of neutron-deficient nuclei $^{233-236}$Am have been studied using a gas-jet coupled on-line isotope separator. A proton orbital of the ground state of ^{235}Am and neutron orbitals in ^{235}Pu have been estimated from a decay scheme of ^{235}Am. The 5^- ground state and the 1^- EC decaying isomer in ^{236}Am and the configuration of two-quasiparticle states in ^{236}Pu have been established. Q_α values of ^{233,235,236}Am have been determined from measured α-particle energies. It was found that evaluated Q_α values were systematically overestimated by 100–200 keV in neutron-deficient Am nuclei.

INTRODUCTION

Experimental studies of heavy and superheavy nuclei have been performed mainly through α-decay spectroscopy due to high sensitivity for α-particle detections. Q_α values extracted from measured α-particle energies determine nuclear mass surface in heavy element region, which reveals stability and shell structure of heavy nuclei. Alpha-decay probabilities and fine structure of α-particle spectra also provide nuclear structure information such as excited states in heavy nuclei, single-particle energies, proton-neutron configurations, and nuclear deformation. On the other hand, γ and conversion electron spectroscopy, which is powerful to study nuclear structure through excited states and electromagnetic properties, is limited to some specific nuclei with large production cross sections and long half-lives of $\gtrsim$30 min because of severe contamination of γ spectra. In order to measure γ rays of these nuclei, it is necessary to isolate the nuclei of interest from a large amount of other reaction products. The nuclei with long half-lives were studied through chemical separations and off-line mass separations [1]. Recently, in-beam γ rays of ^{254}No with ~2 μb were successfully measured through a recoil decay tagging technique [2,3]. In the present work, we have studied neutron-deficient actinide nuclei through EC and α-decay spectroscopy with half-lives $\lesssim$10 min using the JAERI on-line isotope separator (ISOL) coupled to a gas-jet transport system [4].

Neutron-deficient Am, Cm, and Bk nuclei predominantly decay by electron capture, and their α-decay branching ratios are extremely small. Although some short-lived nuclei ^{232,234}Am and ^{238,240}Bk have been studied through observations of EC-delayed fissions due to extremely high sensitivity for detections of fission fragments [5–8], EC and α-decay studies for the nuclei with half-lives <1 h are scarce. In this paper, experimental studies on EC and α decays of $^{233-236}$Am with half-lives of 2–10 min are presented.

EXPERIMENTS

Neutron-deficient americium isotopes were produced by the reaction of ^{233}U(^{6}Li, xn)$^{233-235}$Am and ^{235}U(^{6}Li, $5n$)^{236}Am at the JAERI tandem accelerator facility. A stack of twenty-one uranium targets set in a multiple-target chamber with 5 mm spacings was bombarded with a ^{6}Li beam of about 400 particle-nA intensity. Each target was electrodeposited with an effective thickness of about 100 μg/cm^2. Reaction products recoiling out of the targets were stopped in He gas loaded with PbI$_2$ clusters, and transported into an ion source of the ISOL with gas-jet stream through an 8 m long capillary. Atoms ionized in the surface ionization-type thermal ion source were accelerated with 30 kV and mass-separated with a resolution of $M/\Delta M \sim 800$. The overall efficiency of this ISOL system including an ionization efficiency and a transport one of the gas-jet system was measured to be 10% for ^{143m}Sm produced in the ^{141}Pr(^{6}Li, $4n$) reaction and 0.3% for ^{237}Am in the ^{235}U(^{6}Li, $4n$) reaction [9]. The smaller value for ^{237}Am results from a lower ionization efficiency for Am atoms than that for Sm ones. For γ-ray measurements, the separated ions were implanted into an aluminum-coated Mylar tape in a tape transport system, and periodically transported to a measuring position equipped with two Ge detectors. Gamma-ray singles, γ-γ coincidence, and γ-γ delayed coincidence measurements were performed. For α-decay measurements, a four-position rotating-wheel system was employed. The separated ions were implanted into 10 μg/cm^2 thick PVC/PVAc foils set on the periphery of the wheel. The wheel periodically rotates 90°, which conveys the implanted sources to three consecutive detector stations. Each of the detector stations was equipped with two Si detectors (HAMAMATSU PIN photodiode 18 mm × 18 mm) placed on both sides of the foil to measure α particles with 85% efficiency. Another Si detector was also placed at the implantation station. To determine α/EC branching ratios, a Ge detector was placed at the first detector station. Alpha and γ-ray singles, and α-γ coincidence measurements were performed.

RESULTS

^{233}Am

The nucleus ^{233}Am was identified for the first time in the present experiments through an α-α correlation analysis. The α decay of ^{233}Am is followed by five successive α decays of ^{229}Np ($T_{1/2}$ = 4.0 min) → ^{225}Pa (1.7 s) → ^{221}Ac (52 ms) → ^{217}Fr (22 μs) → ^{213}At (125 ns). Since the last four nuclides decay via 100% α-particle emissions with short half-lives, the α-α correlation events among these five nuclides can be unambiguously identified. Although the ^{229}Np has a substantially long half-life of 4.0 min, extremely low-contaminated α sources from the ISOL allowed us to identify the correlations between the α decays of ^{233}Am and the following α-α correlation events due to low-counting rate conditions of the α detectors. Figure 1 shows an α-particle spectrum constructed from the observed α-α correlation events. The α lines of ^{233}Am and ^{229}Np were clearly distinguished each other, and those of the other nuclides in the decay chain are also seen. The α-particle energy of ^{233}Am was determined to be 6780(17) keV. From the decay curve of the 6780 keV α particles, the half-life of ^{233}Am was determined to be 3.2(8) min. As for the EC decay of ^{233}Am, no Pu KX rays were observed. Based on the detection efficiency for Pu KX rays, the α-branching intensity was estimated as $I_\alpha > 3\%$.

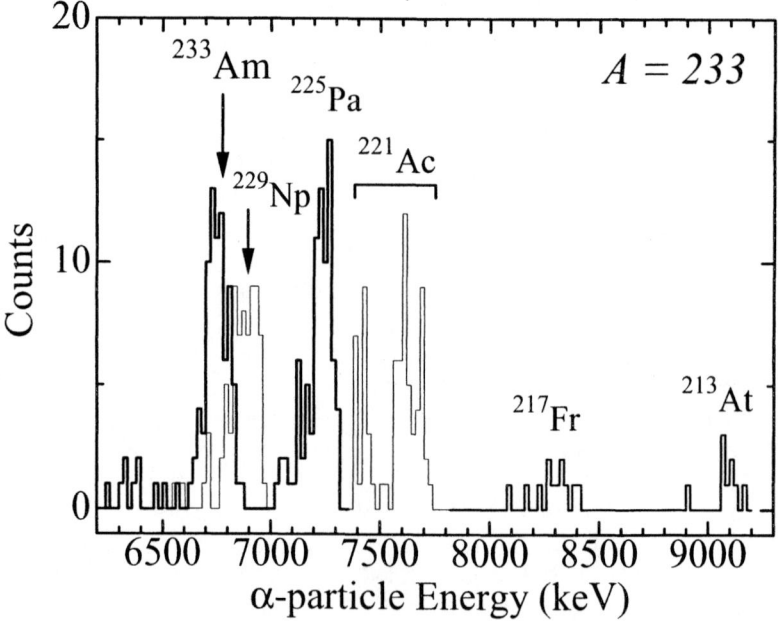

FIGURE 1. α-particle spectrum constructed from α-α correlation events observed in the mass-233 fraction.

^{234}Am

For the α decay of ^{234}Am, Hall et al. [6] reported a 6.46 MeV α group with an α/EC branching ratio of $3.9(12) \times 10^{-4}$ deduced from an α intensity and Pu KX-ray one. The half-life value of ^{234}Am was determined to be 2.32(8) min from the decay of EC-delayed fission events [6]. In the present experiment, Pu KX-ray peaks were weakly observed in the mass-234 fraction, but no α peak was detected at 6.46 MeV. The upper limit of $I_\alpha < 0.04\%$ was deduced for the 6.46 MeV α group.

^{235}Am

The 6457(14) keV α peak was clearly observed in the mass-235 fraction as shown in Fig. 2. Disintegration rate of the observed α counts agreed very well with that of the Pu K_αX rays. Taking a weighted average of those values, we have determined the half-life of ^{235}Am as 10.3(6) min. This half-life value is consistent with the literature value of 15(5) min by Guo et al. [10] extracted from the growth and decay of ^{235}Pu in chemically purified Am fractions. The α-branching intensity of $I_\alpha = 0.40(5)\%$ was derived from the ratio between the observed α and Pu KX-ray intensities.

To establish the energy level in ^{231}Np populated by the α decay of ^{235}Am, an

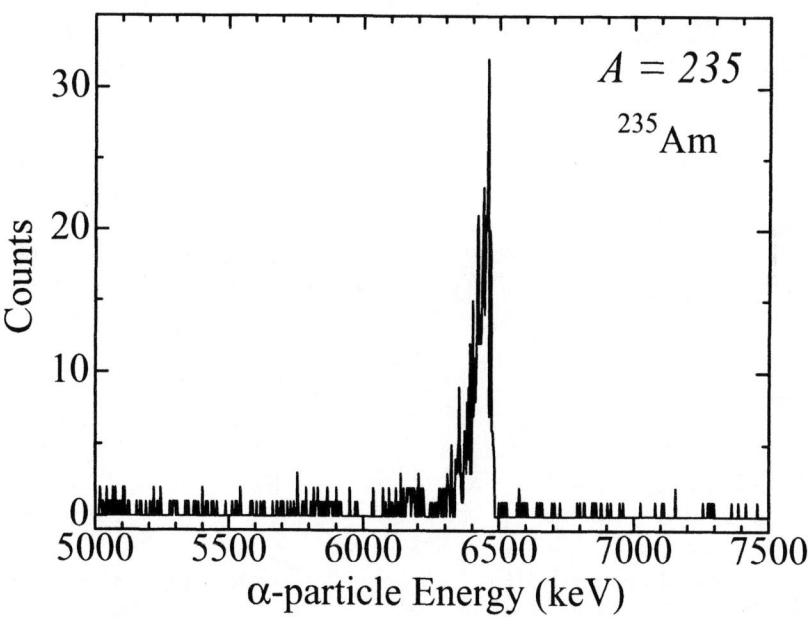

FIGURE 2. α-particle spectrum observed in the mass-235 fraction.

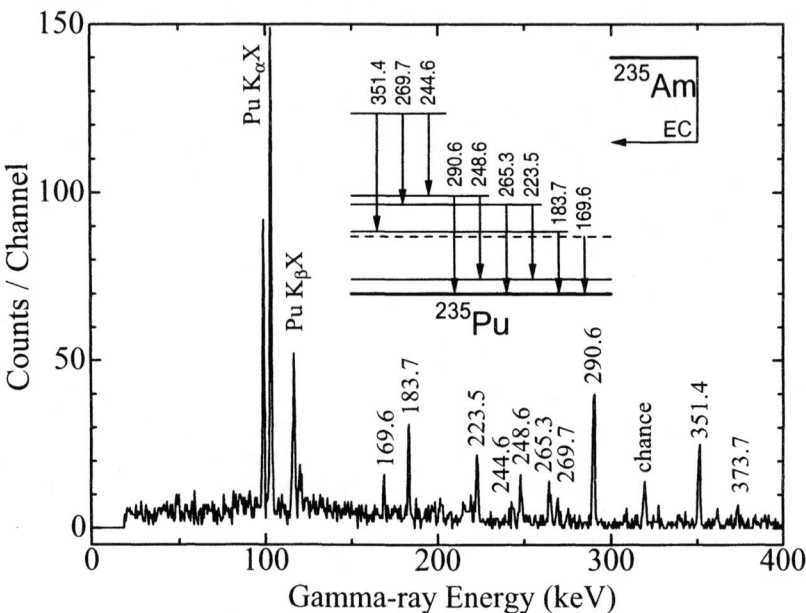

FIGURE 3. γ-ray spectrum in coincidence with Pu $K_\alpha X$ rays associated with the EC decay of ^{235}Am. The inset is a partial decay scheme of ^{235}Am.

α-γ coincidence measurement was performed. For 230 α-particle detections at 6.46 MeV, no α-γ coincidence event was observed. This result leads to an upper limit of the expected $E1$ transition energy of <15 keV between the 5/2+[633] ground state and the 5/2−[523] state to be populated by favored α transitions.

A γ-ray spectrum in coincidence with Pu $K_\alpha X$ rays is shown in Fig. 3. Observed γ lines were attributed to the EC decay of ^{235}Am due to the coincidence with Pu KX rays and the separation of the mass number. Based on γ-γ coincidence relationships, a decay scheme was established as shown in Fig. 3. The <50% EC-branching intensity to the ground state of ^{235}Pu was deduced from an observed Pu KX-ray intensity.

^{236}Am

The α decay of ^{236}Am was previously reported by Hall et al. [11]. They reported a 6.41 MeV α group with an α/EC branching ratio of $4.2(6) \times 10^{-4}$. In the present experiment, however, no 6.41 MeV α peak was observed; the upper limit of the branching intensity for 6.41 MeV α particles was deduced to be $I_\alpha < 2 \times 10^{-5}$. Instead of it, a very weak α group with a few-minute half-life was observed at 6150 keV. The α-branching intensity was deduced to be 4×10^{-5}. We tentatively

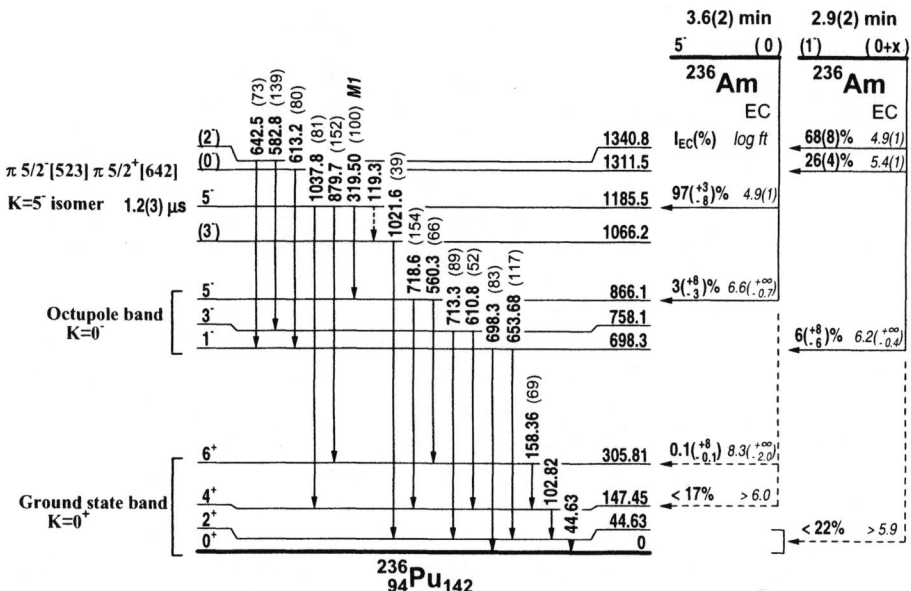

FIGURE 4. A decay scheme of ^{236g,m}Am.

assigned this 6150 keV α group to that associated with the α decay of ^{236}Am.

A decay scheme of ^{236}Am was established as shown in Fig. 4. Only the ground state band with spins up to 16$^+$ has been known in ^{236}Pu [12]. The 698, 758, and 866 keV levels are considered to be the 1$^-$, 3$^-$, and 5$^-$ states in the $K^\pi = 0^-$ octupole band which typically appears at low excitation energy in light actinide nuclei. The 1186 keV 5$^-$ state with a 1.2(3) μs half-life is a K isomer with $K^\pi = 5^-$. It was found that there are two EC-decaying states in ^{236}Am; one is a high-spin state with a 3.6(2) min half-life, and the other is a low-spin state with a 2.9(2) min half-life. These half-life values were extracted from decay curves of γ rays depopulating high-spin states in ^{236}Pu and low-spin ones.

DISCUSSIONS

Nilsson orbitals and proton-neutron configurations

The proton orbital of the ground state of ^{235}Am is expected to be $\pi 5/2^-[523]$ or $\pi 5/2^+[642]$. In fact, the 95th proton of the ground state of $^{237-244}$Am occupies the $\pi 5/2^-[523]$ orbital, and the $\pi 5/2^+[642]$ orbital lies close to the Fermi surface. The $\pi 5/2^+[642]$ hole state is located at 206 keV in ^{241}Am and 187 keV in ^{239}Am, and is expected to lower in energy with decreasing neutron number corresponding to decreasing deformation. For the ground state of ^{235}Am, the present experimental results exclude the $\pi 5/2^+[642]$ assignment as follows. Around this neutron-deficient actinide region, most of EC and β^- transitions, not only first-forbidden but also allowed ones, show $\log ft \gtrsim 5.9$, and only the transitions between the $\pi 5/2^+[642]$ and the $\nu 5/2^+[633]$ orbitals indicate small $\log ft$ values of 5.2–5.9 [12]. The EC transition from the ground state of ^{235}Pu to the $\pi 5/2^+[642]$ state in ^{235}Np shows a $\log ft = 5.4$, indicating that the neutron orbital of the ground state of ^{235}Pu is $\nu 5/2^+[633]$. If the ground state of ^{235}Am is the $\pi 5/2^+[642]$ state, the EC transition to the ground state of ^{235}Pu also shows a small $\log ft$ value. The experimental limit of $\log ft > 5.8$ contradicts this expectation. Thus, the proton orbital of the ground state of ^{235}Am is considered as the $\pi 5/2^-[523]$ orbital.

The EC decays of $^{237-240}$Am are characterized by the fast $\pi 5/2^-[523] \to \nu 5/2^+[622]$ transitions with constant $\log ft$ values of 6.0–6.2. The observed 535 keV level in ^{235}Pu is a candidate for the $\nu 5/2^+[622]$ state because of the large EC feeding to this level. The 41.9 keV level in ^{235}Pu is most likely the $7/2^+$ state in the $5/2^+[633]$ band. The 291 and 265 keV levels are considered as the $5/2^-[752]$ state and the $5/2^+$ state in the $3/2^+[631]$ band, respectively, owing to observed γ branching ratios. The energy spacing between the 184 keV level and the tentatively established 170 keV level is consistent with that between the $1/2^+$ and $3/2^+$ states in the $1/2^+[631]$ band. These spin assignments are tentative, but it is concluded that there are no other Nilsson states below 170 keV in ^{235}Pu.

According to systematics of energy positions of the Nilsson orbitals, the 95th proton and the 141st neutron of the ground state of ^{236}Am are expected to occupy the $\pi 5/2^-[523]$ and the $\nu 5/2^+[633]$ orbitals, respectively. The experimental results of the EC decay of ^{235}Am gave us reliable knowledge to predict this configuration. The $\pi 5/2^-[523]\nu 5/2^+[633]$ configuration leads to the 5^- assignment for the EC-decaying high-spin state in ^{236}Am, and this 5^- state is considered as the ground state according to the Gallagher and Moszkowski coupling rule [13]. The candidate for the low-spin isomer is the 0^- state or its signature partner 1^- state with $K^\pi = 0^-$ and the $\pi 5/2^-[523]\nu 5/2^+[633]$ configuration.

The EC transitions to the 1186, 1312, and 1341 keV states in ^{236}Pu show small $\log ft$ values of 4.9 and 5.4, suggesting that these transitions mainly arise from the $\pi 5/2^+[642] \to \nu 5/2^+[633]$ transition. Since the ^{236}Am^{g,m} have the $\pi 5/2^-[523]\nu 5/2^+[633]$ configuration and the occupied $\pi 5/2^+[642]$ orbital, the $\pi 5/2^+[642] \to \nu 5/2^+[633]$ transition generates a $\pi 5/2^-[523]\pi 5/2^+[642]$ configuration, hence the 1186, 1312, and 1341 keV states in ^{236}Pu are assigned to the $\pi 5/2^-[523]\pi 5/2^+[642]$ two-quasiparticle states. The $\pi 5/2^-[523]\pi 5/2^+[642]$ two-quasiparticle states were also reported in ^{240}Pu at 1309, 1411, and 1438 keV with spin and parities of 5^-, 0^-, and 2^-, respectively [12]. The 1309 keV level is the $K^\pi = 5^-$ $(\pi 5/2^-[523]\pi 5/2^+[642])_{5^-}$ state, and the 1411 and 1438 keV levels are interpreted as the $(\pi 5/2^-[523]\pi 5/2^+[642])_{0^-}$ state and its 2^- rotational band member, respectively [14]. The 1312 and 1341 keV levels in ^{236}Pu are also interpreted as the $(\pi 5/2^-[523]\pi 5/2^+[642])_{0^-}$ state and its 2^- rotational band member, respectively. This interpretation leads to the 1^- assignment for the low-spin isomer in ^{236}Am.

α-decay partial half-lives and Q_α values

The α-decay partial half-lives T_α for ^{233,235}Am are plotted in Fig. 5 as a function of released α-decay energy E_α^* which is a measured α-particle energy corrected by its recoil energy and electron screening [15]. T_α values for the ground state to ground state transitions in even-even Pu isotopes and those for the most prominent α transitions in odd-mass $^{237-243}$Am are also plotted. For ground state to ground state transitions in even-even nuclei, a semi-empirical relations between T_α and E_α^* is known [15]. A fitted curve in Fig. 5 for Pu isotopes represents this relation. T_α values for Am isotopes also exhibit this relation and lie close to the fitted curve for Pu isotopes within a factor of 2.5, indicating that these α transitions in Am isotopes are favored transitions with hindrance factors $\leq$2.5. The T_α values for ^{233,235}Am also clearly indicate that these α decays are favored transitions. This means that these α decays populate the same Nilsson state as that of the ground state of the parent nucleus, that is, the $\pi 5/2^-$[523] state in Np isotopes. The α-γ coincidence measurements revealed that the energy position of the $\pi 5/2^-$[523]

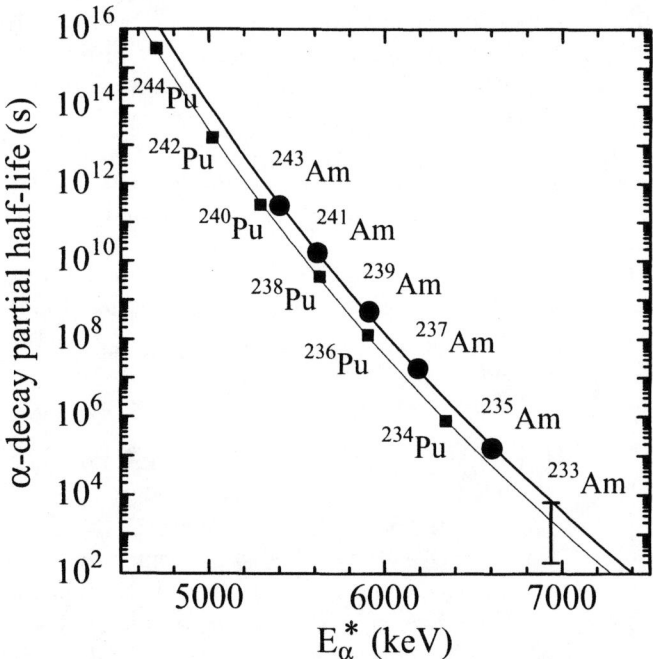

FIGURE 5. α-decay partial half-lives as a function of E_α^*. The partial half-lives for the ground state to ground state transitions in even-even Pu isotopes and the most prominent α transitions in odd-mass $^{233-243}$Am are plotted.

TABLE 1. Comparison between experimental and evaluated Q_α values for $^{233-236}$Am. The evaluated values are taken from systematic predictions by Audi et al. [16]

	^{233}Am	^{234}Am	^{235}Am	^{236}Am
Q_α(sys.) [keV]	7105(240)	6875(220)	6705(220)	6395(140)
Q_α(exp.) [keV]	6898(17)		6569(14)	~6286(40)
ΔQ_α(sys–exp)	+207 keV		+136 keV	+109 keV

state in Np isotopes decreases with neutron number; 60, 49, and <15 keV in ^{237}Np, ^{235}Np, and ^{231}Np, respectively. This lowering indicates the decreasing deformation with neutron number, and suggests that the $\pi 5/2^-[523]$ state becomes the ground state at ^{229}Np.

For ^{236}Am, although observed α group is considered as an admixture of α decays of ^{236g,m}Am, measured α-decay intensity is consistent with the interpretation that these α decays are favored transitions; the α decay of the 5^- ground state populates the $(\pi 5/2^-[523]\pi 5/2^+[642])_{5^-}$ state in ^{232}Np, and that of the 1^- isomer also populates the $(\pi 5/2^-[523]\pi 5/2^+[642])_{1^-}$ state. The T_α value of the 6.46 MeV α transition of ^{234}Am deduced from the Hall's data [6] also lies near the line for the favored transitions. However, the 6.46 MeV is the same energy as the α transition of ^{235}Am. Hall et al. did not separate the ^{234}Am and ^{235}Am nuclei in their experiments. Considering the production cross sections of ^{234,235}Am and the α-decay intensity of ^{235}Am, we concluded that Hall et al. mistook α particles of ^{235}Am for those associated with ^{234}Am.

Since the level energies of the $\pi 5/2^-[523]$ states in ^{229,231}Np are expected within experimental uncertainties of E_α values and the energy of the $(\pi 5/2^-[523]\pi 5/2^+[642])_{5^-}$ state in ^{232}Np is roughly estimated as <60 keV, we assumed here these level energies are zero for ^{229,231}Np and 30±30 keV for ^{232}Np to deduce the Q_α values of ^{233,235,236}Am. In Table 1, the deduced Q_α values are compared with systematic predictions evaluated by Audi et al. [16]. It was found that the evaluated Q_α values are systematically overestimated by 100-200 keV in neutron-deficient Am isotopes.

IONIZATION OF SUPERHEAVY ELEMENTS

Finally, we mention the potential of an ion-source type ISOL to study superheavy elements. To detect superheavy elements, highly efficient separation techniques and highly sensitive detection techniques are essential. If the ionization efficiency of the ISOL is improved, detections of superheavy elements may become possible. The ISOL is able to provide extremely low-contaminated and low-counting rate mass-separated sources which achieves unambiguous mass identification of the superheavy elements and extremely high sensitive detecting condition. Although the elements 104–110 are difficult to be ionized by the ion source of the ISOL because these elements are expected as hardly volatile elements, the elements 111–116 are probably able to be ionized efficiently. This raises a possibility to identify the mass number of the element-114. Although the limit of the lifetime of nuclei to be detected using the ISOL is of the order of 1 s, the element-114 has long-lived isotopes [17]. Moreover, the element-119, which is expected to be an alkali metal, is ionized with 100% efficiency by using the present surface ionization-type ion source, and the elements 120, 121, $\cdots$ are also expected to show high ionization efficiencies. We think the ISOL is one of the most effective apparatus to identify the element-119. Furthermore, the ionization of the elements-119, 120, 121, $\cdots$ gives information on chemical properties of these elements. A high ionization efficiency proves that the element-119 is really an alkali metal, and the oxidation number of the elements-120 and 121 can be determined through the formations of metal-fluoride ions or monoxide ions in the ion source and their identifications by the mass-separations. The realization of productions of elements ≥ 119 is highly desired to study both nuclear structure and chemical properties of superheavy elements.

REFERENCES

1. Ahmad, I., Chasman, R.R., and Fields, P.R., *Phys. Rev. C* **61**, 044301 (2000).
2. Reiter, P. *et al.*, *Phys. Rev. Lett.* **82**, 509-512 (1999).
3. Leino, M. *et al.*, *Eur. Phys. J. A* **6**, 63-69 (1999).
4. Tsukada, K. *et al.*, *Phys. Rev. C* **57**, 2057-2060 (1998).
5. Hall, H.L. *et al.*, *Phys. Rev. C* **42**, 1480-1488 (1990).
6. Hall, H.L. *et al.*, *Phys. Rev. C* **41**, 618-630 (1990).
7. Kreek, S.A. *et al.*, *Phys. Rev. C* **49**, 1859-1866 (1994).
8. Gangrskii, Yu.P. *et al.*, *Yad. Fiz.* **31**, 306-317 (1980). [*Sov. J. Nucl. Phys.* **31**, 162-168 (1980).]
9. Sakama, M. *et al.*, JAERI-Review 98-017, 37-38 (1998).
10. Guo, J. *et al.*, *Z. Phys. A* **355**, 111-112 (1996).
11. Hall, H.L., *Ph.D. thesis*, University of California, Report No. LBL-27878 (1989).
12. *Table of Isotopes*, 8th ed., edited by Firestone, R.B., and Shirley, V.S., Wiley, New York, 1996.
13. Gallagher, Jr., C.J., and Moszkowski, S.A., *Phys. Rev.* **111**, 1282-1290 (1958).
14. Hseuh, H.-C. *et al.*, *Phys. Rev. C* **23**, 1217-1227 (1981).
15. Rasmussen, J.O., *Alpha-, Beta- and Gamma-ray Spectroscopy*, edited by Siegbahn, K., North-Holland, Amsterdam, 1966, pp. 701-743.
16. Audi, G. *et al.*, *Nucl. Phys. A* **624**, 1-124 (1997).
17. Oganessian, Yu.Ts. *et al.*, *Nature* (London), **400**, 242-245 (1999).

Gas Chromatographic Studies of the Heaviest Elements

Annett Vahle

Forschungszentrum Rossendorf, Institut für Radiochemie
for a FZR – PSI Villigen – Universität Bern – FLNR Dubna – GSI Darmstadt – TU Dresden –
Universität Mainz – LBNL collaboration

Abstract. Recent experiments are reviewed which used on-line isothermal gas chromatography techniques for studying chemical properties of transactinide elements. Elements 106 (Sg) and 107 (Bh) could assigned to groups 6 and 7 of the Periodic Table. The chemical characterization of elements 108 (Hs) and 112 has been prepared. A first attempt to chemically identify element 112 has been unsuccessful. Nevertheless, the chemical isolation of transactinide elements with $Z > 107$ appears feasible. It could also serve to confirm the discovery of SHE decaying to nuclides which have previously been unknown.

INTRODUCTION

The interest in the chemistry of the transactinide elements lies in the fact that near the end of the periodic table so called "relativistic effects" play an important role in determining the chemistry of the heaviest elements. Deviations from the regularities of the periodic table will result from the strong influence of these effects. Electron configurations different from those known for the lighter homologs may occur as well as unusual oxidation states and radii.

Recent studies of the chemical properties of Rf and Db have demonstrated that the chemical properties of the heaviest elements cannot be reliably predicted from simple extrapolations of the regularities in the groups and periods of the periodic table. Nevertheless, extrapolations of chemical properties are invaluable as a basis for the interpretation of experimental results.

Extensive relativistic quantum-chemical calculations of molecules combined with fundamental physicochemical considerations of the interaction of these molecules with their chemical environment allow quite detailed predictions of the chemical properties of the heaviest elements and their compounds. The experimental verification of these predictions is highly desirable, but also demanding.

The difficulties in studying the chemistry of the transactinides lie in their short half-lives and their extremely low production rates. Chemical studies can only be performed on a "atom-at-a-time" scale. Separation techniques to isolate the heaviest elements from interfering compounds have to be fast, efficient and applicable to single atoms. The high reaction rates in the gas phase and the possibility of using continuously operating chromatography systems lend gas phase chemistry the required efficiency and speed.

EXPERIMENTAL TECHNIQUES

A first attempt to isolate a volatile compound of a transactinide element by means of frontal gas chromatography was performed as early as 1966 [1].

In later experiments the same Dubna group used an improved technique, the so-called thermochromatography (TC) method [2], to determine the adsorption behavior of Rf-, Db-, and recently also Sg-halides and/or oxyhalides [2 - 4]. In thermochromatography, a longitudinal, negative temperature gradient is established along the chromatography column in the flow direction of the carrier gas. Volatile species are deposited in the column according their volatility and form distinct deposition zones. The very high speed at which the production of volatile species and their separation occurs and the high overall efficiency are the main advantages of this method. However, this technique is not suited for real-time detection of the nuclear decay of separated species and the determination of their half-lives.

In order to overcome these shortcomings the so-called OLGA (**O**n-**L**ine **G**as Chromatography **A**pparatus) technique was developed [5, 6]. The most advanced version, OLGA III [7], as it was used for the study of element 107 (Bh) is shown in Fig. 1.

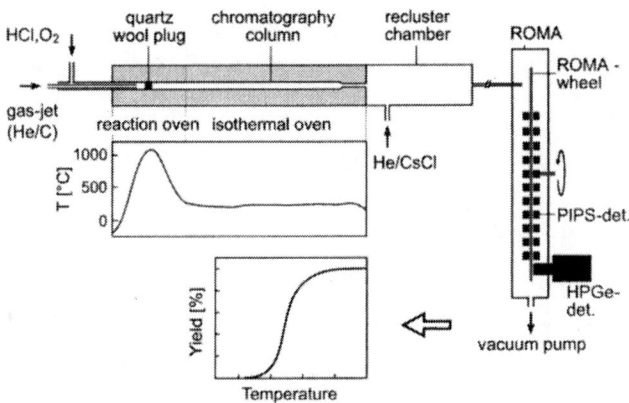

FIGURE 1. The OLGA III system used for the study of Bh

In the target chamber - not shown here - nuclear reaction products, recoiling from the thin target, are thermalized in He loaded with carbon aerosol particles. Attached to the surface of the aerosol particles the reaction products are continuously transported through a capillary to the OLGA set-up.

In the first section of the chromatography column, the reaction products are stopped on a quartz wool plug heated to 1050 °C. Reactive gases (HCl and O_2) are added in order to form volatile oxychlorides as well as to oxidize the carbon particles.

The second part of the column serves as the isothermal chromatography section, where the chromatographic separation of the formed species is achieved. The retention time of these compounds in the column depends on their interaction with the column surface at the chosen isothermal temperature. Only molecules formed with radionuclides having a longer nuclear life-time than the chemical retention time pass the chromatography set-up. By measuring the yields at different isothermal temperatures, the retention times and thus ΔH°_{ads} can be determined.

Volatile species leaving the column are attached to new aerosol particles in the re-cluster chamber for transport to the detection system, a rotating wheel (or a moving tape system in previous experiments).

Here, the aerosol particles are impacted in vacuum on thin foils mounted along the circumference of a rotating wheel and subsequently stepped in front of several **p**assivated **i**on-implanted **p**lanar **s**ilicon (PIPS) detectors which register α-particles and SF events in an event-by-event mode. This allows to search for correlated decay chains and it ensure the unambiguous detection of the separated nuclides. The chemical speciation, however, remains undetermined. Indirect evidence is possible, if data predicted on the base of thermodynamic arguments and model calculations correspond with data evaluated from experimental results.

At present, using the OLGA technique it is possible to study nuclides with $T_{1/2} \geq 5$ s, which can be produced in reactions with $\sigma \geq 5$ pb.

GAS PHASE STUDIES OF ELEMENTS WITH Z ≥ 106

For a long time only isotopes of elements 104 (Rf) and 105 (Db) with $T_{1/2} \geq 5$ s were known. Compounds of these elements have been studied extensively [8 - 12]. Partly, Rf and Db behave like typical group 4 and 5 elements, but also evidence for the strong influence of relativistic effects in the chemistry of Rf and Db was found.

This review is restricted to more recent experiments with even heavier elements.

Element 106 (Seaborgium)

First attempts to chemically characterize Sg were reported by the Dubna group in 1993 [13]. They used their on-line TC equipment to study the short-lived 0.9 s nuclide ^{263}Sg produced in the ^{249}Cf (^{18}O, 4n) reaction. SF tracks observed in the column were attributed to the decay of ^{263}Sg based on chemical arguments [14, 15, 4]. This work has been met with skepticism concerning both the assignment of the observed SF activity and as well as the chemical state of Sg- and W-oxychlorides [16].

In 1993, two new isotopes of Sg, ^{265}Sg and ^{266}Sg, with half-lives of 7.4 s and 21 s at production cross sections of about 240 and 25 pb were discovered [17, 18]. Two years later, Sg could be unambiguously chemically identified for the first time by an international collaboration of radiochemists [19]. The OLGA technique was used to separate Sg produced in the reaction ^{22}Ne + ^{248}Cm as volatile oxychloride compound from interfering by-products.

In a second experiment series it could be shown that this Sg compound, SgO_2Cl_2, is even or less volatile like the corresponding W compound as it was expected from theoretical predictions. As first thermochemical property of a Sg compound ΔH°_{ads} (SgO_2Cl_2) was determined to be -98 kJ/mol [20]. Sg was found to behave like a typical group 6 element in this chemical system and could be assigned to group 6 of the periodic table.

High-temperature gas chromatography was used to study a second Sg compound in 1998. It was proven by identification of only two ^{266}Sg atoms after separation at 1000 °C that Sg forms a low volatile Sg-oxide hydroxide, $SgO_2(OH)_2$, as typical for group 6 elements [21].

Element 107 (Bohrium)

For a long time, ^{264}Bh was the longest-lived known isotope of element 107 (Bh) with a half-live of only 440 ms, too short for chemical studies. Early experiments aiming at the production of longer-lived Bh isotopes and the chemical identification of Bh failed [22, 23]. Only in 1999 a new, neutron-rich isotope, ^{267}Bh, was discovered in the reaction ^{249}Bk (^{22}Ne, 4n) [24]. The half-live of 17 s and the production cross section of about 50 pb make ^{267}Bh an ideal candidate for gas phase chemical studies.

With oxidizing chlorinating gases Bh should form a volatile oxychloride, presumably BhO_3Cl. This compound was predicted to be more stable and less volatile than ReO_3Cl or TcO_3Cl from both relativistic Density-Functional calculations [25] and classical extrapolations [26].

In August/September 1999 a mixed target of ^{249}Bk and ^{159}Tb was irradiated with ^{22}Ne particles at a beam energy of 119 MeV for a time period for about 4 weeks. ^{267}Bh was chemically isolated as volatile oxychloride using OLGA III. ^{176}Re which was produced simultaneously in the reaction ^{159}Tb(^{22}Ne, 5n) served as a yield monitor for the chemical separation process.

Experiments at three isothermal temperatures were conducted. Four correlated decay chains (N_R = 1.3), attributed to the decay of ^{267}Bh, were detected at 180 °C. Based on the measured cross section of about 50 pb for the production reaction an overall yield of the whole separation and detection process of 7 %, or a sensitivity limit of 8 pb was deduced for this temperature.

At 150 °C and 75 °C two (N_R = 0.1) respectively no events were registered at a similar sensitivity as for 180 °C, whereas ^{169}ReO$_3$Cl still passed the isothermal part of the column with about 80 % relative yield at 75 °C.

The adsorption properties of BhO_3Cl were quantified using a microscopic model of the adsorption process [27]. The standard adsorption enthalpy of BhO_3Cl was calculated to be $\Delta H^°_{ads} = -75^{+6}_{-7}$ kJ/mol. This value compares well with estimated values of -78 ± 5 kJ/mol obtained for BhO_3Cl applying a physisorption model adjusted to the experimental adsorption enthalpies of TcO_3Cl and ReO_3Cl [25], as well as with -74 ± 12 kJ/mol deduced from thermodynamic extrapolations [26].

Employing an empirical correlation between the standard adsorption enthalpy on quartz and the standard sublimation enthalpy [20] it is possible to estimate the sublimation enthalpy of macroscopic amounts of BhO_3Cl to be $\Delta H^°_{subl} = 89^{+18}_{-17}$ kJ/mol.

In model experiments it was found that TcO_3Cl was too volatile to be adsorbed on CsCl particles after chemical separation. Therefore, $FeCl_2$ aerosol particles, presumably acting as reducing surface, were used instead of CsCl. The fact that Bh was transported with CsCl aerosol particles like Re confirm the stronger similarity of Bh to Re compared to Tc.

It was concluded from the good correspondence of theoretical predictions and experimental results that Bh belongs to group 7 of the periodic table. Obviously, and similar to the group 6 element Sg, relativistic effects are not strongly expressed in the investigated Bh-oxychloride compound [28].

Element 108 (Hassium)

The most obvious chemical compound of Hs suitable for its chemical separation is the tetroxide, HsO_4, which is likely to be highly volatile and well-suited for chemical studies. First attempts to isolate HsO_4 were performed in the eighties, but were unsuccessful [29 - 31].

In 1996 a new isotope of element 108 (Hs), ^{269}Hs, was discovered [32] with a half-life sufficiently long for chemical studies. However, calculated cross sections in the $^{26}Mg + {}^{248}Cm$ reaction to produce ^{269}Hs (and/or ^{270}Hs) are on the order of only a few pb [16]. A very efficient chemistry device with a high overall yield is required to envisage chemistry experiments at this level. Therefore, new devices were developed by several groups. These devices work on the same principle, but experimental set-ups and detection systems differ from each other in detail.

Here, the apparatus for In-situ volatilization and on-line detection of transactinide elements (IVO) is represented as an example [33]. Considerable improvements compared to OLGA are the direct coupling of target chamber and gas chromatography and the possibility to deposit the volatile, separated species on the detector surface by diffusive adsorption.

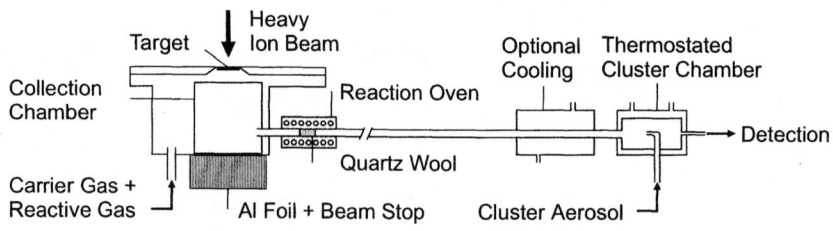

FIGURE 2. The IVO set-up

IVO consists of a recoil chamber, flushed with a gas where the reaction products are thermalized. If required, a small admixture of reactive gas provides the in-situ formation of volatile compounds. Volatile molecules or elements are swept through a quartz chromatography column which reaches into the recoil chamber.

The first few cm of the quartz column are heated in the reaction oven to 600 °C - 800 °C to increase the diffusion of less volatile, unwanted by-products to the wall. They are retained by irreversible adsorption. The elevated temperature also favors the formation of volatile compounds. If required, the chromatography column can be cooled to measure yield curves.

In the cluster chamber, molecules that pass through the column can be attached to aerosol particles and will be transported to the detection system. It's also possible to replace the cluster chamber and to deposit the volatile separated species directly on the detector surface by diffusive adsorption.

IVO was tested with OsO_4, which should behave similar to HsO_4. As carrier gas He with 10 % O_2 was used. The chemical yield of the system was determined to be about 50 %. Po, often produced in heavy element experiments from impurities of the target and the target assembly and strongly interfering with the α-spectroscopic detection of transactinides, was not volatile under the chosen experimental conditions. A decontamination factor of at least $2 \cdot 10^4$ was achieved.

With this encouraging results the chemical separation of Hs as HsO_4 seems to be feasible within the near future.

Elements 109 - 111

At the moment, chemists do not aim at the chemical characterization of these elements. The longest-lived known isotopes of elements 109 and 111 are ^{268}Mt and ^{272}111 with half-lives of only 70 ms and 1.5 ms, respectively, far too short for chemical studies. An isotope of element 110 which should be suitable for chemical studies was discovered in 1999 as decay product of ^{289}114 [34, 35]. ^{281}110 decays by emission of an α-particle with $T_{1/2} \approx 1$ min. However, this nuclide has never been produced directly and chemical studies have not been prepared.

Elements 112 + 114

The last two years have seen a remarkable progress in the synthesis of superheavy nuclei near the theoretically predicted shell closures at Z = 114 and N = 184. New isotopes of elements 114 and 116 and their subsequent decay chains were observed at FLNR [34 - 36]:

^{289}114 (α, 30 s) → ^{285}112 (α,15 min) → ^{281}110 (α, 1.6 min)→ ^{277}108 (SF, 16 min)
^{292}116 (α, 47 ms) → ^{288}114(α, 2.4 s) → ^{284}112 (α, 53.9 s) → ^{280}110 (SF, 6.9 s)
$\qquad\qquad\qquad\quad$ ^{288}114(α, 0.8 s) → ^{284}112 (α, 10.3 s) → ^{280}110 (SF, 14.3 s)
$\qquad\qquad\qquad\quad$ ^{288}114(α, 4.6 s) → ^{284}112 (α, 18.0 s) → ^{280}110 (SF, 7.4 s)
^{287}114(α, 1.3 s) → ^{283}112 (SF, 9.3 min)
^{287}114(α, 14 s) → ^{283}112 (SF, 3.8 min)

Two nuclei of ^{283}112 were also obtained directly: ^{283}112 (SF, 3.0 and 0.9 min) [37].

These results open up new possibilities to study the chemical properties of superheavy elements (SHE). In addition, chemical identification of the proton number is very important, because all the members of the above decay chains end in previously unknown nuclei.

Element 112 and to some extend also element 114 are predicted to be highly volatile noble metals [38]. They should be similar or even more volatile than Hg and can be transported in gas-jet systems without the help of aerosol particles. As a consequence of the high volatility, however, it might be difficult to adsorb such atoms onto surfaces under vacuum, a prerequisite for assaying α-particle spectra.

Therefore, predictions have been made about the interaction of SHE with several metal surfaces [39]. The idea is, that similar to the amalgamation processes of Hg, also elements such as 112 and 114 are predicted to have a high tendency to form strong intermetallic bonds and could be adsorbed directly on detector surfaces which are coated by thin layers of appropriate metals.

As a crude estimate the adsorption enthalpy of at least -80 kJ/mol is needed to fix an atom on a surface for several minutes. Optimal backings for final SHE samples should be Pd or Cu since they guarantee relatively high adsorption enthalpies of the volatile SHE and both of these metals can easily be kept free from oxide layers.

A study of the chemisorption properties of Hg on various metal surfaces in pure He at room temperature showed that Hg was adsorbed quantitatively on Pd, Au, and Pt. For Cu, Ti, Fe, Ni, Cd, Pb the adsorption efficiency was less than 5- 15 % [40].

A flow-through detection array was constructed at FLNR [40]. A pair of square PIPS detectors coated with thin layers of Au or Pd ($\sim$ 40 µg/cm^2) are installed in each chamber. The working area of a detector is 18 x 18 mm, the distance between the detectors is 1 mm, which allow to obtain a high detection efficiency (80 %) with a resolution of about 100 keV for 6 MeV α-particles. The residence time of the gas in each detector chamber is a fraction of a second.

A first attempt to chemically identify element 112 was performed at the Dubna U-400 cyclotron in January 2000 [40]. ^{283}112 and ^{185}Hg were produced by bombarding a mixed ^{238}U$_3$O$_8$/^{nat}Nd target with ^{48}Ca ions. The recoils were thermalized in pure He at atmospheric pressure and transported through a 25 m long teflon capillary to the detection apparatus. The transportation time was 25 s at a gas flow rate of 500 ml/min. The detection apparatus consisted of 8 detection chambers in series coated with layers of Au (1 - 6) or Pd (7 - 8). The chambers were positioned inside an assemblage of 84 ^{3}He-filled detectors (in a moderator) to register prompt fission neutrons.

If element 112 behave chemically like Hg and all efficiencies measured for Hg held also for element 112, the detection of some 3 SF events within 10 days beam time at an integral beam dose of $7 \cdot 10^{17}$ could be expected, taking into account the cross section given in [37]. During the irradiation no SF events could be detected.

This experiment did not give an unambiguous answer as to the physical and chemical properties of element 112 - either the chemical properties of element 112 are more Rn-like, or the statistics of the experiment didn't allow us to register the element 112. However, the experiment showed the possibility to chemically identify nuclides produced with very low cross sections of only a few pb.

Further experiments will follow in the next future. It is planned to at least double the beam dose and to upgrade the detector system for measurements of α-decays and SF events in gas using a special ionization chamber.

However, the detection of a SF product after chemical separation is problematic, since only small contamination in the procedure might cause events in the detectors from SF-decaying actinides. Therefore, a second approach [41] proposes to study another isotope of element 112 decaying by emission of an α-particle with a half-life > 5 s. ^{284}112 should be well-suited but there is presently no reaction known to produce this isotope in a direct way. ^{284}112 can be obtained as daughter nuclide of ^{288}114 produced in the reaction ^{244}Pu (^{48}Ca, 4n) with a cross section of only about 1 pb. The detection of 3 α-SF decay chains per 15 day bombardment is expected if the sensitivity can be improved by a gain factor of 10 compared to the chemical separation of Bh. This objective should be reached by transport in gas-jet systems without the help of aerosol particles and by direct diffusive adsorption of separated species on covered detector surfaces. A rotating target wheel should allow to increase the beam intensity.

In conclusion, prospects to characterize the SHE with $Z = 112$ and $Z = 114$ chemically within the next years appears feasible. The chemical isolation of SHE could also serve to confirm the discovery of SHE decaying to nuclides which have previously been unknown.

REFERENCES

1. Zvara, I. et al., *At. Energ.* **21**, 83 (1966).
2. Zvara, I. et al., *Inorg.Nucl.Chem. Lett.* **7**, 1109 (1971).
3. Zvara, I. et al., *Soviet Radiochemistry* **18**, 371 (1976).
4. Zvara, I. et al., *Radiochim. Acta* **81**, 179 (1998).
5. Gäggeler, H. et al., *Radiochim. Acta* **38**, 103 (1985).
6. Gäggeler, H. et al., *Nucl. Instr. Meth.* **A309**, 201 (1991).
7. Türler, A., *Radiochim. Acta* **72**, 7 (1996).
8. Türler, A. et al., *J. Radioanal. Nucl.Chem.* **160**, 327 (1992).
9. Kadkhodayan et al., *Radiochim. Acta* **72**, 169 (1996).
10. Türler, A. et al., *J. Alloys Comp.* **271-273**, 287 (1998).
11. Gäggeler, H. et al., *Radiochim. Acta* **57**, 93 (1992).
12. Türler, A. et al., *Radiochim. Acta* **73**, 55 (1996).
13. Timokhin, S. N. et al., "Chemical Identification of Element 106 by the Thermochromatographic Method", Proceedings of the International School-Seminar on Heavy Ion Physics, Dubna, 1993, Vol. 1, p. 204.
14. Timokhin, S. N. et al., *J. Radioanal. Nucl.Chem., Letters* **212**, 31 (1996).
15. Yakushev, A. B. et al., *J. Radioanal. Nucl.Chem.* **205**, 63 (1996).
16. Kratz, J. V., "Chemical Properties of the Transactinide Elements," in *Heavy Elements and Related New Phenomena*, edited by W. Greiner and R. K. Gupta, Singapore: World Scientific, 1999, Vol. 1, pp. 129-193.
17. Lazarev, Yu. A. et al., *Phys. Rev. Lett.* **73**, 624 (1994).
18. Türler, A. et al., *Phys. Rev. C* **57**, 158 (1998).
19. Schädel, M. et al., *Nature* **388**, 55 (1997).
20. Türler, A. et al., *Angew. Chem. Int. Ed.* **38**, 138 (1999).
21. Hübener, S. et al., FZR Institute of Radiochemistry. Annual Report 1998, FZR-247, p. 67 (1999).
22. Zvara, I. et al., *Sov. Radiochem.* **26**, 72 (1984).
23. Schädel, M. et al., *Radiochim. Acta* **68**, 7 (1995).
24. Wilk, P. et al., submitted to *Phys. Rev. C*.
25. Pershina, V., and Bastug, T., submitted to *J. Chem. Phys. A*.
26. Eichler, R., PSI internal report, TM-18-00-04 (2000).
27. Zvara, I., *Radiochim. Acta* **38**, 95 (1985).
28. Eichler, R. et al., submitted to *Nature*.
29. Zhuikov, B. L. et al., Report JINR P7-86-322, p. 15 (1986).
30. Dougan, R. J. et al., LLNL Annual Report FY 87, UCAR 10062/87 (1987) 4.
31. Hulet, E. K. et al., LLNL Annual Report FY 87, UCAR 10062/87 (1987) 4.
32. Hofmann, S. et al., *Z. Phys. A* **354**, 229 (1996).
33. Düllmann, C. et al., submitted to *Nucl. Instr. Meth. A*.
34. Oganessian, Yu. Ts. et al., *Phys. Rev. Lett.* **83**, 3154 (1994).
35. Oganessian, Yu. Ts. et al., *Nature* **400**, 242 (1999).
36. Oganessian, Yu. Ts. et al., submitted to *Phys. Rev. C*.

37. Oganessian, Yu. Ts. et al., *Eur.* Phys. J. A **5**, 63 (1994).
38. Eichler, B., Report JINR P12-9454, (1976).
39. Eichler, B., and Rossbach, H., *Radiochim. Acta* **33**, 121 (1983).
40. Yakushev, A. B. et al., "First Attempt to Chemical Identify Element 112", Extended Abstract, 5th Int. Conf. on Nuclear and Radiochemistry (NRC5), Pontresina, Sept. 2000.
41. Gäggeler, H. et al., *UNILAC proposal (2000).*

The Performance of Mean-field Models for Superheavy Elements

P.-G. Reinhard,[a] M. Bender,[b] T. Bürvenich,[c] T. Cornelius,[c] P. Fleischer,[a] J.A. Maruhn[c]

[a] Inst.f.Theor.Physik, Univ. Erlangen, D-91054 Erlangen, Germany
[b] Gesellschaft für Schwerionenforschung, D-64921 Darmstadt, Germany
[c] Inst.f.Theor.Physik, Univ. Frankfurt, D-60054 Frankfurt, Germany

Abstract. We concentrate on the two most widely used mean-field models, Skyrme-Hartree-Fock and relativistic mean field. We discuss their performance in view of the extended range of experimental information in terms of a guided tour through selected examples of success, failures and subsequent improvement of the models drawn from comparison with key observables as isotopic shifts, trends of binding energies or single-particle spectra. Particular attention is paid to the extrapolative power with respect to superheavy elements (SHE). It turns out that the most important aspect is here the actual level sequence as implied by the given mean-field models. We will compare the performance of available parametrisations and discuss briefly possible improvements.

I INTRODUCTION

It is more than three decades ago that speculations on the possible existence of an island of shell-stabilized superheavy elements (SHE) [1,2] have motivated heavy-ion physics as a new field of nuclear research. In the meantime, it turned out that just the production of these new SHE is one of the most tedious tasks in the field of exotic nuclei. But untiring search over decades has now reached the point where the region of new magic shells has been accessed in recent experiments [3,4]. The newly developed and the coming experimental facilities produce more and more new isotopes. First glimpses of elements at $Z=116$ [5] and $Z=118$ [6] have been reported. SHE are thus a topic of current interest. This does also inspire new theoretical surveys because the extrapolation of existing nuclear models to huge nuclei constitutes, on the one hand, a very demanding problem and delivers, on the other hand, a world of new benchmarks for improving the models.

Early theoretical estimates were performed on the basis of the microscopic-macroscopic (mic-mac) approach [1,2] which has developed much further over the years [7,8]. One step deeper towards microscopic description are the self-consistent mean-field models which employ effective energy-density functionals. The two most widely used mean-field models in nuclear physics are the (non-relativistic) Skyrme-Hartree-Fock (SHF) approach and the relativistic mean-field model (RMF). Both models rely on a theoretically motivated and phenomenologically adjusted effective energy-density functional. The reliability of the actual parametrisations has developed very much over the past decades. With 6–10 free parameters a very good reproduction of nuclear ground-state properties for the stable elements is achieved [9–12]. The average quality of the energy still stays somewhat beyond the mic-mac approaches. But mean-field models aim to cover a much broader range of observables up to dynamical features as giant resonances [13]. There is one more crucial aspect: The mic-mac approach, although generally successful, requires preconceived knowledge about the expected nuclear shapes and potentials, which fades away when stepping into unknown regions. Self-consistent calculations, on the other hand, provide potentials and shapes in an unprejudiced manner. This should make mean-field models more robust in daring extrapolations. Nonetheless, SHE are very sensitive to the detailed level structure and thus also challenge the predictive power of mean-field models. It is the aim of this contribution to summarise briefly the state of the art.

II FRAMEWORK

Nuclear matter is a dense Fermi liquid. Thus a fully microscopic description is extremely demanding and for finite nuclei impossible. Practicable microscopic approaches rely on effective energy-density functionals which are the nuclear analogue of the famous electronic density functional theory [14]. The typical procedure in the nuclear case is even more phenomenological. Many-body theory serves merely to motivate a reasonable ansatz for the functional. The actual parameters of the ansatz are then determined phenomenologically by a fit to a selection of relevant observables (energy, radii, formfactor, $l*s$-splitting).

The SHF functional is sketched as follows:

$$E = \int d^3r \, \{ \underbrace{b_0 \rho^2 + b_3 \rho^{2+\alpha}}_{\text{zero-range density-dep.}} + \underbrace{b_2 \rho \Delta \rho}_{\text{gradient corr. (surface prop.)}} + \underbrace{b_1 \rho \tau}_{\text{kinetic}} + \underbrace{b_4 \rho \nabla J}_{\text{spin-orbit}} \}$$

isoscalar: $E/A, \rho, K$ $\quad a_{surf}$ $\quad m^*/m$

isovector: $a_{asy}, \frac{d}{d\rho} a_{asy}$ $\quad a_{asy,surf}$ $\quad \kappa_{sumrule}$ $\quad$ spin-orbit splitting

The leading term is the density dependence parametrised here as a two-body attractive and density-dependent repulsive term. This allows already to fix all relevant nuclear bulk properties. Finite systems require a fine-tuning of surface properties which is achieved by the gradient correction term (much similar as in electronic functionals [14]). Moreover, the strong dressing of nucleons in matter call for an effective mass which is adjusted by the kinetic correction term. Last not least, a strong $l*s$-splitting is crucial for a correct description of the nuclear mean field. This is guaranteed by the term including the spin-orbit current $\mathcal{J}$. The above presentation is a bit abbreviated. The terms as they stand represent the isoscalar part of the functional. Each term needs to be complemented by its isovector analogue, e.g. for the two-body term $\rho^2 \longrightarrow (\rho_n - \rho_p)^2$.

The nice feature of the Skyrme functional is that each term has an immediate physical interpretation, as indicated in the above sketch. The isoscalar density dependent part is related to bulk binding E/A, equilibrium density ρ, and incompressibility K. The isovector part complements this by asymmetry energy a_{asy} and its slope with density. The gradient corrections determine naturally the surface energies. The kinetic terms adjust the isoscalar effective mass and the sum rule enhancement factor κ. No bulk property can be associated with the spin-orbit term. It is related to the single-particle spectrum of finite nuclei.

The RMF can be sketched as follows:

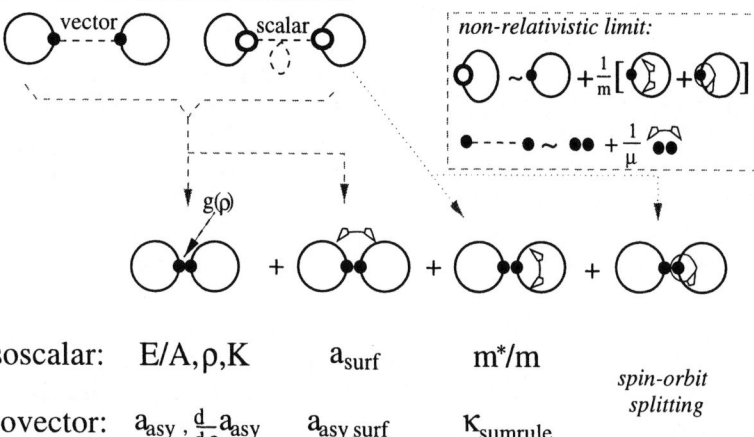

isoscalar: $E/A, \rho, K$ a_{surf} m^*/m spin-orbit splitting

isovector: $a_{\text{asy}}, \frac{d}{d\rho}a_{\text{asy}}$ $a_{\text{asy,surf}}$ κ_{sumrule}

The basic functional is summarised in the two left upper diagrams. It looks much simpler. One takes an obvious selection of meson fields with varied scalar↔vector character complemented by corresponding isovector terms. Some density dependence is required in an effective functional. It is implemented here as a nonlinear self-coupling of the scalar field (indicated by the meson loop in the scalar diagram). Thus the concept of the RMF seems to be much more naturally emerging from a field theoretical perspective. But the many-body aspects are not so transparent in that scheme, particularly the relation to nuclear bulk properties. This relation can be established by considering the non-relativistic limit. The right upper box indicates the necessary expansions. The scalar density delivers as leading term the normal density and as v/c-corrections the kinetic as well as the spin-orbit term. The finite range of the mesons is expanded as zero-range coupling and gradient correction. Inserting these approximations yields the functional as represented by the diagrams in the second line of the sketch. It looks almost identical to the Skyrme functional. The interpretation in terms of bulk properties then proceeds as before. There is one aspect, however, which is very hard to map: the form of the density dependence. The mechanism in the RMF goes through non-linear meson coupling and is much different from the SHF with its straightforward expansion in powers of density ρ. A thorough comparison of the density dependences is still a task for future research.

There is one more variant of the RMF which tries to stay closer to SHF right from the beginning. It is an RMF where the meson fields are eliminated in favour of a direct point-coupling of the various densities [15]. We will also consider results from that model.

There exists world of different parametrisations for SHF as well as RMF. We confine the discussion to a few well adjusted and typical sets. For SHF we consider the parametrisations SkM* [16], SkP [17], SkI3, SkI4 [11], and SLy4 or SLy6 [10]. The force SkP uses effective mass $m^*/m = 1$ and is designed to allow a self-consistent treatment of pairing. The other forces all have smaller effective masses around $m^*/m = 0.7 - 0.8$. The force SkM* was first to deliver acceptable incompressibility and fission properties. The forces SLy4/6 stem from an attempt to cover properties of pure neutron matter together with normal nuclear ground state properties. The forces SkI3/4 employ a spin-orbit force with isovector freedom to simulate the relativistic spin-orbit structure. SkI3 contains a fixed isovector part exactly analogous to the RMF, whereas SkI4 is adjusted allowing free variation of the isovector spin-orbit force. The modified spin-orbit force has a strong effect on the spectral distribution in heavy nuclei and thus for the predictions of SHE. For the RMF we consider the parametrisations NL-Z [18] or NL-Z2 [19] and NL3 [12]. The forces NL-Z and NL-Z2 comes from fits with the choice of observables much similar to those of SkI3 and SkI4 where particularly the charge formfactor was taken care of. NL3 is fitted without looking at the formfactor but with taking more care about the isovector trends. For the PRMF we consider the parametrisation from Los Alamos [15], called here P-LA, and a recent fit with data as in SkI3/4 and NL-Z2, called P-Fx.

All models are complemented by BCS pairing with matrix elements deduced form a zero-range pairing force.

III RESULTS AND DISCUSSION

Fig. 1 summarises the performance of the various forces for normal nuclei between oxygen and lead. The reproduction of energy is very good in all cases with an average error safely below 0.5%. Differences can be seen for the parameters of the charge distribution. The four forces with particularly low error in surface thickness σ are those which included the formfactor in the fit. This indicates that this is a feature which can be accommodated by each model but which does not necessarily emerge from a fit without these data points. The pattern for the diffraction radius R

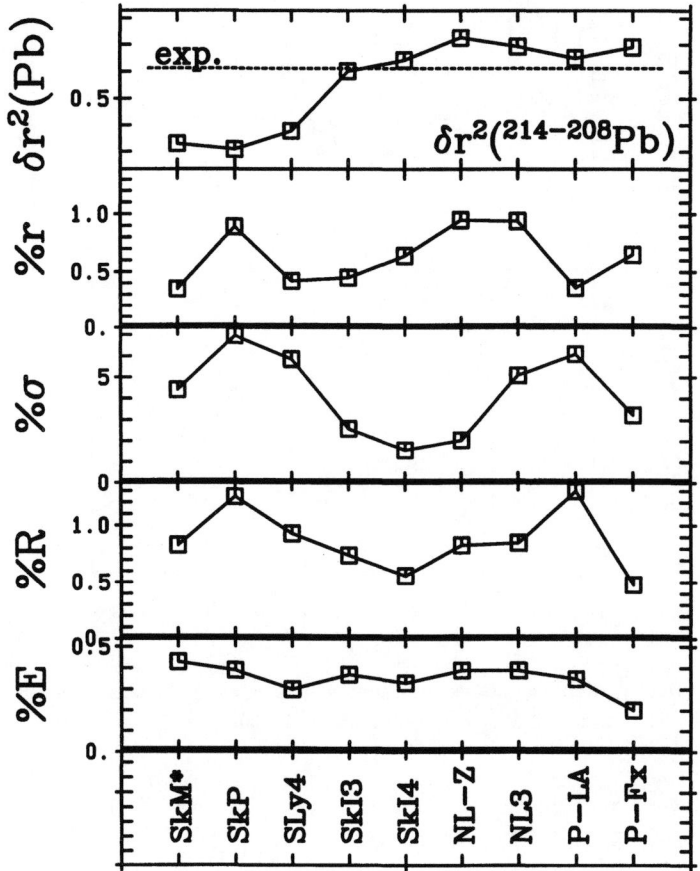

FIGURE 1. Relative error of energy E, diffraction radius R, surface thickness σ, and r.m.s. radius r for the selection of forces. The uppermost panel shows the isotopic shift of the radius in Pb and the experimental value is indicated by a horizontal dotted line.

are similar to those of σ which again is related to having the formfactor in the fit or not. Nonetheless nearly all forces provide a good description of R. The r.m.s. radii are also generally very well reproduced with errors safely below 1%. Thus far all forces perform about equally well for nuclear bulk properties (with understandable difference in the quality of σ). A rather recent key feature is provided by the isotopic shift of the r.m.s. radius (uppermost panel). All relativistic model reproduce that more or less automatically correct. All conventional Skyrme models underestimate the value by factor two. Only SkI3 and SkI4 which employ an isovector generalised spin-orbit force perform as well as the RMF. This, of course, has consequences on

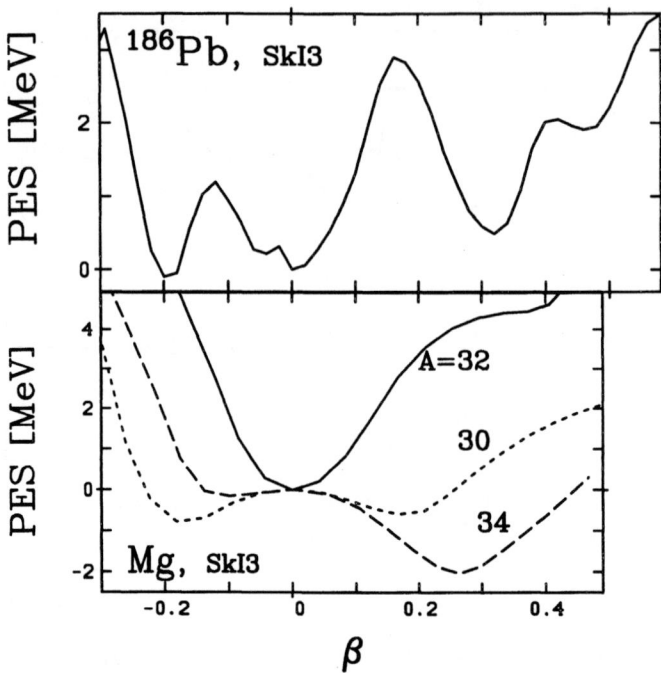

FIGURE 2. Deformation energies versus dimensionless quadrupole deformation β for a variety of exotic nuclei and computed with various. forces as indicated.

the actual level structure and thus on the extrapolation to SHE.

It is a general feature of all exotic nuclei to exhibit a pronounced shape coexistence [20,21]. This is demonstrated in fig. 2 for a few examples from light and heavy nuclei. Nearly all examples show a strong competition between different deformation minima. The actual ground state will then be a superposition of these competing configurations. Pure mean-field calculations have to be interpreted with care. A particularly impressive example is ^{186}Pb which shows five local minima in the range of deformation accessible by the ground state fluctuations of β. This has consequences on the excitation spectrum. In fact, experiments have revealed the very unusual feature that the lowest states are a sequence of two low-lying 0^+ states before the first 2^+ state comes [22]. Our calculations yield a 0^+ spectrum with 0.3 and 0.8 MeV which agrees qualitatively with the experimental findings.

We now turn to results from SHE. Before presenting the new material, we recall basic results from earlier publications. The various forces which have comparable quality in normal nuclei (see fig. 1) deliver different quality for known SHE [23]. But there remains a good handful of forces, RMF as well as SHF, which perform very well even for SHE. The interesting question where the next spherical doubly-

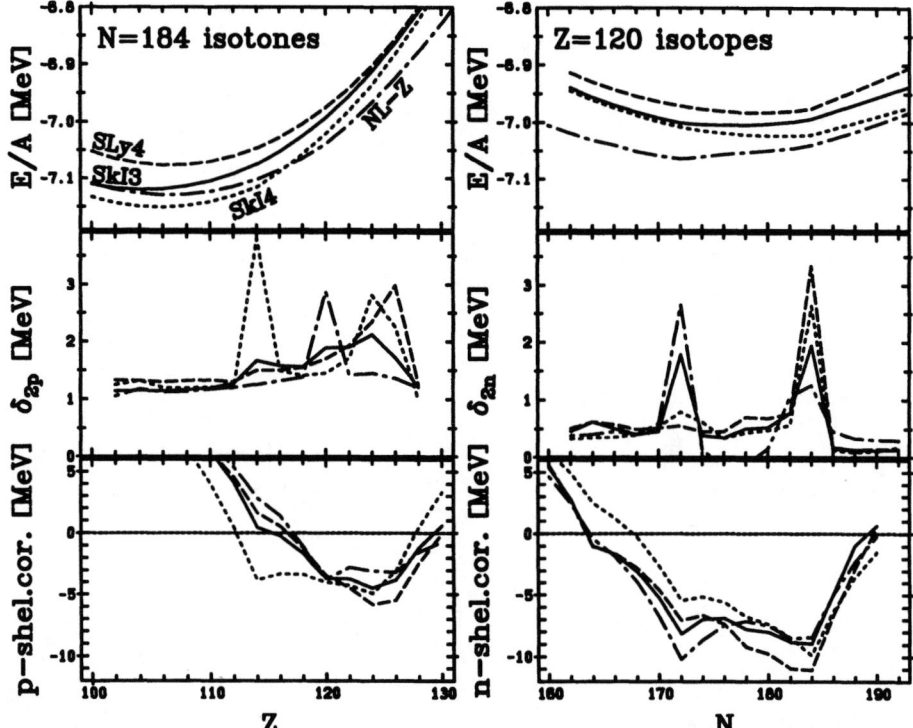

FIGURE 3. Shell effects for a superheavy isotonic and isotopic chain and for various forces. The lowest panel shows the Strutinsky shell correction and the middle panel the two-nucleon shell gap as measures for shell closure. The upper panel complements the information by the binding energies as such.

magic nuclei appear receives an ambiguous answer from the mean-field models. The predictions of the various forces differ, particularly what the proton shell closure is concerned (leaving here the choice Z=114, Z=120 and Z=126). It turns out that these predictions depend sensitively on the single-particle structure of the models, particularly on the spin-orbit splittings [19]. Those force forces which produce best spin-orbit splittings in known nuclei predict Z=120 as next magic proton shell.

Fig. 3 takes up the question of shell effects in SHE. We consider two different measures for "magicitiy", the two-nucleon shell gaps, $\delta_{2n} = E(N+2) - 2E(N) + E(N-2)$ and the analogous δ_{2p}, and the Strutinsky shell correction energy whose computa-

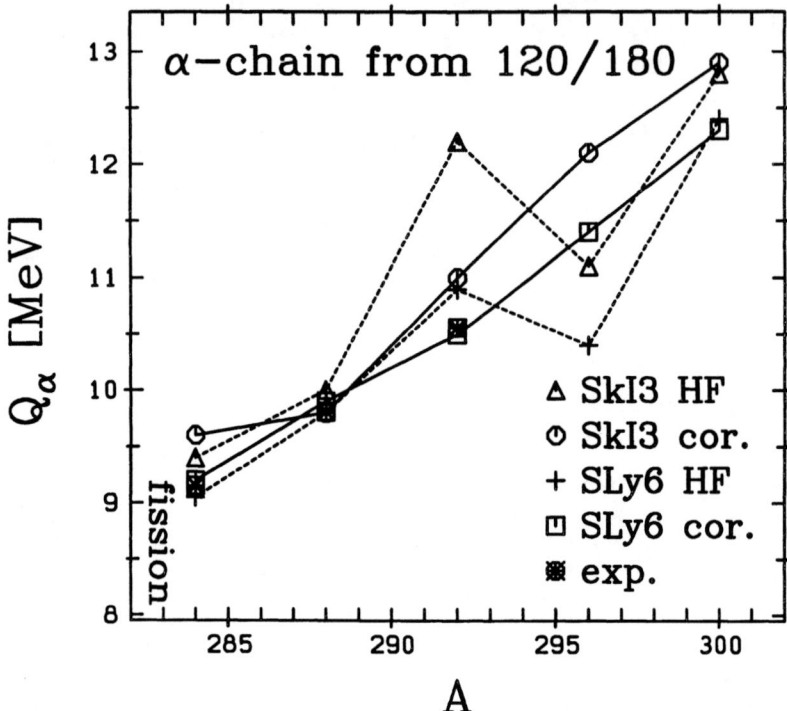

FIGURE 4. Q_α values for the chain of SHE starting at 120/180. Two different forces are considered as indicated. And for each force the pure SHF results is shown as well as a computation including quadrupole-collective ground state correlations.

tion requires special care for SHE [24]. Magic shell closures are distinguished by peaks in the shell gaps. One see in the left middle panel of the figure immediately the different predictions for proton shells along the isotonic chain, whereas the two neutron shells at N=172 and N=184 seem to stick out unambiguously. At second glance, we realize that the δ_{2p} are very small anyway. This means that proton shells will not be very pronounced. In fact, they are at the fringe of what could be a shell closure. It is thus worthwhile to look at the alternative measure in terms of the Strutinsky shell correction. Negative values indicate additional binding from shell effects, for SHE the so called shell stabilisation of an else wise unstable system. Pronounced spiky minima indicate a shell closure. We see in the lower panels of fig. 3 that all forces predict qualitatively the same feature, namely a broad minimum landscape with rather flat spikes. The good news is that there is indeed an island of stability induced by shell effects. The bad news is that there will be no

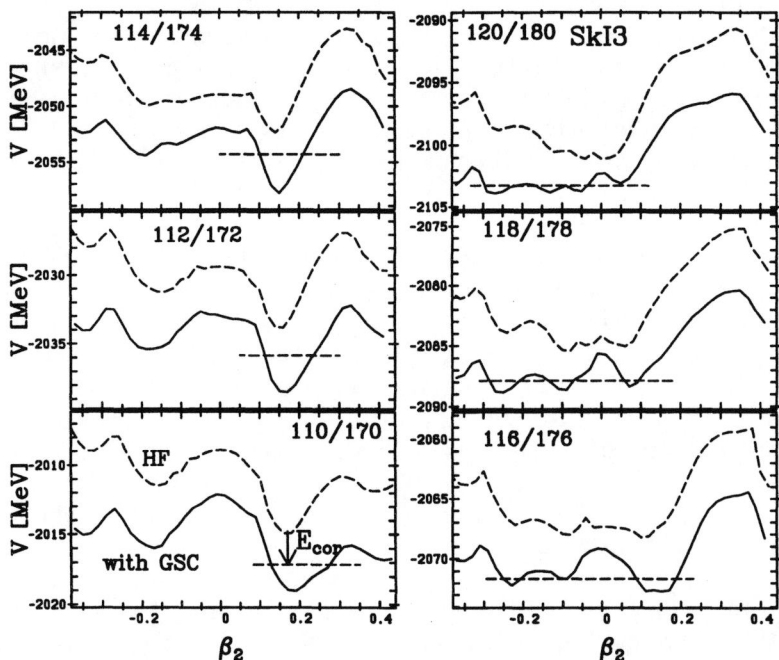

FIGURE 5. Deformation energies versus dimensionless quadrupole momentum for the nuclei along the α-decay chain from 120/180 and computed with SkI3. The dashed line shows the pure SHF results and the full line the energy after subtraction of the collective zero-point energies. The dashed horizontal lines indicate the position of the collective grounds state.

pronounced shell closures. Almost any nucleus in that region is sufficiently stabilised, but none stick truly out amongst the others. A final look at the binding energies as such shows that the regime of lowest binding energies does not necessarily agree with largest shell stabilisation, particularly along the isotonic chain. Only the shell stabilised, but less well bound, elements will be sufficiently stable against fission.

Just recently, one has measured a long α-chain of SHE starting 116/176 [5]. Fig. 4 shows theoretical on this chain (in fact on the larger chain starting from 120/180). The pure SHF results agree nicely with data for the lower two elements. There is a strange kink at 116/176 which does not comply with the data. We recall the lesson learned from fig. 2 that exotic nuclei (and SHE belong to this brand) show shape coexistence and that pure mean-field models may be insufficient. This is indeed the case here. We have computed the collective ground state in the deformation plane (see e.g. [25]) and deduced the quadrupole correlation energy from that. The correlated results are also shown in fig. 4. These are free from such kinky trends and agree nicely with the experimental values where available. The example demonstrates that detailed predictions on SHE will require inclusion of ground state correlations. It also hints that correlations may wipe out any visible signs of shell closure.

The strong impact of correlations seen in fig. 4 suggest pronounced shape coexistence in SHE. This is checked in fig. 5 where we show the potential energy surfaces for the nuclei along the chain. It is obvious that the larger (thus more exotic) nuclei hove strong shape isomerism. The SHF minimum as such is meaningless and one is compelled to consider the whole collective ground state correlations. The mechanism for that can be read off from the figure. One first has to subtract from the given energy surface the spurious collective motion (quadrupole, rotation, centre-of-mass), the zero-point energies. This yields a much lowered energy surface (from dashed to full lines in the plot). Within the correct energy surface, one than computes the collective dynamics. The ground state lies above the absolute potential minimum. The energetic position of the collective ground state then represents the correlated ground state energy. One sees from the figures that the correlation effect on the energy can be much different, depending on the softness of the energy surface. We have learned from the previous fig. 4 that the correlation effect smoothen the trends as predicted by the pure mean-field models.

IV CONCLUSIONS

We have provided a quick tour through the structure of mean-field models, their general features and particularly their predictions for superheavy elements (SHE). The relativistic mean-field model (RMF) as well as non-relativistic Skyrme-Hartree-Fock (SHF) has been discussed. Both approaches deliver an excellent description of normal nuclei. Exotic nuclei add new aspects with useful new selection criteria, as e.g. the isotopic shifts in the lead region. The SHE are probably the most demanding extrapolation because these nuclei exist from a subtle interplay between the general trends from bulk properties and shell stabilisation related detailed single-particle levels. It was found that the question of doubly magic SHE dissolves in tiny energy differences and non-negligible correlation effects. There is a broad range of shell stabilised elements without a truly pronounced magic shell closure. The studies on the α-decay chain around 116/176 shows that mean-field models provide reliable estimates of the Q_α values. The details, however, are to wiggly. The problem is resolved by invoking collective ground state correlations. It turns out that the heavier SHE in that chain have extremely soft deformation energy surfaces which induces considerable correlation energies. Taking these into account provides a nice agreement with the experimental values and trends of the Q_α.

Acknowledgement: This work was supported by Bundesministerium für Bildung und Forschung (BMBF), Project No. 06 ER 808 and by the Gesellschaft für Schwerionenforschung (GSI).

REFERENCES

1. Mosel, U., Greiner, W., Z. Phys. **222**, 261 (1969).
2. Nilsson, S.G., et al, Nucl. Phys. **A131**, 1 (1969).
3. Hofmann, S., Rep. Prog. Phys. **61**, 639 (1998).
4. Lazarev, Yu. A., et al., Phys. Rev. **C54**, 620 (1996).
5. Oganessian, Yu.Ts. at al, private communication (2000).
6. Ninov, V., et al, Phys. Rev. Lett. **83**, 1104 (1999).
7. Patyk, Z., and Sobiczewski, A., Nucl. Phys. **A533**, 132 (1991).
8. Möller, P., Nix, J. R., Nucl. Phys. **A549**, 84, (1992); J. Phys. **G 20**, 1681, (1994).
9. Reinhard, P.-G., Rep. Prog. Phys. **52**, 439 (1989).
10. Chabanat, E., et al., Nucl. Phys. **A635**, 231 (1998), **A643**, 441 (1998).
11. Reinhard, P.-G., and Flocard, H., Nucl. Phys. **A584**, 467 (1995).
12. Lalazissis, G.A., König, J., Ring, P., Phys. Rev. C **55**, (1997) 540.
13. Reinhard, P.-G. Nucl. Phys. **A649** (1999) 305c
14. Gross, E.K.U., Dreizler, R.M., *Density functional theory*, Springer, Berlin 1990.
15. Nikolaus, B.A., et al. Phys. Rev. C **46**, 1757 (1992).
16. Bartel, J. et al., Nucl. Phys. **A386**, 79 (1982).
17. Dobaczewski, J., Flocard, H., and Treiner, J., Nucl. Phys. **A422**, 103 (1984).
18. Rufa, M., et al., Phys. Rev. C **38**, 390 (1989).
19. Bender, M., et al., Phys.Rev. C **60**, 34304 (1999).
20. Reinhard, P.-G., et al., Phys.Rev. C **60**, 14316 (1999).
21. Reinhard, P.-G., et al., Hyperf. Int. **127**, 13 (2000)
22. Andreyev, A.N., et al. Nature **405**, 430 (2000).
23. Rutz, K., et al., Phys. Rev. C **56**, 238 (1997)
24. Kruppa, A.T., et al., Phys. Rev C **61**, 034313 (2000)
25. Reinhard, P.-G., et al., RIKEN Review **26**, 23 (2000)

Decay properties of heavy and superheavy nuclei predicted by nuclear mass formulas

H. Koura

Advanced Research Institute for Science and Engineering, Waseda University
3-4-1 Okubo, Shinjuku-ku, Tokyo 169-8555, JAPAN

Abstract. The nuclear mass formula which was recently constructed by the author and his collaborators is used to study heavy and superheavy nuclei. The most distinctive characteristic of this formula is that the shell energy of a deformed nucleus is expressed as an appropriate mixture of spherical shell energies added to an average deformation energy. This mass formula gives ground-state masses and shapes of nuclei ranging from light nuclei ($Z \geq 2$, $N \geq 2$) to superheavy nuclei. In the region of superheavy nuclei, the shell energies of this mass formula show a few magic numbers, which are not so pronounced as in ^{132}Sn and ^{208}Pb. The α-decay Q-values are calculated, and the α-decay half-lives are estimated with use of a phenomenological formula. These results are compared with experimental data and other predictions. The potential energy surface for fission can be calculated by the same method as used for obtaining the shell energies. Based on these energy surfaces, the spontaneous fission is discussed for some selected superheavy nuclei.

I INTRODUCTION

The author's group has recently developed a new method of determining nuclear shell energies which are to be incorporated into a mass formula [1-4]. In this method the shell energy of a deformed nucleus is calculated from shell energies of neighboring spherical nuclei by mixing them with appropriate weights. Details of this method and the final mass formula was reported in Ref. [5].

In response to the recent increase of interest in superheavy elements we give some results predicted by the new mass formula. In Section II, we outline the new method of constructing the mass formula. In Section III, we give some results of the new mass formula for superheavy nuclei, and also compare some decay properties with other predictions and experimental results.

II MASS FORMULA

A Single-particle potentials

We first calculate shell energies for the neutron groups and for the proton groups in spherical nuclei using an extreme single-particle model; we refer to them as crude shell energies. We use a spherical single-particle potential proposed recently [6,7]. The essential point of this potential is in its central part, which is given by

$$V_{\text{cen}}(r) = V_0 \frac{1}{\{1+\exp\left[(r-R_v)/a_v\right]\}^{a_v/\kappa}} \left\{1 + V_{\text{dp}}\frac{1}{1+\exp\left[-(r-R_v)/a_v\right]}\right\}. \quad (1)$$

Here, κ is a parameter governing the behavior at large distances, and V_{dp} is introduced to form a dip in the surface region. When $\kappa = a_v$ and $V_{\text{dp}} = 0$, this potential reduces to the Woods-Saxon potential. The potential parameters V_0, V_{dp}, R_v, a_v, κ are taken to be smooth functions of Z and N. The spin-orbit part has the ordinary form associated with the Woods-Saxon potential. The charge symmetry is imposed on these potentials. In calculating the Coulomb energy of a single proton, special consideration is given to the partial deviation of the proton from the single-particle motion. This potential reproduces fairly well the experimental single-particle levels of 15 doubly-magic and magic-submagic nuclides in a wide nuclidic region ranging from ^{4}He, ^{8}He to ^{208}Pb.

B Crude and refined spherical shell energies

Once the single-particle potential of the nucleus (Z, N) is prepared, we put n neutrons or n protons in it from its bottom. Then the sum of the single-particle energies, which is denoted by $E_{\text{nsp}}(n; Z, N)$ (or $E_{\text{psp}}(n; Z, N)$), is a function of n, Z and N. For the purpose of extracting the deviations from a general tendency in this sum, we subtract from the sum a smooth function $\overline{E}_{\text{nsp}}(n; Z, N)$ (or $\overline{E}_{\text{psp}}(n; Z, N)$) which represents the general tendency of $E_{\text{nsp}}(n; Z, N)$ (or $E_{\text{psp}}(n; Z, N)$). Then, the deviations are given as

$$E_{i\text{fl}}(n; Z, N) = E_{i\text{sp}}(n; Z, N) - \overline{E}_{i\text{sp}}(n; Z, N), \quad (i=\text{n}, \text{p}). \quad (2)$$

The subtraction of the smooth function $\overline{E}_{i\text{sp}}(n; Z, N)$ is made in two steps. We first subtract the Thomas-Fermi energy, and further subtract a smooth function of n, Z and N. With these deviations, we obtain the crude shell energies as

$$E_{\text{ncr}}(Z, N) = E_{\text{nfl}}(N; Z, N), \quad E_{\text{pcr}}(Z, N) = E_{\text{pfl}}(Z; Z, N). \quad (3)$$

For details, see Ref. [5].

Next, we modify these crude shell energies by taking into account the BCS-type pairing, and also make some phenomenological reduction of the shell energies; we refer to the neutron and proton shell energies thus obtained as refined spherical shell energies. In order to include the pairing effect, we take a weighted average of the crude shell energies of neighboring nuclei with certain weights related to the occupation probabilities of the single-particle levels in the BCS theory. It is likely that the simple single-particle plus pairing model is not sufficient to take full account of the configuration mixing. The remaining configuration mixing will probably reduce the magnitudes of the shell energies. This effect is simply represented by a multiplication of the shell energies by a reduction factor μ. Then we obtain the refined spherical neutron and proton shell energies $E_{ns}(Z, N)$ and $E_{ps}(Z, N)$.

For a spherical nucleus the nuclear shell energy is simply the sum of the refined spherical neutron and proton shell energies:

$$E_{0s}(Z, N) = E_{ns}(Z, N) + E_{ps}(Z, N). \tag{4}$$

C Deformation

The shell energy of a deformed nucleus is expressed as the sum of two parts: the intrinsic shell energy $E_{in}(Z, N)$ and the average deformation energy $\overline{E_{def}}(Z, N)$. As for the method, we only give a sketch of it. (See Ref. [5] in details.) We assume that the intrinsic shell energy of a deformed nucleus is expressed as a superposition of the proton and neutron shell energies of some spherical nuclei. The weights in this superposition are obtained by decomposing the deformed nucleus

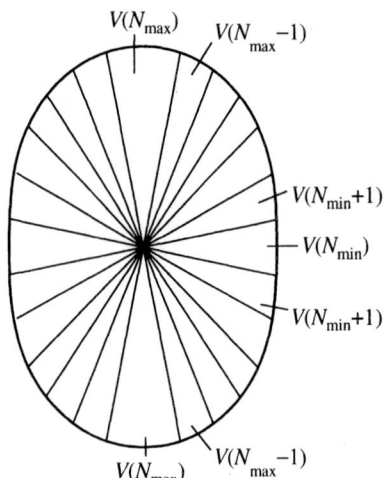

FIGURE 1. A schematic illustration of the decomposition of a deformed nucleus into fractional spherical nuclei. $V(N)$ corresponds to the fraction of the spherical nucleus with the neutron number N.

into fractional spherical nuclei according to the radii as illustrated in Fig. 1 (There is some complication in the correspondence between the radii of the deformed and spherical nuclei [5].)

As the contributions to the average deformation energy, we consider three kinds, the changes of the surface energy and the Coulomb energy, and an energy to favor prolate shapes to obtain the observed dominance of prolate deformation.

We limit the deformation to axially and reflectionally symmetric shapes and assume the same shape for the neutron group and proton group. We use uniform neutron and proton distributions with a sharp-cut surface. Then the nuclear shape is described by the following radii as a function of the polar angle θ:

$$R(\theta) = \frac{R_0}{\lambda} \left[1 + \alpha_2 P_2(\cos\theta) + \alpha_4 P_4(\cos\theta) + \alpha_6 P_6(\cos\theta) + \cdots \right], \quad (5)$$

where $P_{2i}(\cos\theta)$ ($i = 1, 2, \ldots$) are the Legendre polynomials and α_{2i} are parameters to specify the shape. Furthermore, R_0 is the radius of the sphere whose volume is equal to that of the deformed nucleus under consideration, and this volume conservation is guaranteed by the denominator λ. We take the expansion in Eq. (5) down to the $P_6(\cos\theta)$ term.

The shell energy of the nucleus (Z, N) is obtained by minimizing the sum of the intrinsic shell energy and the average deformation energy:

$$E_{\rm sh}(Z, N) = \min_{\alpha_2, \alpha_4, \alpha_6} \left[E_{\rm in}(Z, N) + \overline{E_{\rm def}}(Z, N) \right]. \quad (6)$$

The deformation parameters α_2, α_4 and α_6 giving the minimum energy specify the shape of the ground state. We show the shell energies in Fig. 2.

D Mass formula

The composition of our mass formula is the same as that of the TUYY formula [8]. It consists of three parts as

$$M(Z, N) = M_{\rm g}(Z, N) + M_{\rm eo}(Z, N) + M_{\rm sh}(Z, N), \quad (7)$$

where $M_{\rm g}(Z, N)$ is the term representing the gross feature of the nuclear mass surface, $M_{\rm eo}(Z, N)$ is the even-odd term, and $M_{\rm sh}(Z, N)$ is the shell term for which we use the shell energies obtained in the last section. The functional forms of the gross term and even-odd term are given in Ref. [5].

In order to determine the values of the parameters in the gross and even-odd terms we compare the calculated masses with the masses in the Audi-Wapstra95 [9] excluding the systematics values and also excluding the nuclides with $Z = 0, 1$ and/or $N = 0, 1$. Then, 1835 masses are available. The root-mean-square (RMS) deviation of this formula from experimental data is 680.2 keV. We refer to this formula as KUTY hereafter. In Fig. 3, deviations of calculated masses from experimental data are roughly shown.

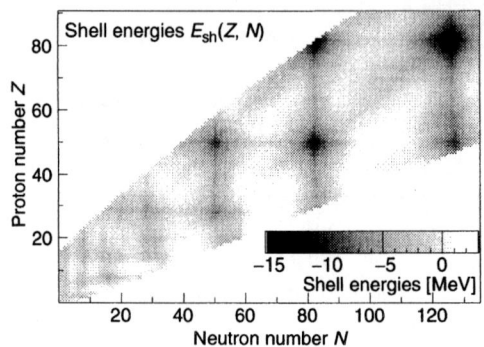

FIGURE 2. Calculated shell energy [5].

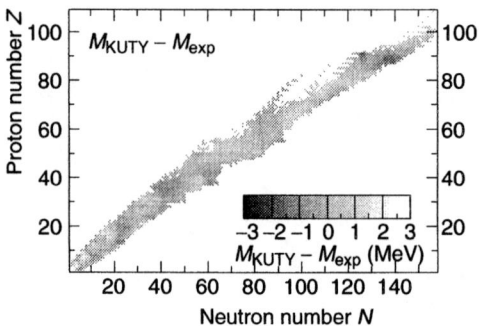

FIGURE 3. Calculated masses [5] minus experimental masses.

III APPLICATION TO SUPERHEAVY ELEMENTS

We give some results of the new mass formula relevant to superheavy elements (SHE). We also compare our predicted quantities with those of two recent mass formulas, the Finite Range Droplet Model (FRDM) formula [10] and the Extended Thomas-Fermi plus Strutinsky Integral (ETFSI) formula [11].

A Single-particle levels

We first show in Figs. 4 and 5 the spherical single-particle levels in two superheavy nuclides $^{310}126$ and $^{298}114$ (these single-particle levels were used as factors to construct the mass formula rather than deduced from the mass formula. The neutron gap at $N = 184$ is 2.49 MeV for $^{310}126$, and 2.50 MeV for $^{298}114$. The proton gap at $Z = 114$ is 1.88 MeV for $^{298}114$, and that at $Z = 126$ is 1.77 MeV for $^{310}126$. These values are considerably smaller than the magic gaps in the known doubly-magic nuclide ^{208}Pb, where the $N = 126$ neutron gap is 3.43 MeV and the

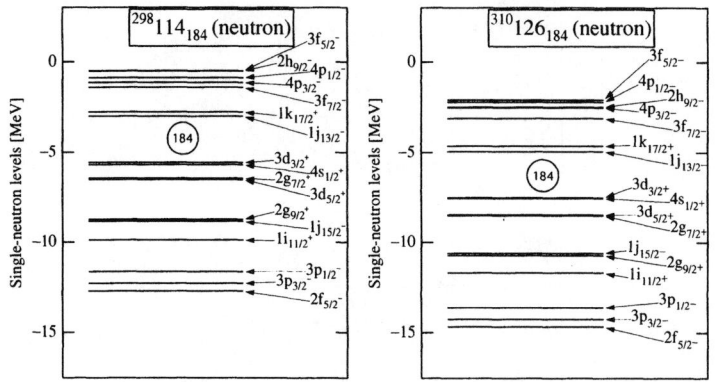

FIGURE 4. Calculated neutron single-particle levels in $^{298}114$ and $^{310}126$ [6].

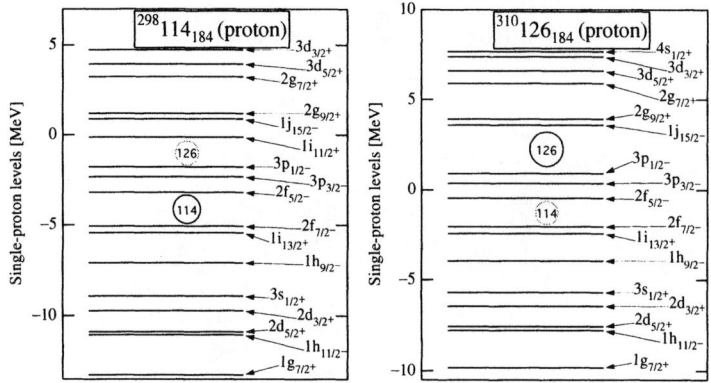

FIGURE 5. Calculated proton single-particle levels in $^{298}114$ and $^{310}126$ [6].

$Z = 82$ proton gap is 4.21 MeV. This fact suggests that the magicities in SHE are less pronounced than in ^{208}Pb (and also in ^{132}Sn).

B Shell energies of SHE

We show in Fig. 6 the shell energies of our formula in the superheavy region. In this figure we see that the alleged magicity at $Z = 114$ is not so remarkable, while the nucleus $^{310}126$ is doubly-magic although its double-magicity is weaker than those of ^{132}Sn and ^{208}Pb as expected.

According to our mass formula, the β-stability line roughly goes straight from ^{257}Fm to $^{297}113$.

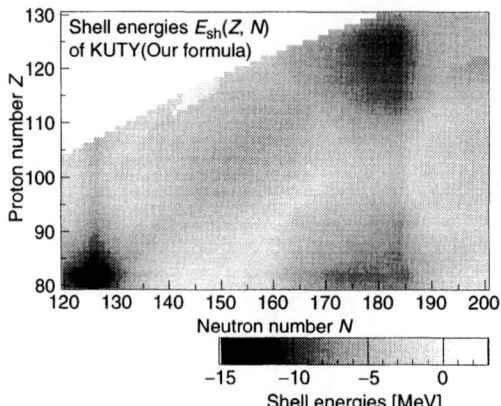

FIGURE 6. Shell energy of KUTY formula [5].

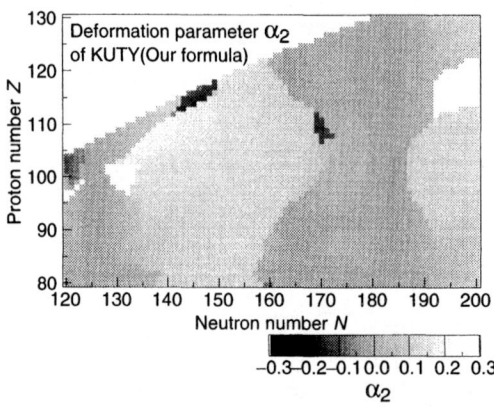

FIGURE 7. Deformation parameter α_2 of KUTY formula [5].

C Nuclear shapes

We show the deformation parameter α_2 in Fig. 7. This figure suggests that many nuclides around $N = 184$ have spherical shapes.

D Q_α and T_α of SHE

The main decay mode of known heaviest elements is α-decay. We calculate α-decay Q-values, Q_α, from the mass formula, and also estimate the α-decay half-lives T_α with use of the phenomenological formula by Viola and Seaborg [13],

$$\log T_\alpha(Z, N) = (aZ + b)/\sqrt{Q_\alpha} + (cZ + d), \qquad (8)$$

with

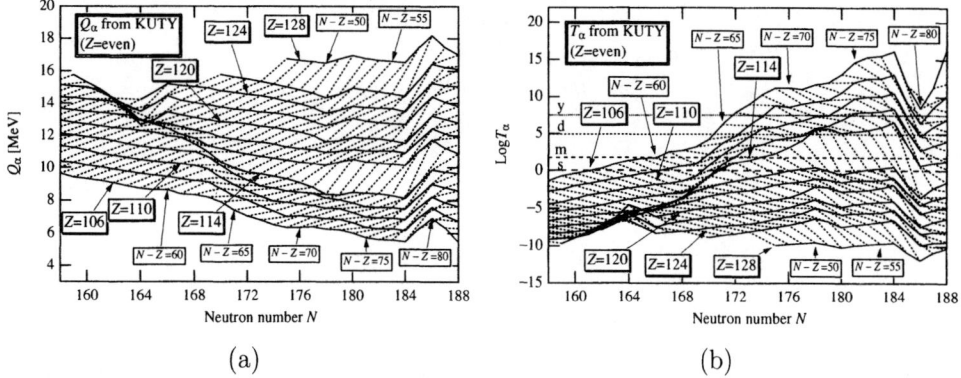

FIGURE 8. Q_α (a) and T_α (b) of superheavy elements by KUTY formula [5] for even Z. The solid lines connect isotopes and dotted lines connect α-decay chains.

$$a = 1.66175, \quad b = -8.5166, \quad c = -0.20228, \quad d = -33.9069, \tag{9}$$

which were determined from comparison with the α-decay data of even-even nuclei [14]. We show Q_α and T_α in Fig. 8. In Fig. 8(a), our α-decay Q-values present a feature of magicity at $Z = 114$ and at $Z = 126$ as relatively wide gaps between isotope lines, while a similar figure with use of FRDM (not shown here) has a larger gap only at $Z = 114$, and that with use of ETFSI shows no gap. Our α-decay half-lives T_α depend on nuclides rather moderately, or regularly, compared with the other two predictions.

We also compare Q_α and T_α with the recent experimental data in the vicinity of the nuclide $^{288}114$ [15–18]. Table 1 shows these data. Note that we tentatively identified the quantity $Q_\alpha = [A/(A-4)]E_\alpha$, in which E_α is the reported α-particle energy and A the mass number of the parent nucleus, as the experimental ground-state Q-value. The differences between the experimental values and the theoretical ones are within 700 keV, not unexpected from the general deviations of the calculated Q_α-values from the Q_α-values of known lighter nuclides. Comparison of the Q_α-values is also made in Fig. 9, which makes the tendencies of the experimental and calculated Q_α-values clearer.

TABLE 1. Recent experimental data and some theoretical results of Q_α.

Nuclide	Experiment	KUTY [5]	ETFSI [11]	FRDM [10]	Smolańczuk [12]
$^{284}112_{172}$	9.30 [17]	9.22	8.88	8.70	–
$^{287}114_{173}$	10.44 [16]	9.83	10.01	9.31	–
$^{288}114_{174}$	9.98 [17]	9.47	9.84	9.17	–
$^{289}114_{175}$	9.85 [15]	9.31	9.61	8.87	–
$^{292}116_{176}$	10.71 [18]	10.33	10.21	10.83	11.07

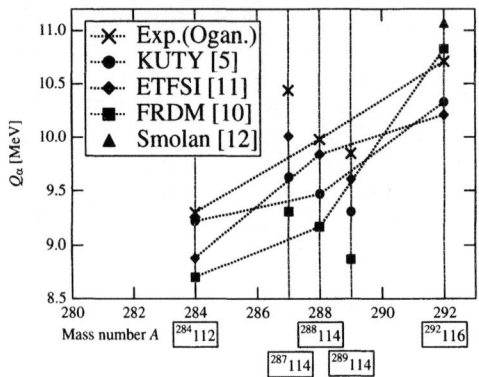

FIGURE 9. Recent experimental data and corresponding theoretical values of Q_α.

E Spontaneous fission

Although the mass formula is concerned only with the equilibrium shapes of nuclei, we can use the same method to calculate the nuclear energies in other shapes. We show in Figs. 10–12 the energy surfaces against the nuclear deformation for some superheavy nuclei. For the nuclide $^{283}112$, the height of the fission barrier is only about 3 MeV and its width is relatively narrow, and the spontaneous fission half-life is expected to be rather short. On the contrary, for the nuclides $^{293}118$ and $^{289}110$, the fission barrier heights are about 7 MeV and about 5 MeV, respectively,

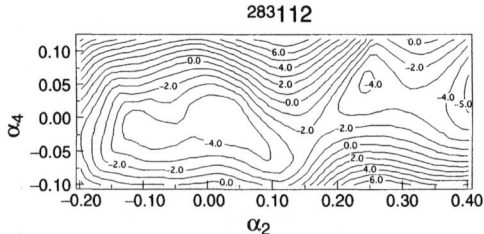

FIGURE 10. Calculated energy surface of $^{283}112$.

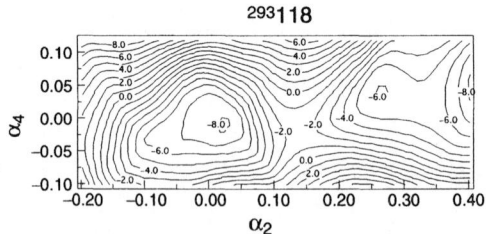

FIGURE 11. Calculated energy surface of $^{293}118$.

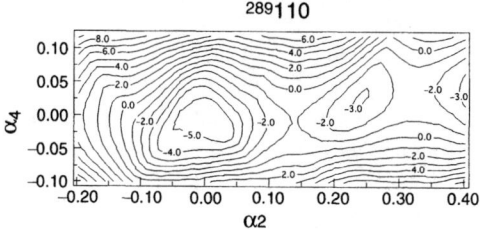

FIGURE 12. Calculated energy surface of $^{289}110$.

and their widths are fairly wide. Therefore, the spontaneous fission of these nuclides is expected to have very long partial half-lives, much longer than the α-decay half-lives. By the way we have chosen $^{289}110$ as the nuclide expected to have the longest half-life among the β-stable superheavy nuclides reachable by $\beta^+\varepsilon$ decays. The α-decay half-life of $^{289}110$ is estimated to be about 1 year.

F Conclusion

With the aid of our mass formula, we could form a picture of the decays of superheavy nuclei. Further theoretical as well as experimental investigations are necessary to make this picture more accurate.

REFERENCES

1. H. Koura, M. Uno, T. Tachibana and M. Yamada, *Technical Report of Advanced Research Center for Science and Engineering, Waseda University*, **No.95-25**, (1995).
2. M. Uno, H. Koura, T. Tachibana and M. Yamada, *International Conference on Exotic Nuclei and Atomic Masses (ENAM 95)*, ed. M. de Saint Simon and O. Sorlin, (Editions Frontieres, 1995), p. 159.
3. H. Koura, T. Tachibana, M. Uno and M. Yamada, *JAERI-Conf* **96-008**, 284 (1996).
4. H. Koura, M. Uno, T. Tachibana and M. Yamada, *Exotic Nuclei and Atomic Masses (ENAM 98)*, ed. B.M. Sherill, D.J. Morrissey and C.N. Davids, (AIP Conference Proceedings 455, 1998), p. 114.
5. H. Koura, M. Uno, T. Tachibana and M. Yamada, *Nucl. Phys.* **A674**, 44 (2000).
6. H. Koura and M. Yamada, *Nucl. Phys.* **A671**, 96 (2000).
7. H. Koura and M. Yamada, *Technical Report of Advanced Research Institute for Science and Engineering, Waseda University*, **No.99-14**, (1999).
8. T. Tachibana, M. Uno, M. Yamada and S. Yamada, *Atomic Data and Nuclear Data Tables* **39**, 251 (1988).
9. G. Audi and A.H. Wapstra, *Nucl. Phys.* **A595**, 409 (1995).
10. P. Möller, J.R. Nix, W.D. Myers and W.J. Swiatecki, *Atomic Data and Nuclear Data Tables* **59**, 185 (1995).

11. Y. Aboussir, J.M. Pearson, A.K. Dutta and F. Tondeur, *Atomic Data and Nuclear Data Tables* **61**, 127 (1995).
12. R. Smolańczuk, *Phys. Rev.* **C56**, 812 (1997).
13. V.E. Viola, Jr. and G.T. Seaborg, *J. Inorg. Nucl. Chem.* **28**, 741 (1966).
14. Z. Patyk, J. Skalski, A. Sobiczewski and S. Ćwiok, *Nucl. Phys.* **A502**, 591c (1989).
15. Yu.Ts. Oganessian, et al., *Phys. Rev. Lett.* **83**, 3154 (1999).
16. Yu.Ts. Oganessian, et al., *Nature* **400**, 242 (1999).
17. Yu.Ts. Oganessian, et al., *Phys. Rev.* **C62**, 041604(R) (2000).
18. Yu.Ts. Oganessian, et al., preprint.

Microscopic description of damped collective motion

Shuhei Yamaji[1], Helmut Hofmann [2] and Fedor A.Ivanyuk[3]

1) Cyclotron Center, Riken, Wako, Saitama, 351-01, Japan

2)Technischen Universität München, D-85747 Garching, Germany

3) Institute for Nuclear Research, 03028 Kiev, Ukraine

Abstract

The damping mechanism of the damped collective motion is studied microscopically on the basis of the linear response theory.

As for isovector modes the giant dipole resonance built on highly excited states in ^{208}Pb is considered. The width due to the two-body collisional damping is found to be able to reproduce well the temperature dependence of the observed width.

As for isoscalar modes, the slow collective motion like fission is studied. The transport coefficients, friction γ, inertia M and local stiffness C are computed along a fission path of ^{224}Th. The calculated effective damping rate $\eta = \gamma/(2\sqrt{M|C|})$ at the saddle is in qualitative agreement with the observed one which increases markedly with temperature. More realistic description of solving the transport equation is necessary to get the quantitative agreement. The approximate functional form of the friction is provided for that.

Moreover, it is noticed that the effect of shell structure on the friction should be taken into account as well as the shell correction energy in the theoretical estimate of the production of supper heavy elements based on the transport equation.

1 Introduction

The dissipation of nuclear matter and its evolution as a function of temperature is one of the fundamental properties of nuclear matter.

In case of fast collective motion of isovector mode, the damping mechanism of the giant dipole resonances as a function of spin and temperature has been highly debated. The reviews are given, for example, in [1], [2]. The theoretical models which are used at present to explain the temperature dependence are the two-body collisional damping model (see, for example, [3] for the collisional damping to 2p-2h states in Landau-Vlasov equation, [4] - for the collisional damping to 1p-1h + 1 phonons states, [5] - for the collisional damping to 2 phonons states) and the thermal fluctuation model (see, for example, [6]). Whether the temperature dependence arises from collisional damping of nucleons or thermal fluctuations of of nuclear potential landscape is still unclear, although many experimental studies have been done (see, for example, [7], [8]). The study of the giant dipole resonance based on the transport theory [9], in which the collisional damping to 2p-2h states are treated, will shed a light on this debate.

For isoscalar modes at finite excitations the best information available at present comes from experiments on fission accompanied by the emission of light particles or gamma rays. Nowadays it seems possible to get the information on the effective damping rate η and its temperature dependence [10],[11],[12]. They find an η which increases markedly with T. Such behavior is hard to understand within the macroscopic models (See, for example, [13]).

In the present work, we apply the linear response theory, first, to the study of isovector dipole resonance in ^{208}Pb in order to check how well it works. Then, we apply it to the case of fission of ^{224}Th [15], [16] and compare the calculated effective damping rate at the saddle with the observed one, which can be extracted from [12]. Finally, we discuss the temperature dependence of microscopically calculated transport coefficients and suggest some simple approximations which can be used in dynamical codes based on Langevin equation(for example, [14]).

2 The linear response approach to collective motion

In this section we will outline briefly outline only basic theoretical features, the details can be found in[17].

We assume to have a Hamiltonian $\hat{H}(\hat{x}_i, \hat{p}_i, Q)$ which depends on deformation through the shape variable Q. For the sake of simplicity we just take one such degree of freedom. However, it does not suffice to restrict this $\hat{H}(\hat{x}_i, \hat{p}_i, Q)$ to the deformed shell model, for which one would have (for A particles)

$$\hat{H}_{sm}(\hat{x}_i, \hat{p}_i, Q) = \sum_{l=1}^{A} \hat{h}(\hat{\vec{x}}_l, \hat{\vec{p}}_l, Q) \quad , \tag{2.1}$$

where $\hat{h}(\hat{\vec{x}}_l, \hat{\vec{p}}_l, Q)$ stands for the dynamics of particle l. First of all, the expectation value of $H_{sm}(\hat{x}_i, \hat{p}_i, Q)$ does not represent correctly the system's total energy. Secondly, as it stands this Hamiltonian would not account for effects of collisions. Following [22] the first deficiency is cured by the Strutinsky method. The second issue can be taken care of by adding the effects of collisions when treating dynamical forces.

Since the nucleus is isolated, the total energy $E_{tot}(t) = \langle \hat{H}(\hat{x}_i, \hat{p}_i, Q) \rangle$ must be a *constant of motion*. Hence the equation of motion for the $Q(t)$ can be constructed from the energy conservation:

$$0 = \frac{d}{dt} E_{tot} = \dot{Q} \langle \frac{\partial \hat{H}(\hat{x}_i, \hat{p}_i, Q)}{\partial Q} \rangle_t \equiv \dot{Q} \langle \hat{F}(\hat{x}_i, \hat{p}_i, Q) \rangle_t \qquad (2.2)$$

2.1 Local linearization

One may evaluate the *intrinsic* quantity $\langle \hat{F}(\hat{x}_i, \hat{p}_i, Q) \rangle_t$ by effectively using the expansion of the Hamiltonian

$$\hat{H}(\hat{x}_i, \hat{p}_i, Q) = \hat{H}(\hat{x}_i, \hat{p}_i, Q_0) + (Q - Q_0)\hat{F}(\hat{x}_i, \hat{p}_i, Q_0) + \frac{1}{2}(Q - Q_0)^2 \langle \frac{\partial^2 \hat{H}}{\partial Q^2}(Q_0) \rangle_{Q_0, T_0}^{qs}, \qquad (2.3)$$

for the case that Q stays in the neighborhood of some properly chosen Q_0 for some ("microscopically") large time interval δt. The second order term is approximated by a *static* density operator $\hat{\rho}_{qs}$ which is determined by $\hat{H}(\hat{x}_i, \hat{p}_i, Q_0)$. The $\hat{\rho}_{qs}$ is meant to represent a thermal equilibrium at Q_0. The only coupling term between collective and intrinsic motion is then given by the term of first order in $Q - Q_0$.

It can be shown (see e.g.[17]) that (2.2) leads to the following form of the local equation of motion:

$$k^{-1} q(t) + \int_{-\infty}^{\infty} \tilde{\chi}(s) q(t-s) ds = 0. \qquad (2.4)$$

Here $q = Q - Q_m$ measures the deviation of the actual Q from the center of the oscillator approximating the true potential in the neighborhood of Q_0. The $\tilde{\chi}$ is the causal response function associated to the dynamics of the nuclear "property" $\langle \hat{F} \rangle$. The coupling constant k is expressed as [17]

$$-k^{-1} = \langle \frac{\partial^2 \hat{H}}{\partial Q^2}(Q_0) \rangle_{Q_0, T_0}^{qs} + (\chi(0) - \chi^{ad}) \simeq \frac{\partial^2 f}{\partial Q_0^2} + \chi(0) \qquad (2.5)$$

with $\chi(0)$ being the static response and χ^{ad} being the adiabatic susceptibility.

2.2 Transport coefficients from the collective response

To solve (2.4), we introduce a (hypothetical) external force $\tilde{f}_{ext}(t) \hat{F}$ and evaluate how the system responds to it. One can define the collective response function $\chi_{coll}(\omega)$ via $\delta \langle \hat{F} \rangle_\omega = -\chi_{coll}(\omega) f_{ext}(\omega)$. It can be brought to the form [9]

$$\chi_{coll}(\omega) = \frac{\chi(\omega)}{1 + k\chi(\omega)} \qquad (2.6)$$

which is known to be standard for the case of zero temperature (see e.g. [22] and [26]).

The dissipative part of $\chi_{coll}(\omega)$ represents the distribution of strength over various possible local modes, which exhibit themselves as individual peaks. The corresponding "dispersion relation" or secular equation

$$\chi(\omega) + 1/k = 0 \qquad (2.7)$$

is easily recognized to come from the Fourier transform of (2.4). A solution of (2.7) leads to complex frequencies ω_ν, which actually come in pairs $\omega_\nu^\pm = \pm \mathcal{E}_\nu - i\Gamma_\nu/2$. We approximate the response associated of the low frequency mode by that of a damped oscillator with the transport coefficients M, γ and C.

$$\begin{aligned}(\chi_{coll}(\omega))^{-1}\delta <F>_\omega &= -f_{ext}(\omega) \\ &\Downarrow \\ (\chi_{osc}(\omega))^{-1}\delta <F>_\omega &\equiv k^2(-M\omega^2 - \gamma i\omega + C)\delta <F>_\omega = -f_{ext}(\omega),\end{aligned} \qquad (2.8)$$

2.3 Collisional damping of nucleonic motion

Let us turn to the damping mechanism used in our theory. To evaluate the damping mechanism fully for some given $V_{res}^{(2)}$ would be too a tremendous task. We therefore use a scheme borrowed from the way one would treat the effects of collisions in time dependent mean field theories like ETDHF or its classic versions as given by the Landau-Vlasov equation with two-body collisions. In this paper we will just state the final expressions referring to [17].

The Fourier transform of the dissipative part of the intrinsic response function finally can be written as

$$\chi''(\omega) = \int \frac{d\hbar\Omega}{4\pi} \left(n(\Omega-\omega/2) - n(\Omega+\omega/2)\right) \sum_{jk} |F_{jk}|^2 \varrho_k(\Omega-\omega/2)\varrho_j(\Omega+\omega/2) \qquad (2.9)$$

Here, $n(x)$ is the Fermi function determining the occupation of the single particle levels $|k>$. The latter are the eigenstates of the Hamiltonian $\hat{h}(\hat{\vec{x}},\hat{\vec{p}},Q_0)$ with corresponding energies e_k. The $\varrho_k(\omega)$ represents the distribution of the single particle strength over more complicated states. It is here that the effects of collisions come into play. A finite $V_{res}^{(2)}$ gives rise to finite self-energies for which both real and imaginary parts are considered according to the formulae $\Sigma(\omega,T) = \Sigma'(\omega,T) - i\Gamma(\omega,T)/2$, with

$$\Gamma(\omega,T) = \frac{1}{\Gamma_0} \frac{(\hbar\omega - \mu)^2 + \pi^2 T^2}{1 + \frac{1}{c^2}[(\hbar\omega-\mu)^2 + \pi^2 T^2]} \qquad (2.10)$$

and μ being the chemical potential. Then the $\varrho_k(\omega)$ becomes

$$\varrho_k(\omega) = \frac{\Gamma(\omega,T)}{(\hbar\omega - e_k - \Sigma'(\omega,T))^2 + \left(\frac{\Gamma(\omega,T)}{2}\right)^2} \qquad (2.11)$$

In (2.10) the $1/\Gamma_0$ represents the strength of the "collisions", viz the coupling to more complicated states. The cut-off parameter c allows one to account for the fact that the imaginary part of the self-energy does not increase indefinitely when the excitations get away from the Fermi surface. In the present calculation we choose $\Gamma_0 = 33 MeV$ and c=20 MeV like in [18].

For not too small temperature the friction in (2.8) can be written in so called zero frequency limit as $\gamma = \frac{\partial \chi''(\omega)}{\partial \omega}|_{\omega=0}$ or

$$\gamma = -\int \frac{d\hbar\Omega}{4\pi} \frac{\partial n(\Omega)}{\partial \Omega} \sum_{jk} |F_{jk}|^2 \varrho_k(\Omega) \varrho_j(\Omega). \qquad (2.12)$$

2.4 The single-particle Hamiltonian

The single particle Hamiltonian $\hat{h}(\hat{x}, \hat{p}, Q)$ of (2.1) is chosen to be given by the two-center shell model of [19] and [20]:

$$\hat{h} = -\frac{\hbar^2 \nabla^2}{2m} + \hat{V}(\hat{\rho}, \hat{z}) + \hat{V}_{ls}(\hat{x}, \hat{l}, \hat{s}) + \hat{V}_{l^2}(\hat{x}, \hat{l}) \qquad (2.13)$$

Here V is the two-center potential in cylindrical coordinates with m, $\hat{s}$ and $\hat{l}$ being the nucleons' mass and the operators for spin and angular momentum, respectively. A detailed description of the construction of V can be found in [19] and [20]. As basic shape parameters, for the symmetric case, there are the distance $z_0 = z_2 - z_1$ between the centers z_i of the two potential wells, the neck-parameter ϵ and the fragment deformation parameter δ. The momentum-dependent part in (2.13) consists of the spin orbit-coupling term $\hat{V}_{ls}$ and l^2-term $\hat{V}_{l^2}$. For them the angular momentum $\hat{l}$ is described in the stretched coordinates. The strengths κ_i of ls-term and $\kappa_i \mu_i$ of l^2-term are taken from [21].

2.5 The collective coordinates, the operator $\hat{F}$ and the coupling constant k

For the description of isovector dipole resonance, we follow the discussion of Ref.[22]. Since neutrons and protons oscillate against each other in opposite phase, the density of nucleons with isospin projection t_3 may be parametrized by

$$\hat{n}(r, \theta, \phi, t_3, Q) = n_0 f(r + 2t_3 Q R_0 Y_{10}(\theta, \phi)) \frac{N(t_3)}{A}, \qquad (2.14)$$

where $n_0 f(r)$ is the density distribution of nucleons with spherical shape and $N(t_3)$ stands for the neutron number for neutrons and the proton number for protons. The first-order change in the above density is

$$\delta \hat{n}(r, \theta, \phi, t_3, Q) = Qm\omega_0^2(2t_3)\frac{N(t_3)}{A} n_0 2r Y_{10}(\theta, \phi))\theta(r - R_0) \qquad (2.15)$$

for the harmonic oscillator approximation to the independent-particle model, where $n_0 = A/(\frac{4\pi}{3}R_0^3)$. After folding $\delta\hat{n}$ by a zero-range isovector two-body interaction $V_0\vec{t_1}\cdot\vec{t_2}\delta(\vec{r_1}-\vec{r_2})$, one obtains the corresponding change in the single particle potential, which yields the expression for the operator $\hat{F}$ as

$$\hat{F} = 2t_3 r Y_{10} U_0^t A m \omega_0^2, \tag{2.16}$$

where $U_0^t = n_0 V_0/A$ is the strength of the isospin potential and $U_0^t = (96/A)$ MeV is used for the numerical analysis. The coupling constant k is directly calculated from Eq.(2.5) by assuming $\chi(0) = \chi^{ad}$

$$k^{-1} = \frac{1}{\pi} U_0^t A^2 (m\omega_0^2)^2 <r^2> \tag{2.17}$$

For the description of fission, we use a fission coordinate $Q = R_{12}/(2R_0)$ along a fission path as the collective coordinate, where R_{12} measures the relative distance between the centers of mass of the two fragments, and R_0 stands for the radius of the spherical configuration. We identify the fission path with a line of minimal potential energy specified by two parameters z_0 and δ. The neck parameter, which is sensitive to specifications of the scission configuration, is fixed to be $\epsilon = 0.4$. For the fission variable Q the operator $\hat{F}$ is defined as

$$\hat{F} = \frac{\partial \hat{h}}{\partial Q} = \left(\frac{\partial Q}{\partial z_0} + \frac{\partial \delta}{\partial z_0}\frac{\partial Q}{\partial \delta}\right)^{-1} \left(\frac{\partial \hat{h}}{\partial z_0} + \frac{\partial \delta}{\partial z_0}\frac{\partial \hat{h}}{\partial \delta}\right) \tag{2.18}$$

where the derivative $\partial\delta/\partial z_0$ is to be taken along the fission path $\delta = \delta(z_0)$.

For the collective response function (2.6), we need the coupling constant k as given by (2.5). This expression can be evaluated from the free energy f and the static response $\chi(0)$. The free energy $f(Q,T)$ will be written as a sum of Coulomb and surface energies plus the shell correction part:

$$f(Q,T) = f_{Coul}(Q,T) + f_{surf}(Q,T) + f_{sc}(Q,T). \tag{2.19}$$

The Coulomb and surface energies is approximated by $f_{Coul}(Q,T) = f_{Coul}(Q,T=0)(1-\alpha T^2)$ and $f_{surf}(Q,T) = f_{surf}(Q,T=0)(1-\beta T^2)$ with the values of α and β taken to be 0.000763 and 0.00553 MeV^{-2} [23]. The free energies $f_{Coul}(Q,T=0)$ and $f_{surf}(Q,T=0)$ have been evaluated according to [24]. In Table III of [24], there are several sets of parameters. Here, we choose the set with the values of $a = 0.65 fm$, $a_s = 21.836 MeV$, $K_s = 3.48$ corresponding to the radius parameter $r_0 = 1.2 fm$. For the shell correction we assume the form (c.f.[26]): $f_{sc}(Q,T) = f_{sc}(Q,T=0)\tau/\sinh\tau$, with $\tau = (2\pi^2 T)/(\hbar\omega_0)$. The shell correction $f_{sc}(Q,T=0)$ is evaluated according to [25].

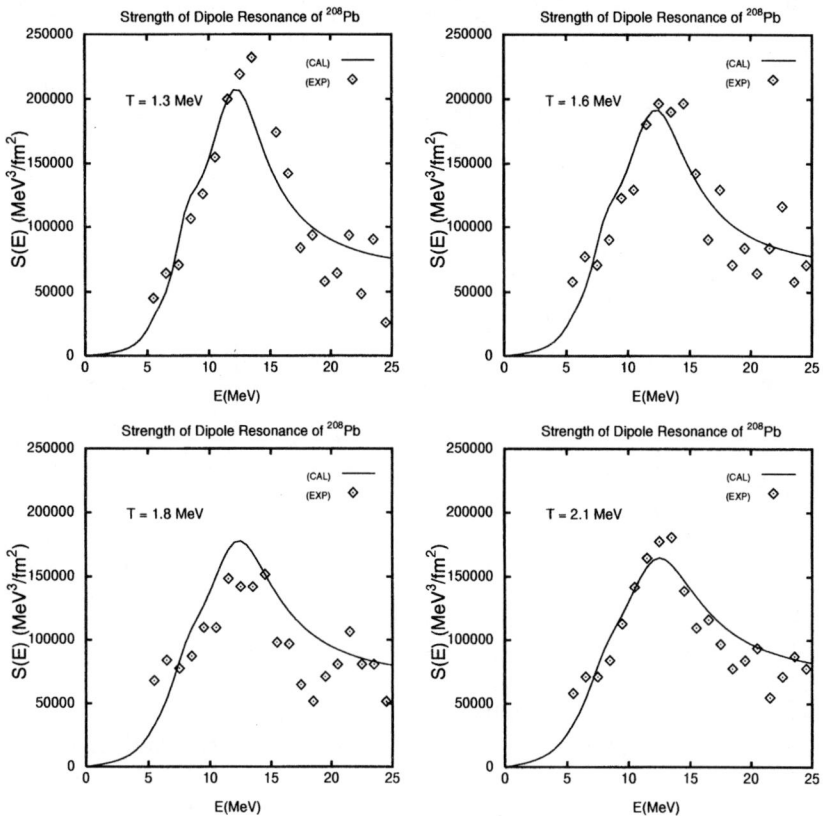

Figure 1: The calculated strength function as function of the energy E is compared with the observed strength at the temperatures $T = 1.3, 1.6, 1.8$ and 2.1 MeV.

3 Numerical results

We, first, discuss the numerical results for the case of isovector dipole resonance. We evaluate the strength function $S(E)$, which is the imaginary part of the collective response function $\chi_{coll}(E)$ in (2.6). The calculated strength functions $S(E)$ for ^{208}Pb at T= 1.3, 1.6, 1.8 and 2.1 MeV are compared with the experimental ones [8] in Fig.1. Since the experimental results in [8] are given in arbitrary units, they are multiplied by a common factor 64,500 in the figures to match to the calculated results. From these figures, one can see that the increase of width as a function of temperature is well explained by the two-body collisional damping.

Next, we have evaluated transport coefficients such as friction γ, inertia M and local stiffness C, as a function of temperature, along the fission path for the symmetric fission of ^{224}Th and the quantities $\varpi = \sqrt{|C|/M}$, $\beta = \gamma/M$ and $\eta = \gamma/2\sqrt{M|C|}$ at the minimum and saddle points. It was found that, in the values of η and β at the barrier and the minimum as well as their variation with T do not differ much from each other. The details were given in [15]. The reduction $\left(\sqrt{1+\eta_s^2} - \eta_s\right)$ in

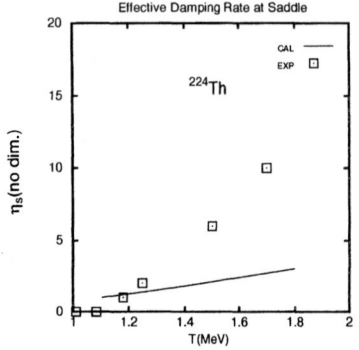

Figure 2: The calculated and experimental effective damping rates η_s at the saddle as function of temperature.

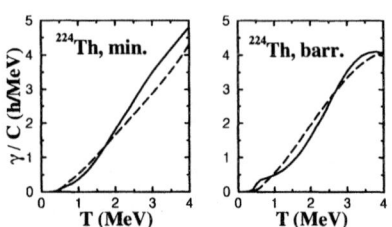

Figure 3: The relaxation time τ_{coll} for collective motion at the potential minimum and at the barrier: the microscopic result (solid curve) compared to the approximation (3.1)-(3.3) (dotted curve).

fission width in Kramers' formula enhances particles and γ-rays emission from the interior. The experimental η_s is determined by the CASCADE calculations [12] in order to reproduce the enhancement of the γ-ray multiplicities over nondissipative CASCADE calculations. The calculated η_s is compared with the experimental one in Fig.2. The calculated η_s is in qualitative agreement with the experimental one, which increases rapidly with temperature. However, in these experimental studies they tried to parameterize a more complicated issue by the single quantity η_s. Such a procedure is very appealing for reasons of simplicity. But this is dangerous. More realistic description by solving the transport equation (see, for example, [14]) may be necessary. In order to perform such kind of dynamical analysis, it is necessary to account for the temperature dependence of the damping parameter. The increase of damping with the temperature cannot be obtained within macroscopic models like [13]. It can can be given by our microscopic model [9]. However the microscopic calculations the transport coefficients especially for the multi-dimensional case are rather time-consuming. Therefore, it is desirable to obtain some simple functional form for the temperature dependence. Here, we study the dependence of the ratios $\gamma/|C|$ and γ/M on the temperature [27].

In Fig.3 we plot the relaxation time $\tau_{\text{coll}} = \gamma/|C|$ for motion in a (locally harmonic) potential as function of T.

The fully drawn lines show the microscopic results. The dashed curves are calculated by the functional form with a cut-off parameter c_{macro}:

$$\tau_{\text{coll}} = \frac{\gamma}{|C|} \approx \frac{2}{\hbar \Gamma_0} \frac{\pi^2 T^2}{1 + \pi^2 T^2 / c_{\text{macro}}^2} \tag{3.1}$$

One should expect that this ratio reaches a macroscopic limit at larger temperatures, which for this formula is obtained above $T_{\text{h.T}} \simeq c_{\text{macro}}/\pi$. For this reason the c_{macro} may be fixed to warrant this feature, namely

$$\tau_{\text{coll}}|_{T_{\text{h.T}}} = \left.\frac{\gamma(T)}{|C(T)|}\right|_{T_{\text{h.T}}} \approx \frac{\gamma_{\text{wall}}/2}{|C_{\text{LDM}}(T)|} \tag{3.2}$$

which is achieved for

$$c_{\text{macro}}^2 = \frac{\Gamma_0}{2\hbar} \frac{\gamma_{\text{wall}}/2}{|C_{\text{LDM}}(T)|} \tag{3.3}$$

Here we accounted for results obtained by several previous numerical calculations, see e.g.[15], [16], [17]. They showed that the value of friction at large T is somewhat below the wall formula.

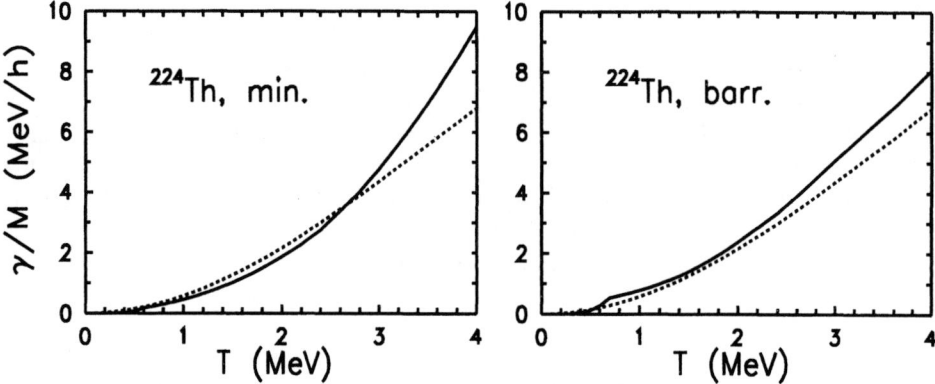

Figure 4: The inverse relaxation time $1/\tau_{\text{kin}} \equiv \gamma/M$ as function of temperature for nucleus ^{224}Th: the microscopic result (solid curve) compared to the approximation (3.4) (dotted curve).

The ratio γ/M determines the inverse relaxation time to the Maxwell distribution. In Fig.4 we show it on the left hand panel as function of T. In this case the dashed curve represents the estimate given in [27], namely

$$\gamma/M \approx 2\Gamma_{\text{sp}}(\mu, T)/\hbar = \frac{2}{\Gamma_0} \frac{\pi^2 T^2/\hbar}{1 + \pi^2 T^2/c^2} \approx \frac{0.6 T^2}{1 + T^2/40} MeV/\hbar \quad (T \text{ in} MeV). \tag{3.4}$$

The approximations (3.1) and (3.4) to the *microscopic* quantities $\gamma/|C|$ and γ/M are expressed in terms of *macroscopic* quantities like C_{LDM} or γ_{wall} which are often used in dynamical calculations. So, Eqs. (3.1) and (3.4) supply a "cheap" way to account for the correct temperature dependence of the transport coefficients.

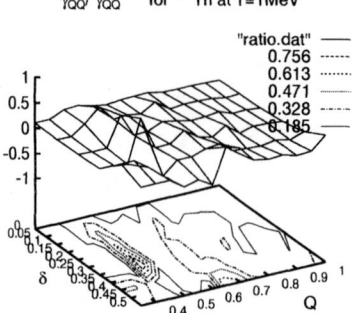
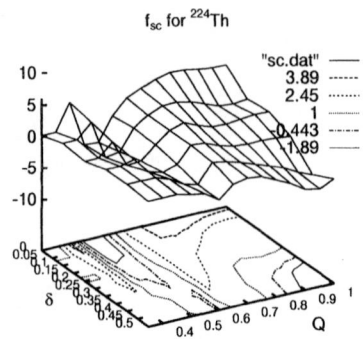

Figure 5: The calculated ratio $\gamma_{QQ}/\gamma_{QQ}^{w}$ for T= 1 MeV. The suffix w corresponds to wall formula. γ_{QQ} stands for the diagonal friction for Q-motion.

Figure 6: The shell correction energy f_{sc} as function of deformations Q and δ.

We, finally, would like to notice the importance of shell effect on friction, when we estimate the production cross section of supper heavy elements by solving the transport equation. The effect of shell correction energy has been taken into account and is found to be important to predict the cross section [28]. We want to point out the importance of shell effect on friction, too. In Fig.5, we show the ratio of the microscopic friction to the macroscopic wall friction $\gamma_{QQ}/\gamma_{QQ}^{w}$ as function of deformations Q and δ at temperature T = 1 MeV. In Fig.6, we show the shell correction energy. Comparing these two figures, we see the shell effect in the friction as well as in the shell correction energy. We have negative shell correction energy with large absolute value for the case of big energy gap at the Fermi surface in the single particle level structure. For such case, we expect small value of the ratio $\gamma_{QQ}/\gamma_{QQ}^{w}$. This can be guessed from the expression of friction γ in (2.12). In Fig. 6, the shell correction energy is minimum at Q = 0.55, δ = 0.0, where we have small value of ratio in Fig. 5. When the shell correction energy is positive and large, we expect large values of friction. This occurs near the saddle at Q = 0.75 and δ = 0.4.

4 Summary

The damping mechanism of the collective motion is studied microscopically on the basis of the linear response theory.

For isovector modes, the strength functions of the giant dipole resonance calculated on the basis of the two-body collisional damping reproduce well the experimental ones.

For isoscalar modes, the calculated effective damping rate at the saddle for the fission of ^{224}Th is in qualitative agreement with the observed one which increases markedly with temperature. To get quantitative agreement, it may be necessary to treat the fission more realistically and to solve the transport equation such as Langevin or Fokker-Planck equation. For that, the temperature dependence of transport coefficients is studied. We suggest a functional form of the temperature dependence, which improves the macroscopic coefficients.

We notice that the shell effect on friction should not be neglected as well as the shell effect of the total energy, when we estimate the production cross section of supper heavy elements theoretically.

Finally, we would like to mention the two assumptions in the treatment of damping due to the residual two-body interaction $\hat{V}_{res}^{(2)}(\hat{x}_i, \hat{p}_i)$.

First, to calculate the response function, we neglect the contribution of the particle-hole effective interaction to damping for which we have not found a simple treatment. The neglected correction was argued to have an opposite phase to the contribution of the renormalization of the independent particle or hole Green function and to reduce appreciably the damping width Γ within the approximate treatment of 2p-2h configurations by 1p-1h + 1 vibration configurations [4]. But the renormalized particle or hole Green function in the present theory does not represent the exact one body Green function. It represents the single particle motion in the intrinsic system only, which should not contain collective effects [29]. Therefore, the correction for the particle-hole interaction may be reasonably neglected, based on the discussion that it is of a higher power of density compared to the correction due to the renormalization of independent particle or hole Green function in the infinite matter.

Second, we take the single particle width $\Gamma(\omega, T)$ like the one in the infinite matter, which depends only on the single particle energy and temperature. However, the dependence of the single particle width on other properties of the single particle levels may be neglected, if one is interested in the gross structure of the damping.

References

[1] K.A.Snover, Annu. Rev. Nucl. Part. Sci. **36** (1986) 545

[2] J.J.Gaardhoeje, Annu. Rev. Nucl. Part. Sci. **42** (1992) 483

[3] V.Baran, M.Colonna, M.Di Toro et al , Nucl. Phys. **A599** (1996) 29c

[4] P.Dontai, N.Giovanardi, P.F.Bortignon and R.A.Broglia, Phys. Lett. **B383**(1996)15

[5] N.D.Dang, K.Tanabe and A.Arima, Phys. Rev. Lett. **80** (1998) 4145

[6] W.E.Ormand, P.F.Bortignon, R.A.Broglia and A.Bracco, Nucl. Phys. **A614** (1997) 217

[7] G.Gervais, M.Thoennessen and W.E.Ormand, Phys. Rev. **C58** (1998) R1377

[8] T.Baumann et al., Nucl. Phys. **A635** (1998) 428

[9] S. Yamaji, H. Hofmann and R. Samhammer, Nucl. Phys. **A475** (1988) 487

[10] P. Paul and M. Thoennessen, Ann. Rev. Part. Nucl. Sci. **44** (1994) 65

[11] D.J. Hofman, B.B.Back and P.Paul, Phys. Rev. **C51** (1995) 2597

[12] I.Dioszegi, N.P.Shaw, I.Mazumdar et al , Phys. Rev. **C61** (2000) 024613

[13] J.Blocki, Y.Boneh, J.R.Nix, A.J.Sierk, W.J. Swiatecki, Ann. Phys. **113** (1978) 330

[14] T.Wada, Y.Abe and N.Carjan, Phys. Rev. Lett.**70** (1993) 3538

[15] S. Yamaji, F.A.Ivanyuk and H. Hofmann, Nucl. Phys. **A612** (1997) 1

[16] F.A.Ivanyuk, H.Hofmann, V.V.Pashkevich and S.Yamaji, Phys. Rev. **C55**(1997)1730

[17] H. Hofmann, Phys.Rep. **284** (1997) 137

[18] P.J. Siemens, A.S. Jensen and H. Hofmann, Nucl. Phys. **A409** (1983) 135

[19] J. Maruhn and W. Greiner, Z.f.Phys. **251** (1972) 431

[20] S. Yamaji and A. Iwamoto, Z.f.Phys. **A313** (1983) 161

[21] D. Scharnweber, W. Greiner and U. Mosel, Nucl.Phys. **A164** (1971) 257

[22] P.J. Siemens and A.S. Jensen, "Elements of Nuclei: Many-Body Physics with the Strong Interaction", Addison and Wesley, 1987

[23] C. Guet, E. Strumberger and M. Brack, Phys.Lett. **B205** (1988) 427

[24] H.J. Krappe, J.R. Nix and A. Sierk, Phys.Rev. **C20** (1979) 992

[25] M. Bolsterli, E.O. Fiset, J.R. Nix and J.L. Norton, Phys. Rev. **C5** (1972) 1050

[26] A. Bohr and B.R. Mottelson, Nuclear Structure, Vol.II (Benjamin, London, 1957)

[27] H.Hofmann, F.A.Ivanyuk and S.Yamaji, to be published

[28] T.Wada,T.Tokuda, K.Okazaki, M.Ohta, Y.Aritomo and Y.Abe, Proc.4th Int. Conf. on Dynamical Aspect of Nuclear Fission, Slovak, Republic,(19-23,Oct,1998) 77

[29] R.Alkofer, H.Hofmann and P.J.Siemens, Nucl. Phys. **A476** (1988) 213

Reaction Theory for the Synthesis of Superheavy Elements

T. Wada, Y. Aritomo[†], T. Ichikawa, M. Ohta, and Y. Abe[‡]

Department of Physics, Konan University, 8-9-1 Okamoto, Kobe 658-8501, Japan
E-mail: wada@konan-u.ac.jp
[†]*Flerov Laboratory of Nuclear Reactions, JINR, Dubna, Moscow region 141980, Russia*
[‡]*Yukawa Institute for Theoretical Physics, Kyoto University, Kyoto 606-8502, Japan*

Abstract. The dynamical process of synthesizing superheavy elements is studied on the basis of the fluctuation-dissipative dynamics. The whole process is divided into three stages, i.e., the approaching stage, the formation stage, and the surviving stage. For the study of the formation stage, a three-dimensional Langevin equation is used from the contact of two nuclei to calculate the competition between the complete fusion and the quasi-fission. We estimate the effects of the nuclear deformation on the fusion probability. The results are consistent with the experimental results. For the description of the surviving stage, a statistical model is used to estimate the competition between the fission and the evaporation process. In this stage, it is very important to include the temperature-dependent shell correction to the fission barrier and the collective enhancement of the level density. From the study of the isotope dependence of the production cross section, the survival probability is found to be very sensitive to the separation energy of neutron. The results show the importance of the use of neutron-rich beams and targets.

I INTRODUCTION

The search for the superheavy elements had an exciting year of 1999. After the production of $Z = 112$ at GSI in 1996 [1], there had been a time of silence until FLNR in Dubna reported the production of $Z = 112$ with hot fusion reaction with ^{48}Ca bombarding on ^{238}U [2]. Successively, in 1999, they performed experiments with Pu targets, ^{48}Ca on ^{242}Pu and on ^{244}Pu, to produce the new element $Z = 114$ [3]. LBL in Berkeley reported the production of $Z = 118$ with the cold fusion reaction with ^{86}Kr bombarding on ^{208}Pb [4]. The production cross section was reported to be as large as 2 pb that was a few orders of magnitude larger than the value expected from the extrapolation of the systematic data of GSI with the cold fusion reaction [5]. The new experiments stimulated the further study in other laboratories like GANIL and RIKEN.

Understanding of the reaction mechanism of the synthesis of superheavy elements is necessary for the quantitative estimation of the cross section. For this purpose, we propose to use the framework of the fluctuation-dissipation dynamics, which is expressed by the Kramers (Fokker-Planck) equation or by the Langevin equation. For simplicity, we assume that the formation and the decay of the compound nuclei are independent even for the massive system like superheavy elements. With this approximation, we can express the evaporation residue cross section as the product of the fusion probability and the survival probability.

The fusion stage is further divided into two stages: sticking probability (before contact of projectile and target) and compound-nucleus-formation probability (after contact). In the preliminary analysis [6], we employ the one-dimensional Smoluchowski equation in the elongation degree of freedom for the symmetric channel taking into account the temperature dependent shell correction energy. We pointed out a new possibility of synthesizing the superheavy element with relatively hot fusion reaction. We also performed a three-dimensional calculation including mass-asymmetric degree of freedom and obtained the cross section for ^{48}Ca + ^{244}Pu reaction before the Dubna experiment [7]. In the formation stage, the problem of fusion-hindrance plays an important role. Our framework is an extension of the extra-push model [8] by Swiatecki to include the fluctuation of the motion.

The survival probability depends strongly on the excitation energy of the compound nucleus in the case of superheavy element synthesis. The survival probability decreases rapidly as the excitation energy gets higher, because the fission barrier due to the shell correction energy disappears. In this sense, the excitation energy dependence of the shell correction energy is very important in estimating the survival probability for superheavy elements. The survival stage is treated with the statistical model including the modification due to the effect of the dissipation. We discuss the isotope dependence of the survival probability in connection with the separation energy of neutrons.

In Sec. 2, we discuss the phenomenon of fusion hindrance in connection with the macroscopic model. In Sec. 3, we factorize the evaporation residue cross section into three factors: sticking probability, formation probability, and survival probability. Section 4 is devoted to the formation probability. And Sec. 5 is for the survival probability. Summary and conclusions are given in Sec. 6.

II FUSION HINDRANCE

The hindrance of fusion is observed in the symmetric-like fusion reactions forming $Z > 80$ nuclei [9], where Z denotes the atomic number of the fused system. Several interpretations have been given for this phenomenon [10]. Our interpretation is based on the work by Swiatecki and his coworkers [8]. He considered that the passage of the one-dimensional potential barrier (Coulomb barrier) might not be sufficient for fusion to occur. By introducing a neck and a mass-asymmetry degree of freedom, the further dynamical evolution of the system after being captured inside the one-dimensional potential barrier was considered under the dissipation. Since the dissipation phenomena inherent in these macroscopic models are necessarily connected with fluctuation in the dynamic evolution of the nuclear system, it is necessary to include the fluctuation and to solve the fluctuation-dissipation dynamics in this multi-dimensional space.

For light systems ($Z < 80$), fusion occurs when the two nuclei come into contact overcoming the incident Coulomb barrier. Because the Coulomb repulsion between two nuclei is rather small, the attractive nuclear force makes the system to form a spherical compound nucleus and fusion occurs with large probability. This situation is described successfully by the Bass model [11].

For heavy systems ($Z > 80$), the Coulomb repulsion becomes strong and starts to compete with nuclear attraction even after the contact of two nuclei. Therefore, to go to the spherical region, the system needs extra energy after the contact.

The difference between the two cases lies in the relative location of the ridge and the contact configuration. In the case of light systems, the ridge lies outside of the contact configuration in the multi-dimensional deformation space. Therefore, once the projectile and the target contact it is automatically goes to the spherical shape. On the other hand, in the case of the heavy systems, the ridge lies inside of the contact configuration and we need an extra energy after the contact to overcome this ridge to come into the spherical region. This is the basic idea of the extra-push model by Swiatecki [8].

The relative location of the ridge and the contact configuration is essentially determined by the strength of the Coulomb force between the projectile and the target. In this sense, the charge product, or more precisely the effective fissility [12] determines the situation. The critical value for the onset of the fusion hindrance is inferred as $x_{\text{eff}} \approx 0.75$. Suppose the system has the effective fissility a little smaller than this critical value. If the projectile and the target is spherical, there is no or very weak fusion-hindrance. But on the other hand, if one of them has prolate deformation, the distance between the charge centers depends on the orientation of the deformed nuclei. If they approach in tip configuration (the spherical nucleus colliding at the tip of the deformed nucleus), the distance is farther and it locates outside of the ridge; that mean the system suffers the fusion-hindrance.

Recently, Ikezoe and his collaborators performed the experiments using the deformed targets, ^{60}Ni + ^{154}Sm ($x_{\text{eff}} = 0.735$) [13] and ^{76}Gd + ^{150}Nd ($x_{\text{eff}} = 0.749$) [14]. They found the necessity of the extra-extra-push energy for the system to form the spherical compound nuclei if they collide in tip configuration. The extra-extra-push energy they obtained from the experiment is 20MeV for ^{60}Ni + ^{154}Sm and 14MeV for ^{76}Gd + ^{150}Nd. This effect is very important because in ^{48}Ca + ^{244}Pu experiment, the effective fissility is $x_{\text{eff}} = 0.75$ and ^{244}Pu is prolate deformed. If we collide the two nuclei in tip configuration, the Coulomb barrier is low and we can make them contact with lower incident energy, that means lower excitation energy of the compound system. This is very preferable in view of the small survival probability. But the loss due to the fusion-hindrance has to be taken into account in estimating the total gain (or loss).

III EVAPORATION RESIDUE CROSS SECTION

The reaction process of the production of the superheavy elements can be divided into three stages and can be written in the following formula,

$$\sigma_{EV} = \frac{\pi\hbar^2}{2\mu E} \sum_{\ell=0} (2\ell+1) f_\ell p_\ell w_\ell, \quad (1)$$

where f_ℓ denotes the sticking probability that describes the process before the contact of projectile and target nuclei and also describes the barrier penetration in the case of sub-barrier fusion, p_ℓ the formation probability that describes the competition between complete fusion and quasi-fission, and w_ℓ the survival probability that describes the competition between fission and particle evaporation.

The formation probability of a compound system and the survival probability in de-excitation are governed by very different physics. Formation mainly depends on the growing Coulomb forces in the process, which strongly depends on the mass-asymmetry of the colliding systems. On the other hand, the de-excitation losses are governed by the height of the fission barrier in the compound nucleus.

Because of the difference in the time scale in these two processes, we assume that we can factorize the whole process like in Eq. (1).

IV FORMATION PROBABILITY

We adopt the three-dimensional nuclear deformation space with the two-center parametrization [15]. The neck parameter ϵ is fixed to 1.0 in the present calculation, deformation parameters δ_1 and δ_2 of the colliding nuclei are taken to be equal, *i.e.*, $\delta_1 = \delta_2 = \delta$, and we use the asymmetry parameter $\alpha = (A_1 - A_2)/(A_1 + A_2)$ where $A_{1,2}$ denotes the mass number of target/projectile nucleus. We treat z_0 (distance between two potential centers), δ and α as the two collective parameters to be described by the Langevin equation. The multi-dimensional Langevin equation is given in the following form,

$$\frac{dq_i}{dt} = \left(m^{-1}\right)_{ij} p_j,$$
$$\frac{dp_i}{dt} = -\frac{\partial V}{\partial q_i} - \frac{1}{2}\frac{\partial}{\partial q_i}\left(m^{-1}\right)_{jk} p_j p_k - \gamma_{ij}\left(m^{-1}\right)_{jk} p_k + g_{ij} R_j(t), \quad (2)$$

where V is the potential energy, m_{ij} and γ_{ij} are the shape-dependent collective inertia and dissipation tensors, respectively. The normalized random force $R_i(t)$, is assumed to be a white noise, *i.e.*, $\langle R_i(t) \rangle = 0$ and $\langle R_i(t_1) R_j(t_2) \rangle = 2\delta_{ij}\delta(t_1 - t_2)$. The strength of the random force g_{ij} is given by $\gamma_{ij} T = g_{ik} g_{jk}$, where T is the temperature of the compound nucleus calculated from the excitation energy as $E_x = aT^2$ with a denoting the level density parameter. The potential is calculated as the sum of a generalized surface energy [16], Coulomb energy, and the centrifugal

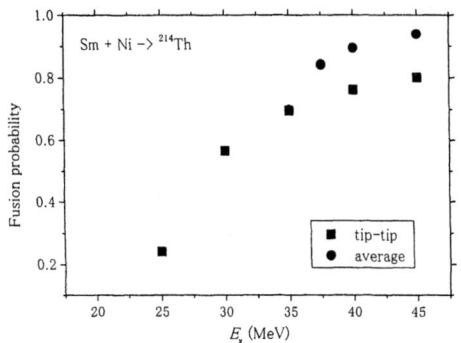 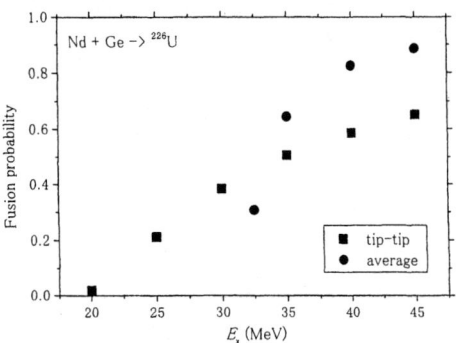

FIGURE 1. Fusion probability for ^{60}Ni + ^{154}Sm system. Squares denote the results for tip-to-tip configuration with $\delta = 0.2$. Circles denote the results for spherical case.

FIGURE 2. Fusion probability for ^{76}Gd + ^{150}Nd system. Squares denote the results for tip-to-tip configuration with $\delta = 0.2$. Circles denote the results for spherical case.

energy with the moment of inertia of the rigid body. Hydrodynamical inertia tensor is adopted with the Werner-Wheeler approximation for the velocity field, and the wall-and-window one-body dissipation [17] is adopted for the dissipation tensor. At $t=0$, each trajectory starts from $z_0 = 1.6 R_0$ and $\delta = 0$ corresponding to the contact configuration with the initial velocity in the z_0 direction, where R_0 denotes the radius of the spherical compound nucleus. Excitation energy of the composite system E_x is calculated for each trajectory as,

$$E_x = E_0 - \frac{1}{2}\left(m^{-1}\right)_{ij} p_i p_j - V(q), \quad (3)$$

where E_0 is given as $E_0 = E_{cm} - Q$ with Q denoting the Q-value of this reaction and E_{cm} the incident energy in center-of-mass frame. We count the number of trajectories that come into the spherical region and divide it with the total number of trajectories to obtain the fusion probability.

In the calculation, for simplicity, we only calculate the central collision, *i.e.* $\ell = 0$. Figure 1 shows the resulting fusion probability for the case of ^{60}Ni + ^{154}Sm system and Fig. 2 for ^{76}Gd + ^{150}Nd system, respectively. Both figures have the same tendency. First we calculate the fusion probability for the initial deformation $\delta = 0$. Because both systems have the effective fissility that is very close to the critical value, we expect a very week or no fusion-hindrance. In the figures, circles denote the results with the initial deformation $\delta = 0$. As you see, the fusion-hindrance is very week in this case. And the fusion probability falls down to zero at the barrier energy that is around 30-35 MeV. Then we calculate the deformed case. For simplicity, we take the same deformation $\delta = 0.2$ for both the target and the projectile and collide them with tip-to-tip configuration. In the figures, squares denote the results with the deformation and with tip-to-tip configuration. Because the barrier is low for this case, the fusion probability extends to the lower energies, but because of the fusion hindrance, the value is small. We obtain the extra-extra-push energy of around 10-15 MeV for the deformed case that is consistent with the experimental findings.

From this results, we can speculate that with the tip configuration, ^{48}Ca + ^{244}Pu system can fuse with smaller excitation energy than 35MeV. With smaller excitation energy, we can expect larger survival probability, however, because of the fusion-hindrance, the fusion probability will be smaller. To obtain a quantitative conclusion, we need to perform a systematic study so that we can fix the parameters like the strength of the friction.

V SURVIVAL PROBABILITY

While the fusion process is governed by nuclear dynamics, the de-excitation process is primarily determined by the statistical weight of the different de-exciation channels as treated by the statistical model [18,19].

The de-excitation of an excited highly fissile nucleus by neutron emission is governed essentially by the i-fold product of the Γ_n/Γ_f ratios of the i single steps in the neutron evaporation cascade. When we employ the Bohr-Wheeler formula for fission width, the ratio Γ_n/Γ_f is expressed as,

$$\frac{\Gamma_n}{\Gamma_f} = \frac{4mR_0^2}{\hbar^2} \frac{\int_0^{E^*-B_n} \epsilon\rho(E^* - B_n - \epsilon)d\epsilon}{\int_0^{E^*-B_f} \rho(E^* - B_f - K)dK} \qquad (4)$$

where $\rho(E)$ denotes the nuclear level density, E^* the intrinsic excitation energy of the compound nucleus, B_n the neutron separation energy, and B_f the fission barrier. ϵ is the kinetic energy of emitted neutron and K is the kinetic energy of fission fragments at saddle. The level density depends exponentially on the entropy. For a Fermi gas, the entropy $S = 2\sqrt{aE^*}$ where a is the level density parameter which is related to the single-particle level density at the Fermi surface.

In the case of the high excitation energy comparing with the neutron separation energy and the fission barrier height, the integration can be done approximately, and we obtain the following expression for Γ_n/Γ_f as,

$$\frac{\Gamma_n}{\Gamma_f} = A_0 \exp\left[2a_n^{1/2}(E^* - B_n)^{1/2} - 2a_f^{1/2}(E^* - B_f)^{1/2}\right], \qquad (5)$$

where

$$A_0 = \frac{4mR_0^2 a_f(E^* - B_n)}{\hbar^2[2a_f^{1/2}(E^* - B_f)^{1/2} - 1]}. \qquad (6)$$

The fission barrier B_f is the sum of the barrier calculated from macroscopic morels B_{LD} and the shell correction δU for the ground state and saddle point configurations,

$$B_f = B_{LD} + \delta U. \tag{7}$$

For superheavy elements the macroscopic barrier B_{LD} is small and can be neglected compared to δU. This microscopic correction to the binding energy depends on the shape of the nucleus and it is restricted to a narrow region of deformation parameters. The occupation probability of the single-particle levels around the Fermi surface varies gradually when the nuclear temperature increases. This is why the effects of shells and pairing correlations on the level density are expected to decrease and to vanish with increasing excitation energy. Due to this temperature smearing of microscopic effects in the level density, the shell correction energy obtained with Strutinsky method is washed out. Ignatyuk et al. suggested that the shell effects are damped exponentially [20]

$$\delta U = \delta U_0 \exp(-E^*/E_d), \tag{8}$$

where δU_0 denotes the shell correction energy at $E^* = 0$.

There is another factor we need to take into account when we consider this process. That is the collective enhancement factor of the level density [21]. Collective excitations, as rotation and vibration, enhance the level density up to excitation energies characterized by a critical energy E_{crit}. The enhancement factor K_{coll} of level density depends on the nuclear structure of the nucleus. K_{coll} can be expressed as a function of the quadrupole deformation parameter β_2,

$$K_{\text{coll}} = \frac{T}{\hbar^2} J_0 \left[1 + \sqrt{\frac{5}{16\pi}} \beta_2 + \frac{45}{28\pi} \beta_2^2 \right], \tag{9}$$

where $J_0 = (2/5) A m R_0^2$. One should note that the value of Γ_n is determined by the level density of the ground state deformation, whereas the value of Γ_f is determined by the level density of the saddle point that is usually highly deformed. Therefore, this collective enhancement effect may enhance fission decay, especially for the case that the ground state has a spherical shape.

The effect of the fluctuation-dissipation dynamics is taken into account in the statistical model in terms of the Kramers factor [22] K_{Kr},

$$K_{\text{Kr}} = \frac{\hbar \omega_1}{T} \left(\sqrt{1 + x^2} - x \right) \tag{10}$$

where ω_1 is the parameter depending on the curvature of the energy surface at the saddle point, $x = \gamma/2\omega_0$ is the reduced friction coefficient γ divided by ω_0, the curvature of the energy surface at the ground state deformation, and T is the nuclear temperature.

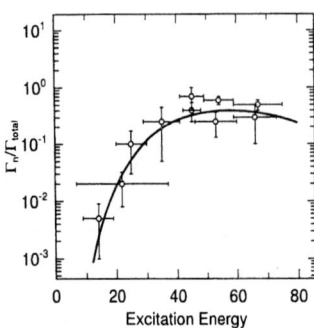
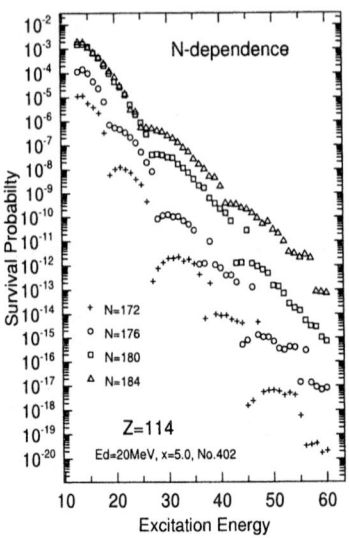

FIGURE 3. Theoretical fit to the experimental $\Gamma_n/(\Gamma_n + \Gamma_f)$ for $Z = 102$. In the calculation, the neutron number is fixed to $N = 154$, the shell dumping factor $E_D = 20$MeV, and the friction parameter $x = 5.0$.

FIGURE 4. The neutron number dependence of the survival probability for the compound nuclei with $Z = 114$. $N = 178$ corresponds to the reaction system ^{48}Ca+^{244}Pu.

Including those effects, the ratio Γ_n/Γ_f now reads,

$$\frac{\Gamma_n}{\Gamma_f} = \frac{K_{\text{coll}}(gr.st.)}{K_{\text{coll}}(sad)K_{\text{Kr}}} A_0 \exp\left[2\sqrt{a_n E_n^*} - 2\sqrt{a_f E_f^*}\right], \qquad (11)$$

where $E_n^* = E^* - B_n$ and $E_f^* = E^* - B_f$.

In order to evaluate Γ_n/Γ_f, the neutron separation energy B_n, the shell correction energy δU and the quadrupole deformation β_2 are estimated by using the mass table of Möller et al [23]. The deformation for the saddle shape is taken from the liquid drop model calculation [24]. The shell-damping factor E_D is fixed to 20 MeV.

The level density parameter at ground state deformation a_n is assumed to have the following form,

$$a_n = \frac{A}{10}\left\{1 + \frac{\delta U_0(gr.st.)}{E_n^*}\left[1 - \exp\left(-\frac{E_n^*}{E_d}\right)\right]\right\} \qquad (12)$$

The level density parameter at saddle point deformation a_f, on the other hand, is treated as an adjustable parameter and we determine the energy dependence of a_f by fitting the experimental $\Gamma_n/(\Gamma_n + \Gamma_f)$ for $Z = 102$ as is shown in Fig. 3.

By using the theoretical formula for Γ_n/Γ_f given above, we can simulate the neutron evaporation cascade. Starting with the excitation energy E^*, we estimate stochastically the neutron kinetic energy ϵ by assuming the neutron spectrum emerging from the compound nucleus with temperature T as $\epsilon^{1/2}\exp(-\epsilon/T)$, and calculate $\Gamma_n^{(1)}/\Gamma_{tot}^{(1)} = \Gamma_n^{(1)}/(\Gamma_n^{(1)} + \Gamma_f^{(1)})$. Then we set new excitation energy as $E^{(1)*} = E^* - \epsilon - B_n^{(1)}$. If the energy $E^{(1)*}$ is above the particle threshold, we repeat the same process to get $\Gamma_n^{(2)}/\Gamma_{tot}^{(2)}$ and $E^{(2)*} = E^{(1)*} - \epsilon - B_n^{(2)}$ until the excitation energy becomes below the particle threshold. For the case of excitation energy below the particle threshold but above the fission barrier, the important decay channel is replaced by the γ emission and the fission.

The detailed analysis can be found in Ref. [25], here we only show one result concerning the isotope dependence of the evaporation residue cross section. Figure 4 shows the neutron number dependence of the survival probability for the compound nucleus with $Z = 114$. The neutron number varies from $N = 172$ to 184 and the corresponding variation in the average neutron separation energy is from 7MeV to 5MeV.

As can be seen from the figure, at the excitation energy about 30MeV that corresponds to (HI, $3n$) reaction, we see the difference of 5-6 orders of magnitude depending on the neutron number. It can be said that the neutron richness by three units enhances the survival probability about an order of magnitude in the case of compound nucleus with $Z = 114$.

VI SUMMARY AND DISCUSSIONS

The year 1999 was an exciting year for nuclear physics in the sense that it showed us the possibility to reach the island of stability that has been searched for over 30 years. New elements claimed to be produced but yet to be confirmed by further experiments. The location of the center of the island of stability will be determined from the lives and the binding energies of the newly found nuclides.

Fluctuation-dissipation dynamics is applied for the study of the reaction mechanism of the synthesis of the superheavy elements. The most interesting feature of this reaction is that there is no macroscopic barrier for fission and the nuclei are stabilized against fission merely by the shell correction energy which depends strongly on the excitation energy. The formation probability of the compound nuclei after the contact of projectile and target is calculated with a three-dimensional Langevin equation. With this framework, we can treat the problem of the fusion hindrance in a proper way, that is necessary to treat the competition between fusion and quasi-fission reaction.

The survival probability of the superheavy compound nuclei is estimated by treating the decaying process by the statistical model. Competition between neutron evaporation and fission is the most important one. The essential ingredient is the level density parameter and it is necessary to study its excitation energy dependence due to the shell correction and to the collective enhancement. The neutron evaporation width is very sensitive to the binding energy of neutrons. The use of the neutron-rich beam/target is desirable for the larger survival probability.

ACKNOWLEDGEMENTS

The author would like to thank Dr. S. Yamaji for providing us with a code to calculate the energy surface, mass and friction. This work was supported in part by the Grant-in-Aid for Scientific Reseach (No. 10640287) of the Ministry of Education, Science, Sports and Culture in Japan.

REFERENCES

1. S. Hofmann et al., Z. Phys. **A354** (1996) 229.
2. Yu.Ts. Oganessian et al., JINR Report No.E7-98-212, 1988.
3. Yu.Ts. Oganessian et al., Nature (London) **400** (1999) 242;
   Yu.Ts. Oganessian et al., Phys. Rev. Lett. **83** (1999) 3154.
4. V. Ninov et al., Phys. Rev. Lett. **83** (1999) 1104.
5. S. Hofmann, Rep. Prog. Phys. **61** (1998) 639.
6. Y. Aritomo, T. Wada, M. Ohta, and Y. Abe, Phys. Rev., **C55** (1997) R1011;
   Y. Aritomo, T. Wada, M. Ohta, and Y. Abe, Phys. Rev., **C59** (1999) 796.
7. T. Wada et al., Proc. 4th Int. Conf. on Dynamical Aspects of Nuclear Fission (World Scientific, Singapore, 2000) p. 77.
8. W.J. Swiatecki, Nucl. Phys., **A376** (1982) 275.
9. K.-H. Schmidt and W. Morawek, Rep. Prog. Phys., **54** (1994) 949.
10. W. Reisdorf, *Lecture Notes in Physics*, **317** (Springer, 1988) p.26;
    N.V. Antonenko et al., Phys. Lett., **B319** (1993) 425;
    C.E. Aguiar et al., Nucl. Phys., **A517** (1990) 205.
11. R. Bass, *Nuclear Reactions with Heavy Ions* (Springer, 1980).
12. S. Bjørnholm and W.J. Swiatecki, Nucl. Phys., **A391** (1982) 471.
13. S. Mitsuoka, H. Ikezoe, K. Nishio, and J. Lu, Phys. Rev., **C62** (2000) 054603.
14. K. Nishio, H. Ikezoe, S. Mitsuoka, and J. Lu, Phys. Rev., **C62** (2000) 014602.
15. J. Maruhn and W. Greiner, Z. Phys., **251** (1972) 431;
    S. Suekane et al., JAERI-memo 5918 (1974);
    K. Sato, S. Yamaji, K. Harada, and S. Yoshida, Z. Phys. **A290** (1979) 145.
16. H.J. Krappe, J.R. Nix, and A.J. Sierk, Phys. Rev., **C20** (1979) 992.
17. J.R. Nix and A.J. Sierk, Nucl. Phys., **A428** (1984) 161c.
18. V.F. Weisskopf, Phys. Rev., **52** (1937) 295.
19. N. Bohr and J.A. Wheeler, Phys. Rev., **56** (1939) 426.
20. A.V. Ignatyuk et al., Sov. J. Nucl. Phys., **21** (1975) 255.
21. A. Junghans et al., Nucl. Phys. **A629** (1998) 635.
22. H.A. Kramers, Physica (Utrecht), **7** (1940) 284.
23. P. Möller, J.R. Nix, W.D. Myers and W.J. Swiatecki, Atomic Mass Nucl. Data Table, **59** (1995) 185.
24. S. Cohen and W.J. Swiatecki, Ann. Phys. **22** 406 (1963).
25. M. Ohta, Proc. of Workshop "Fusion Dynamics at the Extreme" (World Scientific, Singapore) to be published.

Fusion and Quasifission in Collisions of Heavy Nuclei

G.G.Adamian[1,2,3], N.V.Antonenko[1,2] A.Diaz Torres[1,4], W.Scheid[1]
and Yu.M.Tchuvil'sky[5]

[1] *Institut für Theoretische Physik der Justus-Liebig-Universität, D-35392 Giessen, Germany*
[2] *Joint Institute for Nuclear Research, 141980 Dubna, Russia*
[3] *Institute of Nuclear Physics, Tashkent 702132, Uzbekistan*
[4] *Department of Physics University of Surrey, Guildford, Surrey, UK*
[5] *Institute of Nuclear Physics, Moscow State University Moscow 119899, Russia*

Abstract. It is shown that the compound nucleus is formed by a transfer of nucleons (dinuclear system concept) and not by a melting of the nuclei. The experimental evaporation residue cross sections and the mass distributions of quasifission products in fusion reactions leading to the production of heavy and superheavy nuclei are well reproduced in the dinuclear system model.

INTRODUCTION

The existing fusion models are distinguished by the choice of the relevant collective degree of freedom which is mainly responsible for the complete fusion. Many models assume a melting of the nuclei along the internuclear distance R (or the elongation λ of system) [1,2]. In the dinuclear system (DNS) concept [3–5] the compound nucleus is reached by a series of transfers of nucleons from the light nucleus to the heavy one. The dynamics of the DNS is considered as a combined diffusion in the degrees of freedom of the mass asymmetry $\eta = (A_1 - A_2)/(A_1 + A_2)$ (A_1 and A_2 are the mass numbers the DNS nuclei). Here, $\eta=0$ and $|\eta|=1$ mean a symmetric DNS and fused system, respectively. In present paper we check whether these two different types of models of fusion allows us to describe the available experimental data. What is important in fusion of heavy nuclei: melting or nucleon transfers?

ADIABATIC TREATMENT OF FUSION IN λ

Models describing the fusion process as an internuclear melting of nuclei usually use adiabatic potential energy surfaces (PES). We apply the microscopic two-center shell model with the following coordinates **q**: elongation $\lambda = \ell/(2R_0)$, where ℓ is the

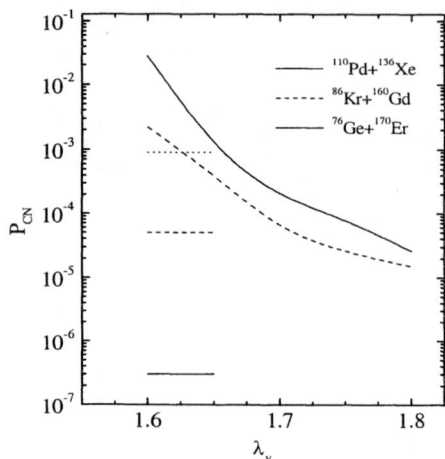

FIGURE 1. Dependence of the fusion probability P_{CN} on the starting value λ_v for various reactions leading to the same compound nucleus ^{246}Fm. P_{CN} extracted from the experimental data are given by horizontal lines.

length of the system and $2R_0$ the diameter of compound system, neck parameter ε with $\varepsilon=0$ showing no neck and $\varepsilon \approx 1$ showing necked-in shapes, mass asymmetry η and deformation parameters $\beta_i = a_i/b_i$ which are ratios of semiaxes of the clusters $i=1$ and 2. The adiabatic potential is obtained as

$$U(\lambda, \varepsilon, \eta, \eta_z, \beta_i) = U_{LDM} + \delta U_{shell}, \qquad (1)$$

where U_{LDM} is the liquid drop potential, δU_{shell} the shell correction part originating from the TCSM. In addition to the PES the masses $M_{q_i q_j} = M_{q_i q_j}^{WW}$ are calculated with Werner-Wheeler (or hydrodynamic) approximation and the friction coefficients are found with the expression $\gamma_{q_i q_j} = \Gamma M_{q_i q_j}/\hbar$ (Γ is an average width of single particle states).

The fusion takes place with a scenario of the macroscopic dynamical model (including fluctuations): 1) With the hydrodynamical masses M_{ij}^{WW} neck fastly grows after the nuclei touch each other and the united system falls into the fission-type valley in a short time of $(3-4) \times 10^{-22}$s. 2) The compound nucleus is formed due to the diffusion process to smaller elongation λ (or relative distance R) in this valley. The necessary condition for the compound nucleus formation is that the fusing system passes inside the fission saddle point at $\lambda = \lambda_{sd}$.

In order to calculate the fusion probability we started from a value of $\lambda = \lambda_v$ in the fission valley obtained from the dynamical calculation of the descent into this valley. The fusion probability is determined by the leakage $\Lambda_{fus}^\lambda(t)$ through the barrier in λ which separates the strongly deformed configuration with equilibrated large neck and the compound nucleus:

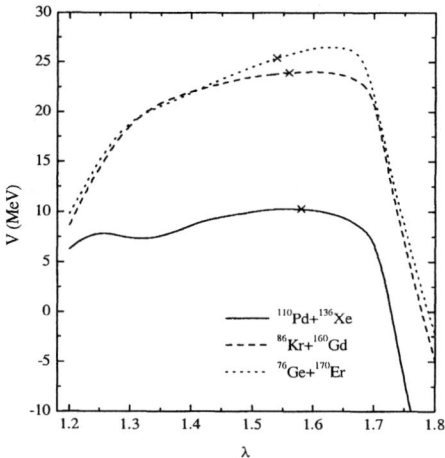

FIGURE 2. Adiabatic potential of different combinations leading to ^{246}Fm as function of λ for a fixed value of $\varepsilon=0.75$. The crosses denote the touching configurations.

$$P_{CN} = \int_0^{t_0} \Lambda_{fus}^\lambda(t)dt, \qquad (2)$$

where t_0 is the lifetime of the system and $\Lambda_{fus}^\lambda \sim \exp(-B_{fus}^\lambda/T)$ (Kramers expression). Here, $B_{fus}^\lambda = U(\lambda_{sd}) - U(\lambda_v)$ is the barrier for fusion in λ and $T = \sqrt{E^*/a}$ the local thermodynamic temperature with the excitation energy E^* of the system. For nearly symmetric reactions ^{100}Mo+^{100}Mo, ^{100}Mo+^{110}Pd, ^{110}Pd+^{110}Pd and others, the fusion probabilities in λ are much larger [6] than the values found from experimental data. While the experimental fusion probability decreases with mass asymmetry in the entrance channel [7], the calculated data give the opposite tendentious (Fig. 1). Experimental evidence for a hindrance of fusion has been raised mainly from impossibility to produce fermium evaporation residues with nearly symmetric projectile-target combinations. The adiabatic consideration of fusion in λ mostly gives the wrong dependence of the fusion probability on the isotope composition of the colliding nuclei [6]. The qualitative and quantitative contradictions obtained with the adiabatic scenario of the fusion result the existence of additional hindrance for the fast growth of the neck and the motion to smaller λ.

MICROSCOPIC INERTIA TENSOR

The hydrodynamical mass for the neck coordinate is small and allows to increase the neck of the DNS rapidly to a shape of a strongly deformed compound nucleus. However, crossings of single particle levels lead to a large inertia of the system which hinders the growth of the neck. Using the single-particle spectrum ϵ_α and wave functions $|\alpha>$ of both the adiabatic and diabatic TCSM, one can obtain the mass parameters with an extended cranking formula [8]

$$M_{ij}^{cr} = \hbar^2 \sum_{\alpha,\beta} \frac{<\alpha|\frac{\partial \hat{H}}{\partial q_i}|\beta><\beta|\frac{\partial \hat{H}}{\partial q_j}|\alpha>}{(\epsilon_\beta - \epsilon_\alpha)^2 + \frac{1}{4}(\Gamma_\beta + \Gamma_\alpha)^2} \frac{n_\alpha - n_\beta}{\epsilon_\beta - \epsilon_\alpha}, \qquad (3)$$

where Γ_α and Γ_β are the widths of the single particle states. The main contribution to M_{ij}^{cr} arises from the diagonal elements at $\epsilon_\beta \to \epsilon_\alpha$. Comparing our results with M_{ij}^{WW} obtained in the Werner-Wheeler approximation for a touching configuration with excitation energy 30 MeV (T=1.3 MeV), we find $M_{\lambda\lambda}^{cr} = M_{\lambda\lambda}^{WW}$, $M_{\varepsilon\varepsilon}^{cr} \approx$ (20-30)$M_{\varepsilon\varepsilon}^{WW}$ and $M_{\lambda\varepsilon}^{cr} \approx 0.4\ M_{\lambda\varepsilon}^{WW}$, practically independent of the mass number of the system [6,8]. As result we conclude that the initial neck coordinate is nearly kept due to the large microscopical mass parameter for a time comparable with the reaction time. Since the mass of the neck degree of freedom is large, the neck parameter can be taken approximately fixed ($\varepsilon = 0.75$) during the fusion. As in previous case, the adiabatic treatment of fusion in λ (at the fixed neck parameter) yields fusion probabilities P_{CN} which are considerably overestimated in comparison to the experimental data. The reason of this overestimate is clearly seen in Fig. 2. In contrast to experiment the hindrance of fusion in these potentials is practically absent ($P_{CN} \approx 1$) because there is no internal fusion barrier for the motion to smaller elongations [8].

EFFECT OF STRUCTURAL FORBIDDENNESS OF FUSION IN λ

Since the adiabatic PES (with and without fixed neck parameter) is not adequate for the description of fusion, the question is arised how fast is the transition between an initially diabatic (sudden or frozen density) and adiabatic regimes during the fusion process. The main question is the validity of the use of the adiabatic PES from the beginning of the fusion process.

Algebraic description

Let us first to see the static picture. It is known that after the formation of the DNS there is large structural forbiddenness effect for the motion to smaller elongations (melting) in the heavy cluster system [9]. This effect is because of the difference created by the realizing Pauli principle between the compound state and the heavy dinuclear cluster state. For the fusion channel $A_1 + A_2 \to A$, with the wave function

$$\Psi_{A_1+A_2} = \hat{A}_{as}\{\Psi_{A_1}\Psi_{A_2}\phi(R)\}, \qquad (4)$$

TABLE 1. Experimental and calculated minimal values of the energy thresholds ΔE_{min}, at which fusion is possible, are compared with the B_λ and q (see text).

System	ΔE_{min} (MeV) exp. [7]	ΔE_{min} (MeV) DNS model [3,9]	q	B_λ (MeV)
^{40}Ar+^{206}Pb	-0.5±3	0	16	105
^{76}Ge+^{170}Er	10±5	8	22	144
^{86}Kr+^{160}Gd	≥15.7	11.5	20	131
^{110}Pd+^{136}Xe	≥23.5	15	26	170
^{96}Zr+^{124}Sn	6.5±3	5	28	183
^{90}Zr+^{90}Zr	0.0	0.0	30	196
^{100}Mo+^{100}Mo	7.2	6.0	30	196
^{110}Pd+^{110}Pd		12.0	30	196

where Ψ_{A_i} are the cluster wave functions which have the lowest oscillator quanta numbers N_{A_i} allowed by Pauli principle, $\phi(R)$ is the function of their relative motion along collision axis z and $\hat{A}_{as}$ is operator of antisymmetrization, the minimal oscillator quanta $N_{A_1+A_2}$ is determined by application of the SU(3) group theory. The fact is that $\Psi_{A_1+A_2}$ and Ψ_A belong to different SU(3) (or Sp(2,R) for deformed nuclei A_1 and A_2) representations. We have to construct the U(3) irreducible representation $[f_z f_y f_x]$ for $\Psi_{A_1+A_2}$ which possesses by minimal sum $f_x + f_y + f_z$ and does not vanish with antisymmetrization. The proper procedure for the construction of the vector of highest weight of this representation is the use of the vector of highest weight for the wave function of the relative motion and the vectors of lowest weights for the wave functions of the fragments which possess by the maximal value of the sum $f_x^{(1)} + f_y^{(1)} + f_x^{(2)} + f_y^{(2)}$ (upper index denotes the fragment). The resulting $N_{A_1+A_2}$ value turns out to be essentially large than N_A if the mass of the lighter fragment is rather large. Therefore, in heavy ion physics we obtained the generalization of the Talmi-Moshinsky rule. The minimal difference $q = N_{A_1+A_2} - N_A$ is referred to the degree of structural forbiddenness for fusion channel $A_1 + A_2 \to A$. The wave function $\Psi_{A_1+A_2}$ of the DNS has nonvanishing overlap integral with wave function Ψ_A of the compound nucleus if $N_{A_1+A_2} \geq N_A + q$. The forbiddenness effect is model independent and the SU(3) approach is only the simple method to determine it. The quantity $B_{fus}^\lambda = \hbar\omega q$ ($\hbar\omega = 41\text{MeV}A^{-1/3}$) is a qualitative estimation of the minimal energy thresholds for the fusion in relative distance R degree of freedom at fixed mass asymmetry η. If the excitation energy of system is much smaller than the value of B_{fus}^λ, the fusion in λ is strongly forbidden. The dependence of q on η is not monotonic, the q-value is maximal for almost symmetric combinations and it decreases with increasing η. For very asymmetric combinations, the structural forbiddeness practically disappears. Therefore, the main reason for removing the structural forbiddenness in λ–channel is the coupling between cluster channels with different mass asymmetries. Since the absolute values of q (or B_{fus}^λ) are large for not very asymmetric DNS, there is a strong hindrance for the evolution of these DNS to smaller λ due to the large energy barrier between the initial DNS and

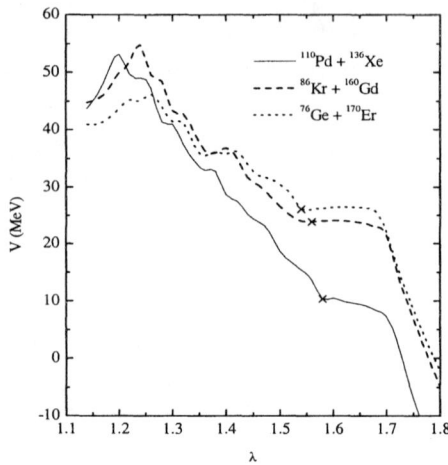

FIGURE 3. Dynamical diabatic potential. The notations are the same as in Fig. 2.

compound shapes (Table 1) [9].

Dynamical diabatic description

It is important to study whether the system has time for destroying the "memory" on the structural forbiddenness and reorganizing the density of the system for the transition from the sudden or diabatic potential $U_{di}(\lambda)$ to the adiabatic potential $U_{ad}(\lambda)$ [10]:

$$V(\lambda,t) = U_{di}(\lambda)\exp(-\int_0^t \frac{dt}{\tau(\lambda,t)}) + U_{ad}(\lambda)[1 - \exp(-\int_0^t \frac{dt}{\tau(\lambda,t)})]. \qquad (5)$$

A time-dependent relaxation time τ is microscopically calculated using TCSM. The dynamical diabatic potential $V(\lambda,t)$ is the relevant tool to measure the structural forbiddenness. This potential shows a strong increase with decreasing elongation λ. Their repulsive character hinders the DNS to melt into the compound nucleus.

The dynamical diabatic potential at the time of the lifetime of the initial DNS (Fig. 3) has a very large fusion barrier in λ and, correspondingly, the fusion probability in λ is negligible for combinations leading to ^{246}Fm. It should be noted that dynamical potentials were calculated using the minimally possible relaxation time for the transition between diabatic and adiabatic potentials [10]. The calculated energy thresholds for the complete fusion in the $\lambda-$ and $\eta-$channels lead to the conclusion that the DNS evolution to the compound nucleus proceeds in the mass asymmetry degree of freedom. For example, the average fusion barriers B^*_{fus} in mass asymmetry are about 10, 12 and 15 MeV for the reactions ^{76}Ge+^{170}Er

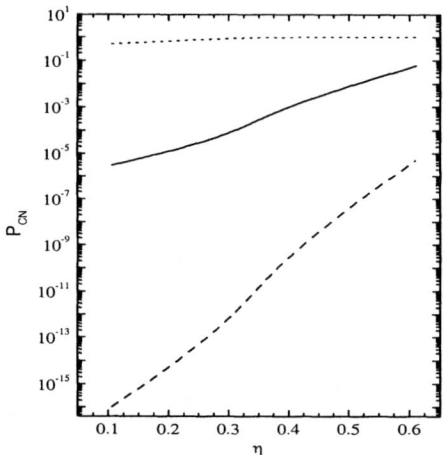

FIGURE 4. Fusion probability P_{CN} in the reactions leading to ^{246}Fm with excitation energy 30 MeV as a function of the mass asymmetry in the entrance channel. The result of the adiabatic treatment of the fusion in λ is presented by the dotted line. The upper limit of the fusion probability in λ in the dynamical diabatic treatment is presented by the dashed line. The fusion probability in the η channel at closed fusion channel in λ is presented by the solid line and is in a good agreement with the experiment [7].

(η=0.4), ^{86}Kr+^{160}Gd (η=0.3) and ^{110}Pd+^{136}Xe (η=0.1), respectively [8]. The fusion barrier in λ is about 3–4 times larger than the fusion barrier B^{η}_{fus} in mass asymmetry. As is shown in Fig. 4, the fusion probability P_{CN} in η strongly increases with mass asymmetry in the entrance channel. The same behaviour was experimentally established [7]. For the reactions ^{40}Ar+^{206}Pb and ^{76}Ge+^{170}Er, the values of P_{CN} in η yield a good agreement with experimental data from evaporation residue cross sections.

In compound systems heavier than ^{246}Fm the difference between the fusion barriers and fusion probabilities in both λ- and η- channels is even larger [10]. Our analysis with the diabatic dynamics demonstrates that a structural forbiddenness exists for a direct motion of the nuclei to smaller internuclear distances during the fusion process. Fusion of heavy nuclei along the internuclear distance in the coordinates R or λ is practically closed. These facts strongly support our standpoint that the correct model of fusion of heavy nuclei is the dinuclear system model where fusion is described by the transfer of nucleons.

CROSS SECTIONS FOR SUPERHEAVY NUCLEI

The available experimental data [11,12] on the evaporation residue cross section $\sigma_{ER} = \sigma_{cap} P_{CN} W_{sur}$ of lead- and actinide-based fusion reactions can be well reproduced with the dinuclear system model with exception of the reaction

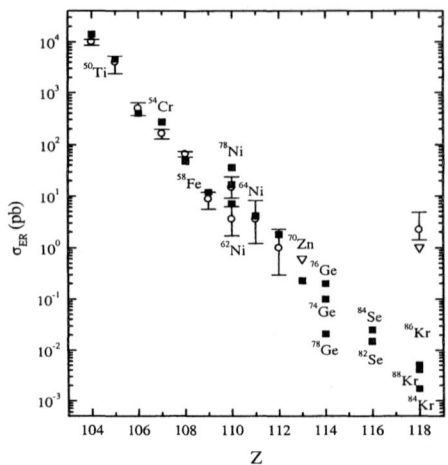

FIGURE 5. Evaporation residue cross sections for ^{208}Pb- and ^{209}Bi-based fusion. The full squares and the open circles are calculated and experimental data, respectively. The open triangles give experimental upper limits (GSI).

^{86}Kr+^{208}Pb→^{293}118+1n [4] (Fig 5). In recent experiments at GSI, GANIL, RIKEN the element 118 was not detected and value of σ_{ER} was estimated as less than 1 pb in agreement with our calculations. The strong decrease of σ_{ER} with the charge number of the fused system is caused by the decrease of the probability P_{CN} for forming the compound nucleus due to a strong competition between complete fusion and quasifission (decay of the DNS). For example, for ^{70}Zn+^{208}Pb→^{277}112+1n we found P_{CN}=1× 10^{-6}, σ_{ER}=1.8 pb (exp. 1.0 pb) and for ^{86}Kr+^{208}Pb→^{293}118+1n we obtained P_{CN} = 1.5× 10^{-10}, σ_{ER}=5.1 fb [4]. The surviving probability W_{sur} varies moderately with the shell structure of the compound nucleus. In the Pb-based reactions with neutron-rich nuclei ^{70,74,78}Ni, ^{80}Zn,^{86}Ge and ^{92}Se the increase of W_{sur} with the number of neutrons could be compensated by decreasing P_{CN}. For example, in the ^{62}Ni+^{208}Pb reaction the yield of the Z=110 element is comparable with the yields in the ^{70,74}Ni+^{208}Pb reactions [5].

In the actinide-based reactions ^{48}Ca+^{232}Th, ^{238}U, ^{242,244}Pu, ^{248}Cm, ^{249}Cf the P_{CN} decrease with increasing Z, but they are larger than in the Pb-based reactions. For ^{48}Ca+^{244}Pu→^{289,288}114+3n,4n, P_{CN} is about 10^5 times larger than in ^{76}Ge+^{208}Pb→^{283}114+1n. The gain in fusion and capture probabilities for actinide-based reactions with respect to cold fusion reactions is not compensated by the loss in the survival probability of the compound nucleus. So, our comparison of the formation cross section of element Z=114 in the Pb- and actinide-based reactions shows that the latter one is larger ($\sigma_{3n,4n} \approx 1$ pb) [3,4]. The σ_{ER} in the ^{48}Ca+^{248}Cm,^{249}Cf reactions are smaller than the experimental $\sigma_{ER} \approx 1$ pb in the ^{48}Ca+^{244}Pu reaction [11].

QUASIFISSION PROCESS IN DINUCLEAR SYSTEM

In order to support the DNS mechanism of fusion, it is very crucial and important to describe also other experimental observables correctly, like mass and charge distributions of the quasifission products which accompany the fusion process. The competition between complete fusion and quasifission is extremely important in the DNS evolution. In the DNS concept the quasifission process is simultaneously described as the evolution of the DNS in the charge (mass) asymmetry coordinate Z (A) by nucleon transfer between the nuclei and the decay of the DNS into the direction of increasing R (or elongation λ). So, the quasifission process delivers products with different η. A diffusion process leads to the exchange of nucleons between the two touching fragments, thus generating a time-dependent distribution in the charge (mass) asymmetry of the DNS. This process is described by a master-equation for the probability $P_Z(t)$ to find the DNS at the time t in the configuration with charge number Z of the light nucleus [13]:

$$\frac{\partial P_Z(t)}{\partial t} = \Delta^{(-)}_{(Z+1)} P_{(Z+1)}(t) + \Delta^{(+)}_{(Z-1)} P_{(Z-1)}(t) - (\Delta^{(+)}_Z + \Delta^{(-)}_Z + \Lambda^{qf}_Z) P_Z(t), \quad (6)$$

where $P_Z(0) = \delta_{ZZ_i}$ and the microscopically calculated transport coefficients $\Delta^{(\pm)}_Z$ characterize the probability rate of the proton transfer from a heavy to a light nucleus $\Delta^{(+)}_Z$ or in opposite direction $\Delta^{(-)}_Z$. In the calculations with (6) the isotopic composition of the nuclei forming the DNS is chosen from the condition of a $N/Z-$equilibrium in the system. The coefficient $\Lambda^{qf}_Z \sim \exp(-B^\lambda_{qf}/\Theta)$ is the rate of the DNS decay probability in R (or λ). The decaying DNS has to overcome the potential barrier $B^\lambda_{qf}(Z)$ [3] which value coincides with the depth of the pocket in the nucleus-nucleus potential as a function of R. The height B^λ_{qf} of this barrier monotonically decreases with increasing Z because the increasing Coulomb repulsion leads to very shallow pockets in the nucleus-nucleus potential for near symmetric configurations. The excitation energy (temperature $\Theta(Z)$) of the DNS increases with decreasing mass (charge) asymmetry. The measurable charge (mass) yield for quasifission can be expressed by the product of the formation probability $P_Z(t)$ of the DNS configuration with charge (mass) asymmetry Z and the decay probability in λ determined by the Kramers rate Λ^{qf}_Z:

$$Y_Z(t_0) = \Lambda^{qf}_Z \int_0^{t_0} P_Z(t) dt. \quad (7)$$

Here, t_0 is the time of reaction which is determined by solving the equation $\sum_Z \Lambda^{qf}_Z \int_0^{t_0} P_Z(t) dt \approx 1$.

In Fig. 6 the distributions reveal large widths. The masses of products are substantially different from the initial target–projectile masses and symmetric fragments can be formed with a quite large cross section. The main peak of the charge

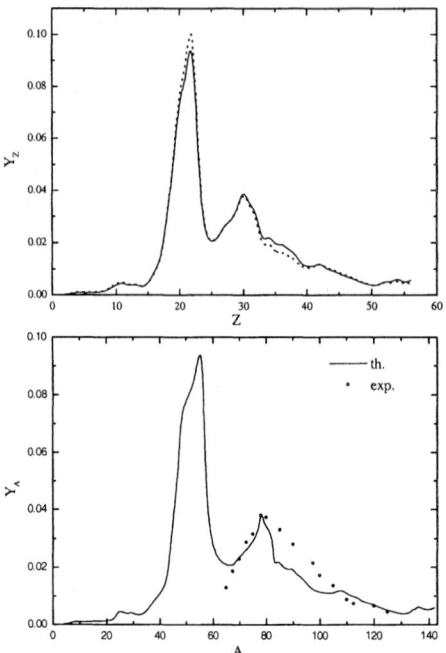

FIGURE 6. Charge (upper part) and mass (lower part) yields of the quasifission products as a function of Z and A of the light fragment, respectively, for the hot fusion reaction ^{48}Ca+^{238}U and an excitation energy of the initial DNS of E*= 5 (dotted curve) and 15 (solid curve) MeV. The experimental data [4] are shown by solid points for E*= 10 MeV corresponding to an excitation energy of the compound nucleus of 33 MeV.

and mass distributions is around the initial configuration of the DNS. The maxima and minima in the yields correlate with minima and maxima in the DNS potential energy, respectively. In the reactions ^{48}Ca+^{238}U and ^{48}Ca+^{244}Pu the maximum yield of the quasifission fragments occurs around the nucleus ^{208}Pb for the heavy fragment where the DNS potential energy has a transparent minimum. The calculated mass distributions of quasifission products are in good agreement with the available experimental data [14]. It should be noted that, near the initial combination, quasifission events overlapping with the products of deep inelastic collisions were not taken into account in the experimental data because of the difficulties to discriminate them. The decrease of quasifission cross section $\sigma_{qf}(A) - \sigma_{cap}Y_A(t_0)$ and increase of asymmetry of mass distribution with decreasing bombarding energy under the Coulomb barrier can be mainly explained by the reduction of the DNS formation cross section σ_{cap} at the approaching stage of collision.

Taking the deformation of the nuclei in the DNS in the ^{48}Ca+^{238}U reaction into account, we find that the calculated average total kinetic energy <TKE>=240 MeV of the quasifission products with $70 < A < 120$ is in agreement with the experi-

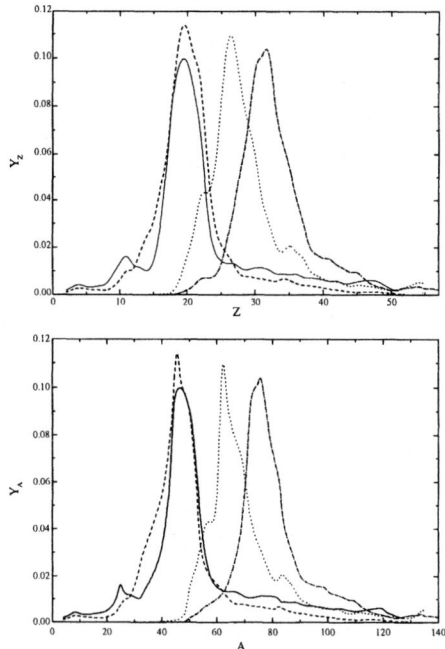

FIGURE 7. The same as in Fig. 6, but for the cold fusion reactions $^{48}\text{Ca}+^{208}\text{Pb}\rightarrow^{256}102$ (solid curves), $^{50}\text{Ti}+^{208}\text{Pb}\rightarrow^{258}104$ (dashed curves), $^{64}\text{Ni}+^{208}\text{Pb}\rightarrow^{272}110$ (dotted curves) and $^{76}\text{Ge}+^{208}\text{Pb}\rightarrow^{284}114$ (dashed-dotted curves).

mental data [15] and the systematics of Viola et al. Considering the fluctuations of the Coulomb interaction due to the fluctuations of the DNS charge asymmetry (at fixed mass asymmetry), the fluctuations of the quadrupole deformation parameters of the DNS nuclei and fluctuations of the bending mode in the DNS, it is possible to explain the large variance of the TKE distribution as a function of the mass numbers of fragments.

The calculations of quasifission distributions of charge and mass for the ^{208}Pb-based cold fusion reactions leading to the synthesis of the elements $^{258}104$, $^{272}110$ and $^{284}114$ (Fig. 7) are important for the planned quasifission experiments in many laboratories. It is seen that in the Pb-based reactions the charge and mass distributions have no pronounced maxima besides the main peak around the entrance DNS. This structure of the mass (charge) distribution of the quasifission products well corresponds to the peculiarities of the DNS potential energy which has the consequence that the DNS consisting of such strongly bound nuclei like ^{208}Pb for $Z > Z_i$ are absent.

SUMMARY

The main conclusions are: 1) The DNS nuclei are hindered to melt together by a variation in the relative distance. This occurs due to the relatively large inertia of the neck degree of freedom and structural forbiddenness effects. The comparison of the calculated energy thresholds for the complete fusion in different relevant collective variables shows that the dinuclear system prefers to evolve in the mass asymmetry coordinate by nucleon transfer to the compound nucleus. 2) The evaporation residue cross sections for heavy and superheavy elements and quasifission products of hot fusion reactions are correctly described within the DNS model. This facts support the idea that the fusion process proceeds along the mass asymmetry degree of freedom. 3) For the cold fusion, reactions the quasifission products are practically associated with fragmentations near the initial DNS. 4) The quasifission process is the main factor suppressing the complete fusion of heavy nuclei. In these reactions the fusion-fission and fusion events are much smaller than the events of the production of certain quasifission products.

G.G.A. is grateful for the support of the Alexander von Humboldt-Stiftung. This work was supported in part by DFG and RFBR.

REFERENCES

1. A. Sandulescu, R.K. Gupta, W. Scheid, W. Greiner, *Phys. Lett. B*, **60** 225 (1976).
2. W.J. Swiatecki, *Phys. Scripta*, **24** 113 (1981); Y.Aritomo, T.Wada, M.Ohta and Y.Abe, *Phys. Rev. C*, **59** 796 (1999).
3. V.V. Volkov, *Izv. AN SSSR ser.fiz.*, **50** 1879 (1986); N.V. Antonenko, E.A. Cherepanov, A.K. Nasirov, V.B. Permjakov, V.V. Volkov, *Phys. Lett. B*, **319** 425 (1993); *Phys. Rev. C*, **51** 2635 (1995); E.A. Cherepanov, preprint JINR, E7-99-27(1999); R.V.Jolos, A.K.Nasirov, A.I.Muminov, *Eur. Phys. J. A*, **4** 245 (1999).
4. G.G.Adamian, N.V.Antonenko, W.Scheid and V.V.Volkov, *Nucl. Phys. A*, **633** 409 (1998); **678** 24 (2000);*Nuovo Cimento A*, **110** 1143 (1997).
5. G.G. Adamian, N.V. Antonenko, A. Diaz-Torres, W. Scheid, *Nucl. Phys. A*, **678** 24 (2000).
6. G.G. Adamian, N.V. Antonenko, S.P. Ivanova, W. Scheid, *Nucl. Phys. A*, **646** 29 (1999).
7. H.Gäggeler et al., *Z. Phys. A*, **316** 291 (1984).
8. G.G. Adamian, N.V. Antonenko, A. Diaz-Torres, W. Scheid, *Nucl. Phys. A*, **671** 233 (2000).
9. G.G. Adamian, N.V. Antonenko, Yu.M. Tchuvil'sky, *Phys. Lett. B*, **451** 289 (1999).
10. A. Diaz-Torres et al., *Phys. Lett. B*, **481** 228 (2000).
11. Yu.Ts.Oganessian et al., Phys. Rev. Lett. 83 (1999) 3154.
12. S.Hofmann and G.Münzenberg, *Rev. Mod. Phys.*, (2000) (to be published).
13. A. Diaz-Torres et al., *Nucl. Phys. A*, in print (2000).
14. M.G. Itkis et al., 7th *Int. Conf. on Clustering Aspects of Nuclear Structure and Dynamics*, Rab, Croatia (1999) p.386; preprint JINR, E15-99-248(1999).
15. W.Q.Shen et al., *Phys. Rev. C*, **36** (1987) 115.

Production of Superheavy Elements in Cold Fusion Reactions

V.Yu. Denisov

GSI-Darmstadt, Planckstrasse 1, D-64291 Darmstadt, Germany
Institute for Nuclear Research, Prospect Nauki 47, 252028 Kiev, Ukraine

Abstract. The cold fusion reactions leading to superheavy elements with Z=104-116 has been discussed in our model recently [5]. Presently we shortly discuss our model and extend our consideration to fusion reactions (^{86}Kr, ^{87}Rb, ^{88}Sr) + ^{208}Pb and ^{86}Kr + ^{209}Bi leading to elements with Z=118-120. The available experimental cross-section data for the reactions are well described.

INTRODUCTION

The synthesis of superheavy elements (SHEs) was and still is an outstanding research object. The properties of SHEs are studied, both theoretically as well as experimentally [1-17]. In two series of experiments the heaviest elements from 107 to 109 and from 110 to 112 were synthesized at GSI (Darmstadt) by using cold fusion [1-3]. In cold fusion, SHEs are produced by reactions of the type X + (Pb,Bi) → SHE + 1n at subbarrier energies. The excitation energy of a compound nucleus formed by cold fusion is low, ≈ 10–20 MeV. It was measured that the center-of-mass kinetic energy of reaction partners leading to elements with $Z \leq 112$ corresponds to the fusion barrier or is even less [2]. The cross-section for the synthesis of SHEs is very small and decreases rapidly with increasing atomic number. The experimental investigation of an excitation function for the SHE production becomes increasingly difficult due to the very small cross-sections and narrow width of the excitation function [1-3]. The full width at half maximum of the excitation function is about 4 MeV [1-3].

An appropriate theoretical models should reproduce all observed phenomena. In the Ref. [5] we present a model for the description of measured excitation functions for the SHE production. The maximum position and the width of the excitation function for cold fusion reactions X+^{208}Pb,^{209}Bi leading to elements with Z=104-112 are well described in [5], see also Figs 1,2. Within our approach [5] the process of the SHE formation goes through three stages: (*i.*) The capture of two spherical nuclei and the formation of a common shape of the two touching nuclei. Low-energy surface vibrations and a transfer of few nucleons are taken into account on the first step of a reaction. (*ii.*) The formation of a spherical or near spherical compound nucleus. (*iii.*) The surviving of an excited compound nucleus during evaporation of neutrons and γ-ray emission which compete with fission. A lowering of the fission barrier was taken into account, which arises from a reduction of shell effects at increasing excitation energy of the compound nucleus.

One of the heaviest systems studied experimentally over a wider range of excitation energy is ^{58}Fe+^{208}Pb $\rightarrow^{265}$Hs* + n [1], the data are shown in Fig. 1. The experimental data are compared with several modifications of our model. In the simplest case, using tunneling through a one-dimensional barrier and the WKB method, the results strongly underestimate the experimental fusion cross-sections. Better agreement is obtained, when the neutron transfer channels from lead to

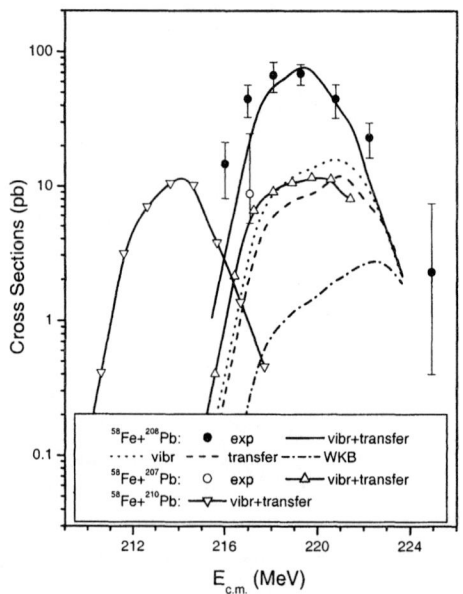

FIGURE 1. Calculated excitation functions for the reactions ^{58}Fe + ^{207,208,210}Pb $\rightarrow^{264,265,267}$Hs + n. The continuous curve shows the results for the reaction ^{58}Fe + ^{208}Pb $\rightarrow^{265}$Hs + n taking into account both the low-energy 2^+ and 3^- vibrations and the neutrons transfer channels. The dotted and the dashed curve shows the results for considering solely the 2^+ and 3^- vibrations and the neutron transfer channels, respectively. The result of the one-dimensional WKB approach is shown by the dash-dotted curve. The data obtained for the reaction ^{58}Fe + ^{207}Pb $\rightarrow^{264}$Hs + n are represented by ($\triangle$) and those for ^{58}Fe + ^{210}Pb $\rightarrow^{267}$Hs + n by ($\triangledown$). In both cases only the results including vibrations and transfer are shown. The relations taking into account the channels separately are similar as in the case of ^{58}Fe + ^{208}Pb. The experimental data shown here are taken from [1-4].

iron are taken into account. Similarly, the cross-sections increase by including in the calculations the low-energy 2^+ and 3^- surface vibrational excitations of both projectile and target. The best results are obtained by considering transfer and vibrations simultaneously. The value of parameters and other details are presented in Ref. [5].

In our model [5] we have adjusted and other parameters, which are taken from experimental data and other calculations. Note that we able to describe data for reactions ^{58}Fe+^{207}Pb $\rightarrow^{264}$Hs* + n (see Fig. 1) and ^{58}Fe+^{209}Bi $\rightarrow^{266}$Mt* + n (see Fig. 11 in [5]) by using the same fitting parameters which we fixed for reaction ^{58}Fe+^{208}Pb $\rightarrow^{265}$Hs* + n.

The results of the calculations for reactions between the projectiles ^{66,68,70}Zn and targets ^{207,208,210}Pb are presented in Fig. 2. The penetration through the inner barrier (the second stage of the reaction), which taken place for these reactions during shape evolution from two touching configuration of colliding nuclei to spherical or

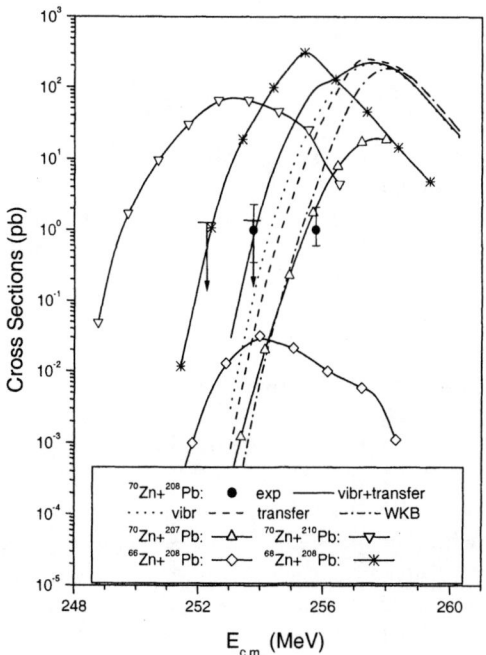

FIGURE 2. Calculated excitation functions for the reactions ^{66,68,70}Zn + ^{207,208,210}Pb $\rightarrow$ 112 + n. The notation for the reactions ^{70}Zn + ^{207,208,210}Pb corresponds to ^{58}Fe + ^{207,208,210}Pb in Fig. 1. The insert explains the assignment of the reactions to the symbols. The experimental data shown here are taken from [1-3].

near spherical compound nucleus [5], is important for these reactions, because the intrinsic excitation energies E^*_{CN} of the compound nucleus $^{278}112$ ($E^*_{CN} \approx 10$ MeV at collision energy $E_{cm} = 253.8$ MeV and $E^*_{CN} \approx 12$ MeV at $E_{cm} = 255.8$ MeV [3]) are less then the value of inner barrier ($B_{Sph} \approx 13.5$ MeV [5]). Our calculations for reaction ^{70}Zn + ^{208}Pb $\to ^{277}112$ + n are well agree with experimental data at collision energy 253.8 MeV and underestimated data obtained in May 2000 at energy 255.8 MeV.

REACTIONS LEADING TO ELEMENTS WITH $Z=118-120$

Recently element 118 is formed in reaction ^{86}Kr+^{208}Pb $\to ^{293}118$+n in Berkeley [6]. This discovery stimulates various studies of reaction mechanisms leading to SHE formation. Note that the same reaction has been studied in GSI (Darmstadt, Germany) [3], GANIL (Caen, France) [15] and RIKEN (Wako, Japan) [16], but the no events related to the formation of element $^{293}118$ was observed. The discussion of some experimental results can be found in [3]. Due to these undefinite experimental situation the theoretical studies of the reaction ^{86}Kr+^{208}Pb $\to ^{293}118$+n is very important. It is interesting to investigate reaction ^{86}Kr+^{208}Pb $\to ^{293}118$+n in the framework of our model [5]. In this part we study cold fusion reactions leading to elements 118-120. Note that the reactions leading to elements with $Z=118-120$ could be measured on various facilities in the near future. Therefore below we discuss only new results obtained for the SHE production cross section in collisions: (^{86}Kr, ^{87}Rb, ^{88}Sr) + ^{208}Pb, and ^{86}Kr + ^{209}Bi.

As pointed in Introduction, the formation of a common shape of the two touching nuclei is taken place on the first stage of reaction in our model [5]. The projectile and target penetrate through the outer barrier (capture barrier) formed by both Coulomb and nuclear ion-ion interactions. The capture stage is very important for the formation of the SHE with Z$\leq$112 in cold fusion reaction in our model. However, for heavier systems with 114$\leq$Z$\leq$120 this process is not so important, because the height of a capture barrier is smaller than the height of an inner barrier B_{Sph} for our parametrization of ion-ion interaction. The inner barrier is taken place on the way from the two-touching shape, related to the projectile and target, to the near spherical compound nucleus, see for details [5,9,10]. The potential energy related to Coulomb and nuclear forces between the separate ions is smaller than collision energies used in cold fusion reactions for synthesis of elements with 114$\leq$Z$\leq$120. Therefore, the inner barrier plays a key role and the capture barrier is not essential for the formation of the elements 114 or heavier in our model.

Note that similar ratio between collision energy and height of the interaction potential between two separate ions used for cold fusion SHE production is also found in Ref. [17] by using proximity potential and in Refs. [2-3] by using Bass potential.

The shape of a fusing system after formation of a touching configuration is elongated, asymmetric and laced. From such a configuration the system develops further to a near spherical compound nucleus and later to the ground-state, which can be deformed or spherical. We propose that this evolution of the shape is described by smooth reduction of parameters p and q in shape parametrization [5]

$$R(\vartheta) = R(p,q)[1 + p \sum_{\ell=2,even}^{N} \beta_\ell Y_{\ell 0}(\vartheta) + q \sum_{\ell=3,odd}^{N} \beta_l Y_{\ell 0}(\vartheta)], \qquad (1)$$

where p and q are equal to 1 at a touching point of two spherical colliding ions and the deformation parameters β_ℓ are fixed at a touching point. The values of the deformation parameters for SHEs in the ground-state are $p, q \approx 0$. The

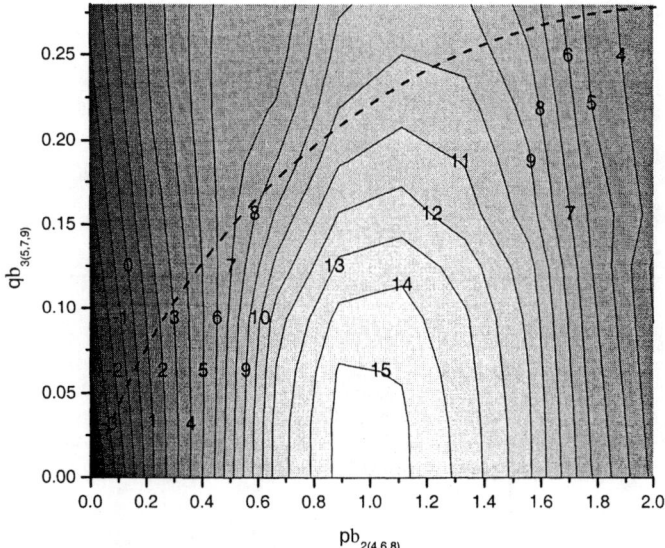

FIGURE 3. Potential energy surface as a function of the deformation parameters β for cold fusion reaction ^{86}Kr+^{208}Pb→^{294}118. The touching configuration of the spherical projectile and target nucleus close to the upper right corner and that of the ground-state close to bottom left. The dash line is a tunneling trajectory which is draw by eye assuming that all deformations are monotonously changes during motion to equilibrium compound nucleus shape. The ratio between even and odd deformation parameters is fixed, see text for details. The contour lines are drawn in steps of 1 MeV, the maximum values and several others are given in MeV on graph.

parameters p and q ($0 \leq p \leq 1$, $0 \leq q \leq 1$) are connected to elongated and asymmetrical degrees of freedom of a nuclear shape during the formation of the spherical compound nucleus, respectively. The radius $R(p,q)$ depends on p and q because of the volume conservation.

The potential energy surface for the nucleus $^{294}118$ formed in reaction ^{86}Kr + ^{208}Pb is presented in Fig. 3. Although the shape parametrization (1) is rather rough at a touching point, it is possible to study the inner barrier, which has to be crossed during the process of sphere formation, as shown in Fig. 3. The potential energy surface is related to the sum of macroscopic and shell correction energies.

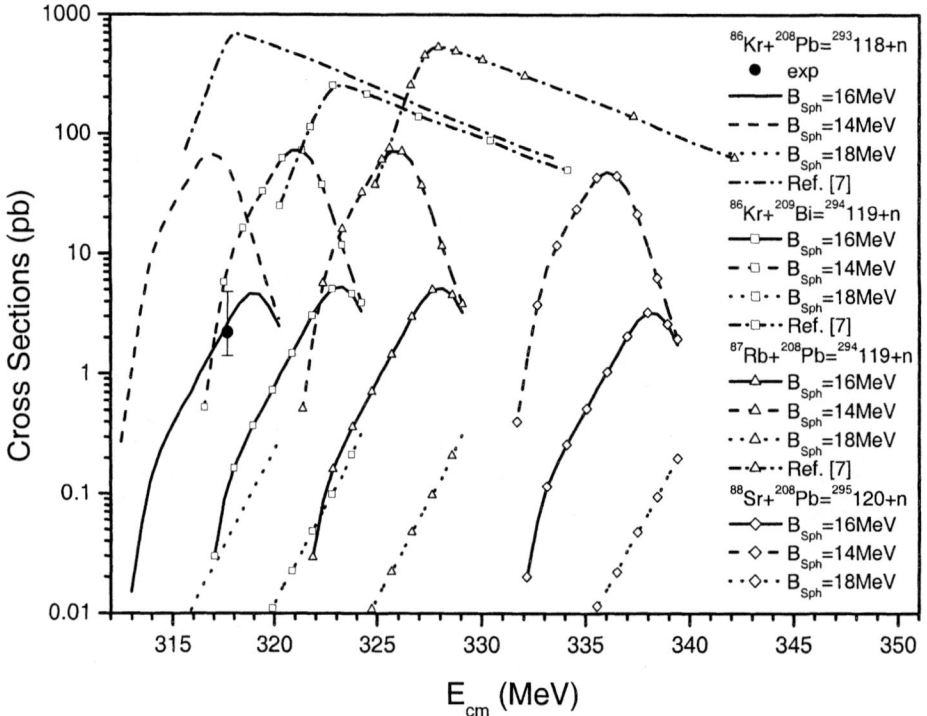

FIGURE 4. Calculated excitation functions for reactions ^{86}Kr + ^{208}Pb $\to ^{293}118$ + n, ^{86}Kr + ^{208}Bi $\to ^{294}119$ + n, ^{87}Rb + ^{208}Pb $\to ^{294}119$ + n and ^{88}Sr+^{208}Pb $\to ^{195}120$+n. A decreasing (increasing) of the barrier B_{Sph} by 2 MeV (see text) increases (decreases) the cross-section considerably. The experimental data obtained for the reaction ^{86}Kr + ^{208}Pb $\to ^{293}118$ + n in [6] are represented by filled dot. The excitation functions for reactions ^{86}Kr + ^{208}Pb $\to ^{293}118$ + n, ^{86}Kr + ^{208}Bi $\to ^{294}119$ + n, ^{87}Rb + ^{208}Pb $\to ^{294}119$ + n from the Ref. [7] are also presented.

The potential energy surface is a function of parameters p and q in the deformation space $p\beta_2, q\beta_3, p\beta_4, q\beta_5, ..., q\beta_9$.

The transmission coefficient for the barrier penetration during the shape evolution from the touching projectile and target nuclei to the near spherical shape of a compound nucleus was estimated in the Hill-Wheeler approximation for various collision energies E.

After the first two stages of the reaction, the transmission through the capture (before touching of ions) and the inner (after touching of ions) barriers, a compound nucleus with an number A is formed with equilibrium deformation near sphericity and excitation energy E^*. The dominant decay modes of a heavy compound nucleus at low excitation energy, what is the case in cold fusion, are the fission and the evaporation of neutrons. The residue $A - 1$ can still be excited after neutron emission. Again, it may fission or cool down by emission of γ's. We take into account all these processes. In excited nuclei the fission barrier B^* is reduced due to both the washing out effect of the shell correction and the centrifugal potential [5]. These two effects are very important for the correct description of SHE production cross sections.

We start discussion with the excitation function calculated for reaction ^{86}Kr+^{208}Pb $\to$ ^{293}118+n, because experimental data exist for this reaction [6]. The experimental data are compared with result of our calculations in Fig. 4. We make calculations of the SHE production cross sections for different values of the inner barrier $B_{Sph} = 14$, 16 and 18 MeV. The value $B_{Sph} \approx 14$ MeV is close to the value obtained in other model [9,10]. The value $B_{Sph} \approx 16$ MeV is near to evaluated by using Fig. 3. The reduction of the height of the inner barrier by 2 MeV from 16 to 14 MeV or from 18 to 16 MeV increases the production cross section by the one order of magnitude.

As pointed in Introduction, in our model [5] we have adjusted and other parameters, which are taken from other calculations or experimental data. Here we apply the same values of fitted parameters as for elements 112 in [5]. The other parameters are taken from [18-25]. The values of static fission barrier for the even-even compound nuclei are taken from [7]. We evaluate fission barriers for odd and odd-odd nuclei by using fission barriers for the nearest even-even nuclei. The parameters used in our calculation for this reaction and other reactions considered below are listed in Table 1.

The good description of the experimental cross section is obtained for reaction leading to elements 104-112 in [5] and for reaction ^{86}Kr+^{208}Pb $\to$ ^{293}118+n in calculation with $B_{Sph} = 16$ MeV in Fig. 4. The result of calculation for this reaction obtained in Ref. [7] is also presented in Fig. 4. The excitation function obtained in Ref. [7] overestimates experimental data and has larger width than that obtained in our model. The cross section 0.0051 pb for reaction ^{86}Kr+^{208}Pb $\to$ ^{293}118 + n obtained in Ref. [11] underestimates strongly experimental value $2.2^{+2.6}_{-0.8}$ pb [6]. The production cross sections 0.5 pb for reaction ^{86}Kr+^{208}Pb $\to$ ^{293}118 + n evaluated in Ref. [12] is close to the experimental data.

Now we consider reactions leading to elements 119-120.

TABLE 1. The parameters of ground-state and saddle-point properties of the compound nucleus (CN) and of evaporation residues (SHEs, after neutron emission). E_{nbe}^{CN} is the binding energy of the CN [19], δE_{shel}^{CN} (B_{CN}) and δE_{shel}^{SHE} (B_{SHE}) are the ground state shell corrections (fission barrier heights) of the CN and SHE respectively, β_{2gs} and β_{2sadl} are the ground state and saddle point deformations of the CN and SHE, E_{nsep}^{CN} is the neutron separation energy in the CN, T_{sf} is the spontaneous-fission lifetime of the CN in the ground-state, γ is the shell-correction damping parameter [5] and c_V is the strength factor of the nuclear heavy ions interaction potential [5].

Colliding ions	^{86}Kr+^{208}Pb	^{87}Rb+^{208}Pb	^{86}Kr+^{209}Bi	^{88}Sr+^{208}Pb
CN	^{294}118	^{294}118	^{294}118	^{294}118
E_{nbe}^{CN} (MeV)	-2081.80	-2081.48	-2081.48	-2081.85
δE_{shel}^{CN} (MeV)	-7.67	-7.67	-7.67	-6.77
β_{2gs}	0.070	0.0735	0.0735	0.07
B_{CN} (MeV)	5.45	5.30	5.30	5.15
β_{2sadl}	0.375	0.3675	0.3675	0.3675
E_{nsep}^{CN} (MeV)	7.72	7.67	7.67	7.73
$\log_{10} T_{sf}{}^a$	3.13	0.985	0.985	-1.16
SHEs	^{293}118	^{294}119	^{294}119	^{295}120
δE_{shel}^{SHE} (MeV)	-7.82	-7.89	-7.89	-7.25
B_{SHE} (MeV)	5.18	5.125	5.125	5.0
γ (MeV)$^{-1}$	0.13	0.13	0.13	0.13
c_V	1.05	1.05	1.05	1.05

[a] The T_{sf} is in sec.

The same isotope of element 119 may be formed in two different reactions ^{86}Kr+^{209}Bi $\to$ ^{294}119+n and ^{87}Rb+^{208}Pb $\to$ ^{294}119+n. Therefore, it is very important for future experiments to know which reaction has a larger cross section. We present results for both reactions in Fig. 4. The cross sections of both these reactions are similar, because the SHE production is determinated on the second and third stages of the reaction. We make calculation for the same values of inner barrier as for reaction ^{86}Kr+^{208}Pb $\to$ ^{293}118+n. The compound nuclei formed in collisions ^{86}Kr+^{209}Bi and ^{87}Rb+^{208}Pb are the same, therefore the fission, neutron and γ decay properties are the same for both reactions at the equal excitation energies of compound nuclei. The difference between two these reactions is related only to binding energies of colliding ions. Due to this, maxima of excitation functions for these reactions occur at different collision energies, see Fig. 4.

The production cross sections for reactions ^{86}Kr+^{209}Bi $\to$ ^{294}119+n and ^{87}Rb+^{208}Pb $\to$ ^{294}119+n obtained in Ref. [7] are again much lager than results of our model, see Fig. 4.

The models described in Refs. [5,7,11–13] assume very different mechanisms of the SHE formation. Note that all these models describe well production cross sections for the SHEs with Z$\leq$ 112 in cold fusion reaction. But the models in

[5,7,11–13] predict very different results for the SHE with $Z \geq 114$, see Fig. 4 and Refs. [3,5,7,11–13]. Such a difference is not surprising. The models discussed in Refs. [11–13] use a specific trajectory of the compound nucleus formation related to the concept of a dinuclear system. Within this concept the shape evolution from two touching nuclei to a spherical compound nucleus takes into account only asymmetric shape degree of freedom near touching point, because colliding ions stop near touching point, and afterward the subbarrier multinucleon transfer from projectile (light ion) to target (heavy ion) takes place. Such a mechanism of the SHE formation does not take into account the neck formation and smooth evolution of both symmetric and asymmetric shape degree of freedom as suggested in Fig. 3. Note, that the subbarrier multinucleon transfer at touching point considered in [11–13] should takes a long time due to high value of the barrier, but the existence of a dinuclear system during such a long tunneling process is questionable. The model proposed in Ref. [7] ignores both nuclear interaction between ions during the collision and shape evolution processes.

The excitation functions of cold fusion reactions leading to element 120 obtained in our model for three different values of the inner barrier are also shown in Fig. 4. The tendency observed for reactions leading to elements 118 and 119 are also kept for reaction $^{88}Sr+^{208}Pb \rightarrow {}^{295}120+n$, see Fig. 4. The production cross section for the element 120 is smaller than for elements 118-119 due to smaller values of both fission barriers and ground state shell corrections, see Table 1.

The main difference between reactions considered in this section is related to the properties of colliding nuclei, binding energies and decay properties. Unfortunately, the estimation of the inner barrier penetration is not very precise in our model. Due to this we can not see any difference associated with the inner barrier penetration process in various reactions.

CONCLUSIONS

The theoretical description of reaction processes leading to fusion of SHEs is a difficult task. The nuclei are located at the edge of stability, and the reactions are dominated by shell structure effects in projectile, target as well as in compound nucleus. In addition, shell structure is required for the calculation of the binding energy at large deformation in order to determine the inner barrier and fission barrier.

We plan to improve our model by comparison with experimental data, which will be available in the nearest future.

Acknowledgments

The author would like to thank S. Hofmann, V. Ninov and W. Nörenberg for useful and stimulated discussions. He acknowledges gratefully support from GSI.

REFERENCES

1. G. Münzenberg, *Rep. Prog. Phys.* **51**, 57 (1988).
2. S. Hofmann, *Rep. Prog. Phys.* **61**, 373 (1998).
3. S. Hofmann, G. Münzenberg, *Rev. Mod. Phys.* **72** 733 (2000).
4. V. Ninov, (private communications).
5. V.Yu. Denisov, S. Hofmann, *Phys. Rev.* **C61**, 034606 (2000); *Acta Phys. Polonica* **B31**, 479 (2000).
6. V. Ninov, K. E. Gregorich, W. Loveland, A. Ghiorso, D. C. Hoffman, D. M. Lee, H. Nitsche, W. J. Swiatecki, U. W. Kirbach, C. A. Laue, J. L. Adams, J. B. Patin, D. A. Shaughnessy, D. A. Strellis, and P. A. Wilk, *Phys. Rev. Lett.* **83**, 1104 (1999).
7. R. Smolanczuk, *Phys. Rev.* **C56**, 812 (1997); Phys. Rev. **59**, 2634 (1999); *Phys. Rev.* **C60**, 021301 (1999); *Phys. Rev.* **C 61**, 011601 (R) (2000).
8. Yu.Ts. Oganessian, A.S. Iljinov, A.G. Demin, S.P. Tretyakova, *Nucl. Phys.* **A239**, 353 (1975).
9. P. Möller, J.R. Nix, P. Ambruster, S. Hofmann, G. Münzenberg, *Z. Phys.* **A359**, 251 (1997), and private communications.
10. P. Möller, J.R. Nix, In Proc. Tours Symposium on Nuclear Physics III, Tours, France, September 2-5, 1997; Los Alamos National Laboratory Preprint No. LA-UR-97-3580.
11. G.G. Adamian, N.V. Antonenko, W. Scheid, V.V. Volkov, *Nucl. Phys.* **A633**, 409 (1998); G.G. Adamian, N.V. Antonenko, V.V. Volkov, E.A. Cherepanov, W. Scheid, *Bull. Rus. Acad. Sci. Phys.* **63**, 693 (1999); G.G. Adamian, N.V. Antonenko, W. Scheid, LANL-Preprint-nucl-th/9911078.
12. E. Cherepanov, *Pramana* **53**, 619 (1999).
13. R.V. Jolos, A.I. Muminov, A.K. Nasirov, *Eur. Phys. J.* **A4**, 245 (1999); G.S. Giardina, S. Hofmann, A.I. Muminov, A.K. Nasirov, *Eur. Phys. J.* **A8**, 205 (2000).
14. Y. Aritomo, T. Wada, M. Ohta, Y. Abe, *Phys. Rev.* **C59**, 796 (1999).
15. J. Peter, *Talk on International Workshop Fusion Dynamics at the Extremes*, JINR, Dubna, Russia, May 25 -27, 2000.
16. K. Morita, *Talk on International Workshop Fusion Dynamics at the Extremes*, JINR, Dubna, Russia, May 25 -27, 2000.
17. W.D. Myers, W.J. Swiatecki, *Acta Phys. Polonica* **B31**, 1471 (2000).
18. G. Audi, A.H. Wapstra, *Nucl. Phys.* **A595**, 409 (1995).
19. W.D. Myers, W.J. Swiatecki, *Nucl. Phys.* **A601**, 141 (1996).
20. P. Möller, J.R. Nix, W.D. Myers, W.J. Swiatecki, *At. Data Nucl. Data Tables* **59**, 185 (1995).
21. S. Raman, C.W. Nestor, S. Kahane, K.H. Bhatt, *At. Data Nucl. Data Tabl.* **42**, 1 (1989).
22. R.H. Spear, *At. Data Nucl. Data Tabl.* **42**, 55 (1989).
23. M.A. Kennedy, P.D. Cottle, K.W. Kemper, *Phys. Rev.* **C46**, 1811 (1992);
24. J.P.M.G. Melsen, P.J. Van Hall, S.D. Wassenaar, O.J. Poppema, G.J. Nigh, S.S. Klein, *Nucl. Phys.* **A376**, 183 (1982).
25. M.J. Martin, *Nucl. Data Sheets* **63**, 723 (1991).

The Limit of Nuclear Deformation and Fission Properties of Heavy and Superheavy Elements

Y. L. Zhao*, I. Nishinaka†, Y. Nagame†,
K. Sueki* and H. Nakahara*

Graduate School of Science, Tokyo Metropolitan University, Tokyo 192-0397, Japan
†*Japan Atomic Energy Research Institute, Tokai-Mura, Ibaraki 319-1195, Japan*

Abstract. The degrees of the deformation of nuclei at the scission points in the symmetric and asymmetric fission are experimentally determined. They are found to be constant, respectively, in a wide range of A_f among the asymmetric fission and among the symmetric fission of highly excited nuclei although in the latter for the spontaneous fission and low energy induced fission, the degree of deformation decreases gradually for $A_f > 245$ until $A_f \sim 260$. The constancy of the degree of the scission deformation in each fission mode leads to a TKE systematic formula for each fission mode which is comparable to Viola's TKE formula. Based on the new TKE formulas derived from the deformation parameter of the fissioning nucleus, the extension of the present knowledge to predict the fission properties of the superheavy elements is addressed. It is pointed out that the symmetric and asymmetric fission valleys may merge into one for the superheavy elements around the $A=280$-290 region.

I. INTRODUCTION

The number of elements in the periodic table has been extended up to 118 after the recent experimental reports on the observation of $Z=114$ events by the Dubna group [1] and $Z=116$ and 118 events by the Berkeley group [2]. Nuclear and chemical properties of these superheavy elements are attractive topics [3,4]. But unfortunately, direct experimental exploration of the nuclear properties *e.g.*, fission decay, of such heavy elements as $Z=114$ and 118 is impossible because of the very low production rate of these nuclei. The dynamic deformation motions of the nucleus in fission process depend on various parameters such as fissility, nucleonic structure, temperatures, *etc.* of a nuclear system. In general, the fission properties such as the mass-yield distribution and kinetic energy release vary gradually and systematically with the nucleus in the chart of nuclides. But when the nucleus becomes as heavy as Fm isotopes, dramatic changes in the fission properties were observed [5,6]. Spontaneous fission of these nuclei is found to give extremely sharp mass-yield curves with an abnormally large TKE release which deviates from the prediction of the existing systematics [7].

But when the nuclear fission process was studied via a new viewpoint of the final deformation of the fissioning nucleus, we have found some essential simplicity behind the apparent complexity of the fission phenomena observed [8,9]. Also the seemingly dramatic changes in fission properties observed for very heavy elements are found to be due to the gradual increase of the fragment shell effects for the symmetric mass divisions. They can be systematically understood based on the known knowledge from the fission properties of the relatively lighter niclides [8,9]. In fact, we have so far accumulated a large amount of experimental data for fission from the preactinide through the actinide and up to the transactinide elements. In this paper, an attempt has been made to extend our present knowledge toward the fission properties of the superheavy elements. The results will provide us with the criteria for identifying the superheavy elements if they fission spontaneously.

II. RESULTS OF DATA ANALYSIS AND DISCUSSIONS

2.1. Fine structures in distributions of fission events for compound nuclei with low excitation energy

Some results for proton-induced fission of ^{232}Th [10,11] are shown in Fig. 1. The coincidence method of measuring the kinetic energies and the mass numbers for each pair of complementary fragments was employed in the present work. With a development of heavy ion detectors, this method allowed mass resolution of less than 2 u and the energy resolution of $\sim$2 MeV [12]. The observed TKE distribution, $y(\text{TKE},A_1,A_2)$, as a function of the fragment mass A_1 and A_2 for a given mass split $(A_1+A_2=A_f)$, is given in Figs. 1 by solid circles. The abscissa is the TKE value and the ordinate is the yield (counts) for the fission product (indicated by the mass number in each frame) within a TKE window of 2 MeV.

In Figs. 1, the slopes of the two wings of some of $y(\text{TKE},A_1,A_2)$ curves are quite different. An analysis of these data indicated that the $y(\text{TKE},A_1,A_2)$ curves for mass splits in the very symmetric ($117<A_1<122$) and very asymmetric ($A_1>140$) regions can be well described by a single Gaussian function, whereas those in between them can not.

We have analyzed those fine structures associated with the $y(\text{TKE},A_1,A_2)$ distributions in term of two TKE components, each with a Gaussian shape of different width, and have identified the independent fission paths from the observations of the incident-particle-energy dependence of the yield of each TKE component [12,13]. The results corresponding to the symmetric and asymmetric fission paths, $y(\text{TKE}^{\text{sym}},A_1,A_2)$ and $y(\text{TKE}^{\text{asym}},A_1,A_2)$ distributions, are depicted in Figs. 1 by long-dashed and short-dashed lines, respectively. The sum of both is indicated by solid lines which are in good agreement with the measured data points of solid circles. The distributions of the fission events for the symmetric and asymmetric fission paths in Figs. 1 indicate that in the fission events observed nearby the symmetric mass split, contribution from the asymmetric fission still exist. Similarly, even in the fission events of the asymmetric mass split, contribution from the symmetric fission path is also present.

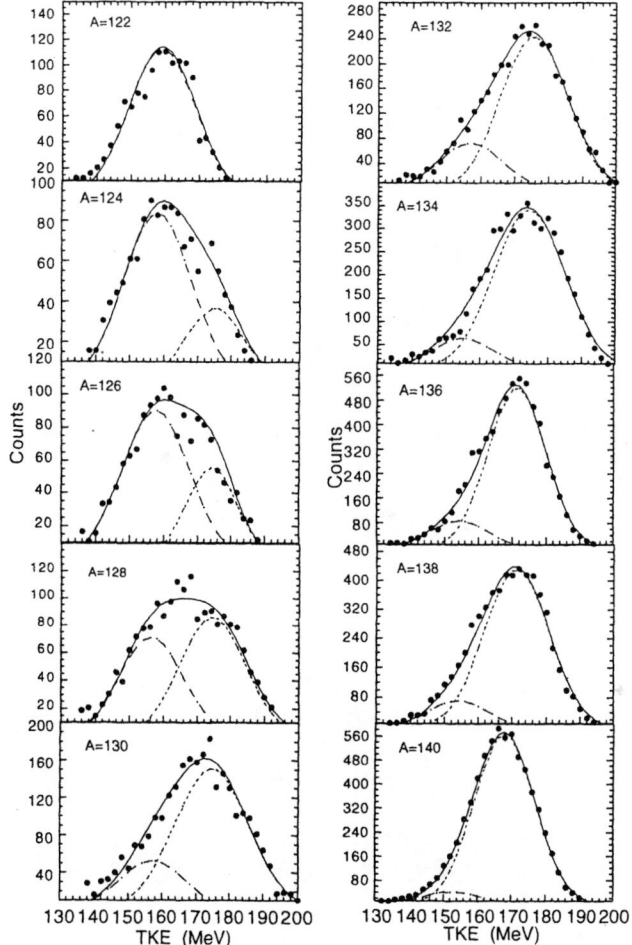

Fig. 1, The observed fine structures (solid circles) associated with the TKE(A_1,A_2) distributions for 11.6 MeV proton Induced fission of ^{232}Th. Solid circles are the data points observed, and the lines were obtained from the two-component analysis.

The scission deformations of the fissioning nuclei experiencing those independent fission paths are shown in Figs. 2. The scission deformation (β) is defined as the distance between the two charge centers estimated from the TKE relative to the sum of the radii of the corresponding spherical nuclei in contact, $\beta=D_0/D$. In other words, β is a physical quantity which describes the shape elongation along the fission coordinate at the instant of the nuclear mass division taking place. The details of the definition can be found in Ref. [8]. The results shown in Figs. 2(a) and 2(b) are the scission deformations and the primary mass-yield distribution, respectively, in the fission after taking each independent fission path of ^{233}Pa fission at the excitation energy around 17 MeV.

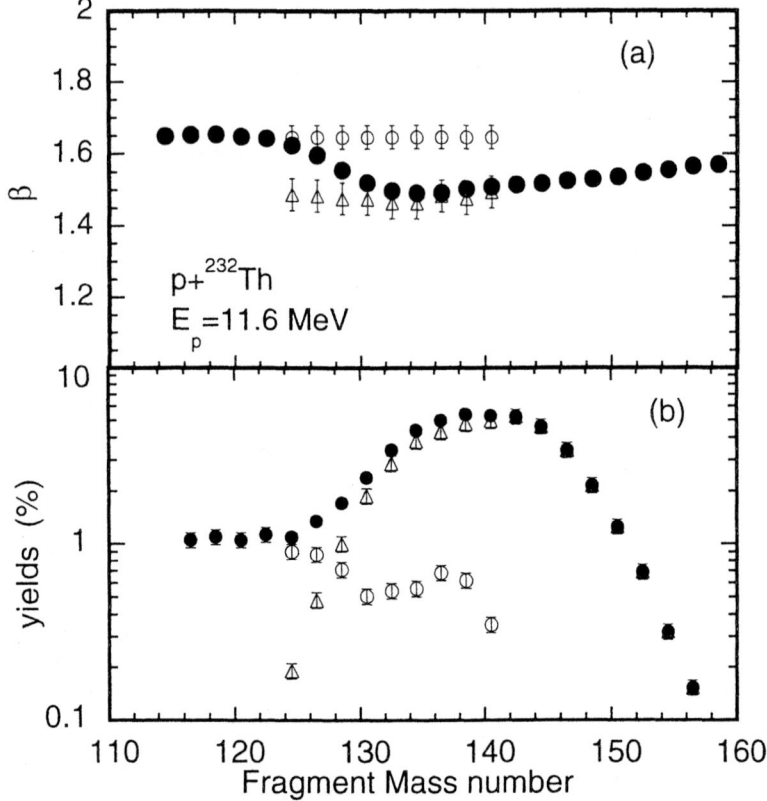

Fig. 2, (a). The dependence of the final shape elongation of the low-energy fission on the way of nuclear mass division, undergoing the mass-symmetric deformation (open circles) and mass-asymmetric deformation (triangles). (b). The corresponding mass-yield distributions of fission fragments indicated by the same symbols with (a), for the symmetric and asymmetric fission paths.

Difference in the β-values for the symmetric and asymmetric fission paths is clearly seen in Fig. 2(a). Solid points in the symmetric region are relatively larger and they are smoothly connected to the open circles which were obtained from the low-TKE components of Figs. 1. The solid circle observed in the asymmetric region ($A > 140$) are relatively smaller as compared to the values for the symmetric path and slightly dependent on the mass split. Open triangles obtained from the high

TKE component seem to lie on the line smoothly extended from the solid circles in the asymmetric region, except for the mass splits with $A_1<132$, where the β_{asym} no longer keeping slightly decreasing nature. This result indicates that the fragment mass $A_1=132$ has the smallest deformation among all fission products. The reason can be the effects of nuclear structures, the spherical shells of $N=82$ and/or $Z=50$.

The important feature is, for a given fissioning system (1) the scission deformations are nearly constant within each fission mode, and (2) the β in the symmetric fission is larger than that for the asymmetric fission. In Fig. 2(b), the observed mass yields are shown by solid circles, they were decomposed into the symmetric (open circles) and asymmetric (open triangles) components by use of the relative intensities of the high-TKE and low-TKE components in the TKE distributions. The details of the analysis method and its physical bases are described in Ref. [12].

2.2. Liquid-drop type deformations of heavy nuclei and their fission properties

The scission deformations in the fission of hot nuclei of ^{235}U ($Z=92$) and ^{252}Fm ($Z=100$) are displayed in Figs. 3. The fission phenomena were observed in the 60 MeV ^{3}He+^{232}Th and 154 MeV ^{20}Ne+^{232}Th reactions, respectively. Fragment mass distributions of the fission of the highly excited nuclei is a single peak with a broad Gaussian-like shape [14], as shown in Fig. 3(c) as an example. For fusion-fission reactions induced by high energy heavy ions, the conditions become quite different from those we have discussed above. The higher energy and larger angular momenta possessed by the compound nuclei would likely destroy the shell structures of nuclei and the deformation motion is dominated by property of liquid drop.

In Figs. 3, the β values deduced from the TKE are approximately independent of the degree of mass asymmetry in mass split, and they are in the region of $\beta=1.65\sim1.60$ which are the value observed for the symmetric fission of the low-energy fission of light actinides.

The effect of nuclear shell structure on the symmetric deformation can be negligible in the symmetric fission of high excitation energy. The β-values for those nuclei undergoing the mass-symmetric fission have been evaluated from a large number of literature data as a function of the mass number of the fissioning nucleus A_f [8], and they are shown in Fig. 4. The shape elongations for all nuclei undergoing the symmetric mass division are found to remain constant of ~1.65. It is nearly independent of the fissioning mass.

The nuclei giving an identical β-value of 1.65 cover a large range of A_f from ~200 to 270 and a large range of excitation energy as indicated in the plot. With the change of the nucleus and its excitation energy, various aspects related to the dynamics [15,16] of the deformation motion of the nucleus, *e.g.*, the potential energy surface, inertia, the dissipative force and the nuclear temperature, *etc.* should be varying. Nevertheless, all of these nuclei end their elongation at the stage of almost the same degree of deformation. This hence raises a further question of what the crucial factors which dominate the maximal deformation of the nucleus are.

Assuming two touching spheroids for the nascent fragments at scission, the degree of deformation of $\beta=1.65$ corresponds to a shape of the major semi-axis to the minor semi-axis of ~ 2.1 to 1. It also corresponds to ~ 2.62-fold of its original diameter of the spherical shape.

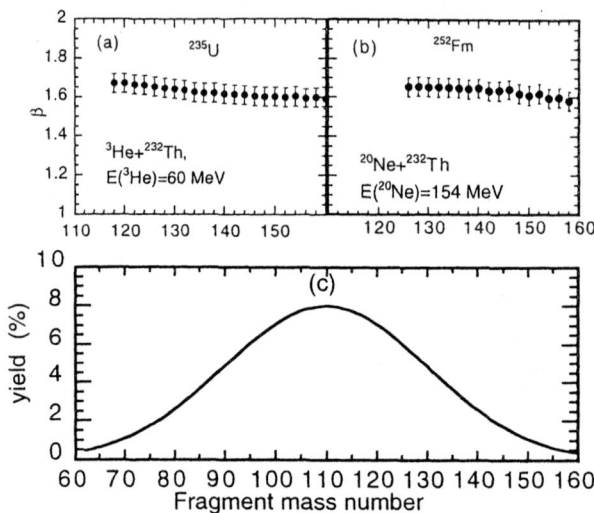

Fig. 3, (a) and (b). The dependence of the final shape elongations of the hot-fusion compound nuclei on the mass split. (c). The corresponding mass-yield distribution of fission fragments.

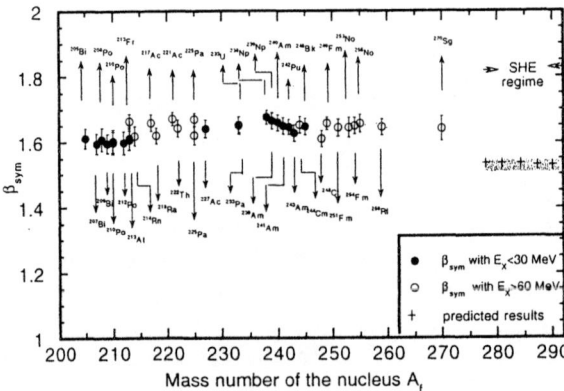

Fig. 4, The final deformation of the nuclei undergoing the symmetric mass division as a function of the mass of the fissioning nucleus A_f.

2.3. Shell-affected deformations of heavy nuclei and their fission properties

The effects of structural shells on the nuclear deformation process are observed in the spontaneous and low-energy induced fission. When a nucleus becomes heavier with increasing the number of protons, it becomes more unstable toward spontaneous fission. In Figs. 5(a) and (b), the shape elongations for spontaneous fission of ^{252}Cf and ^{246}Fm are plotted as a function of the mass split. The asymmetric mass-yield distribution is observed in such spontaneous fission as illustrated in Fig. 5(c). It is very asymmetric with the peak-to-valley ratio of, for instance, ~ 750 for the ^{252}Cf (sf). As observed in other fission processes discussed above, the β-values for the asymmetric fission are still nearly independent of the mass split.

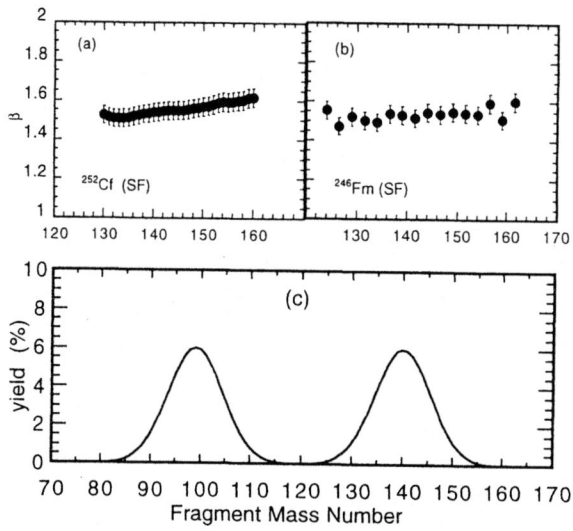

Fig. 5, (a) and (b). The dependence of the final shape elongations of the spontaneous nuclei on the mass split. (c). The corresponding mass-yield distribution of fission fragments.

The shell-effect changing with the nucleonic composition is more explicitly seen in Fig. 6. The β-values for the mass-symmetric deformation leading to the symmetric mass division of $A_1=A_2=A_f/2$ in low-energy and spontaneous fission processes are studied as a function of the fissioning mass A_f. They are found to gradually become smaller with the increase of A_f. It changes from $\beta=1.65$ for ^{244}Cm down to $\beta=1.33$ for nuclei in the $A_f=260$ region. The $\beta=1.65$ is a characteristic value for the liquid-drop type deformation as indicated by the broad dotted line in Fig. 6. This is to say that the mass-symmetric deformation of ^{244}Cm (sf) is still dominated by the liquid-drop like property of the nucleus, and the influence from nuclear shells is very weak. What we can learn from the results in Fig. 6 is that the effects of nuclear shells become stronger for the symmetric fission as the nucleus becomes heavier, as indicated by the decrease of the degree of final deformation of the fissioning nucleus. The $\beta=1.33$ corresponds to the abnormally high TKE components observed in the spontaneous fission of very heavy elements in the ^{258}Fm region. The effect of final fragment shells is explicitly seen here.

The other aspect of the effects of shells on nuclear division process is the mass-asymmetric deformation, for which the β-values are given in Fig. 7. The shape elongation for the nucleus leading to the typical asymmetric mass division producing the fragment mass of $A_1=140$ is used to represent the degree of the asymmetric deformation. One sees that the β for any nucleus undergoing the asymmetric deformation is surprisingly similar. The asymmetric deformation gives a $\beta=1.53\pm0.03$.

Mass asymmetry in fission is related to the effects of nuclear structures. Theoretical calculations of potential energy surface has successfully shown the presence of a fission valley that leads to the asymmetric mass division [17], and it is apparently affected by the shell structures of the final fragments near scission which cause an early neck-in of the scissioning nucleus, thus giving a smaller β value of 1.53.

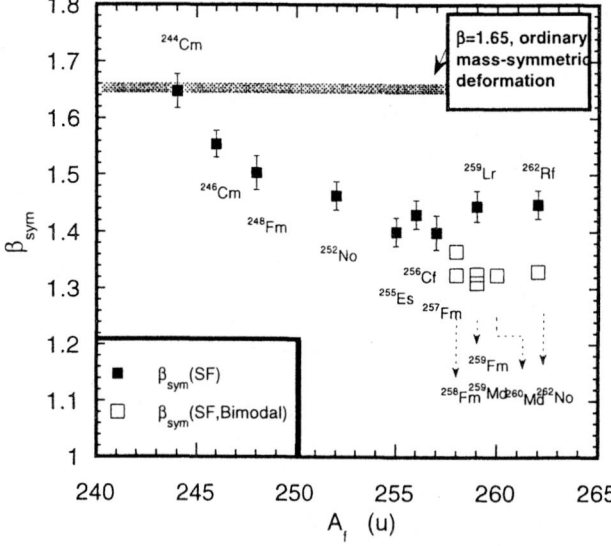

Fig. 6, The effects of nucleonic composition on the final deformation for heavy elements fissioning spontaneously.

Another interesting result in Fig. 7 needs to be noted here. Open circles show the β-value for the lower TKE components observed in the bimodal fission process of very heavy elements [5], giving $\beta\sim1.53$, which is the degree of deformation for the nuclei undergoing the ordinary mass-asymmetric fission as observed in fission of lighter actinides. This hence indicates such a fact that the shell-directed ordinary asymmetric fission is competing with the strongly shell-affected special symmetric fission.

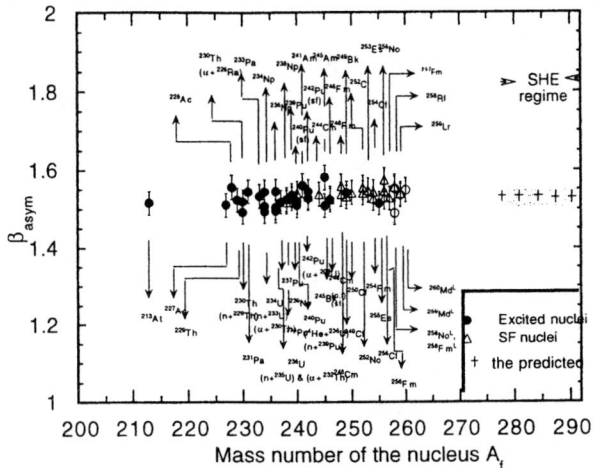

Fig. 7, The final deformation of the nuclei undergoing the asymmetric mass division as a function of the mass of the fissioning nucleus A_f.

III. THE EXPECTED FISSION PROPERTIES OF HEAVY AND SUPERHEAVY ELEMENTS

3.1. The final degree of deformation

In Fig. 7, the lower average TKE observed in the bimodal fission of nuclei around ^{258}Fm gives the β value of 1.53 which is the characteristic value observed for the mass-asymmetric deformation in the fission process of the lighter actinides. This fact strongly suggests the presence of the asymmetric fission path in heavy nuclei. According to the recent dynamical calculation of fission paths in multi-dimensional potential surface for ^{258}Fm, two main paths are present after passing the respective barrier [18]. One proceeds to a symmetric valley with a more compact shape of the scissioning nucleus and the other to a more reflection-asymmetric valley with a more elongated shape at scission. A similar calculation for the ^{254}Fm fission has been done and the results indicated that there were two paths with different thickness of the second barrier. The path leading to the symmetric fission valley has a thicker barrier to penetrate through and reaches the scission with less compact shape compared to the one in ^{258}Fm, and the other leading to an reflection-asymmetric fission valley has a thinner barrier and ends with an elongated scission shape. These results for ^{254}Fm is in conformity with the experimental observations which give a double-humped asymmetric mass yield curve for the spontaneous fission of ^{254}Fm. In the calculations, deformation parameters for the multipolarities of $\lambda=3,5,6$ were fixed for the minimum potential energies. But more recent calculations by Möller and Iwamoto [17] showed the presence of a different saddle for each deformation path to the reflection-symmetric and to the asymmetric fission valley with similar predictions for elongation of the scission shapes.

The degrees of the symmetric and asymmetric deformations for superheavy elements (in $A_f \sim 280$ region) undergoing spontaneous fission are expected from Figs. 4 and 7, respectively, as indicated by the + symbol. Either of them gives the β value of about 1.53, because the asymmetric valley and symmetric valley may merge into one due to the effects of the deformed fragment shells, for instance, the deformed shell of N=88 leading to the well-known mass peak of $A \sim 140$ for the most of fissioning nuclei. In other words, in the region of $A_f \sim 280\text{-}290$, the symmetric fission is again affected by the fragment shells which favor deformed fragments with the mass number in the $A \sim 140$ regime, unlike the case for heavy elements in the $A_f \sim 260$ region where the effects of the fragment shells favor spherical shapes giving fragments with the mass number in the $A \sim 130$ regime.

3.2. The TKE release, New TKE formulas

Applying the deformation parameters (β) to calculate the Coulomb repulsion energy, we have derived new TKE-formulas as a function of the charge and mass of the fissioning nucleus, separately for the symmetric and asymmetric fission. The details of the derivation is available in Ref. [9]. The mass-symmetric deformation of $\beta_{sym}=1.65$ results in a TKE formula for the symmetric fission as,

$$\text{TKE}_{sym} = 0.1173 \times \left(Z_f^2/A_f^{\frac{1}{3}}\right) + 7.5 \quad (\text{MeV}). \tag{1}$$

And the mass-asymmetric deformation of $\beta_{asym}=1.53$ gives a TKE formula for the asymmetric fission process as,

$$\text{TKE}_{asym} = 0.1217 \times \left(Z_f^2/A_f^{\frac{1}{3}}\right) + 3.5 \quad (\text{MeV}). \tag{2}$$

The TKE in fission can be approximated by a linear function of $Z_f^2/A_f^{1/3}$ as proposed in early 1960's [19] by Viola et al. The comparison of the present and Viola's TKE formulas has been made in Ref. [9]. The disparity between the Viola's and the present TKE-formulas is that (1) the former was obtained from a linear fitting of the available experimental data, and the latter was derived from a defined physical quantity, the deformation parameter β of the nucleus; (2) The present ones give the TKE functions separately for the symmetric and asymmetric fission process.

In the region with $A_f = 280 \sim 290$, fission events resulting from the symmetric and asymmetric valleys may not be disentangled, and the concept of the two modes fission in this region may need to be modified, as discussed above. The possibility that the two fission valleys may merge into one leads to the expected property for the $^{293}118$ fission to be the symmetric fission having a broad mass-yield distribution and the average total kinetic energy of about 270-280 MeV (with $\beta=1.55\text{-}1.50$ for the eq.(1)) even though the prediction of Viola's formula gives a TKE of 256 MeV. A similar TKE prediction results in 250-260 MeV for the nucleus of $^{290}114$.

V. SUMMARY

The fission properties of heavy and superheavy elements were studied based on the degree of the deformation in the fission process of nuclei from the preactinide through the actinide and up to transactinide elements. For either of the mass-symmetric and mass-asymmetric deformations, the degree of deformation of the fissioning nuclei near the scission point is constant for a wide region of A_f. For a given nucleus, the degree of deformation is also independent of the mass division within each fission mode. The identical deformation parameter of fissioning nuclei leads a new TKE formula separately for the symmetric and asymmetric fission. It is pointed out that in the superheavy elements around the $A=280-290$ region, the symmetric and asymmetric fission valleys may merge into one.

The present results raise new questions: What are the critical factors that bring about the identical value of the elongation at the point of scission in nuclear fission?!

ACKNOWLEDGEMENTS

One of the authors (Y.L.Z) wishes to thank Prof. M. Ebihara of TMU/Tokyo for assistance. Y.L.Z also wishes to thank Dr. Ken Gregorich of LBNL/Berkeley for discussions on the experimental and theoretical aspects of the production of superheavy elements. The extensive discussions on theoretical results for the fission paths in actinide nuclei with Profs. Ohta and Wada of Konan University, Dr. Yamaji of RIKEN and Prof. Abe of Kyoto University are gratefully acknowledged. This work was partly supported by a grant from the Japan Society for the Promotion of Science (Y.L.Z).

REFERENCES

1. Yu. Ts. Oganessian, A. V. Yeremin, A. G. Popeko, S. L. Bogomolov, G. V. Buklanov, M. L. Chelnokov, V. I. Chepigin, B. N. Gikal, V. A. Gorshkov, G. G. Gulbekian, M. G. Itkis, A. P. Kabachenko, A. Yu. Lavrentev, O. N. Malyshev, J. Rohac, R. N. Sagaidak, S. Hoffman, S. Saro, G. Giardina, and K. Morita, Nature, **400**, 242 (1999).
2. V. Ninov, K. E. Gregorich, W. Loveland, A. Ghiorso, D. C. Hoffman, D. M. Lee, H. Nitsche, W. J. Swiatecki, U. W. Kirbach, C. A. Laue, J. L. Adams, J. B. Patin, D. A. Shaughnessy, D. A. Strellis, and P. A. Wilk, Phys. Rev. Lett. **83**, 1104 (1999).
3. M. Schädel, W. Brüchle, R. dressler, B. Eichler, H. W. Gäggeler, R. G'unther, K.E.Gregorich, D.C. Hoffman, S.Hübener, D.T.Jost, J.V.Kratz, W.Paulus, D.Schumann, S. Timokhin, N. Trautmann, A. Türler, G. Wirth, A. Yakuschev, Nature **388**, 55 (1997).
4. P. Reiter, T. L.Khoo, T.Lauritsen, C.J.Lister, D.Seweryniak, A.A.Sonzogni, I.Ahmad, N.Amzal, P.Bhattacharyya, P.A.Butler, M.P.Carpenter, A.J.Chewter, J.A.Cizewski, C.N.Davids, K.Y.Ding, N.Fotiades, J.P.Greene, P.T.Greenlees, A.Heinz, W.F.henning, R.D.Herzberg, R.V.F.Janssens, G.D.Jones, H.Kankaanpää,

F.G.Kondev, W.Korten, M. Leino, S.Siem, J.Uusitalo, K.Vetter, I.Wiedenhöver, *Phys. Rev. Lett.* **84**, 3542 (2000).
5. E.K. Hulet, J.F. Wild, R.J. Dougan, R.W. Lougheed, J.H. Landrum, A.D. Dougan, M. Schädel, R.L. Hahn, P.A. Baisden, C.M. Henderson, R.J. Dupzyk, K.Sümmerer, and G.R. Bethune, *Phys. Rev. Lett.* **56**, 313 (1986).
6. D. C. Hoffman, and M. R. Lane, *Radiochim. Acta.* **70/71**, 135 (1995).
7. V.E. Viola, K. Kwiatkowski, and M. Walker, *Phys. Rev. C* **31**, 1550 (1985).
8. Y. L. Zhao, I. Nishinaka, Y. Nagame, M. Tanikawa, K. Tsukada, S. Ichikawa, K. Sueki, Y. Oura, H. Ikezoe, S. Mitsuoka, H. Kudo, and H. Nakahara, *Phys. Rev. Lett.* **82**, 3408 (1999).
9. Y. L. Zhao, Y. Nagame, I. Nishinaka, K. Sueki, H. Nakahara, *Phys. Rev. C* **62**, 014612 (2000).
10. Y. Nagame, I. Nishinaka, K. Tsukada, Y. Oura, S. Ichikawa, H. Ikezoe, Y.L. Zhao, K. Sueki, H. Nakahara, M. Tanikawa, T. Ohtsuki, H. Kudo, Y. Hamajima, K. Takamiya, Y.H. Chung, *Phys. Lett. B* **387**, 26 (1996).
11. Y. L. Zhao, Y. Nagame, I. Nishinaka, K.Tsukada, K. Sueki, Y. Oura, H. Nakahara, S. Ichikawa, H. Ikezoe, M.Tanikawa, T. Ohtsuki, and H.Kudo, *J. Alloys and Comp.* **271** , 327 (1998).
12. Y. L. Zhao, I. Nishinaka, Y. Nagame, M. Tanikawa, S. Goto, K. Sueki, K. Tuskada, S. Ichikawa, H. Nakahara, *Phys. Rev. C*, to be published.
13. Y. Nagame, I. Nishinaka, K.Tsukada, Y. Oura, S. Ichikawa, H. Ikezoe, Y. L. Zhao, K. Sueki, H. Nakahara, M.Tanikawa, T. Ohtsuki, H.Kudo, Y. Hamajima, K. Takamiya, and Y. H. Chung, *Radiochim. Acta*, **78**, 3 (1997).
14. M. G. Itkis, N. A. Kondrat'ev, S. I. Mul'gin, V. N. Okolovich, A. Ya. Rusanov, and G. N. Smirenkin, *Sov. J. Nucl. Phys.* **52**(4), 601 (1990).
15. T. Wada, Y. Abe, and N. Carjan, *Phys. Rev. Lett.* **70**, 3538 (1993).
16. Y. Abe, S. Ayik, P.G. Reinhard, and E. Suraud, *Phys. Reports* **275**, 49 (1996).
17. P. Möller and A. Iwamoto, *Phys. Rev. C* **61**, 047602 (2000).
18. S. Ćwiok, P. Rozmej, A. Sobiczewski and Z. Patyk, *Nucl. Phys.* **A491**, 281 (1989).
19. V. E. Viola, and T. Sikkeland, *Phys. Rev.* **130**, 2044 (1963).

Five-Dimensional Potential-Energy Surfaces and Coexisting Fission Modes in Heavy Nuclei

Peter Möller*, David G. Madland*, Arnold J. Sierk*, and
A. Iwamoto†

*Theoretical Division, Los Alamos National Laboratory, Los Alamos, New Mexico 87545, USA
†Department of Materials Sciences, Japan Atomic Energy Research Institute, Tokai-mura, Naka-gun, Ibaraki, 319-1195 Japan

Abstract. We calculate complete fission potential-energy surfaces versus five shape coordinates: elongation, neck diameter, light-fragment deformation, heavy-fragment deformation, and mass asymmetry for even nuclei in the range $82 \leq Z \leq 100$. The potential energy is calculated in terms of the macroscopic-microscopic model with a folded-Yukawa single-particle potential and a Yukawa-plus-exponential macroscopic model in the three-quadratic-surface parameterization. The structure of the calculated energy landscapes exhibits multiple valleys leading to different scission configurations. The properties of these valleys and the saddle-points at the beginning of these valleys can be directly related to bimodal fission properties observed in the radium region, in the light-actinide region, and in the fermium region [1-4]. The rms deviation between calculated and experimental fission-barrier heights is only 1.08 MeV for 31 nuclei from ^{70}Se to ^{252}Cf.

INTRODUCTION

When a heavy nucleus divides into two fragments in nuclear fission, two key aspects of the process have challenged researchers since the discovery of fission more that 60 years ago. First, what is the threshold energy for the reaction and, second, what are the shapes involved in the transition from a single nuclear system to two separated daughter fragment nuclei? These two questions are intimately connected. The energy of a nucleus as a function of shape defines a landscape in a multi-dimensional deformation space. It is the energy of the lowest mountain pass, or saddle-point, in this landscape, connecting the nuclear ground state with the region corresponding to separated fragments that represents the threshold energy of the fission process.

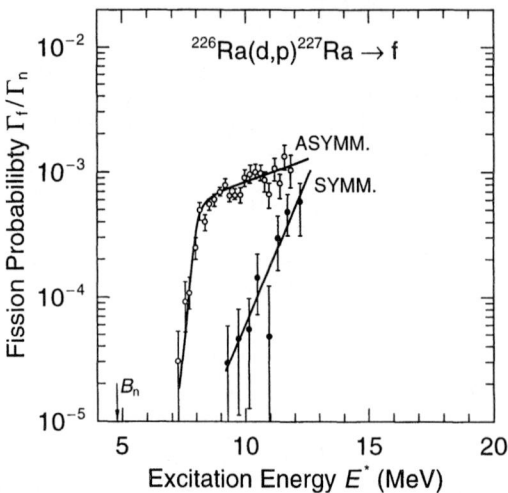

FIGURE 1. Fission probability data show different thresholds for mass-asymmetric and mass-symmetric fission near ^{227}Ra. The figure is based on a figure in Ref. [1]).

However, despite many fission potential-energy-surface calculations over the years certain features of the fission process have remained unexplained. For example:

1. Nuclei near ^{228}Ra exhibit two fission modes. We show in Fig. 1 an example of the extensive data obtained in Ref. [1]). In one mode, with the lower threshold energy, the fragment mass distribution is asymmetric and the fragment total kinetic energy is about 10 MeV higher than in the other, symmetric mode. The kinetic energies indicate that the scission configuration is more compact for the asymmetric mode than for the symmetric mode. From the totality of the data Ref. [1]) concludes: "Thus it seems that after the gross determination of the symmetric or asymmetric character of fission made already at the barrier, the two components follow a different path with no or little overlap in the development from the barrier to the scission configuration."

2. Most actinide nuclei near the line of β-stability undergo mass-asymmetric fission. From Th to Fm the heavy fragment mass is close to 140, with the remainder of the mass in the light fission fragment.

3. Near the far end of the actinide region fission properties change suddenly and sometimes exhibit a two-mode character in the same nucleus. For example, the fragment mass distribution changes abruptly from mass-asymmetric for ^{256}Fm to mass symmetric for ^{258}Fm along with a correlated increase in the fragment total kinetic energy (TKE) by about 35 MeV. But ^{258}Fm also exhibits the asymmetric mode with lower TKE with a small probability: fission of such nuclei is characterized as bimodal.

Over the past decades many calculations based on 1000 or so grid points have been presented. However, to properly describe the evolution of a single nuclear shape into two fragments of different mass and deformation, for example one spherical ^{132}Sn-like fragment and one deformed fragment, we have concluded that complete deformation spaces based on at least five independent shape parameters are required [5]). This leads to multi-million grid-point spaces.

SHAPE PARAMETERIZATION

Because fragment shell effects strongly influence the structure of the fission potential-energy surface long before scission, often in the outer saddle region, it is crucial to include in calculations the nascent-fragment deformations as two independent shape degrees of freedom. In addition, elongation, neck diameter, and mass-asymmetry shape-degrees of freedom are required, at a minimum, to adequately describe the complete fission potential-energy surface. For nascent-fragment deformations we choose spheroidal deformations characterized by Nilsson's quadrupole ϵ parameter. This single fragment-deformation parameter is sufficient because higher-multipole shape-degrees of freedom are usually of lesser importance in the fission-fragment mass region below the rare earths.

The three-quadratic-surface parameterization (3QS) is ideally suited for the above description.[6]) In the 3QS the shape of the nuclear surface is specified in terms of three smoothly joined portions of quadratic surfaces of revolution. Using this parameterization we here construct, calculate, and investigate complete five-dimensional spaces with 2 610 885 grid points as illustrated in Fig. 2.

A common notation used to characterize the fragment mass asymmetry of a fission event is M_H/M_L where M_H and M_L are the masses of the heavy and light fission fragments respectively. For the purpose of grid generation for the potential-energy calculation it is convenient to relate a mass-asymmetry shape degree of freedom for the pre-scission nucleus to the final fission-fragment mass asymmetry in some fashion, although the final mass division, strictly speaking, cannot be determined from the static shapes occurring before scission. However, the exact nature of our definition of mass asymmetry for a single shape has little effect on the calculated saddle-point energies and shapes because our five-dimensional grid covers all of the physically relevant space available to the 3QS parameterization, regardless of how we choose to define a "mass-asymmetry" coordinate. In order to obtain a definition of mass asymmetry that is meaningful close to scission, and equations that are reasonably simple to work with for the purpose of grid-point generation, we define an auxiliary grid mass-asymmetry parameter α_g

$$\alpha_g = \frac{M_1 - M_2}{M_1 + M_2} \quad (1)$$

where M_1 and M_2 are the volumes inside the end-body quadratic surfaces, were they completed to form closed-surface spheroids. Thus

$$\alpha_g = \frac{a_1^2 c_1 - a_2^2 c_2}{a_1^2 c_1 + a_2^2 c_2} \quad (2)$$

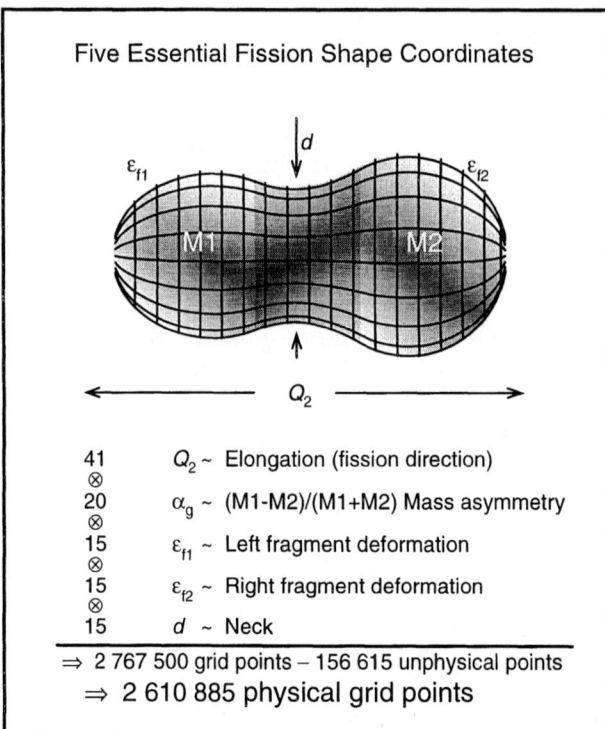

FIGURE 2. Five-dimensional shape parameterization used in our potential-energy calculation. Different shades of gray indicate the three different quadratic surfaces of the shape parameterization used in our calculation. The first derivative is continuous at the intersections of the surfaces. For the nascent spheroidal fragments we characterize the deformations by Nilsson's quadrupole ϵ parameter. Shapes corresponding to certain quadrupole moments do not exist for specific combinations of the other shape parameters. For example, zero quadrupole moment cannot be realized for shapes with very deformed ends. In our grid there exist 156 615 such "unphysical" points. Thus, we are left with 2 610 885 shapes for which we actually calculate the potential energy.

where a denotes the transverse semi-axis and c the semi-symmetry axis of the left (1) and right (2) quadratic surfaces of revolution. With this definition we select 20 coordinate values corresponding to

$$\alpha_g = -0.02\ldots(0.02)\ldots 0.36 \tag{3}$$

We have closely spaced the asymmetry coordinate so that we will be able to spot favorable saddle-point shapes that may not appear in a more sparsely spaced grid. For ^{240}Pu the values 0.00, 0.02, and 0.36 of the mass-asymmetry coordinate α_g correspond to the mass divisions 120/120, 122.4/117.6, and 163.2/76.8, respectively.

Because of the intuitive appeal of the notation M_H/M_L we use it below to characterize the "asymmetry" of a single shape. We then connect M_H and M_L to α_g through

$$M_\text{H} = A\frac{1+\alpha_\text{g}}{2} \quad \text{and} \quad M_\text{L} = A\frac{1-\alpha_\text{g}}{2} \tag{4}$$

for a nucleus with A nucleons. For shapes with a well-developed neck the ratio obtained with this definition can be expected to be close to the final fragment mass-asymmetry ratio. We cannot conveniently use M_1 and M_2 to designate the final fragment mass asymmetries because they do not exactly sum up to the total nuclear volume or mass. Equation (4) simply represents a scaling of M_1 and M_2 so that their sum after scaling adds up to the total mass number A.

ANALYSIS OF FIVE-DIMENSIONAL SPACES

It is a common misconception that the structure of a multi-dimensional potential-energy function can be determined by calculating and displaying the function versus two shape variables, for example, β_2 and β_3 where the function has been "minimized" with respect to additional multipoles such as β_4, β_5, β_6 and β_7.

Figure 3 illustrates, in two dimensions, some of the difficulties that occur in such a search for the relevant fission threshold saddle points in a multidimensional potential-energy landscape. Let Θ represent a coordinate in the fission direction and α all other coordinates and furthermore let the area at $\Theta = -100$ and $\alpha = 6.0$ represent the second minimum in the fission potential energy surface. And let the area at $\Theta = 100$ and $\alpha = -5.0$ represent the valley of separated fission fragments. In many calculations based on "minimizations" with respect to additional coordinates the procedure would now be to increase the fission coordinate Θ by some increment, which for clarity we choose to be as large as 40, while keeping the additional shape-degrees of freedom fixed. This would take us along the initial part of the thin horizontal line to $\Theta = -60$ and $\alpha = 6.0$. Starting from this position the energy would now be minimized for fixed $\Theta = -60$. This would take us down to the large black dot on the dot-dashed line. The process would then be repeated with the result that the energy would be obtained for the succession of black dots on the dot-dashed line. Thus the energy that is obtained is the energy along the dot-dashed line. In our example here all the shape coordinates along this trajectory would vary continuously and no suspicion may arise that the saddle point found, in the region of the white arrow, *is not* the true saddle point, that is, it is not the *lowest* pass leading to the fission valley of separated fragments to the right in the figure. A minimum-energy trajectory should instead take us over the saddle-points identified by the gray arrows in the figure. We can easily "see" this in a two-dimensional case as the example here, but in a higher-dimensional space it takes a clever algorithm to identify all of the relevant structure. This requirement led to the water-flow algorithm described below.

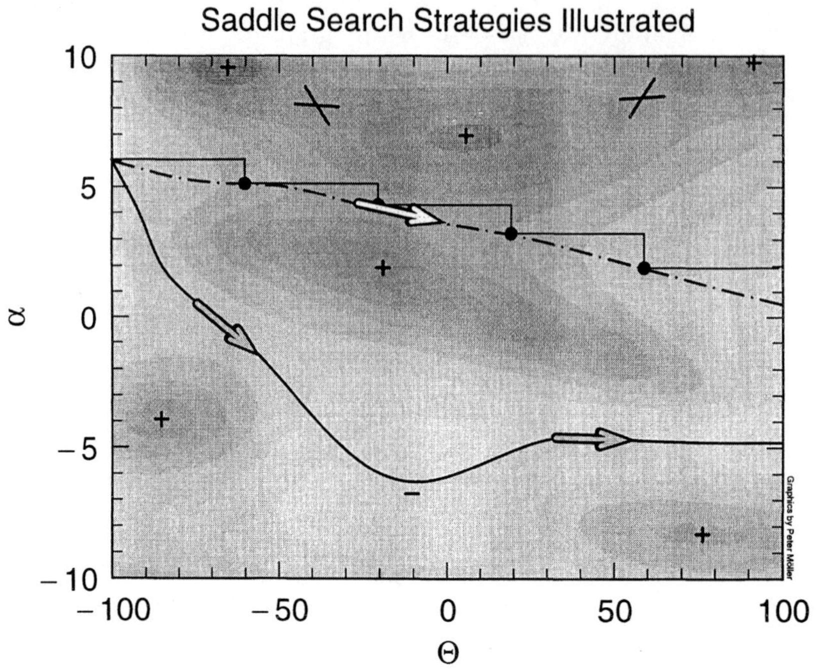

FIGURE 3. Maxima (+), minima (−), and saddle points (arrows or crossed lines) of a two-dimensional function. As discussed in the text it is not possible to obtain a lower-dimensional representation of this surface by "minimizing" with respect to the "additional" α shape-degree of freedom.

A numerical algorithm that locates the saddle points by the criteria that all first derivatives be zero and that the second derivatives have the appropriate signs would locate the saddles marked by crossed lines in the upper part of the figure in addition to the saddles we have already discussed. It would then be difficult to determine if any of these represented the lowest mountain pass between the second minimum and the blue area to the right representing the fission valley of separated fragments. However, as we will show the water-flow method simply bypasses these saddle points.

It is *also* a common misconception that *constrained* self-consistent calculations, for example HF or HFB calculations with Skyrme or Gogny forces[7-9]) automatically take into account all non-constrained variables. For the application to saddle-point determination this is incorrect. A self-consistent calculation constrained in one variable, for example Q_2, would have difficulties similar to those discussed above.

In addition, it is of interest to note that in calculations where the potential energy is displayed as contour diagrams versus two shape variables and in which the energy is minimized with respect to additional multipoles, only relatively few points are required to perform a minimization with respect to, say, 3 additional multipoles, about 30 or so. If the two-dimensional contour diagram is based on 10 by 10 points then only 3 000 points are considered in the calculation. In contrast, we find that to adequately investigate the structure associated with five simultaneous shape-degrees of freedom almost 3 000 000 grid points, that is, 1000 *times* more points than earlier calculations purporting to be multi-dimensional are required.

The technique we use here to investigate the structure of the multidimensional surface is to employ imaginary water flows [5,10] in the calculated 5-dimensional potential energy surface. For example, we imagine that we stepwise flood, in intervals of 1 MeV, the second minimum with water. During the flooding process we check at what water level a preselected "exit" grid point that is clearly in the fission valley near scission gets "wet". When this happens, then the water level has passed the threshold energy level for fission. We can determine the saddle-point energy to desired accuracy by repeating the filling procedure with successively smaller stepwise increases of the water level. In the second such iteration we only need to start the filling procedure at 1 MeV below the level for which the exit point became wet in the first iteration with its 1 MeV stepwise increases in the flooding level, and we then use a 0.1 MeV stepwise increase in the water level. Once the exit point again gets flooded we can again repeat the procedure with a smaller stepwise water-level increase until we have determined the saddle-point energy to desired accuracy. The saddle-point shape can also be obtained from this procedure.

Once the threshold energies for fission have been identified, it is of interest to establish if structure effects in the potential energy provide a mechanism for multi-mode fission, such as the well-known three-peaked mass distribution in ^{228}Ra fission [1]). To look for such structures we ask if there are valleys of distinctly different character running in the fission direction of increasing Q_2. For 10 or more fixed Q_2 values beyond the outer saddle region, we determine all minima in the remaining 4-dimensional space of the two fragment deformations, neck size and mass asymmetry. We find that there are usually two (but sometimes more) distinct valleys in the region beyond the second saddle region, one corresponding to a mass asymmetry α_g of about $[140 - (A - 140)]/A$ and one corresponding to mass symmetry $\alpha_g = 0$. To understand the significance of these valleys it is necessary to study their interconnections in the five-dimensional deformation-energy space.

Variations of the flooding algorithm allow us to determine that separate saddle points provide entries to the two valleys and the respective energies of these saddle points. Once the lowest saddle has been determined we may block the water flow across this saddle by building an imaginary dam across the saddle region. We can also totally block the water flow beyond a selected maximum Q_2. This prevents water from flowing down one valley and up "the back way" into the other valley. To determine the height of the ridge between the two valleys along their entire length we study for each fixed Q_2 the remaining 4-dimensional space in which the two valleys correspond to two minima and the ridge to the saddle separating them. We use the flooding algorithm in four dimensions to localize this saddle/ridge.

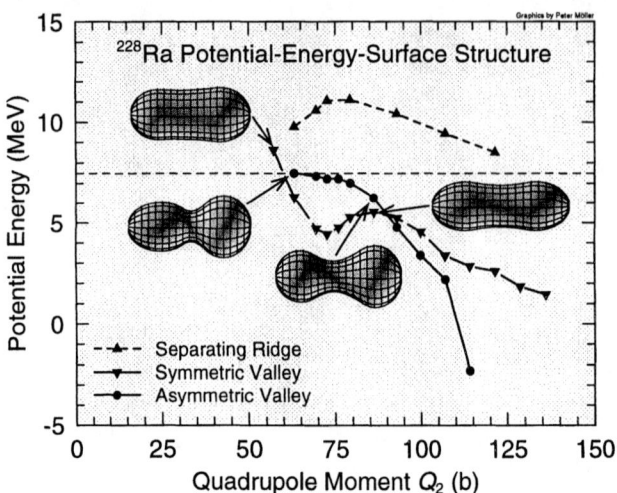

FIGURE 4. Potential-energy valleys and ridges and corresponding nuclear shapes for ^{228}Ra. The first point on each of the two curves with the label "valley" are actually saddle-points at the entrance to the valley that emerges beyond the saddle point. It is of interest to note that the entrance to the symmetric valley is slightly asymmetric. The subsequent points in this valley correspond to symmetric or very nearly symmetric shapes. The entry saddle-point to the symmetric valley is 1.13 MeV higher than the entry saddle-point to the asymmetric valley. The highest point on the separating ridge is 2.47 MeV higher than the symmetric saddle. The thin dashed line represents the threshold energy for fission. All energies are given relative to the spherical macroscopic energy.

RESULTS

We have calculated five-dimensional potential-energy surfaces for 138 even-even nuclei from Pb to Fm. We are currently subjecting these surfaces to various types of imaginary water-flow analyses as discussed in the previous section.

As examples of the structures we have found in the calculated 5-dimensional surfaces we show in Figs. 4 and 5 some fission-valley and separating-ridge features obtained for ^{228}Ra and ^{232}Th. The first point on the fission-valley potential-energy curves in Figs. 4 and 5 is the saddle point for entry into the particular valley. The nuclear shapes corresponding to the saddle points are shown to the left in the figure. Shapes corresponding to the symmetric and mass-asymmetric valleys at $Q_2 = 86$ b are shown to the right. Note that the shape corresponding to the *entry* to the mass-*symmetric* valley is slightly mass-*asymmetric*. The thin dashed line is the calculated threshold potential energy for fission which, to be consistent with the other curves, is given relative to the spherical macroscopic energy.

The calculated structure of the potential-energy surface therefore is consistent with the observed bimodal fission features in this region of nuclei [1,2]. The high ridge separating the two valleys for ^{228}Ra is peaked at 2.47 MeV above the en-

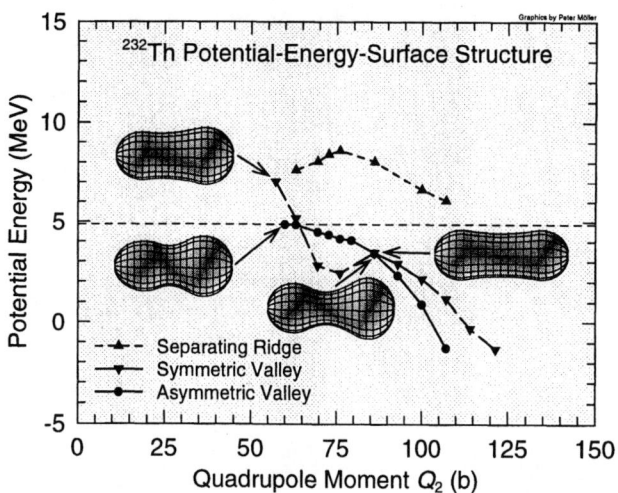

FIGURE 5. Potential-energy valleys and ridges and corresponding nuclear shapes for ^{232}Th. The first point on each of the two curves with the label "valley" are actually saddle-points at the entrance to the valley that emerges beyond the saddle point. It is of interest to note that the entrance to the symmetric valley is slightly asymmetric. The subsequent points in this valley correspond to symmetric or very nearly symmetric shapes. The entry saddle-point to the symmetric valley is 2.17 MeV higher than the entry saddle-point to the asymmetric valley. The highest point on the separating ridge is 1.56 MeV higher than the symmetric saddle. The thin dashed line represents the threshold energy for fission. All energies are given relative to the spherical macroscopic energy.

trance saddle to the symmetric valley. It therefore keeps the mass-symmetric and mass-asymmetric modes well separated up to scission, which is consistent with the experimentally observed data discussed in the introduction. Compare also with Fig. 1. Our results in Fig. 4 are *also* consistent with the observed total fragment kinetic energies which are about 10 MeV higher for asymmetric fission than for symmetric fission for several nuclei in this region[1]).

For ^{232}Th the lower separating ridge, peaked at 1.56 MeV above the entrance saddle to the symmetric valley, allows the symmetric component to partially revert back to the asymmetric valley before scission for ^{232}Th. Therefore, there is only a very weak symmetric fission component in low-energy fission of ^{232}Th. We find that the existence of at least two paths in the five-dimensional surface is a general result for nuclei in this region and we are now studying their relative importance over the large range of nuclei for which we have calculated potential-energy surfaces. We note that experimental fission data in the light-actinide region are currently interpreted in terms of two fission paths, one mass symmetric and the other mass asymmetric. The saddle leading to mass-symmetric division is observed to be 1 to 2 MeV higher than the saddle leading to mass-asymmetric division for nuclei in this region, in excellent agreement with our calculated potential-energy surfaces. Also, the experimental total fragment kinetic energies are higher in asymmetric

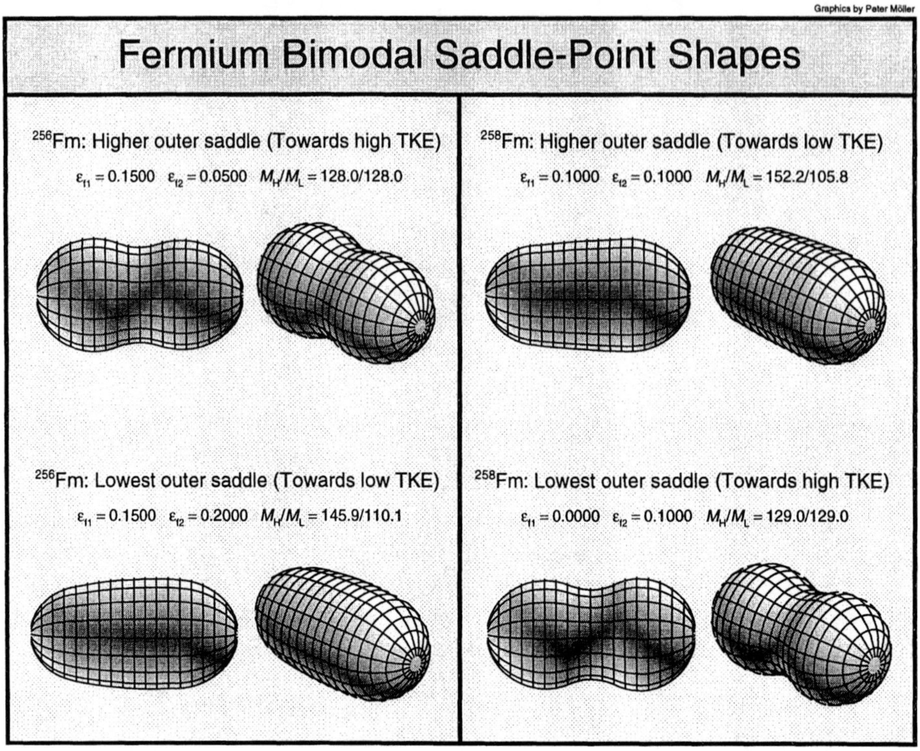

FIGURE 6. Bimodal saddle-point shapes for ^{256}Fm and ^{258}Fm.

fission than in symmetric fission. These observations[2,3] are consistent with the compact and elongated shape configurations that we obtain in the corresponding fission valleys.

Turning to the heavy actinides, nuclei in the region near ^{258}Fm also exhibit bimodal features in fission as discussed in Ref.[4]. We have earlier tentatively identified bimodal structures in calculated *two-dimensional* potential-energy surfaces [11,12], but only now can we verify that these interpretations are still valid when the calculation is taken from two to five dimensions. In the Fm region we have used one of the imaginary water-flow techniques described previously, namely the dam method, to find alternative saddle points that are higher in energy than the lowest threshold saddle point. For ^{256}Fm and ^{258}Fm we find the two distinct classes of saddle points shown in Fig. 6. For ^{256}Fm the shape of the lowest saddle indicates it corresponds to normal, low-TKE fission similar to what is observed in fission of slightly lighter actinides. However, another saddle point exists, which we calculate to be 0.30 MeV higher than the lower saddle point. This may correspond to fission into compact scission configurations with high kinetic energies. For ^{258}Fm the latter type of saddle-point becomes the lowest saddle point. Thus, we reproduce the experimentally observed transition point between asymmetric low-TKE fission and symmetric high-TKE fission as observed experimentally[4].

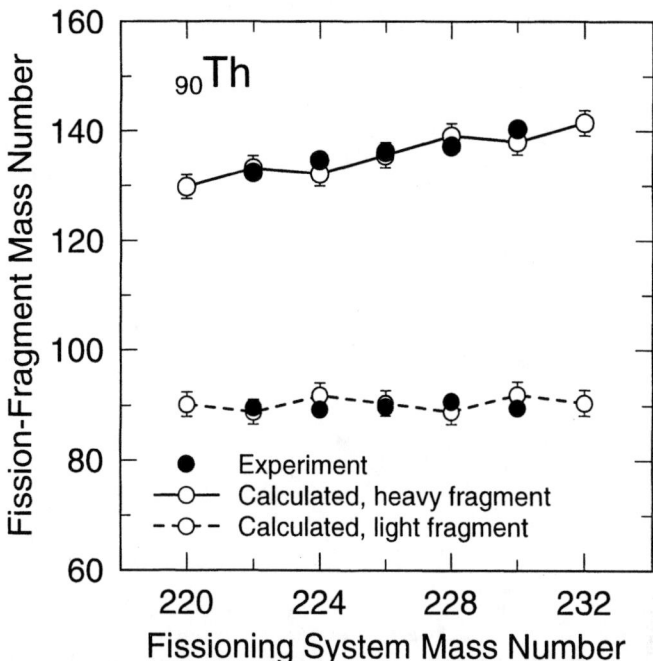

FIGURE 7. Calculated (open symbols) and measured (closed symbols) average mass division in asymmetric fission for a sequence of even isotopes of Th. The error bars on the calculated points correspond to the spacing of mass asymmetry values on the multidimensional shape-coordinate grid. The data is for spontaneous fission when it is available, otherwise data for low-energy induced fission is used. The results reproduce the experimental observation of a heavy fragment at mass number $A \sim 140$ and a light fragment with mass corresponding to the remainder of the original nucleus. However, deviations from this rule of thumb are also reproduced by the calculations.

As pointed out in the Introduction, it is a long-standing observation that in binary fission actinide nuclei preferentially divide into one fragment of about mass 140 and a complementary, smaller fragment of mass $A - 140$, where A is the mass number of the original nucleus. We show in Fig. 7 our calculated results for the fragment masses for the mass-asymmetric valley in fission for seven thorium isotopes. The data are from Refs. [13–15]. For all isotopes we have identified the fission valley corresponding to mass-asymmetric fission at $Q_2 = 99$ b. The value of the mass-asymmetry coordinate α_g at the valley bottom directly yields the mass of the heavy and light fission fragments according to Eq. (4). The "valley floor" corresponds to a local minimum in the four-dimensional space remaining when Q_2 is fixed at a specific value. In Fig. 8 we have plotted the charge asymmetry for some isotopes of Th and U, cf. Ref. [15]. We observe that the charge of the heavy fragment remains constant, whereas there is a substantial variation in the heavy-fragment neutron number along the Th and U isotope chains.

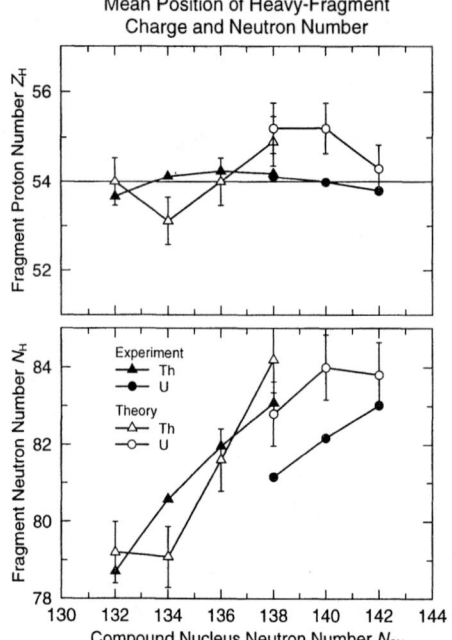

FIGURE 8. Calculated proton number and neutron number of the heavy fission fragment of isotopes of Th and U compared to experiment for the mass-asymmetric fission mode. The error bars correspond to the spacing of the mass-asymmetry shape coordinate.

Our results in this paper are based on calculations with the FRLDM 1992 parameter set [16]). Because of our expanded and more realistic deformation space our calculated barrier heights are systematically lowered relative to earlier, limited-space calculations. To obtain optimum agreement with experiment a readjustment of the FRLDM model parameters is therefore necessary. We have recently performed such a readjustment of the FRLDM model parameters to nuclear ground-state masses and barrier heights in the manner described in Ref. [16]). We obtain an overall barrier rms error of only 1.08 MeV and an overall mass model error of 0.759 MeV. The new calculated and experimental barrier heights are listed in Table 1.

SUMMARY

Our current analysis of the new calculated potential energy landscapes in five dimensions allows us to draw the following conclusions:

1. Multiple fission paths are found for most nuclei in the mass range considered.

2. For radium and light actinide nuclei two paths dominate: one mass-asymmetric and one mass-symmetric. These paths correspond to different fission modes, such as those illustrated in Fig. 1 in the Introduction.

TABLE 1

Calculated "outer" barrier heights compared to experimental barrier data, after readjustment of the macroscopic-model constants. For proton number below 80 the barriers are the macroscopic barriers. A typical experimental barrier uncertainty is 1 MeV.

Z	A	E_B^{exp} (MeV)	E_B^{calc} (MeV)	$E_B^{\text{exp}} - E_B^{\text{calc}}$ (MeV)	Z	A	E_B^{exp} (MeV)	E_B^{calc} (MeV)	$E_B^{\text{exp}} - E_B^{\text{calc}}$ (MeV)
34	70	39.4000	38.4792	0.9208	92	238	5.5000	5.6895	-0.1895
34	76	44.5000	44.7815	-0.2815	92	240	5.5000	6.4866	-0.9866
42	90	40.9200	41.7603	-0.8403	94	236	4.5000	4.5137	-0.0137
42	94	44.6800	45.1974	-0.5174	94	238	5.0000	4.5383	0.4617
42	98	45.8400	47.8580	-2.0180	94	240	5.1500	5.0469	0.1031
80	198	20.4000	22.1022	-1.7022	94	242	5.0500	5.7499	-0.6999
84	210	21.4000	22.2199	-0.8199	94	244	5.0000	6.4935	-1.4935
84	212	19.5000	20.4383	-0.9383	94	246	5.3000	7.1958	-1.8958
88	228	8.1000	7.8328	0.2672	96	242	5.0000	4.4339	0.5661
90	228	6.5000	6.9476	-0.4476	96	244	5.1000	5.1740	-0.0740
90	230	7.0000	6.0608	0.9392	96	246	4.8000	6.0138	-1.2138
90	232	6.2000	5.8163	0.3837	96	248	4.8000	6.6025	-1.8025
90	234	6.5000	5.5551	0.9449	96	250	4.4000	6.1782	-1.7782
92	232	5.4000	4.8495	0.5505	98	250	3.6000	6.0674	-2.4674
92	234	5.5000	5.0518	0.4482	98	252	4.8000	5.8170	-1.0170
92	236	5.6700	5.1745	0.4955					

3. The difference in energy between the symmetric and asymmetric saddle points in our calculated potential-energy surfaces is one to two MeV, which is consistent with the experimentally deduced differences.

4. The shapes we calculate for nuclei evolving in the mass-asymmetric and mass-symmetric valleys are consistent with the total fragment kinetic energies observed for these modes.

5. The long observed mass split in mass-asymmetric fission with an approximately constant heavy fragment mass near $A = 140$ is reproduced in our calculations.

6. Calculated fission-barrier heights agree very well with experimentally extracted barriers for nuclei from ^{70}Se to ^{252}Cf.

Except for Table 1 and the attendant discussion, these results have been obtained in our standard finite-range liquid-drop potential-energy model used for the calculation of nuclear masses. No change in the model or its parameters have been made in the current calculation, relative to its 1992 specification in Ref. [16]).

The calculations on which the results in this paper are based were carried out on the cluster of 4 CPUs at the TANDEM accelerator in JAERI in the winter of 1998–1999 and subsequently on the 140 processor AVALON cluster at Los Alamos. Results of the investigations at JAERI are discussed in Ref. [5]). This research is supported by the US DOE under contract W-7405-ENG-36.

References

1) E. Konecny, H. J. Specht, and J. Weber, Proc. Third IAEA Symp. on the Physics and Chemistry of Fission, Rochester, 1973, vol. II (IAEA, Vienna, 1974) p. 3.

2) Y. Nagame, I. Nishinaka, K. Tsukada, S. Ichikawa, H. Ikezoe, Y. L. Zhao, Y Oura, K. Sueki, H. Nakahara, M. Tanikawa, T. Ohtsuki, K. Takamiya, K. Nakanishi, H. Kudo, Y. Hamajima, and Y. H. Chung, Radiochimica Acta, **78** (1997) 3.

3) Y. L. Zhao, I. Nishinaka, Y. Nagame, M. Tanikawa, K. Tsukada, S. Ichikawa, K. Sueki, Y. Oura, H. Ikezoe, S. Mitsuoka, H. Kudo, T. Ohtsuki, and H. Nakahara, Phys. Rev. Lett. **82** (1999) 3408.

4) E. K. Hulet, J. F. Wild, R. J. Dougan, R. W. Lougheed, J. H. Landrum, A. D. Dougan, M. Schädel, R. L. Hahn, P. A. Baisden, C. M. Henderson, R. J. Dupzyk, K. Sümmerer, and G. R. Bethune, Phys. Rev. Lett. **56** (1986) 313.

5) P. Möller and A. Iwamoto, Phys. Rev. **C 61** (2000) 47602.

6) J. R. Nix, Nucl. Phys. **A130** (1969) 241.

7) S. Åberg, H. Flocard, and W. Nazarewicz, Ann. Rev. Nucl. Sci. **40**, 439 (1990).

8) J. L. Egido, L. M. Robledo, and R. R. Chasman, Phys. Lett. **B393**, 13 (1997).

9) J. F. Berger, M. Girod, and D. Gogny, Nucl. Phys. **A502**, c85 (1989).

10) A. Mamdouh, J. M. Pearson, M. Rayet, and F. Tondeur, Nucl. Phys. **A644** (1998) 389.

11) P. Möller, J. R. Nix, and W. J. Swiatecki, Nucl. Phys. **A469** (1987) 1.

12) P. Möller, J. R. Nix, and W. J. Swiatecki, Nucl. Phys. **A492** (1989) 349.

13) D. C. Hoffman and M. M. Hoffman, Ann. Rev. Nucl. Sci. **24** (1974) 151.

14) L. Dematte, C. Wagemans, R. Barthelemy, P. Dhondt, and A. Deruytter, Nucl. Phys. **A617** (1997) 331.

15) K.-H. Schmidt, S. Steinhäuser, C. Böckstiegel, A. Grewe, A. Heinz, A. R. Junghans, J. Benlliure, H.-G. Clerc, M. de Jong, J. Müller, M. Pfüntzer, and B. Voss, Nucl. Phys. **A665** (2000) 221.

16) P. Möller, J. R. Nix, W. D. Myers, and W. J. Swiatecki, Atomic Data Nucl. Data Tables **59** (1995) 185.

LIST OF PARTICIPANTS

Ackermann, Dieter
Gesellschaft für Schwerionenforschung mbH,
Planckstrabe 1, D- 64291 Darmstadt,
Germany

d.ackermann@gsi.de

Adamian, Gurgen G.
Institut für Theoretische Physik der
Justus- Liebig- Universitiät,
D-35392 Giessen,
Germany

Gurgen.Adamian@physik.nui-giessen.de

Akimune, Hidetoshi
Department of Physics
Konan University
Okamoto 8-9-1, Higashinada
Kobe 658-8501,
Japan

akimune@konan-u.ac.jp

Aliotta, Marialuisa
Lehrstuhl für Experimentalphysik III, Universität Bochum,
D-44780 Bochum,
Germany

marialuisa@ep3.ruhr-uni-bochum.de

Arnould, Marcel
Institut d' Astronomie et d' Astrophysique
Université Libre de Bruxelles Campus
Plaine- CP 226, B- 1050 Brussels,
Belgium

marnould@astro.ulb.ac.be

Asai, Masato
Advanced Science Research Center,
Japan Atomic Energy Research Institute,
Tokai, Ibaraki 319-1195,
Japan

asai@tdmalphl.tokai.jaeri.go.jp

Azaiez, Faisal
Institut de Physique Nucléaire,
IN2P3-CNRS, 91406 Orsay Cedex,
France

azaiez@ipno.in2p3.fr

Blank, Bertram
CEN Bordeaux- Gradignan,
Le Haut- Vigneau,
F-33175 Gradignan Cedex,
France

blank@cenbg.in2p3.fr

Burrows, Adam
Department of Astronomy and Steward Observatory, The University of
Arizona, Tucson, AZ 85721
USA

aburrows@as.arizona.edu

Chubarian, Greg
Cyclotron Institute, Texas A & M
University, College Station, Texas 77843,
USA

Chubarian@leper.tamu.edu

Denisov, Vitali Yu.
GSI- Darmstadt,
Planckstrasse 1, D-64291 Darmstadt,
Germany

denisov@gsi.de

Flocard, Hubert C.
Theory Group, Institut de Physique
Nucléaire F- 91406 Orsay Cedex,
France

flocard@ipno.in2p3.fr

Fukushima, Akira
Department of Physics
Konan University
Okamoto 8-9-1, Higashinada
Kobe 658-8501,
Japan

dn921001@center.konan-u.ac.jp

Goriely, Stephane
Institut d' Astronomie et
d' Astrophysique Université Libre de
Bruxelles Campus de la Plaine, CP 226
1050 Brussels
Belgium

sgoriely@astro.ulb.ac.be

Grawe, Hubert
GSI, Darmstadt, Germany
Planckstrasse 1, D-64291 Darmstadt,
Germany

grawe@axp602.gsi.de

Guerreau, Daniel
GANIL
BP 5027
F-14076 Caen Cedex
France

gurreau@ganil.fr

Guet, Claude
Départment de Recherche Fondamentale
sur la Matiére Condensée,
CEA- Grenoble, 17, rue des Martyrs,
F-38054 Grenoble CEDEX 9,
France

cguet@cea.fr

Heger, Alexander
Astronomy Department, University of
California, Santa Cruz, CA 95064,
USA

alex@ucolick.org

Ichikawa, Takatoshi
Department of Physics
Konan University
Okamoto 8-9-1, Higashinada
Kobe 658-8501,
Japan

dn021001@center.konan-u.ac.jp

Ikezoe, Hiroshi
Advanced Science Research Center,
JAERI, Tokai, Ibaraki 319- 1195,
Japan

ikezoe@popsvr.tokai.jaeri.go.jp

Itahashi, Takahisa
Research Center for Nuclear Physics,
Osaka Univ., Ibaraki, Osaka,
Japan

itahasi@rcnp.osaka-u.ac.jp

Iwamoto, Akira
Advanced Science Research Center
Japan Atomic Energy Research Institute
Tokai, Naka, Ibaraki 319-01,
Japan

iwamoto@hadron01.tokai.jaeri.go.jp

Junker, Matthias
Laboratori Nazionali del Gran Sasso
S.S.17bis km 18+910, 67010 Assergi(AQ),
Italy

junker@lngs.infn.it

Kifonidis, Konstantinos
Max-Planck-Institut für Astrophysik,
Karl-Schwarzschild-Strasse 1,
D-85741 Garching,
Germany

kok@mpa-garching.mpg.de

Koura, Hiroyuki
Advanced Research Institute for Science
and Engineering, Waseda University
3-4-1 Okubo, Shinjuku-ku,
Tokyo 169-8555,
Japan

koura@mn.waseda.ac.jp

Kubo, Satomi
Department of Physics
Konan University
Okamoto 8-9-1, Higashinada
Kobe 658-8501,
Japan

satomi@center.konan-u.ac.jp

Langer, Norbert
Astronomical Institute, Utrecht
University,
The Netherlands

N.Langer@astro.uu.nl

Lewitowicz, Marek
GANIL
BP 5027
F-14076 Caen Cedex
France

lewitowicz@ganil.fr

Leygnier, Jerome
Laboratoire Aimé Cotton, C.N.R.S
UPR 3321, Bât. 505 Université Paris
Sud, 91405 Orsay Cedex,
France

jerome.leygnier@lac.u-psud.fr

Mathews, Grant J.
Center for Astrophysics, Department of
Physics, University of Notre Dame
Notre Dame, IN 46635,
USA

gmathews@nd.edu

Mengoni, Alberto
ENEA, Applied Physics Division, Via
Don Fiammelli 2, I-40129 Bologna,
Italy

Möller, Peter
Theoretical Division
Los Alamos National Laboratory
Los Alamos, New Mexico 87545
USA

moller@moller.lanl.gov

Morimoto, Kouji
The Institute of Physical and Chemical
Research (RIKEN), 2-1, Hirosawa,
Wako-shi, Saitama 351-0198,
Japan

morimoto@postman.riken.go.jp

Morita, Kosuke
The Institute of Physical and Chemical
Research (RIKEN), 2-1, Hirosawa,
Wako-shi, Saitama 351-0198,
Japan

morita@rikaxp.riken.go.jp

Nakamura, Masato
College of Science and Technology,
NIHON University, 7-24-1, Narashino-
dai, Funabashi 274-8501,
Japan

mooming@phys.ge.cst.nihon-u.ac.jp

Oganessian, Yuri Ts.
Flerov Laboratory of Nuclear Reactions,
Joint Institute for Nuclear Research,
141980 Dubna, Moscow region,
Russia

oganessian@sungraph.jinr.ru

Ohta, Masahisa
Department of Physics
Konan University
Okamoto 8-9-1, Higashinada
Kobe 658-8501,
Japan

masaota@konan-u.ac.jp

Ohtsuki, Tsutomu
Laboratory of Nuclear Science, Tohoku
University, Mikamine, Taihaku,
Sendai 982-0826,
Japan

ohtsuki@lns.tohoku.ac.jp

Orr, Nigel A.
Laboratorie de Physique Corpusculaire,
IN2P3- CNRS, ISMRa et Université de
Caen, 14050 Caen Cedex,
France

orr@caelav.in2p3.fr

Reinhard, Paul- Gerhard
Inst.f.Theor.Physik, Univ. Erlangen,
D- 91054 Erlangen,
Germany

reinhard@theorie2.physik.uni-
erlangen.de

Rolfs, Claus
Institut für Physik mit Ionenstrahlen,
Ruhr- Universität Bochum,
D-44780 Bochum,
Germany

rolfs@ep3.ruhr-uni-bochum.de

Saito, Susumu
Department of Physics, Tokyo Institute of
Technology Oh- okayama, Meguro-ku,
Tokyo 152-8551,
Japan

saito@stat.phys.titech.ac.jp

Shibutani, Mizuho
Department of Physics
Konan University
Okamoto 8-9-1, Higashinada
Kobe 658-8501,
Japan

Shinohara, Nobuo
Japan Atomic Energy Research Institute
Department of Materials Science
Research Group for Innovative Nuclear
Science Tokai, Ibaraki 319- 1195,
Japan

shino@popsvr.tokai.jaeri.go.jp

Smith, Michael S.
Physics Division, Oak Ridge National
Laboratory Oak Ridge, TN 37830,
USA

msmith@mail.phy.orni.gov

Stodel, Christelle
GANIL, B.P 5027, F-14076 Caen Cedex 5,
France

stodel@ganil.fr

Strieder, Frank
Institut für Experimentalphysik III[1]
Ruhr- Universität Bochum,
D- 44780 Bochum,
Germany

strieder@ep3.ruhr-uni-bochum.de

Sumiyoshi, Kohsuke
Numazu College of Technology (NCT)
Ooka 3600, Numazu, Shizuoka 410- 8501,
Japan

sumi@la.numazu-ct.ac.jp

Suzuki, Takeshi
Department of Physics,
Niigata University, Niigata 950-2181,
Japan

suzuki@nuexne.sc.niigata-u.ac.jp

Takahashi, Kohji
Max- Planck- Institut für Astronomie
Königstuhl 17, D-69117 Heidelberg,
Germany

takahash@mpia-heidelberg.mpg.de

Takahashi, Noriaki
Osaka Univ. / Osaka Gakuin Univ.
Kishibe – Minami 2-36-1, Suita,
Osaka 564-8511,
Japan

ntakahas@rcnpax.rcnp.osaka-u.ac.jp

Utsunomiya, Hiroaki
Department of Physics
Konan University
Okamoto 8-9-1, Higashinada
Kobe 658-8501,
Japan

hiro@konan-u.ac.jp

Vahle, Annett
Forschungszentrum Rossendorf,
PF 510119, D-01314 Dresden,
Germany

A.Vahle@fz-rossendorf.de

Vogt, Karsten
Institut für Kernphysik, Technische
Universität Darmstadt,
Schlossgartentrasse 9,
D-64289 Darmstadt,
Germany

vogt@ikp.tu-darmstadt.de

Wada, Takahiro
Department of Physics, Konan University,
8-9-1 Okamoto, Kobe 658-8501,
Japan

wada@konan-u.ac.jp

Yabana, Kazuhiro
Institute of Physics, University of
Tsukuba
Tsukuba 305-8571,
Japan

yabana@nucl.ph.tsukuba.ac.jp

Yamagata, Tamio
Department of Physics
Konan University
Okamoto 8-9-1, Higashinada
Kobe 658-8501,
Japan

yamagata@center.konan-u.ac.jp

Yamaji, Shuhei
Cyclotron Center, Riken, Wako,
Saitama 351-01,
Japan
yamajis@rarfaxp.riken.go.jp

Yamasaki, Kaoru
Department of Physics
Konan University
Okamoto 8-9-1, Higashinada
Kobe 658-8501,
Japan

kaoru@konan-u.ac.jp

Yasuhira, Masatomi
Department of Physics, Kyoto University,
Kitashirakawa- Oiwake- cho,
Kyoto 606-8502,
Japan

yasuhira@ruby.scphys.kyoto-u.ac.jp

Zhao, Yu- Liang
Graduate School of Science,
Tokyo Metropolitan University,
Tokyo 192-0397,
Japan

zhao-yuliang@c.metro-u.ac.jp

Author Index

A

Abe, Y., 411
Ackermann, D., 323
Adamian, G. G., 421
Alamanos, N., 344
Aliotta, M., 116
Amar, N., 344
Angélique, J. C., 344
Anne, R., 344
Antonenko, N. V., 421
Aritomo, Y., 411
Arnould, M., 76
Asai, M., 358
Auger, G., 344
Azaiez, F., 269, 287

B

Babilon, M., 137
Bayer, W., 137
Bender, M., 377
Blank, B., 280
Bréchignac, C., 184
Burrows, A., 13
Bürvenich, T., 377

C

Cahuzac, Ph., 184
Casandjian, J. M., 344
Caurier, E., 287
Concina, B., 184
Cornelius, T., 377

D

Daugas, J. M., 287
Dayras, R., 344
Denefleh, K., 137
Denisov, V. Y., 433
de Tourreil, R., 344
Döring, J., 287
Drouart, A., 344

E

Ejiri, H., 127
Enders, J., 137

F

Fahlander, C., 287
Fleischer, P., 377
Flocard, H. C., 232
Fontbonne, J. M., 344

G

Galaviz, D., 137
Gillibert, A., 344
Goriely, S., 53
Górska, M., 287
Goto, S., 358
Grawe, H., 287
Grévy, S., 344
Grzywacz, R., 287
Guerreau, D., 344
Guet, C., 167

H

Haba, H., 358
Hanappe, F., 344
Hartmann, T., 137
Hauschild, K., 287
Heger, A., 3, 44
Herwig, F., 3
Hoffman, R. D., 44
Hofmann, H., 399
Hue, R., 344
Hutter, C., 137

I

Ichikawa, S., 358
Ichikawa, T., 411
Ikezoe, H., 334

Itahashi, T., 127
Ivanyuk, F. A., 399
Iwamoto, A., 455
Iwasa, N., 354

J

Jeong, S. C., 334
Junker, M., 85

K

Kajino, T., 159
Kanungo, R., 354
Käppeler, F., 72
Kato, T., 354
Katori, K., 354
Kawade, K., 358
Kawazoe, Y., 204
Kifonidis, K., 21
Kojima, Y., 358
Komori, M., 127
Koura, H., 388
Kudo, H., 354
Kudomi, N., 127
Kume, K., 127

L

Lalleman, A. S., 344
Langer, N., 3
Lattuada, M., 116
Lecesne, N., 344
Leenhardt, S., 287
Legou, T., 344
Lewitowicz, M., 287, 344
Leygnier, J., 184
Lichtenthäler, R., 344
Liénard, E., 344

M

Madland, D. G., 455
Maruhn, J. A., 377
Maruyama, Y., 204
Masumoto, K., 204

Mathews, G. J., 64
Maunoury, L., 344
Mengoni, A., 72
Miljanic, D., 116
Mitsuoka, S., 334
Mittig, W., 344
Mohr, P., 137
Möller, P., 455
Morimoto, K., 354
Morita, K., 354
Müller, E., 21

N

Nagai, Y., 127
Nagame, Y., 358, 443
Nakahara, H., 358, 443
Nakamura, M., 174
Nishinaka, I., 358, 443
Nishio, K., 334
Nowacki, F., 287

O

Oganessian, Y. T., 303
Ohno, K., 204
Ohsumi, H., 127
Ohta, M., 411
Ohtsuki, T., 204
Orr, N. A., 247, 344
Oura, Y., 358

P

Palacz, M., 287
Pellegriti, M. G., 116
Péter, J., 344
Pfützner, M., 287
Pizzone, R. G., 116
Plagnol, E., 344
Plewa, T., 21
Politi, G., 344

R

Rauscher, T., 44
Reinhard, P.-G., 377
Rejmund, M., 287

Rolfs, C., 116, 145
Rykaczewski, K., 287

S

Saint-Laurent, M. G., 344
Saito, S., 214
Sakama, M., 358
Satou, K., 334
Sawicka, M., 287
Scheid, W., 421
Shibata, M., 358
Shiga, K., 204
Shikano, K., 204
Shinohara, N., 223
Sierk, A. J., 455
Smith, M. S., 102, 159
Sorlin, O., 287
Spitaleri, C., 116
Steckmeyer, J. C., 344
Stodel, C., 344
Strieder, F., 92
Suda, T., 354
Sueki, K., 443
Sugai, I., 354
Sumiyoshi, K., 48
Suzuki, T., 259

T

Takahashi, K., 76
Takahisa, K., 127
Takeuchi, S., 354
Tanihata, I., 354
Tchuvil'sky, Y. M., 421
Tillier, J., 344
Tokanai, F., 354
Toki, H., 127
Torres, A. Diaz, 421
Tsukada, K., 358
Typel, S., 116

U

Uchiyama, K., 354
Utsunomiya, H., 159
Uusitalo, J., 287

V

Vahle, A., 368
Villard, B., 184
Villari, A. C. C., 344
Vogt, K., 137
Volz, S., 137

W

Wada, T., 411
Wakasaya, Y., 354
Wellstein, S., 3
Wieleczko, J. P., 344
Wieloch, A., 344
Wolter, H. H., 116
Woosley, S. E., 44

Y

Yabana, K., 196
Yamaguchi, T., 354
Yamaji, S., 399
Yasuhira, M., 33
Yeremin, A., 354
Yoneda, A., 354
Yoshida, A., 354
Yoshida, S., 127

Z

Zhao, Y. L., 443
Zilges, A., 137